Second Edition

IMAGE ANALYSIS
Methods and Applications

Edited by
Donat-P. Häder, Dr. rer. nat.

CRC Press
Taylor & Francis Group
Boca Raton London New York

CRC Press is an imprint of the
Taylor & Francis Group, an **informa** business

CRC Press
Taylor & Francis Group
6000 Broken Sound Parkway NW, Suite 300
Boca Raton, FL 33487-2742

First issued in paperback 2019

© 2001 by Taylor & Francis Group, LLC
CRC Press is an imprint of Taylor & Francis Group, an Informa business

No claim to original U.S. Government works

ISBN-13: 978-0-8493-0239-8 (hbk)
ISBN-13: 978-0-367-39824-8 (pbk)
Library of Congress Card Number 00-027902

Library of Congress Cataloging-in-Publication Data

Image analysis / editor, Donat-P. Häder.—2nd ed.
  p. cm.
Rev. ed. of: Image analysis in biology.c1992
Includes bibliographical references.
ISBN 0-8493-0239-0 (alk. paper)
1. Microscopy. 2. Image processing. I. Häder, Donat-Peter. II. Image analysis in biology.
QH205.2 .I43 2000
570′.28′2—dc21   00-027902
                                                                                              CIP

Visit the Taylor & Francis Web site at
http://www.taylorandfrancis.com

and the CRC Press Web site at
http://www.crcpress.com

Second Edition
IMAGE ANALYSIS
Methods and Applications

Preface

Automatic image analysis has become an important tool in many fields in biology, medicine, and other sciences. In 1992, we compiled a handbook on *Image Analysis in Biology*, to which a large number of key experts in the field contributed. Since 1992, the development of both software and hardware technology has experienced quantum leaps, which has dramatically improved the versatility of digital image analysis. Therefore, it became necessary to write this book; it is not an updated new edition but rather an almost completely new book featuring the most advanced techniques of image analysis. New authors have agreed to contribute to this volume and write on exciting new topics.

The applications range from the submicroscopic (transmission and scanning electron microscopy) and even atomic level (scanning tunnel microscopy) to the macroscopic scale. Simple techniques include counting and quantifying objects such as leaf areas or sizes of microorganisms. Specific mathematical filters have been developed to enhance the quality of the original image or to extract specific features of interest. More complex programs analyze the form of objects in order to, for example, discriminate cancer cells from normal tissue cells. Three-dimensional analysis of proteins, organelles, or macroscopic objects is even more complex since the number of calculations multiplies with the number of layers in the third dimension.

Real-time applications include cell and organism tracking used to analyze motility, velocity, and the direction of movement of motile organisms. Several techniques have been developed to detect the position of individuals and to follow these objects during a predetermined time interval. These techniques have been optimized to extract movement parameters of a large number of motile objects during recent space experiments.

As the editor, I appreciate very much the contributions of the authors to this volume. Their specialized and diverse expertise in the various aspects of hardware, software, and biological application is reflected in each chapter, all of which are characterized by a high quality standard.

Donat-P. Häder
Erlangen, Germany

Editor

Donat-P. Häder, Dr. rer. nat., is a professor of botany, Department of Botany and Pharmaceutical Biology at the Friedrich-Alexander University at Erlangen, Germany. He received his doctoral degree and his habilitation from the University of Marburg. He held a research associate position at Michigan State University, DOE, East Lansing, Michigan, and was visiting scientist at the Chemistry Department of Texas Tech University, Lubbock, Texas, Consiglio Nazionale delle Ricerche (CNR), Pisa, Italy, and the National Research Lab, Okazaki, Japan.

Professor Häder has worked on the photomovement of microorganisms, the effect of solar ultraviolet radiation on phytoplankton, and is involved in space biology studying the effect of microgravity on motility in flagellates. He was a member of a committee on ecology for the German Ministry for Science and Technology, expert for an Enquete commission of the German Parliament, and is a member of a UNEP commission on the effects of the ozone destruction.

One of the tools for his research activities is a realtime image analysis system developed over the last 15 years. He has published more than 360 original papers and has been involved in 11 books as author, translator, or editor.

Contributors

Shanti J. Aggarwal, Ph.D.
Department of Mechanical Engineering and
  Biomedical Engineering Program
The University of Texas at Austin
Austin, TX

Norio Baba, Ph.D.
Dept. of Electrical Engineering
Kogakuin University
Tokyo, Japan

Keith A. Bartels, Ph.D.
Bioengineering Department
Southwest Research Institute
San Antonio, TX

Luigi Bedini
Istituto di Elaborazione della Informazione
CNR
Pisa, Italy

Daniel R. Beniac, Ph.D.
Department of Medical Biophysics
Faculty of Medicine
University of Toronto and Division of
  Molecular and Structural Biology
Ontario Cancer Institute
Toronto, Ontario, Canada

Alan C. Bovik, Ph.D.
Professor
Department of Electrical and Computer
  Engineering and Biomedical Engineering
The University of Texas at Austin
Austin, TX

Eva-Bettina Bröcker, Prof. Dr.
Cell Migration Laboratory
Department of Dermatology
University of Würzburg
Würzburg, Germany

Axel Budde, Dr.
HaSoTec GmbH
Hardware & Software Technology
Rostock, Germany

Giuliano Colombetti, Dr.
Istituto di Biofisica
CNR
Pisa, Italy

Primo Coltelli
CNR, CNUCE
Pisa, Italy

Gregory J. Czarnota, Ph.D., M.D.
Research Associate
Department of Medical Biophysics
Faculty of Medicine
University of Toronto
Ontario Cancer Institute
Toronto, Ontario, Canada
and
Adjunct Professor
Department of Chemistry and Mathematics
Ryerson Polytechnic University
Toronto, Ontario, Canada

Hervé Delacroix
Professor
Université Pierre et Marie Curie
Centre de Génétique Moléculaire
France

Kenneth R. Diller, Ph.D.
Professor
Department of Mechanical Engineering and
  Biomedical Engineering Program
The University of Texas at Austin
Austin, TX

David B. Dusenbery, Ph.D.
Professor
School of Biology
Georgia Institute of Technology
Atlanta, GA

Mauro Evangelisti
Consorzio Pisa Ricerche
Pisa, Italy

Neil A. Farrow, Ph.D.
Department of Medical Biophysics
Faculty of Medicine
University of Toronto and Division of
  Molecular and Structural Biology
Ontario Cancer Institute
Toronto, Ontario, Canada

Peter Friedl, M.D., Ph.D.
Cell Migration Laboratory
Department of Dermatology
University of Würzburg
Würzburg, Germany

Paolo Gualtieri, Dr.
CNR, Istituto di Biofisica
Pisa, Italy

Donat-P. Häder, Dr. rer. nat.
Professor
Institut für Botanik
  und Pharmazeutische Biologie
Erlangen, Germany

George Harauz, Ph.D.
Professor
Department of Molecular Biology and Genetics
  and Biophysics Interdisciplinary Group
University of Guelph
Guelph, Ontario, Canada

Tsuyoshi Hayakawa, Dr.
Tsukuba Research Laboratory
Hamamatsu Photonics
Hamamatsu City, Japan

Michael E.J. Holwill, Ph.D.
Physics Department
King's College
Strand
London, England

Nak H. Kim, Ph.D.
Hankuk University of Foreign Studies
Seoul, Korea

Michael Lebert, Dr.
Institut für Botanik und
  Pharmazeutische Biologie
Erlangen, Germany

Roberto Marangoni, Ph.D.
Instituto di Biofisica
CNR
Pisa, Italy

Fatemah Merchant, Ph.D.
Perceptive Scientific Instruments Inc.
League City, TX

Kenji Omasa, Prof. Dr.
Graduate School of Agriculture and Life
  Sciences
Department of Biological and Environmental
  Engineering
The University of Tokyo
Bunkyo, Tokyo, Japan

F. Peter Ottensmeyer, Ph.D.
Professor
Department of Medical Biophysics
Faculty of Medicine
University of Toronto and Division of
  Molecular and Structural Biology
Ontario Cancer Institute
Toronto, Ontario, Canada

Michael Radermacher, Dr.
Max-Planck-Institut für Biophysik
Abt. Strukturbiologie
Frankfurt/M., Germany

Peter Richter
Institut für Botanik und
  Pharmazeutische Biologie
Erlangen, Germany

Jean Paul Rigaut, Sc.D., M.D., Ph.D.
Director
Laboratoire d'Analyse d'Images en Pathologie
  Cellulaire (AIPC)
I.U.H., Hopital Saint-Louis
Paris, France

Nico A.M. Schellart, Dr.
Laboratory of Medical Physics
Academic Medical Center
University of Amsterdam
Amsterdam, The Netherlands

Walter Steffen, Dr.
Department of Biological Sciences
Center for Biological Visualization Techniques
Institute for Cell Biology and Biosystems
  Technology
University of Rostock
Rostock, Germany

Harald Tahedl
Institut für Botanik und
  Pharmazeutische Biologie
Erlangen, Germany

Tetsuo Takahashi, Dr.
Professor
School of Materials Science
Japan Advanced Institute of Science and
  Technology
Tatsunokuchi-machi
Nomi-gun
Ishikawa, Japan

Helen C. Taylor, Ph.D.
Physics Department
King's College
Strand
London, England

Anna Tonazzini
Istituto di Elaborazione della Informazione
CNR
Pisa, Italy

Vladimir P. Tychinsky, Prof. Dr.
Moscow State Institute for Radioengineering,
  Electronics and Automation, MIREA
Moscow, Russia

Dieter G. Weiss, Prof. Dr.
Department of Biological Sciences
Institute for Cell Biology and Biosystems
  Technology
University of Rostock
Rostock, Germany

Contents

Microscopy and Ultramicroscopy

Movement Analysis

Introduction

1 Introduction

Donat-P. Häder

Vision by man and machine differ in many respects, and certainly the human visual apparatus and brain are still superior to machine vision.[1,2] However image analysis facilitates quantitative determination and objective interpretation of selected parameters. The quantum leaps in technology as well as new and potent algorithms have made possible the development of powerful image analysis systems. Many systems are now capable of performing realtime analysis of complex tasks.

Since its first experimental prototypes in the 1970s, computerized image analysis has witnessed rapid growth and enormous progress. Digitization of analog images is the basis of image enhancement and the extraction of specific features of interest such as dimensions, areas of a distinct color or a gray shade,[3,4] or movement of objects in a sequence of images.[5,6]

The advent of more powerful hardware and the development of robust and efficient algorithms[7-9] have laid the ground for highly specialized and automatic image analysis, which has found applications in all fields of experimental sciences and medicine.[10-14] Not only the explosion in computer technology but also the development of intelligent and low-noise cameras have dramatically improved image analysis.[15,16] One of the first tasks in image analysis was the improvement and quantification of cellular structures.[17-21] Both light microscopy[22-24] and electron microscopy[25,26] are still important fields for image enhancement and analysis, and even holograms have been used as image sources.[27] With current technology, even single large molecules can be visualized.[28,29] Confocal microscopy,[30] atomic force microscopy,[31] and fluorescence imaging[32-35] would not be possible without computer-aided image analysis. In medicine, one ambitious goal is to apply automatic image analysis systems to recognize and identify specific features[36] in order to reliably determine the quantitative distributions of, for example, blood cells or sediments[37,38] and to identify abnormal cells such as cancer cells.[39-41] Quantitative cell identification requires a high spatial resolution.[42] In addition to morphometric methods,[43] image analysis is used for three-dimensional (3-D) reconstruction of biological structures.[44-47] In ecology, fluorescence imaging is applied to identifying stress situations[48-51] or nutrient deficiency[52] in plants and algae.

This volume is divided into several sections, the first of which describes the methods of image analysis, including 3-D analysis, quantification by laser-scanning confocal microscopy, and quantitative area determination. This section provides a general introduction into the techniques, including a chapter on digital filters, and describes the principles of object detection in digital images as well as single photon imaging.

The second section focuses on the applications of image analysis in microscopy, including contrast enhancement in video images, automatic detection and recognition of live cells, fluorescence imaging, as well as image analysis of bending shapes of eukaryotic flagella and cilia. The third section is devoted to the analysis of subcellular structures in electron micrographs and protein analysis, as well as the reconstruction of crystals and 3-D histological objects from serial section images.

The final section describes the techniques used for movement analysis of microorganisms, bacteria, and animals, which require fast routines to track the organisms in real-time in a sequence of video frames. It also describes the use of tracking methods in bioassays as quantitative tools for ecology and water monitoring.

REFERENCES

1. Poggio, T., Vision by man and machine, *Sci. Am.*, 250, 68, 1984.
2. Blake, A., Real-time seeing machines?, *Nature*, 328, 759, 1987.
3. McMillan, P.J., Yakush, A., Frykman, G., Nava, P.B., and Ras, V.R., Minima equalization: a useful strategy in automatic processing of microscopic images, *J. Microscopy*, 148, 253, 1987.
4. Russ, J.C. and Russ, J.C., Automatic discrimination of features in grey-scale images, *J. Microscopy*, 148, 263, 1987.
5. Turano, T.A., D'Arpa, P., Clark, W.L., and Williams, J.R., A time-lapse, image digitization videomicroscope system based on a mini computer with large peripheral memory, *Comput. Biol. Med.*, 15, 177, 1985.
6. Thurston, G., Jaggi, B., and Palcic, B., Cell motility measurements with an automated microscope system, *Exp. Cell Res.*, 165, 380, 1986.
7. Wampler, J.E. and Kutz, K., Quantitative fluorescence microscopy using photomultiplier tubes and imaging detectors, in *Methods in Cell Biology*, Wang, Y.L., and Taylor, D.L., Eds., Academic Press, San Diego, 1989, 239.
8. Sasov, Y., An integrated PC-based image analysis system for microtomography and quantitative analysis of inner micro-object structure, *J. Microscopy*, 156, 91, 1989.
9. Liter, J.C. and Bülthoff, H.H., An introduction to object recognition, *Z. Naturforsch.*, 53c, 610, 1998.
10. Marshall, E.A. and Pickering, W.M., The computational restoration of autoradiographic images, *Comput. Biol. Med.*, 24, 1, 1994.
11. Dhawan, A.P., A review on biomedical image processing and future trends, *Comput. Meth. Progr. Biomed.*, 31, 141, 1990.
12. Kenny, P.A., Dowsett, D.J., Vernon, D., and Ennis, J.T., A technique for digital image registration used prior to subtraction of lung images in nuclear medicine, *Phys. Med. Biol.*, 35, 679, 1990.
13. Jung, F., Mrowietz, C., Moll, A., Beller, K.D., and Wenzel, E., The influence of ultrasound contrast media that achieve pulmonary passage on the microcirculation in rats: a prospective randomized parallel group study, *Adv. Cardiac Echo-Contr.*, 5, 87, 1997.
14. Baba, N., Satoh, H., and Nakamura, S.I., Serial section image reconstruction by voxel processing, *Bioimages*, 1, 105, 1993.
15. Grimson, W.E.L., The intelligent camera: images of computer vision, *Proc. Natl. Acad. Sci. U.S.A.*, 90, 9791, 1993.
16. Oshiro, M., Cooled CCD vs. intensified cameras for low light video — Applications and relative advantages, *Meth. Cell Biol.*, 56, 45, 1998.
17. Frank, J., Image analysis of single macromolecules, *Electron Microsc. Rev.*, 2, 53, 1989.
18. Holmquest, J., Antonsson, D., Bengtsson, E., Danielsson, P.E., Eriksson, O., Hedblom, T., Martensson, A., Nordin, B., Olsson, T., and Stenkvist, B., TULIPS — The Uppsala-Linkoping image processing system, *Anal. Quant. Cytol.*, 3, 182, 1981.
19. Smith, K.C.A., On-line digital computer techniques in electron microscopy: general introduction, *J. Microscopy*, 127, 3, 1982.
20. Skarnulus, A.J., A computer system for on-line image capture and analysis, *J. Microscopy*, 127, 39, 1982.
21. Spring, K.R., Application of video to light microscopy, in *Membrane Biophysics II, Physical Methods in the Study of Epithelia*, Alan R. Liss, New York, 1982, 15.
22. Inoué, T. and Gliksman, N., Techniques for optimizing microscopy and analysis through digital image processing, *Meth. Cell Biol.*, 56, 63, 1998.
23. Wollmer, W., Application of a small microcomputer to cell image analysis, *Anal. Quant. Cytol. Histol.*, 9, 535, 1987.
24. Weiss, D.G., Video-enhanced contrast microscopy, in *Cell Biology: A Laboratory Handbook*, Academic Press, New York, 1994, 77.
25. Inoue, T., Image processing software for research microscopy: requirements and design of the user interface, in *Electronic Light Microscopy*, Shotton, D., Ed., Wiley-Liss, New York, 1993, 95.
26. Stewart, M., Computer image processing of electron micrographs of biological structures with helical symmetry, *J. Electron Microsc. Tech.*, 9, 325, 1988.
27. Lin, J.A. and Cowley, J.M., Reconstruction from in-line electron holograms by digital processing, *Ultramicroscopy*, 19, 179, 1986.

28. Yonekura, K., Stokes, D.L., Sasabe, H., and Toyoshima, C., The ATP-binding site of Ca-ATPase revealed by electron image analysis, *Biophys. J.*, 72, 997, 1997.

29. Sase, I., Miyata, H., Corrie, J.E.T., Craik, J.S., and Kinosita, K., Jr., Real-time imaging of single fluorophores on moving actin with an epifluorescence microscope, *Biophys. J.*, 69, 323, 1995.

30. Borlinghaus, R. and Gröbler, B., Basic principles of confocal laser scanning microscopy, in *Modern Optics, Electronics and High Precision Techniques in Cell Biology*, Isenberg, G., Ed., Springer, Heidelberg, 1997, 33.

31. Schoenenberger, C.-A., Müller, D.J., and Engel, A., Atomic force microscopy provides molecular details of cell surfaces, in *Modern Optics, Electronics and High Precision Techniques in Cell Biology*, Isenberg, G., Ed., Springer, Heidelberg, 1977, 1.

32. Shaw, P.J., Computer reconstruction in three-dimensional fluorescence microscopy, in *Electron Light Microsc.*, Wiley Liss, New York, 1993, 211.

33. Georgiou, G.N., Ahmet, M.T., Houlton, A., Silver, J., and Cherry, R.J., Measurement of the rate of uptake and subcellular localization of porphyrins in cells using fluorescence digital imaging microscopy, *Photochem. Photobiol.*, 59, 419, 1994.

34. Gentry, B. and Meyer, S., Quantitative mapping of leaf photosynthesis using chlorophyll flurorescence imaging, *Aust. J. Plant Physiol.*, 22, 277, 1994.

35. Käs, J., Guck, J., and Humphrey, D., Dynamics of single protein polymers visualized by fluorescence microscopy, in *Modern Optics, Electronics and High Protein Techniques in Cell Biology*, Isenberg, G., Ed., Springer, Heidelberg, 1997, 101.

36. De Paz, P., Barrio, J.P., and Renau-Piqueras, J., A basic program for determination of numerical density of cytoplasmic compartments. II. Analysis of ellipsoids and cylindrical particles, *Comput. Biol. Med.*, 16, 273, 1986.

37. Koss, L.G., Sherman, A.B., and Adams, S.E., The use of hierarchic classification in the image analysis of a complex cell population. Experience with the sediment of voided urine, *Anal. Quant. Cytol.*, 5, 159, 1983.

38. Preston, K., Jr., High-resolution image analysis, *J. Histochem. Cytochem.*, 34, 67, 1986.

39. Vakil, N. and Everbach, C., Image processing in gastrointestinal endoscopy, *Medical Imaging Systems Techn.*, 5, 101, 1997.

40. Sutherland, K. and Ironside, J.W., Quantifying spongiform change in the brain by image analysis, *Eur. Microsc. Analys.*, 39, 21, 1996.

41. Wittekind, C. and Shulte, E., Computerized morphometric image analysis of cytologic nuclear parameters in breast cancer, *Anal. Quant. Cytol. Histol.*, 9, 480, 1987.

42. Gunzer, U., Aus, H.M., and Harms, H., Letter to the editor, *J. Histochem. Cytochem.*, 35, 705, 1987.

43. Pradère, P. and Thomas, E.L., Image processing of partially periodic lattice images of polymers: the study of crystal defects, *Ultramicroscopy*, 32, 149, 1990.

44. Nierzwicki-Bauer, S.A., Balkwill, D.L., and Stevens, S.E., Jr., Three-dimensional ultrastructure of a unicellular cyanobacterium, *J. Cell Biol.*, 97, 713, 1983.

45. Gras, H.A., 'Hidden line' algorithm for 3D-reconstruction from serial sections — An extension of the NEUREC program package for a microcomputer, *Computer Prog. Biomed.*, 18, 217, 1984.

46. Jimenez, J., Santisteban, A., Carazo, J.M., and Carrascosa, J.L., Computer graphic display method for visualizing three-dimensional biological structures, *Science*, 232, 1113, 1986.

47. Harauz, G. and Flannigan, D., Structure of ribosomes from *Thermomyces lanuginosus* by electron microscopy and image processing, *Biochim. Biophys. Acta*, 1038, 260, 1990.

48. Lichtenthaler, H.K., Lang, M., Sowinska, M., Heisel, F., and Miehe, J.A., Detection of vegetation stress via a new high resolution fluorescence imaging system, *J. Plant Physiol.*, 148, 599, 1996.

49. Lichtenthaler, H.K., Lang, M., Sowinska, M., Summ, P., Heisel, F., and Miehe, J.A., Uptake of the herbicide diuron as visualised by the fluorescence imaging technique, *Bot. Acta*, 110, 158, 1997.

50. Siebke, K. and Weis, E., Imaging of chlorophyll-a-fluorescence in leaves: topography of photosynthetic oscillations in leaves of *Glechoma hederacea*, *Photosynth. Res.*, 45, 225, 1995.

51. Lang, M., Lichtenthaler, H.K., Sowinska, M., Heisel, F., and Miehe, J.A., Fluorescence imaging of water and temperature stress in plant leaves, *J. Plant Physiol.*, 148, 613, 1996.

52. Heisel, F., Sowinska, M., Miehe, J.A., Lang, M., and Lichtenthaler, H.K., Detection of nutrient deficiencies of maize by laser induced fluorescence imaging, *J. Plant Physiol.*, 148, 622, 1996.

Methods of Image Analysis

2 Biological Confocal Reflection Microscopy: Reconstruction of Three-dimensional Extracellular Matrix, Cell Migration, and Matrix Reorganization

Peter Friedl and Eva-Bettina Bröcker

CONTENTS

2.1 INTRODUCTION

Cell motility is a fundamental process in embryonic morphogenesis and tissue homeostasis, including leukocyte recirculation, inflammation, wound healing, and angiogenesis. In addition, cell motility is fundamental to tumor cell invasion and metastasis. In the tissue, cell migration requires coordinated cellular interactions with a three-dimensional (3-D) extracellular tissue matrix. These interactions involve a series of coordinated attachment and detachment events,[1] as well as cellular strategies to overcome the biophysical resistance imposed by the extracellular matrix.[2] With the exception of cell crawling across two-dimensional (2-D) surfaces, it has been difficult to investigate cellular and molecular mechanisms of cell migration in 3-D extracellular matrices, partially because of the lack of dynamic 3-D visualization techniques for the detection of cellular interactions with extracellular matrix.

In the past, 3-D collagen lattices have been used to mimic biochemical and biophysical properties of *in vivo* tissues for studies on cell differentiation and migration, in particular for studies on tumor cell invasion,[3] fibroblast function,[4] and leukocyte trafficking[5] (reviewed in Ref. 6). In this chapter, the technique of confocal reflection contrast in biological research is described, with particular reference to the visualization of tumor cell migration in 3-D collagen lattices, as assessed by 3-D reconstruction and quantification of dynamic aspects of cell-matrix interactions, the redistribution of adhesion molecules relative to fiber insertions, matrix remodeling, and the resulting migratory action. In contrast to widely used 2-D substrata, 3-D extracellular matrix not only requires the cell to attach to matrix ligands by adhesion receptors, such as β1 integrins binding to collagen, fibronectin, and laminin;[7] but in the tissue, cells must overcome the biophysical resistance imposed by the preformed network of matrix fibers, either by adapting the cell shape or by proteolytically reorganizing the matrix structure.[2,6,8,9]

2.2 METHODS TO VISUALIZE CELLULAR INTERACTIONS WITH EXTRACELLULAR MATRIX

The visualization of cell matrix interactions has been largely obtained from fixed samples using transmisssion electron microscopy (TEM) or scanning electron microscopy (SEM).[10] SEM, in combination with immunostaining techniques, allows highly resolved representation of fiber physics, distribution, and composition, as well as receptor distribution of cells interacting with extracellular matrix.[10] However, limitations of SEM lie within the complex series of sample fixation and processing, frequently altering the tertiary structure of proteins as well as a limited visual penetration depth,[10] unless freeze fracture or slicing techniques are used. Last, dynamic events such as cell crawling or tissue remodeling have to be extrapolated from static images using serial sampling of different cells in position and shape.

Several nondestructive techniques have been developed for the detection of cellular dynamics from unfixed samples, including:

1. Phase contrast microscopy
2. Transmission mode differential interference contrast (DIC) microscopy
3. Confocal reflection microscopy in conjunction with immunofluorescence

These techniques support time-series imaging delineating sequential events that occur upon shape change, cell contraction, or endosome trafficking.

Dynamics in living cells can be directly visualized by differential interference microscopy if the substrate is a planar surface.[11] This technique is commonly used for studies on intracellular vesicle trafficking and substrate interaction in cells crawling across a transparent planar surface.[12] The friction of visible light results in enhanced contrast and darkening at areas of close apposition of cell membrane and the underlying substrate. In conjunction with phase contrast or reflection contrast microscopy,[13] dynamic changes of the cell body can be reconstructed in three dimensions[11] up to a resolution of 0.3 μm. This technique, however, is limited to cell migration across 2-D substrates because cell interactions with more complex 3-D substrata (e.g., fibrillar ECM networks), do not give sufficient interference signal for visual detection. Using transmission light, matrix fibers are only detected at very low contrast, thus lacking the possibility of 3-D sectioning and reconstruction. Because of cell crawling and rapid changes in shape, orthotopic simultaneous superimposition of interference signal and fluorescence microscopy is not possible using transmission light optics.

Reflection contrast microscopy is widely used in the polymer research and manufacturing industry, but more recently also in cell biology.[14-16] Reflection contrast is either obtained by conventional light microscopy[17] or by confocal laser-scanning microscopy[18] and can be combined with the detection of fluorescent light.

2.3 CONFOCAL REFLECTION MICROSCOPY AND THREE-DIMENSIONAL RECONSTRUCTION

2.3.1 THEORETICAL ASPECTS OF CONFOCAL REFLECTION

For reflection contrast microscopy,[16] the light is introduced into the sample through the objective and the light reflected back into the objective is detected (Figure 2.1). In confocal microscopy using point lasers, the laser light scanning a sample is scattered by interfaces and/or medium transitions (e.g., from air to glass, glass to medium, medium to collagen fiber, or from medium to cell membrane, Figure 2.1). Light reflected backward into the objective is detected as a bright voxel.

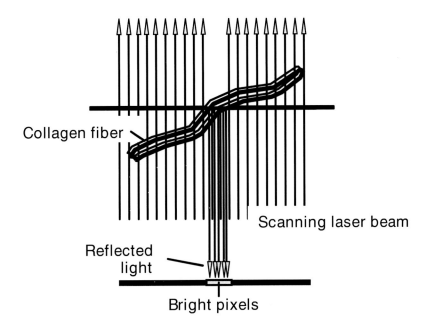

FIGURE 2.1 Principle of confocal reflection contrast. Laser light of 488 and 567 nm is introduced into the sample and light reflected (e.g., by a collagen fiber) is detected as bright voxels.

Differences in contrast largely result from a difference in refractive index (η) between two voxels located in adjacent media, causing the incident ray to deviate from the original path and scatter. Thus, the contrast (C) (i.e., changes in voxel intensity), is proportional to the intensity difference (ΔI) between two phases of the sample divided by the average image brightness $\bar{I}$.

$$C = \frac{\Delta I}{\bar{I}} \tag{2.1}$$

The minimal lateral resolution (d_{min}) of reflection contrast microscopy corresponds to the wavelength (λ) of the light source divided by the numerical aperture of the objective (NA_{obj}) and the numerical aperture of the condenser (NA_{cond}):

$$d_{min} = \frac{\lambda}{NA_{obj} + NA_{cond}} \tag{2.2}$$

approximating 150 to 300 nm for most available confocal systems at high-resolution objectives (NA = 1.3 to 1.4) and the use of ultraviolet and visible light lasers, respectively.

The penetration depth is dependent on the wavelength of the laser light. Light scattering induced by the surrounding medium affects light of lower wavelengths more profoundly than light of longer wavelengths and lower energy. The low reflectance of many living specimens and inhomogeneities of the sample structure lead to a rapid degradation of both the laser beam and the reflection signal. Therefore, the reflectance of biological specimens is frequently improved by the use of reflective dyes or metallic decoration of fixed samples.[18]

2.3.2 CONFOCAL REFLECTION CONTRAST AND 3-D RECONSTRUCTION OF CELL-FREE COLLAGEN LATTICES

Previously, confocal reflection was used for scanning surfaces of biological specimens of high refraction, such as bone and teeth.[16] Because of the low refractance of most other biological specimens, including cells and interstitial tissue, contrast enhancement may be achieved by coating with metal, as previously used for scanning of embryos and cell clusters.[16] In contrast to these previous approaches, in the present study, confocal reflection contrast technology was optimized for hydrated unfixed and unstained 3-D collagen lattices and the cells embedded therein. Analysis was performed using an inverted Leica TCS-4D confocal laser-scanning microscope (Leica, Beusheim, Germany) equipped with a krypton-argon laser. Laser light of 488 and 567 nm was introduced into the sample and the reflected light was passed through a 30/70 beam splitter. A combination of 488 and 567 nm was superior to individual lines at 488 nm (high resolution, low penetration depth) and 567 nm (higher penetration depth at lower resolution and image contrast). For high contrast and resolution, the pinhole was set to 20 to 50 μm (according to previously established optimal confocal discrimination).[19] At a pinhole diameter of 20 to 50 μm, the axial resolutions for consecutive z-scans were approx. 0.7 μm and 1.5 μm, respectively. At lower pinhole diameters, brightness and contrast were degrading; and at diameters greater than 50 μm, the axial resolution for 3-D reconstruction was reduced while the reflection intensity of individual fibers was increased, thus overestimating the actual fiber diameter. Three-dimensional reconstructions of sequential x- and y-sections were obtained from overlay images or calculated as topographical images using the pixel shift approximation[20] of the Leica TCS-4D reconstruction software.

Despite the absence of contrast enhancement by coating with reflective substances, high contrast and image resolution were obtained by confocal reflection scanning of unfixed 3-D hydrated collagen lattices (Figure 2.2). Fibrillar texture, junctions of fibers, and porosity of cell-free collagen lattices as detected from confocal x-y sections (Figure 2.2B, D) strongly corresponded to the aspect obtained in control experiments using scanning electron microscopy (Figure 2.2A, C). High magnification of unfixed lattices resulted in an average fiber diameter of 200 to 350 nm (Figure 2.2D). The overall aspect of fiber texture appeared slightly more nodular by confocal reflection contrast, as compared to smooth and flat fibers using SEM (Figure 2.2A). The use of SEM of fixed and dehydrated lattices resulted in similar fiber characteristics, although the overall fiber diameter was reduced by 10 to 30% as compared to confocal reflection (Figure 2.2C). Furthermore, at high magnification, limitations in sharpness and detail resolution of visible light reflection were apparent (Figure 2.2D).

In cell-free lattices, the fiber orientation in the x- and y- directions follows a random order (Figure 2.2B). In the z-direction, however, collagen fibers were exclusively detected as bright dots from cross-sectioned fibers (Figure 2.2E) in complete absence of vertically oriented fibers, indicating predominant fiber alignment in horizontal order parallel to the underlying cover glass. In x-z sections, similar to x-y scans, fiber detection was possible up to a penetration depth of >100 μm, using the 40x oil immersion objective (NA = 1.3). For the 63x objective (NA = 1.4), fiber detection was possible up to the maximum working distance of approx. 65 μm (not shown). The reflection

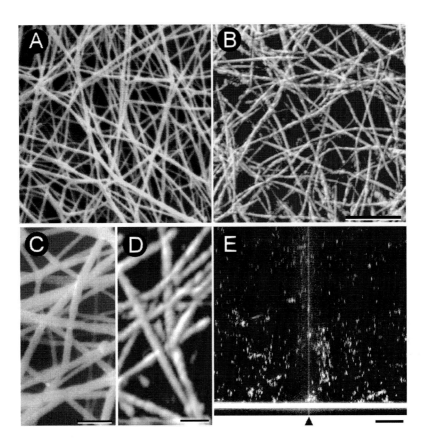

FIGURE 2.2 Confocal reflection imaging of cell-free unfixed 3-D hydrated collagen lattices — comparison with scanning electron microscopy (SEM). Overview and detail of three-dimensionally reconstructed x-y scans obtained from SEM (A, C) or confocal reflection contrast (B, D). x-z section obtained by confocal reflection contrast (E). Collagen matrices were polymerized at a concentration of 1.67 mg/ml as described.[5] For confocal reflection contrast, unfixed lattices were scanned and three-dimensionally reconstructed. The topographical relief (z-coordinate view)[20] was used to reconstruct the 3-D aspect of the fibers (B, D). The central reflex of the objective is indicated (E, arrowhead). The reflection of the cover glass is seen in E (white vertical line). For SEM, collagen lattices were fixed in 6.75% buffered glutaraldehyde (0.1 M phosphate, pH 7.4), rinsed with distilled water, dehydrated in ethanol, incubated in hexamethyl disilazane, critical-point dried, and sputted with 20 to 30 nm gold. Photographs were taken with a Zeiss DSM 962 scanning electron microscope. Bars represent 10 μm (A, B, E) and 1 μm (C, D).

signal provided by individually cross-sectioned fibers was most intense in close apposition to the cover glass (Figure 2.2E, bottom). However, considerable loss of signal intensity and contrast was apparent at higher penetration depth (Figure 2.2E, top). The average pixel intensity/line rapidly degraded with increased penetration depth following a logarithmic function (Figure 2.3). Within the first 20 μm of distance from the cover slip, the reflection intensity was reduced by 80% from the initial mean pixel brightness (Figure 2.3), indicating substantial scattering and refraction of the laser intensity by the aqueous medium and/or fibrils. Consequently, for optimized fiber contrast, a standardized scanning depth of 8 to 20 μm from the cover glass was maintained for the subsequent experiments.

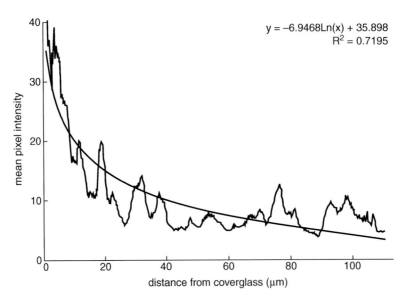

$$y = -6.9468\text{Ln}(x) + 35.898$$
$$R^2 = 0.7195$$

FIGURE 2.3 Changes in mean pixel intensity with increased penetration depth. Densitometry of x-z scan depicted in Figure 2.2E in the vertical direction. The decrease in mean pixel intensity/line in relation to the distance from the cover glass follows a logarithmic function, as indicated.

2.3.3 THREE-DIMENSIONAL RECONSTRUCTION OF CELL-MATRIX INTERACTIONS, CYTOSKELETON, AND ADHESION RECEPTOR DISTRIBUTION

Despite the principal difference of nonfluorescent confocal reflection imaging as compared to fluorescent confocal microscopy (i.e., a confocal reflection signal allows coherent imaging, whereas fluorescence microscopy behaves like an incoherent imaging system),[19] a combination of confocal reflection and fluorescence results in high-resolution sectioning, orthotopic superimposition, and 3-D reconstruction of both signals with little aberration in both x/y- and z-direction.[9] After incorporation into collagen lattices of highly aggressive and metastasizing MV3 melanoma cells (Figure 2.4) or metastatic HT-1080 fibrosarcoma cells (Figure 2.5), the cells undergo profound shape change from initially spheroid and nonmotile to bi- or tripolar and migrating morphologies, as described.[9] This polarization was accompanied by interactions with collagen fibers at the cell edges (Figure 2.4A) and actin polymerization (Figure 2.4B). Upon cell contraction, fibers were reoriented toward the cell body (Figure 2.4A, arrowheads), resulting in nonrandom fiber alignment toward the cell body. If superimposed, F-actin staining showed spike-like clustering at fiber insertions and linear subcortical staining at the center of the cell, which represents the compartment of maximum traction (Figure 2.4A and B).

In MV3 melanoma cells, fiber reorganization and migration are provided by the α2β1 integrin,[9,21] which is the primary collagen receptor in many cell types.[7,22] Ligand binding by β1 integrins induces a series of signaling events mediated by integrin cytoplasmic domains. This results in integrin multimerization, the formation of microscopically detectable clustering followed by the recruitment of cytoskeletal components to the attachment site.[23] The resulting molecular complex consisting of integrins, bound ligand, and cytoskeletal proteins is termed focal adhesion or focal contact.[24] In 3-D collagen lattices, β1 integrin clustering is observed upon interaction with collagen fibers in different cell types upon migration, including melanoma cells,[9] primary fibroblasts,[6] and, as shown here, in migrating HT-1080 fibrosarcoma cells (Figure 2.5). Using pixel shift approximation to rotation stereo,[20] stereo-pairs of both confocal fluorescence of β1 integrin clustering (Figure 2.5, top) and reflection of the matrix architecture (Figure 2.5, bottom) were obtained. In contrast to cells attaching to a planar substrate resulting in flat, spread-out cell morphology and

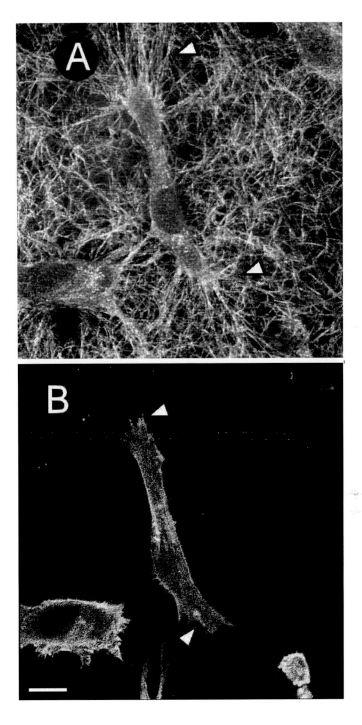

FIGURE 2.4 Polarization of MV3 melanoma cells, interaction with collagen fibers, and actin polymerization. MV3 melanoma cells were incorporated into 3-D collagen lattices. After 18 h, the samples were fixed by 4% depolymerized *para*-formaldehyde and stained with fluorescein-conjugated phalloidin. Sequential *x-y* scans of reflection contrast (A) and fluorescence (B) for the center of the cell ± 2 μm in the *z*-direction were reconstructed (image area 100 × 100 μm). A nonmigrating spheroid cell at the putative onset of polarization is present in the lower left corner.

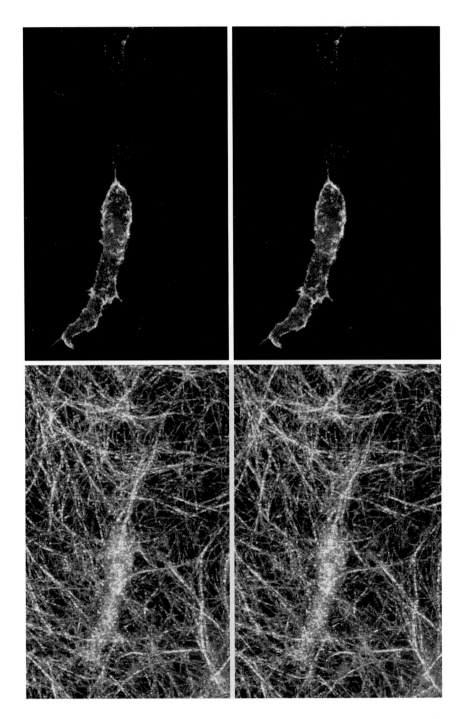

FIGURE 2.5 Three-dimensional reconstruction of a migrating HT-1080 fibrosarcoma cell: clustering of β1 integrins at interactions with collagen fibers. Stereo pairs of β1 integrin staining (top) and confocal reflection image (bottom) of a migrating HT-1080 fibrosarcoma cell in the process of developing new contacts at the leading edge while detaching from the trailing edge and releasing β1 integrins into the lattice. The clustering at interactions to individual fibers and fiber bundles becomes apparent upon superimposition of both channels (not shown). The direction of migration is oriented toward the bottom left. For pixel shift and rotation stereo, the maximum signal criterion was used at an angular difference of ± 5°. Images represent a scanning field of 110 × 70 × 15 μm (in *x*-, *y*-, and *z*-directions, respectively).

integrin clustering toward the basal attachment site, the 3-D matrix substrate supports a tube-like cell shape and three-dimensional β1 integrin clustering (Figure 2.5, top) toward many inserting fibers at the upper and lower side (Figure 2.5, bottom). In addition, β1 integrins were released into the collagen lattice, forming patched and/or string-like deposits along the path of previous migration (Figure 2.5, top). From a technical point of view, both confocal fluorescent and reflection signal were orthotopically represented by the stereo rotation function for 3-D reconstruction.

2.3.4 DYNAMICS IN CELL POLARIZATION AND MIGRATION

In unfixed samples, the dynamic formation and turnover of cell-matrix interactions and receptor distribution was visualized for migrating MV3 melanoma cells (Figure 2.6). Prior to incorporation into the lattice, MV3 cells were labeled with nonblocking anti-CD44 antibody Hermes-3 (kindly obtained from Sirpa Jalkanen, Turku University, Turku, Finland) and secondary lissamine-rhodamine-conjugated Fab-fragments (Jackson Lab., West Grove, PA). CD44 is the primary receptor for hyaluronic acid, which has a hypothesized function in metastatic tumor cell dissemination (reviewed in Ref. 8). In MV3 melanoma cells, CD44 is not directly involved in collagen binding and migration;[21] however, upon melanoma cell migration, substantial amounts of CD44 are detected within the extracellular matrix.[9] To directly show the migration-associated deposition of CD44 into the collagen lattice, confocal time series were obtained from two-channel simultaneous scanning for reflection (Figure 2.6, left) and fluorescence (Figure 2.6, right). To avoid cell damage and related hazardous impact on the migratory action induced by the laser light, the lowest possible laser voltage was combined with an increased pinhole diameter of 150 μm, allowing nondamaging cell observation of up to 1000 single scans (corresponding to 120 scans at eight-fold line averaging for improved signal-to-noise ratio).

The distribution of CD44, in contrast to β1 integrins, was predominantly non-clustered and showed maximal staining toward the trailing edge (Figure 2.6, white asterisks). In the process of cell translocation, CD44 was deposited into the lattices, as detected by the release of fluorescent material from the trailing edge (Figure 2.6, white arrowheads).

Furthermore, besides the shedding of CD44, membrane vesicles containing CD44 were formed at the trailing edge moving toward the cell front, as was apparent from viewing the dynamic time sequence from the video screen (Figure 2.6). Such anterograde endosome trafficking is involved in membrane and receptor recycling in migrating cells required for the flow of membrane and adhesion receptors toward protrusion of the leading edge for the formation of new contacts.[25] The dynamics of cell-fiber interactions at the leading edge, including pulling and fiber traction as well as the deformation of the entire lattice, can be obtained by dynamic sequences (not shown) or by false-color superimposition of sequential scans, showing qualitative aspects of overall matrix deformation in relation to the protrusion of the cell (not shown). The matrix-deforming forces were considerable in these cells, resulting not only in the traction and reorientation of collagen fibers toward the site of attachment at the leading and trailing edges (Figure 2.6, black arrowheads), but also to matrix degradation and the formation of tube-like matrix defects.[26]

2.4 CONCLUSIONS

Although in hydrated biological specimens the refractive indices of various cellular and intercellular structures differ very little from that of the surrounding aqueous milieu,[18] the reflection contrast signal provided by collagen fibrils and, to a lesser extent, cell boundaries is sufficient for sensitive detection up to a scanning depth of 100 μm and a minimal pixel resolution of up to 0.1 μm. Direct comparison of confocal reflection and scanning electron microscopy showed a very similar overall aspect of the matrix architecture, which is in complete accordance with previously published structural data.[10,27] The advantage of gaining direct access to unfixed as well as fixed 3-D matrix structures in a nondestructive manner, as well as the possibility of assessing dynamic changes with

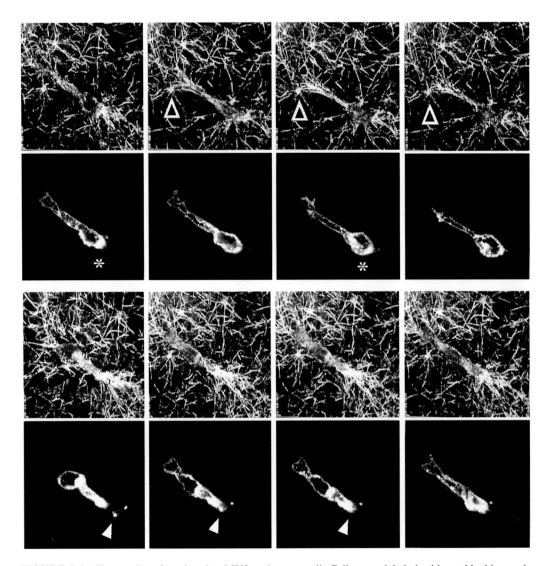

FIGURE 2.6   Time series of a migrating MV3 melanoma cell. Cells were labeled with nonblocking anti-CD44 antibody Hermes-3 prior to incorporation into the collagen lattice. 2 h after matrix polymerization was complete, simultaneous confocal scans were obtained at 15-minute time intervals for the reflection (left) and fluorescence channel (right). Attachment at the leading edge and fiber traction (black arrowhead), accumulation of CD44 toward the trailing edge (asterisks), and the shedding of CD44 into the matrix (white arrowheads) are indicated. Images represent sequential, single x-y scans over a duration of 4 h for a field of 90 × 90 μm. Because of increased pinhole diameter (150 μm) for maximal light detection at low laser power, the reflection of individual collagen fibers appears broader than that obtained for a pinhole diameter of 20 μm (as shown in Figures 2.4 and 2.5).

time in both cells and ECM, allow for sensitive detection of cellular and molecular aspects of cell-matrix interaction, migration, and associated matrix remodeling.

Interestingly, in comparing morphologically and phenotypically distinct cell types using confocal reflection microscopy, marked differences in cellular and molecular migration strategies became apparent. In combination with functional studies on different cell types, including leukocytes, fibroblasts, and tumor cells, confocal reflection microscopy allowed the characterization of differences in migration strategies in 3-D tissues:

1. a slow and integrin-dependent migration of large cells (fibroblasts, some tumor cells) results in profound structural changes in matrix architecture, as opposed to
2. rapid and ameboid migration of small cells such as leukocytes, involving low adhesive forces and dynamic shape change.[6]

In contrast to fibroblast and tumor cells, migration dynamics of leukocytes were 10- to 40-fold faster lacking:

1. integrin clustering,
2. focal contact formation to collagen fibers, and
3. structural remodeling of the matrix architecture.[6]

In conclusion, the combination of confocal reflection contrast and immunofluorescence microscopy will be suitable for novel approaches to investigations in cell biology.

2.5 SUMMARY

The visualization and reconstruction of dynamics in cell-matrix interactions and migration within 3-D extracellular matrices require a high-resolution, continuous-time detection of cells and tissue components. Using confocal reflection microscopy, fixed and unfixed 3-D collagen matrices were visualized and reconstructed at high contrast and a penetration depth of up to 100 µm into the matrix, resulting in a minimal pixel resolution of up to 0.1 µm. Combination with immunostaining or other fluorescence imaging techniques allows a dynamic view of cell-matrix interactions, including (1) cell shape, (2) adhesion receptor distribution, (3) cytoskeletal dynamics, as well as (4) the physical cell interaction with collagen fibers and accompanying structural changes in matrix architecture. Migration associated remodeling of the collagen lattices, as visualized by confocal reflection contrast, includes fiber bending, traction, reorientation, and the deposition of cell surface material into the matrix. In conclusion, real-time confocal reflection microscopy allows detailed investigations of molecular and cellular events in cell migration within 3-D tissue.

ACKNOWLEDGMENTS

This work was supported by the Deutsche Forschungsgemeinschaft (Fr 1155/2-1). The valuable help of Kurt S. Zänker, Georg Krohne, Christian Meyer, Kerstin Maaser, and Martina Joßberger is gratefully acknowledged.

REFERENCES

1. Huttenlocher, A., Ginsberg, M. H., and Horwitz, A. F., Modulation of cell migration by integrin-mediated cytoskeletal linkages and ligand-binding affinity, *J. Cell Biol.*, 134, 1551, 1996.
2. Heino, J., Biology of tumor cell invasion: interplay of cell adhesion and matrix degradation, *Int. J. Cancer*, 65, 717, 1996.
3. Friedl, P., Noble, P. B., Walton, P. A., Laird, D. E., Chauvin, P. J., Tabah, R. J., Black, M., and Zänker, K. S., Migration of coordinated cell clusters in mesenchymal and epithelial cancer explants *in vitro*, *Cancer Res.*, 55, 4557, 1995.
4. Noble, P. B. and Levine, M. D., *Computer-Assisted Analyses of Cell Locomotion and Chemotaxis*, CRC Press, Boca Raton, FL, 1986.
5. Friedl, P., Noble, P. B., and Zänker, K. S., Lymphocyte migration in three-dimensional collagen gels. Comparison of three quantitative methods for analysing cell trajectories, *J. Immunol. Meth.*, 165, 157, 1993.

6. Friedl, P., Zänker, K. S., and Bröcker, E. B., Cell migration strategies in 3-D extracellular matrix: differences in morphology, cell matrix interactions, and integrin function, *Microsc. Res. Tech.*, 43, 369, 1998.

7. Hemler, M. E., VLA proteins in the integrin family: structures, functions, and their role on leukocytes, *Annu. Rev. Immunol.*, 8, 365, 1990.

8. Friedl, P. and Bröcker, E.-B., The biology of cell locomotion within three-dimensional extracellular matrix, *Cell. Mol. Life Sci.*, 57, 41, 2000.

9. Friedl, P., Maaser, K., Klein, C. E., Niggemann, B., Krohne, G., and Zänker, K. S., Migration of highly aggressive MV3 melanoma cells in 3-dimensional collagen lattices results in local matrix reorganization and shedding of $\alpha2$ and $\beta1$ integrins and CD44, *Cancer Res.*, 57, 2061, 1997.

10. Heath, J. P. and Peachey, L. D., Morphology of fibroblasts in collagen gels: a study using 400 keV electron microscopy and computer graphics, *Cell Motil. Cytoskel.*, 14, 382, 1989.

11. Niewöhner, J., Weber, I., Maniak, M., Müller-Tauchenberger, A., and Gerisch, G., Talin-null cells from *Dictyostelium* are strongly defective in adhesion to particle and substrate surfaces and slightly impaired in cytokinesis, *J. Cell Biol.*, 138, 349, 1997.

12. Aletta, J. M. and Greene, L. A., Growth cone configuration and advance: a time-lapse study using video-enhanced differential interference contrast microscopy, *J. Neurosci.*, 8, 1425, 1988.

13. Keith, C. H., Bird, G. J., and Farmer, M. A., Coherent backscatter enhances reflection confocal microscopy, *Biotechniques*, 25, 858, 1998.

14. Cornelese-ten Velde, I., Bonnet, J., Tanke, H. J., and Ploem, J. S., Reflection contrast microscopy. Visualization of peroxidase-generated diaminobenzidine polymer products and its underlying optical phenomena, *Histochemistry*, 89, 141, 1988.

15. Davies, P. F., Robotewskyj, A., and Griem, M. L., Endothelial cell adhesion in real-time. Measurements *in vitro* by tandem scanning confocal image analysis, *J. Clin. Invest.*, 91, 2640, 1993.

16. Boyde, A. and Jones, S. J., Mapping and measuring surfaces using reflection confocal microscopy, in *Handbook of Biological Confocal Microscopy*, Pawley, J. B., Ed., Plenum Press, New York, 1995, 255.

17. Ploem, J. S., Reflection-contrast microscopy as a tool for investigation of the attachment of living cells to a glass surface, in *Mononuclear Phagocytes in Immunity, Infection and Pathology*, Furth, R. van, Ed., Blackwell, Oxford, 1975, 404.

18. Cheng, P. C. and Kriete, A., Image contrast in confocal light microscopy, in *Handbook of Biological Confocal Microscopy*, Pawley, J. B., Ed., Plenum Press, New York, 1995, 281.

19. Wilson, T., The role of the pinhole in confocal imaging system, in *Handbook of Biological Confocal Microscopy*, Pawley, J. B., Ed., Plenum Press, New York, 1995, 167.

20. White, N. S., Visualization systems for multidimensional CLSM images, in *Handbook of Biological Confocal Microscopy*, Pawley, J. B., Ed., Plenum Press, New York, 1995, 211.

21. Maaser, K., Wolf, K., Klein, C. E., Niggemann, B., Zänker, K. S., and Friedl, P., Functional hierarchy of simultaneously expressed adhesion receptors: Integrin $\alpha2\beta1$ but not CD44 mediates MV3 melanoma cell migration and matrix reorganization within three-dimensional hyaluronan-containing collagen lattices, *Mol. Biol. Cell*, 10, 3067, 1999.

22. Klein, C. E., Dressel, D., Steinmacher, T., Mauch, C., Eckes, B., Krieg, T., Bankert, R., and Weber, L., Integrin $\alpha2\beta1$ is upregulated in fibroblasts and highly aggressive melanoma cells in three-dimensional collagen lattices and mediates the reorganization of collagen I fibrils, *J. Cell Biol.*, 115, 1427, 1991.

23. Miyamoto, S., Teramoto, H., Coso, O. A., Gutkind, J. S., Burbelo, P. D., Akiyama, S. K., and Yamada, M., Integrin function: molecular hierarchies of cytoskeletal and signaling molecules, *J. Cell Biol.*, 131, 791, 1999.

24. Burridge, K. and Chrzanowska-Wodicka, M., Focal adhesions, contractility, and signaling, *Annu. Rev. Cell Dev. Biol.*, 12, 463, 1996.

25. Bretscher, M. S., Moving membrane up to the front of migrating cells, *Cell*, 85, 465, 1996.

26. Friedl, P. and Bröcker, E. B., Cancer cell interactions with the extracellular matrix involved in tissue invasion: motility mechanisms beyond the single cell paradigm, in *Extracellular Matrix and Ground Regulation System in Health and Disease*, Heine, H. and Rimpler, M., Eds., Gustav Fischer, Stuttgart, Jena, Lübeck, Ulm, 1997, 7.

27. Friedl, P., Bröcker, E. B., and Zänker, K. S., Integrins, cell matrix interactions and cell migration strategies: fundamental differences in leukocytes and tumor cells, *Cell Adhes. Commun.*, 6, 225, 1998.

3 Automatic Area and Volume Measurements From Digital Biomedical Images

Alan C. Bovik, Shanti J. Aggarwal, Fatemah Merchant, Nak H. Kim, and Kenneth R. Diller

CONTENTS

0-8493-0239-0/00/$0.00+$.50
© 2001 by CRC Press LLC

3.1   INTRODUCTION

Methods for obtaining numerical area data from two-dimensional (2-D) images of biological objects have been studied for nearly 4 decades. More recently, it has become of interest to measure the volumes of three-dimensional (3-D) objects in many applications. In the early 1960s and 1970s, manual methods of area data collection were prevalent, of which the following three are the most reliable. The areas of objects, such as images of cells, multicellular masses, or macroscopic wounds captured in a positive print were measured by:

1. Projecting the outlines of the object using a camera lucida or some similar projection apparatus onto millimeter-ruled graph paper, and then counting the number of squares enclosed within each profile
2. Instead of counting the squares, the areas of traced profiles are measured with the aid of a planimeter
3. Tracing the projected outlines of the object onto a uniform-thickness paper, and subsequently cutting out each profile and weighing it on a microbalance. By comparing the observed weights with samples of a precisely known area, it is possible to obtain a reasonable estimate of the cross-sectional area.

For relatively large specimens such as ova, embryos, and even granulocytes, these techniques are straightforward and reasonably accurate, but are of questionable use for area measurements of small organelles that must be magnified to the limits of optical resolution for viewing. Moreover, apart from the human errors inherent in the manual tracing and interpretation of cellular outline, these approaches are very slow, tedious, and impractical for the measurement of large numbers of cells. The sequence of developing film, positioning the paper or measuring devices, taking the manual measurements, and making computations implies an effort that can be measured in substantial fractions of man-hours, at a minimum. In the case of sequences of images, or 3-D images from which volume is to be measured, the amount of work becomes prohibitive.

Fortunately, rapid technological developments in digital video microscopy, digital video camera technology, dedicated digital image processing hardware and software, and the availability of new generations of powerful, inexpensive desktop workstations have substantially ameliorated these difficulties. Many automated or semi-automated methods are now available that combine simple camera, computer, and software techniques that make it possible to measure object areas or volumes. There are a fair number of area/volume measurement systems that are highly effective for measuring objects that have images of high intrinsic contrast or that have well-defined and easily identified shapes (e.g., that are nearly circular in profile). These systems are often designed for a specific area/volume measurement application, where domain-specific information about the object and background can be used to simplify software algorithms and possibly lead to increased speed. Recently, however, more general techniques have been developed that effectively measure the areas of objects having more complex or tortuous boundaries, nonuniform distributions of optical density (gray level), or images of objects having low contrast, partial transparency, or diffuse detail.

The purpose of this chapter is to review the fundamental elements of workstation-based image area and volume analysis techniques, and their applications to various biological fields. Representative examples are given that are generally extensible to a large variety of objects imaged in the various disciplines. Examples of generically powerful image processing software operations and general considerations involved in area computation in biomedical image analysis are also discussed.

3.2　BIOMEDICAL IMAGE ANALYSIS

This section briefly reviews the types of information that can be used to identify image regions, the areas or volumes of which are to be measured, and more generally gives an overview of the kinds of preprocessing techniques currently available for determining the boundaries of objects or object regions. Video technology, and in particular digital video microscopy, coupled with computer image analysis presents an exciting tool for the fast and accurate analysis of biological structures. The combination of the two technologies allows one to extract the useful information hidden in low-contrast, dark, dynamically changing biological scenes by various well-known techniques such as noise reduction, contrast enhancement, feature extraction, and other manipulations that use the principles of classical image processing, image analysis and mathematical morphology. Some of the most useful features that can be automatically measured from images of a biological scene are those that describe the size and shape of the individual object within the image; for example, object area or volume, object perimeter or surface, degree of circularity or sphericity, aspect ratio, and rectangularity. Local measurements such as boundary or surface curvature can also be very useful. Of course, caution must be applied in the design and implementation of hardware and algorithms for image analysis, since it is possible to generate artifacts not present in the original scene by improper arrangements in the lighting conditions, imaging setup, or software.

Although image analyzers have gained considerable acceptance as analytical tools, their usefulness for quantitative image analysis in biological applications is still limited. Fully automated systems that allow unsupervised measurement of a wide variety of numerical object parameters, and for generic objects of variable shapes and reflectance/transmission properties, are not yet available. The main problem in realizing a fast, efficient, fully automated system resides first in the quality of the images obtained, which are often of low contrast (particularly micrographs) and can present highly complex structures, and second, in the ability of the software system to decompose image data into information-rich primitives or on the basis of specific physical object properties. Many powerful image analysis and object recognition techniques, such as true color processing[1] and image texture analysis,[2,3] have remained relatively unexploited in the analysis of biomedical scenes. Color image analysis has obvious utility for the segregation or recognition of objects based on their chromatic emissive, reflectance, or transmission properties. Biomedical objects often have distinctive natural color, or in the case of stained slides, a truly differentiating color that could aid in better feature or object extraction. Texture analysis can provide important data about the local distribution of gray levels relative to those in surrounding areas, as illustrated by the following example. If one is interested in the therapeutic value of a drug as a wrinkle preventive cream, then one viable approach for evaluating the results could proceed by determining the skin area that is occupied by discernable wrinkles before and after application of the cream. Traditional image analyzing techniques often found in biomedical image analysis systems will provide the ability to measure the relative local lightness and darkness of the sample, but likely will not be able to differentiate between the object classes (wrinkled vs. smooth) since the overall irradiance from the skin types may be the same. Automated texture analysis, on the other hand, can indicate changes in the texture pattern relative to the surrounding areas and, therefore, how the sample is responding to the cream.

The first stage in many biomedical image analysis systems is an object isolation process in which a subject of interest (e.g., a cell, tumor, or bone) is separated from a background of varying optical complexity. Image segmentation techniques devised for this purpose can be broadly divided into three categories:[4,5]

1. Edge- or contour-based approaches, in which sustained intensity changes, or edges, are marked (usually with a discrete derivative operator)
2. Direct region segmentation approaches, in which the image is partitioned according to some homogeneity criterion
3. Gray-level thresholding, where bright (or dark) objects are separated from the background by the use of an appropriate threshold value

Contour-based approaches are primarily used in applications where the object has a relatively smooth (low curvature) boundary,[6-8] or when the object has a simple, invariant shape.[9] Region growing techniques (e.g., split-and-merge) can be useful when the local gray-level statistics of the object are dissimilar to those of the background.[10] Simple thresholding techniques are most adequate for cases where the object or objects of interest have a gray-level distribution that is markedly offset from that of the background. While each of these techniques has been used successfully for many specific applications, they may be difficult to incorporate into a generic imaging system able to deal with a broad range of input images, since most of these methods have a limited range of applicability; that is, they can be applied only to images with certain characteristics or types of intrinsic information that the algorithm assumes, and they also tend to not always be reliable, as they may misclassify a small portion of the region or boundary if the algorithm is not carefully defined.

The following section focuses primarily on the problem of measuring the area or volume after some specific segmentation or boundary-finding technique has been applied. As a vehicle for demonstrating fundamental aspects of the shape correction and area/volume calculation problem, a very simple, but often effective object extraction process is assumed, based on simple thresholding. This is done because the emphasis of this chapter is on automated area and volume measurement, rather than on preparing the images (or the specimens) obtained for the area or volume measurement process. However, the principles described are highly generic, and feature, in particular, methods for ameliorating the types of difficulties or errors that commonly occur with any kind of region segregation technique.

3.3   SEPARATION OF OBJECT FROM BACKGROUND

Once an image has been digitized, it is typically processed through software in several stages. In the first stage, the image is segmented into object and background using a simple gray-level thresholding technique.[4] This is implemented as follows. The first stage of the algorithm often involves the smoothing of the image using, for example, a moving average filter, or possibly a Gaussian-weighted moving average filter. This stage is optional and is used only if the image is noisy, or if the object and/or background have substantial local nonuniformities. In general, the averaging operation results in a distribution of the image gray levels that is more distinctive.[4] In simple terms, the intensity value of the center pixel of the window is replaced by the average intensity (or weighted average) of its neighbors within the window.

For example, the implementation of a two- or three-dimensional (2-D or 3-D) 5×5 (or $5 \times 5 \times 5$) square (or cubic) moving average can be described in mathematical terms that can be easily converted into computer code. The following multi-index notation is used here: $\mathbf{i} = (i, j)$ for 2-D images, and $\mathbf{i} = (i, j, k)$ for 3-D images. Likewise, the notation $0 \leq \mathbf{i} \leq N - 1$ will mean $0 \leq i, j \leq N - 1$ in the 2-D case, and $0 \leq i, j, k \leq N - 1$ in the 3-D case. Assume that the original image is given by $\mathbf{I} = [I(\mathbf{i}); 0 \leq \mathbf{i} \leq N - 1]$. Then, the 5×5 square (or $5 \times 5 \times 5$ cubic) moving average smoothed image $\mathbf{I}_{ave} = [I_{ave}(\mathbf{i}); 0 \leq \mathbf{i} \leq N - 1]$ has elements given by

$$I_{ave}(\mathbf{i}) = \frac{1}{K} \sum_{|\mathbf{n}| \leq 2} X_{\mathbf{i}-\mathbf{n}} \qquad (3.1)$$

for $0 \leq \mathbf{i} \leq N - 1$, where $K = 25$ in 2-D, and $K = 125$ in 3-D, and where the multi-index $\mathbf{n} = (n, m, p)$. It may be observed that the averaging operation at the boundaries of the image (i.e., wherever $\min(\mathbf{i}) < 2$ or $\max(\mathbf{i}) > N - 2$), is not well defined, since the indices over which the average is computed will not all fall within the prescribed range $0 \leq \mathbf{i} \leq N - 1$. This may be easily ameliorated by replicating pixels, i.e., assigning gray levels from the closest pixels to the undefined pixel elements.

A simple gray-level thresholding operation is then applied to the averaged image in order to obtain a first attempt at separating the wound region from the background. This is a simple example of the general technique of segmentation that is appropriate for the example application considered here, since the image objects of interest generally have an average intensity that is significantly different from the average background intensity. Of course, in other applications, the segmentation may proceed on the basis of color separation, textural differences, or some other physically meaningful criterion. One assumes here, however, that the result of the segmentation is a binary image or map indicating the likely locations of object pixels vs. background pixels. This binary map will, of course, generally contain errors resulting from local perturbations of the image from the model assumed. For a determined threshold level τ that falls within the range of allowable gray-level values (usually coded as integers in the range 0 to 255 corresponding to 8 bits of resolution), the thresholded binary image $\mathbf{J} = [J(\mathbf{i}); 0 \leq \mathbf{i} \leq N - 1]$ is defined as

$$J(\mathbf{i}) = \begin{cases} '1' \; ; \; \text{if } I^{ave}(\mathbf{i}) \geq \tau \\ '0' \; ; \; \text{if } I^{ave}(\mathbf{i}) < \tau \end{cases} \tag{3.2}$$

where '1' denotes a binary value indicating that the brightness exceeds the threshold, and hence is likely an object pixel; and the binary digit '0' denotes low-intensity background pixels. For display purposes, it can either be assumed that the binary value '1' corresponds to BLACK and the binary value '0' to WHITE, or vice versa. The selection may depend on whether the application is such that the objects appear as dark regions on a light background, or vice versa, or it may simply depend on which gives a better visual impression.

Many of the binary assignments made by the simple thresholding above may be erroneous, owing to noise, imperfections in the imaging process, or nonuniformities in the intensities of object or background; hence the need for the region and shape correction procedures outlined next. Because the image to be operated on is now binary- or logical-valued, the correction procedures can be conveniently defined as highly efficient logical operations.

In order to determine an appropriate threshold τ, the gray level distribution or histogram of the image must be measured. The histogram $H_I(k); 0 \leq k \leq K - 1$ ($K = 256$ in most cases) of an arbitrary gray-level 2-D or 3-D image $\mathbf{I}$ is a one-dimensional plot or graph of the frequency of occurrence of each image gray level k in $\mathbf{I}$: $H_I(k) = n$ if $\mathbf{I}$ contains exactly n occurrences of gray-level k for each $k = 0,\ldots, K - 1$. The threshold operation is most efficacious if the histogram $H_I(k)$ of the image $\mathbf{I}$ contains two well-separated peaks or modes, corresponding to object and background pixel gray levels. Such a "bimodal" histogram has the appearance depicted in Figure 3.1.

Because the area of the object may vary over a wide range (typically 100 to 10,000 pixels in 2-D images, and 1000 to 1,000,000 pixels in 3-D images), and because there tends to be a large variation in the distribution of gray levels within a typical input image, automatic determination of the threshold is generally quite difficult. Although automated threshold selection criteria have been proposed,[4] some manual intervention is generally required in order to select the best possible threshold. In the current implementation, the gray-level histogram of the image is displayed on the monitor and a threshold is set manually. The output at this stage is the thresholded binary image. To this image, a simple region correction procedure is applied in order to determine the principal object regions of interest. A smoothing or shape correction operation based on concepts of mathematical morphology is then applied prior to measuring the area of the object of interest.

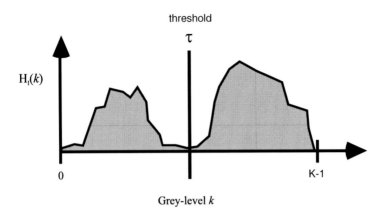

FIGURE 3.1   Appearance of a typical bimodal histogram and one possible threshold location.

3.4   REGION CORRECTION IN THE THRESHOLDED IMAGE

A binary (2-D) image resulting from thresholding a gray-level image typically has an appearance similar to that shown in Figure 3.2; that is, there will usually be misclassifications of both object pixels and background pixels, due to random variations in surface reflectance or absorption, extraneous objects, dust, or image noise. In this section, an algorithm is described that effectively corrects (eliminates) misclassified small regions. The result of the algorithm is a new binary image where all small misclassified regions lying entirely outside the object area are removed using a variant of the "blob coloring" algorithm.[5] Subsequently, all small misclassified regions lying entirely within the object are removed using a dual of the blob coloring algorithm. This algorithm first assigns a region number (or "color") to each nonzero pixel so that all pixels belonging to a same four-connected region will have the same region number. This number-assigning procedure can be done in a single scan of the entire image, as summarized below.

FIGURE 3.2   Typical appearance of a thresholded binary image prior to region correction and boundary smoothing.

3.4.1   BLOB COLORING ALGORITHM

Given a binary (thresholded) image $\mathbf{J} = [J(\mathbf{i}); 0 \leq \mathbf{i} \leq N - 1]$, define a (non-binary) array $\mathbf{R} = [R(\mathbf{i}); 0 \leq \mathbf{i} \leq N - 1]$ where $R(\mathbf{i})$ is the region number of the binary-valued pixel $J(\mathbf{i})$, according to the following pseudo-code. In the code, denote the set of 2-D pixel coordinates $\mathbf{I}_{prev} = \{(i-1, j), (i, j-1)\}$ or 3-D pixel coordinates $\mathbf{I}_{prev} = \{(i-1, j, k), (i, j-1, k), (i, j, k-1)\}$ to denote the set of coordinates that are immediately adjacent (not along diagonals) to $\mathbf{i}$.

```
begin
initialize R: R(i) = 0 for 0 ≤ i ≤ N – 1;
k = 1;
while (0 ≤ i ≤ N – 1) do
begin
if J(i) = '1' then begin
if (J(j) = '0' for all j ∈ Iprev) then
begin
R(i) = k;
k = k + 1;
end;
while (j ∈ Iprev) do
if (J(j) = '1' and R(i) = 0) then
R(i) = R(j);
if (J(j) = '1' and R(i) ≠ 0) then
record R(i) and R(j) as equivalent region numbers;
end
end
end
end.
```

The output of this procedure is the array $\mathbf{R}$ containing all information regarding the connectedness of pixels having the same binary values; that is, all pixels in $\mathbf{J}$ belonging to the same connected region and having the same binary value are assigned the same region number. Using this information, the total number of regions and the area of each region can easily be found. As the implementation of the system described here is designed, for simplicity of exposition, to process images containing only a single object to be measured, only the region with the largest area or volume (largest number of pixels) is retained. All other regions are classified as noise and are thus removed from the input binary image according to the following simple strategy. First, the number of the region having the largest area is identified using the area information acquired in the preceding procedure. This is a trivial operation on the region number array R. The minor region deleting procedure can then be done in another scan of the image as listed below.

3.4.2 MINOR REGION REMOVAL ALGORITHM

Given a binary (thresholded) image $\mathbf{J} = [J(\mathbf{i}); 0 \le \mathbf{i} \le N - 1]$, its region number array $\mathbf{R} = [R(\mathbf{i}); 0 \le \mathbf{i} \le N - 1]$ computed by the previous algorithm, and the number m of the largest region, then delete potentially misclassified pixels according to the following pseudo-code:

```
begin
while (0 ≤ i ≤ N – 1) do
begin
if J(i) = '1' then
if R(i) ≠ m then J(i) = '0';
end
end.
```

Figure 3.3 depicts the result of the blob coloring algorithm and the subsequent removal of minor regions from the binary thresholded 2-D image result in Figure 3.3. Of course, the above code can be modified to instead allow the largest M regions to be preserved, if $M \geq 1$ is found to be a meaningful count of the expected number of objects. However, in perhaps the majority of applications, a single object, having the largest area, will be the region desired to be extracted. For example, in all application cases of the wound area measurement system encountered thus far, currently numbering in the thousands, the above assumption that the object has the largest area has been valid.

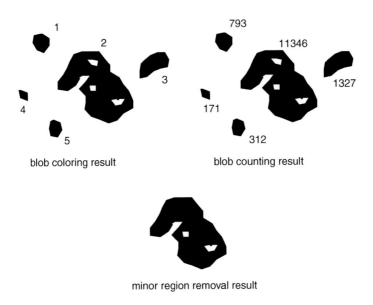

blob coloring result   blob counting result

minor region removal result

FIGURE 3.3   Result of blob coloring and minor region removal algorithms applied to the binary thresholded image of Figure 3.2.

It can be observed from Figure 3.3 that one application of the minor region removal algorithm does not remove those misclassified regions that lie entirely within the object of interest. However, a second application of the blob coloring/counting and minor region removal algorithms to the *logical complement* $\mathbf{J}^c$ of the binary thresholded image $\mathbf{J}$ will accomplish this in trivial matter. The complement image has elements defined according to $\mathbf{J}^c(\mathbf{i}) = $ '1' if $\mathbf{J}(\mathbf{i}) = $ '0' and $\mathbf{J}^c(\mathbf{i}, \mathbf{j}) = $ '0' if $\mathbf{J}(\mathbf{i}, \mathbf{j}) = $ '1'. In the complemented image, the background can be expected to be the largest "blob." Once the region counting and minor region removal has been applied to the complement image, the complement of the resulting corrected image is taken (Figure 3.4). The result of the two passes of these region correction algorithms is a binary image containing a single connected object or BLACK (WHITE) region against a WHITE (BLACK) background. Generally, the number of pixels contained within this corrected image will yield a fairly reasonable estimate of the true object area. However, it is possible to further improve the efficacy of the area measurement system by applying a final smoothing algorithm to the region-corrected binary image. This final step has the effect of correcting misclassifications occurring near the boundary of the object/background regions, which are not amenable to the algorithms described above. For example, the final result of Figure 3.4 contains a significant gulf or bay that may possibly be part of the object region. Errors of this type occur frequently in applications where the boundaries of objects are diffuse, as in many microscopic images.

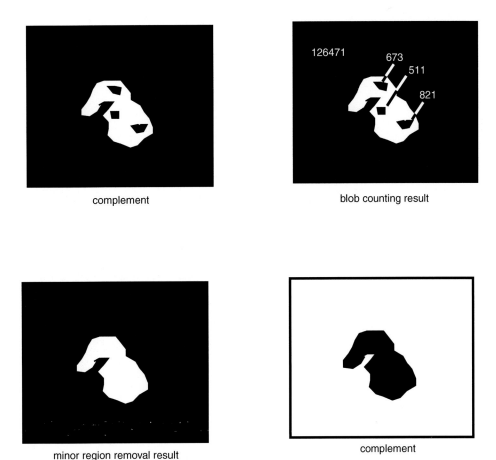

FIGURE 3.4 Dual of blob coloring/counting and minor region removal algorithms applied to the partially corrected image of Figure 3.3.

3.5   BOUNDARY SMOOTHING BY MORPHOLOGICAL FILTERING

While the gray-level thresholding algorithms and region correction algorithms are simple and fast, a common problem occurs when the object or the background has a varying gray level: the detected object boundary may be corroded, containing breaks, gulfs, or peninsulas not corresponding to the physically correct object boundaries. In this section, **morphological filters** are presented as a simple, elegant, and very effective solution to these problems.

   Mathematical morphology, originally developed as an analytical device for the quantification of the geometric structure of objects, has been applied to image processing problems by many authors.[2,13,14,32-37] The techniques and operations of mathematical morphology have the advantages of computational simplicity and intuitive physical meaning. Generally, **binary** morphology is a collection of operations for modifying the shapes of objects in binary images. There is also a more general collection of operations, collectively termed **gray-scale** morphology, but for purposes here the simple binary technique suffices; these can all be defined in terms if simple binary or logical operations.

More specifically, a morphological operation is an interaction between the objects in a binary image and a *window* or *structuring element*. The window, which is usually of simple shape and local support, interacts with the image objects, transforming their shapes in a meaningful way. The type of interaction that occurs between objects and window defines the type of morphological operation occurring. The shape of the window determines the precise way in which the morphological operation modifies the shapes of objects.

3.5.1 WINDOWS

A window **B** defines a geometric relationship between local pixels in an image; it is a device for collecting image coordinates on a local basis. Mathematically, a window **B** is a set of $2M + 1$ image coordinate shifts $\mathbf{m}_i = (m_i, n_i)$ in 2-D or $\mathbf{m}_i = (m_i, n_i, p_i)$ in 3-D, usually symmetrically centered around the zero-shift **0** (hence the odd number of elements, $2M + 1$):

$$\mathbf{B} = \{\mathbf{m}_1, ..., \mathbf{m}^{2M+1}\} \tag{3.3}$$

Figure 3.5 depicts a few 2-D windows as geometric pixel relationships. These may also be expressed as sets of image coordinate shifts as follows:

$$\mathbf{B} = \text{SQUARE}(2M + 1) = \{\mathbf{m}: -N \leq \mathbf{m} \leq N\} \tag{3.4}$$

where $2M + 1 = (2N + 1)^2$, and

$$\mathbf{B} = \text{CROSS}_{2\text{-D}}(2M + 1) = \{(m, 0): -M \leq m \leq M\} \cup \{(0, n): -M \leq n \leq M\} \tag{3.5}$$

Three-dimensional windows can likewise be described; for example, a 3-D cubic window

$$\mathbf{B} = \text{CUBE}(2M + 1) = \{\mathbf{m}: -N \leq \mathbf{m} \leq N\} \tag{3.6}$$

which in our notation is the same as the 2-D SQUARE, except that $2M + 1 = (2N + 1)^3$. It is depicted in Figure 3.6. A 3-D cross-shaped window is given by

$$\begin{aligned}
\mathbf{B} = \text{CROSS}_{3-\text{D}}(2M + 1) \\
= \{(m, 0, 0): -M \leq m \leq M\} \cup \{(0, n, 0): -M \leq n \leq M\} \cup \{(0, 0, p): -M \leq p \leq M\}
\end{aligned} \tag{3.7}$$

and is also depicted in Figure 3.6. It should be noted that windows of similar span may have very different area/volumes or pixel coverages, and hence very different smoothing power. For example, the two 3-D windows in Figure 3.6 have the same span of 5 pixels along each perpendicular dimension, but their pixel coverage differs by nearly an order of magnitude.

3.5.2 WINDOWED SETS

As mentioned above, a window **B** is a device for collecting a set of image pixels according to a specified geometric relationship defined by the shape and size of **B**. This is accomplished at each image coordinate **i** of the binary image **j**, resulting in the indexed **windowed sets**

$$\mathbf{B}\cdot\mathbf{J}(\mathbf{i}) = \{J(\mathbf{i} - \mathbf{m}); \mathbf{m} \in \mathbf{B}\} \tag{3.8}$$

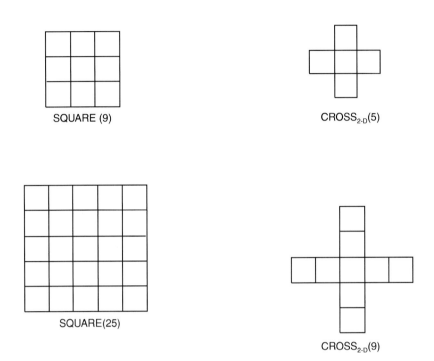

FIGURE 3.5 Typical 2-D windows SQUARE($2M + 1$) and CROSS($2M + 1$), where the numbers in parentheses indicate the number of coordinate shifts in the window.

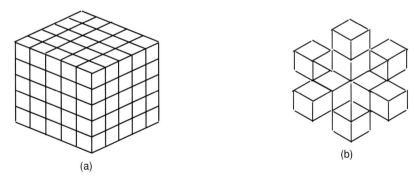

FIGURE 3.6 A 3-D window: CROSS($2M + 1$).

Thus, the windowed set **B·J(i)** has the same geometric interpretation as the set of image pixels in **J** that are covered by (or coincide with) the window **B** when it is centered on the image at coordinate **i**. Thus,

$$\text{SQUARE}(2M + 1) \cdot \mathbf{J}(\mathbf{i}) = \{J(\mathbf{i} - \mathbf{m}): -N \leq \mathbf{m} \leq N\} \tag{3.9}$$

where $2M + 1 = (2N + 1)^2$,

$$\text{CUBE}(2M + 1)\cdot\mathbf{J}(\mathbf{i}) = \{J(\mathbf{i} - \mathbf{m}): -N \leq \mathbf{m} \leq N\} \tag{3.10}$$

where $2M + 1 = (2N + 1)^3$,

$$\text{CROSS}_{2\text{-D}}(2M + 1)\cdot\mathbf{J}(\mathbf{i}) = \{J(i - m, 0): -M \leq m \leq M\} \cup \{J(0, j - n): -M \leq n \leq M\} \tag{3.11}$$

and

$$\text{CROSS}_{3\text{-D}}(2M + 1)\cdot\mathbf{J}(\mathbf{i})$$
$$= \{J(i - m, 0, 0): -M \leq m \leq M\} \cup \{J(0, j - n, 0): -M \leq n \leq M\} \cup$$
$$\{J(0, 0, k - p): -M \leq p \leq M\} \tag{3.12}$$

Using this notation makes it possible to define any kind of binary morphological operation on an image **J**, given a window **B**. While windows **B** other than SQUARE, CUBE, CROSS$_{2\text{-D}}$, and CROSS$_{3\text{-D}}$ can be defined, these are often useful because they are easily indexed sequentially in a computer program, and because they approximate a circular relationship between pixels. An approximately circular shape is desirable because it is usually reasonable to require a window to interact with objects in a manner that is approximately invariant to the local orientations of the object boundaries.

It should be observed that the windowed sets are not well-defined at the boundaries of the image because the coordinate shifts define pixels to be collected that lie outside of the image (i.e., outside of the prescribed range $0 \leq \mathbf{i} \leq N - 1$). As with the average filtering operation defined earlier, this is easily ameliorated by replicating pixels, that is, assigning binary values from the closest pixels to the undefined pixel elements.

3.5.3 Morphological Operations

A morphological operation F on a windowed set of image pixels is a binary or Boolean function of the pixels indexed within the window:

$$J_F(\mathbf{i}) = F[\mathbf{B}\cdot\mathbf{J}(\mathbf{i})] = F\{J(\mathbf{i} - \mathbf{m}); \mathbf{m} \in \mathbf{B}\} \tag{3.13}$$

If the morphological operation F is applied at every pixel coordinate $0 \leq \mathbf{i} \leq N - 1$, then an output or morphologically filtered image $\mathbf{J}_F$ results:

$$\mathbf{J}_F = F[\mathbf{J}, \mathbf{B}] = [J_F(\mathbf{i}); 0 \leq \mathbf{i} \leq N - 1], \tag{3.14}$$

where the elements $J_F(\mathbf{i})$ are given immediately above. The resulting image $\mathbf{J}_F$ is the result of the morphological operation F applied to the image **J**, **with respect to** the window **B**. The emphases in the preceding sentence are important, since the window used determines to some extent the nature of the output.

While there are many other interesting morphological operations, for our purposes, four basic morphological operations — **erosion, dilation, opening**, and **closing** — will suffice. Interested readers may consult Serra.[2]

3.5.3.1 Dilation

Given a binary 2-D or 3-D image **J** and a window or structuring element **B**, the dilation of **J** relative to **B** is defined as

$$\mathbf{J}_{\text{dilation}} = \text{DILATE}(\mathbf{J}, \ \mathbf{B}) \tag{3.15}$$

with elements

$$J_{\text{dilation}}(\mathbf{i}) = \text{OR}[\mathbf{B} \cdot \mathbf{J}(\mathbf{i})], \tag{3.16}$$

where OR denotes the logical or Boolean operation

$$\text{OR}\left(X_1, \ X_2 \dots, \ X_{2M+1}\right) = \begin{cases} \text{'0'} ; \ \text{if } X_1 = X_2 = X_3 = \cdots X_{2M+1} = \text{'0'} \\ \text{'1'} ; \ \text{otherwise} \end{cases} \tag{3.17}$$

for arbitrary binary variables $X_1, X_2, \dots, X_{2M+1}$. Thus, at each image coordinate $\mathbf{i}$, the dilated image $J_{\text{dilation}}(\mathbf{i})$ will take logical value '0' if and only if the windowed set $\mathbf{B} \cdot \mathbf{J}(\mathbf{i})$ contains no logical elements '1', and logical value '1' otherwise. This operation is so termed because it tends to increase the sizes of object or BLACK regions (with logical value '1') in a binary image in a predictable way, as depicted in 2-D in Figure 3.7.

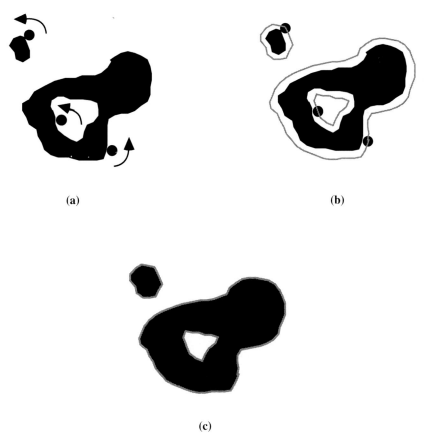

(a)   **(b)**

(c)

FIGURE 3.7 Qualitative 2-D description of the DILATE operation. Since the DILATE operation has a local effect only where the window $\mathbf{B}$ covers both WHITE and BLACK pixels (i.e., near or at object boundaries), it is useful to think of (a) the (approximately circular) window as rolling outside each boundary of BLACK objects in the image, (b) the center of $\mathbf{B}$ thus tracing a set of contours, which (c) form the boundaries of the DILATED image. In 3-D, one can instead imagine a spherical structuring element being rolled over the entire surface of a 3-D binary object, the center of which traces a 3-D surface which is the 3-D dilation.

3.5.3.2 Erosion

Given a binary image **J** and a 2-D or 3-D window or structuring element **B**, the erosion of **J** relative to **B** is defined as

$$\mathbf{J}_{\text{erosion}} = \text{ERODE}(\mathbf{J}, \mathbf{B}) \tag{3.18}$$

with elements

$$J_{\text{erosion}}(\mathbf{i}) = \text{AND}[\mathbf{B} \cdot \mathbf{J}(\mathbf{i})], \tag{3.19}$$

where AND denotes the logical or Boolean operation

$$\text{AND}\left(X_1, X_2 \ldots, X_{2M+1}\right) = \begin{cases} \text{'0'} \text{ ; if } X_1 = X_2 = X_3 = \cdots X_{2M+1} = \text{ '0'} \\ \text{'1' ; otherwise} \end{cases} \tag{3.20}$$

for arbitrary binary variables $X_1, X_2, ..., X_{2M+1}$. Thus, at each image coordinate **i**, the eroded image $J_{\text{erosion}}(\mathbf{i})$ will take logical value '1' if and only if the windowed set $\mathbf{B} \cdot \mathbf{J}(\mathbf{i})$ contains only logical elements '1', and logical value '0' otherwise. This operation is so termed because it tends to decrease the sizes of object or BLACK regions (with logical value '1') in a binary image in a predictable way, as depicted in 2-D in Figure 3.8.

3.5.3.3 Qualitative Properties of Dilation and Erosion

Generally, DILATE removes object holes of sufficiently small size (relative to the window **B**), and gulfs or bays of insufficient width, as depicted in Figure 3.9. Likewise, ERODE removes objects of insufficient size, and peninsulas of insufficient width, as depicted in Figure 3.10. From these properties, it can be deduced that the DILATION of the BLACK objects in an image is equivalent to the EROSION of the WHITE background; likewise, the EROSION of the BLACK objects in an image is equivalent to the DILATION of the WHITE background. Thus,

$$\text{DILATE}(\mathbf{J}, \mathbf{B}) = [\text{ERODE}(\mathbf{J}^c, \mathbf{B})]^c \tag{3.21}$$

and

$$\text{ERODE}(\mathbf{J}, \mathbf{B}) = [\text{DILATE}(\mathbf{J}^c, \mathbf{B})]^c \tag{3.22}$$

where 'c' denotes image complementation. Hence, ERODE and DILATE are dual operations with respect to image complementation. However, it is important to observe that erosion and dilation are only approximate inverses of one another; DILATING an ERODED image only rarely yields the original image. In particular, DILATE cannot recreate peninsulas eliminated by EROSION, nor can it recreate small objects eliminated by EROSION. Likewise, ERODING a DILATED image only rarely yields the original image. In particular, EROSION cannot recreate holes filled by DILATION, nor can it recreate gulfs or bays filled by DILATION. Nevertheless, certain operations defined as concatenations of EROSION and DILATION are in themselves very useful morphological operations, as described next.

3.5.3.4 Opening and Closing

Given a 2-D or 3-D binary image **J** and a window or structuring element **B**, the **opening** and **closing** of **J** relative to **B** are, respectively, defined as

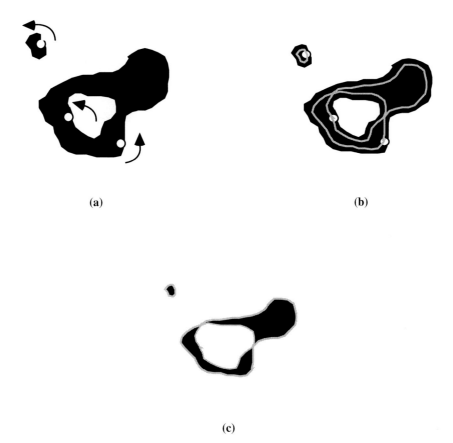

(a)   **(b)**

(c)

FIGURE 3.8   Qualitative 2-D description of the ERODE operation. Since the ERODE operation has a local effect only where the window **B** covers both WHITE and BLACK pixels (i.e., near or at object boundaries), it is useful to think of (a) the (approximately circular) window as rolling inside each boundary of BLACK objects in the image, (b) the center of **B** thus tracing a set of contours, which (c) form the boundaries of the ERODED image. In 3-D, one can instead imagine a spherical structuring element being rolled over the entire **inner** surface of a 3-D binary object, the center of which traces a 3-D surface which is the 3-D dilation.

$$\mathbf{J}_{open} = \text{OPEN}(\mathbf{J}, \mathbf{B}) = \text{DILATE}\,[\text{ERODE}(\mathbf{J}, \mathbf{B}), \mathbf{B}] \tag{3.23}$$

and

$$\mathbf{J}_{close} = \text{CLOSE}(\mathbf{J}, \mathbf{B}) = \text{ERODE}\,[\text{DILATE}(\mathbf{J}, \mathbf{B}), \mathbf{B}] \tag{3.24}$$

Thus, the operation OPEN is the operation DILATE (relative to **B**) applied to the result of the operation ERODE (relative to **B**). Similarly, the operation OPEN is the operation ERODE (relative to **B**) applied to the result of the operation DILATE (relative to **B**). It should be noted that ERODE and DILATE do not commute to each other, that is, it is usually true that

$$\text{OPEN}(\mathbf{J}, \mathbf{B}) \neq \text{CLOSE}(\mathbf{J}, \mathbf{B}) \tag{3.25}$$

Hence, the OPEN and CLOSE operations are not the same. In fact, an interesting property of OPEN and CLOSE is that, for any window **B**,

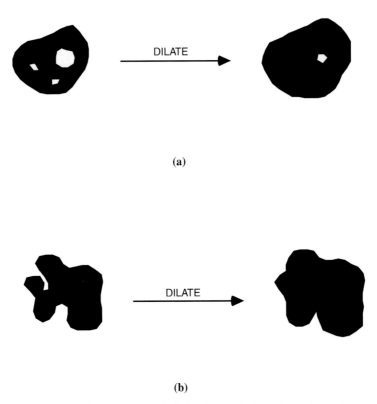

FIGURE 3.9 Qualitative properties of DILATE in 2-D, which fills (a) object holes and (b) gulfs or bays of insufficient size (relative to the size of the window **B**). In 3-D, the effect is the same; only 3-D surface holes and gaps are filled by the DILATE operation.

$$OPEN(\mathbf{J}, \mathbf{B}) \le \mathbf{J} \le CLOSE(\mathbf{J}, \mathbf{B}) \tag{3.26}$$

where for any binary images $\mathbf{J}_1$ and $\mathbf{J}_2$, $\mathbf{J}_1 \le \mathbf{J}_2$ denotes that $J_1(\mathbf{i}) = \text{'1'}$ implies that $J_2(\mathbf{i}) = \text{'1'}$ for any $\mathbf{i}$. Thus, the CLOSE operation must always expand object (BLACK) regions, whereas the OPEN operation must always decrease object (BLACK) regions. However, the concatenations OPEN and CLOSE generally leave the overall (gross) outlines of objects unchanged where there are no extraneous objects, holes, gulfs, or peninsulas. Thus, OPEN and CLOSE are effective object shape smoothing operations.

3.5.3.5 Qualitative Properties of OPEN and CLOSE

In applying morphological operators, one can choose a structuring element of any shape or size. However, most objects arising in biomedical applications have a relatively smooth (low curvature) boundary, at least at the scale of observation. Thus, if it is desired to smooth a boundary using the CLOSE operation, the window **B** should have a smooth boundary and should also be approximately isotropic, since it should not be biased to any particular direction. The best candidate satisfying these conditions is a circle (in 2-D) or a sphere (in 3-D). The effects of applying CLOSE with an approximately circular/spherical window **B** are as follows:

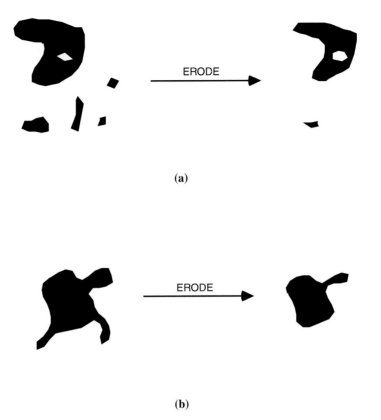

(a)

(b)

FIGURE 3.10   Qualitative properties of ERODE, which removes (a) extraneous objects and (b) peninsulas of insufficient size (relative to the size of the window B). In 3-D, the effect is the same, only 3-D surface objects and peduncles are filled by the DILATE operation.

- No object contour will have a point with a positive curvature greater than $1/r$, where r is the radius of **B**.
- Every gap with width narrower than the diameter of **B** will be filled.
- Every hole with maximum width less than the diameter of **B** will be filled.

Figure 3.11 depicts these properties; the effects of OPEN may be deduced from the fact that OPEN is the dual of CLOSE with respect to image complementation. Note that the above properties include all effects of CLOSE in the case when the image has a single object. If the image has two or more objects, some other effects, such as merging of two separate objects, can arise. Also, note that the size of the structuring element takes an important role in the closing operation, since all of the above three properties are affected by it. In our implementation examples, the radius of the circle/sphere is chosen interactively by a human operator, depending on the state of the thresholded image.

3.5.3.6   OPEN-CLOS, CLOS-OPEN, and Majority Filter

The astute reader has probably observed that the morphological smoothing filters considered thus far (OPEN and CLOSE) each have a **bias** in their respective responses. The OPEN operation removes only "BLACK" noise such as extraneous objects or peninsula, whereas the CLOSE

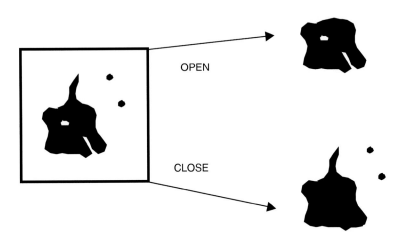

FIGURE 3.11 Depiction of qualitative properties of OPEN and CLOSE.

operation only removes "WHITE" noise such as extraneous holes or gaps. Naturally, it is of interest to know whether it is possible to construct filters that remove both kinds of noise since both types often coexist in the same image.

One approach is to cascade the operations of CLOSE and OPEN. In fact, one can define the CLOS-OPEN and the OPEN-CLOS morphological filters in precisely this way: given a 2-D or 3-D binary image $\mathbf{J}$ and a window or structuring element $\mathbf{B}$, the **open-clos** and **clos-open** of $\mathbf{J}$, with respect to $\mathbf{B}$, are defined as

$$\mathbf{J}_{\text{open-clos}} = \text{OPEN-CLOS}(\mathbf{J}, \mathbf{B}) = \text{OPEN } [\text{CLOSE}(\mathbf{J}, \mathbf{B}), \mathbf{B}] \qquad (3.27)$$

and

$$\mathbf{J}_{\text{clos-open}} = \text{CLOSE-OPEN}(\mathbf{J}, \mathbf{B}) = \text{CLOSE } [\text{OPEN}(\mathbf{J}, \mathbf{B}), \mathbf{B}] \qquad (3.28)$$

Thus, OPEN-CLOS is the OPEN applied to the result of the CLOSE (both relative to $\mathbf{B}$). Similarly, CLOS-OPEN is the CLOSE applied to the result of the OPEN (both relative to $\mathbf{B}$). Both operations can be defined in 2-D, 3-D (or higher dimensions) by appropriate selection of a window or structuring element $\mathbf{B}$ having the same dimensionality as the image being analyzed. The results of the OPEN-CLOS and the CLOS-OPEN operations tend to be quite similar when applied to the same image. Both operations smooth without bias; both "BLACK" noise and "WHITE" noise are removed efficaciously. The sizes of objects are little affected since dilations are always countered with erosions, and vice versa. However, while CLOS-OPEN and OPEN-CLOS are quite similar, they are not mathematically identical, and the result of processing with them will generally be slightly different, although often qualitatively identical. One difference between these filters is that OPEN-CLOS often tends to link neighboring objects together (if they are separated by distances less than the diameter of the window), while the CLOS-OPEN tends to link neighboring holes together. Figure 3.12 illustrates these differences. Usually, these effects are quite spatially limited, but they may be noticeable if there are speckled regions containing clusters of closely spaced BLACK or WHITE blobs of small size.

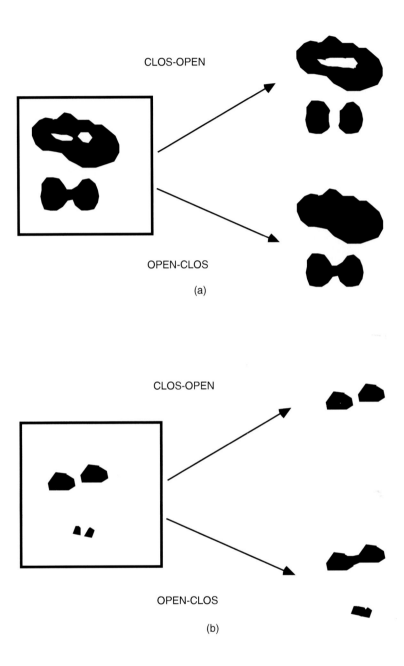

FIGURE 3.12 Depiction of differences between OPEN-CLOS and CLOS-OPEN morphological operations. (a) The CLOS-OPEN tends to link neighboring holes together, if they are separated by a distance less than the window diameter. The OPEN-CLOS does not. (b) The OPEN-CLOS tends to link neighboring objects together, if they are separated by a distance less than the window diameter. The CLOS-OPEN does not.

Another effective operator, which has properties similar to both the OPEN-CLOS and CLOS-OPEN, is the binary **majority filter**. This filter is not composed of concatenations of EROSIONS and DILATIONS, but nevertheless has a simple binary processing description. Given a binary image

J and a 2-D or 3-D window or structuring element **B**, the majority-filtered version of **J** relative to **B** is defined as

$$J_{majority} = \text{MAJORITY}(\mathbf{J}, \mathbf{B}) \tag{3.29}$$

with elements

$$J_{majority}(\mathbf{i}) = \text{MAJ}[\mathbf{B} \cdot \mathbf{J}(\mathbf{i})], \tag{3.30}$$

where MAJ denotes the logical or Boolean operation

$$\text{MAJ}(X_1, X_2 \dots, X_{2M+1}) = \begin{cases} \text{'0' ; if a majority of } X_1, X_2 \dots, X_{2M+1} = \text{are '0'} \\ \text{'1' ; if a majority of } X_1, X_2 \dots, X_{2M+1} = \text{are '1'} \end{cases} \tag{3.31}$$

for arbitrary binary variables $X_1, X_2, \dots, X_{2M+1}$. Thus, at each image coordinate **i**, the filtered image $J_{majority}(\mathbf{i})$ will take logical value '1' if and only if the windowed set $\mathbf{B} \cdot \mathbf{J}(\mathbf{i})$ contains a majority of elements '1', and logical value '0' otherwise. This operation is also called the **median filter** because the majority of a sample is also its median if '1' and '0' are given relative rank. Also, the median filter that is defined on gray-scale images reduces to the majority filter on binary images.

Like the CLOS-OPEN and OPEN-CLOS, the MAJORITY filter removes both objects and holes of too-small size, as well as both gaps (bays) and peninsulas of too-narrow width, as depicted in Figure 3.13. Hence, the MAJORITY filter is also an unbiased smoother. However, the smoothing power of MAJORITY is about half that of OPEN-CLOS and CLOS-OPEN, meaning that a window of about twice the area is needed to fill a hole or gap as compared to the other morphological operators. Like the other operators, MAJORITY does not change the sizes of objects by much. In fact, the MAJORITY is its own dual filter since for any Boolean set **X**:

$$\text{MAJ}(\mathbf{X}^c) = [\text{MAJ}(\mathbf{X})]^c \tag{3.32}$$

The MAJORITY filter is somewhat more complicated than the CLOS-OPEN and the OPEN-CLOS to implement, but it offers similar advantages and simplicity of concept. Figure 3.14 illustrates the various morphological operations that have been introduced here for region correction as applied to an actual 2-D micrograph that was digitized and thresholded. The thresholded image contains numerous defects of the variety described here, as well as a generally poorly connected part of the largest object (near the top).

As an example in 3-D, Figure 3.15 depicts a *Taraxacum officinale* (Compositae) pollen grain that was imaged with a Zeiss laser-scanning confocal microscope operating in sectioning mode, at 63x magnification and 30x zoom (1890x total), using an oil immersion slide with numerical aperture NA = 1.4. The specimen was illuminated with a HeNe laser. The longest actual physical dimension of the specimen was about 38 microns. This image was made available and processed by Dr. Keith Bartels.[30] The original (gray-scale) data, which is of high contrast under the laser illuminant, is not shown because it was quite noisy and cluttered, and because of the difficulty in displaying 3-D intensity data. The thresholded image also contains a considerable amount of extraneous structure. However, median filtering using CUBE(125) (a 5 × 5 × 5 cube-shaped window) produced a very smooth object/surface, with few artifacts remaining.

3.5.3.7 Implementation Aspects

It is of interest to consider the implementation aspects of morphological filters. Although morpho-logical operations can be implemented in a straightforward manner using simple logical operations

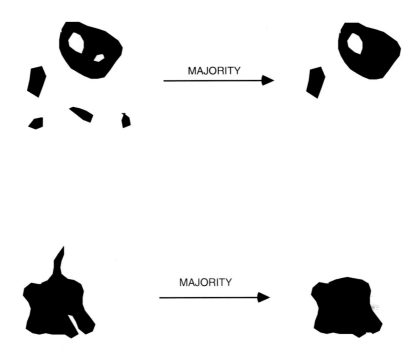

FIGURE 3.13   Depiction of the effect of binary MAJORITY filtering on too-small objects, holes, gaps, and peninsulas. Object size is generally not changed.

defined at each image coordinate, the direct implementation of morphological operations may require excessive computation time. For example, if in 2-D the circle radius (implemented as a 2-D SQUARE, for example) is equal to 5, which is typical for many applications, then the area of the region of support of **B** is approximately equal to $3.14 \times 5^2 \approx 78$ pixels. For an image containing N^2 pixels, the total number of required logical operations is approximately $78 \times N^2$ for either ERODE or DILATE. If a 3-D CUBE window is used, then the computation is multiplied by another factor of 5. The amount of computation can be reduced considerably using the following property: as the position of the window **B** is incremented along a line from coordinate **i** to an adjacent coordinate **j**, a large portion of the windowed set **B·J(i)** will coincide with the windowed set **B·J(j)**; only the elements not contained in the overlap will be different. Hence, instead of recomputing the entire operation, the previous value computed, along with the updated region, can be used. This can be accomplished simply by, at each coordinate, separately storing the AND or OR of the elements contained within the overlap region **B·J(i)** $\cap$ **B·J(j)**. Using this property, it is possible to reduce the amount of computation to a fraction of a direct implementation.

3.6   APPLICATION CONSIDERATIONS AND EXAMPLES

3.6.1   WHEN TO USE BLOB COLORING VS. MORPHOLOGICAL FILTERING

Various applications demand different processing regimens to achieve the best results. In some cases, either blob coloring/removal or morphological filtering is used; in other cases, both are used. Blob coloring/removal can be used if it is desired to remove extraneous objects and holes from

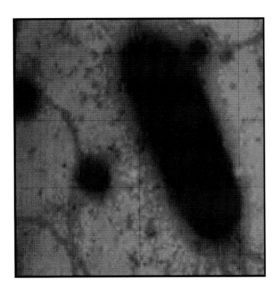

(a)

FIGURE 3.14 Examples of morphological filters. In all cases the window used is B = SQUARE(9). (a) Original 2-D micrograph containing three large objects, (b) thresholded to create binary image **I**, (c) DILATE(**I**, **B**), (d) ERODE(**I**, **B**), (e) CLOSE(**I**, **B**), (f) OPEN(**I**, **B**), (g) OPEN-CLOS(**I**, **B**), (h) CLOS-OPEN (**I**, **B**), (i) MAJORITY(**I**, **B**). Notice the difference between (e), (f), and (g), particularly near the top of the large object, where there are many scattered BLACK and WHITE SPOTS.

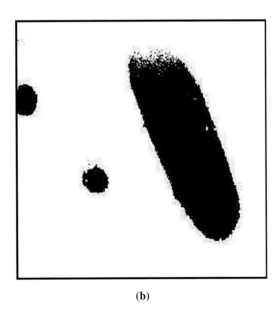

(b)

FIGURE 3.14 (Continued)   (b) Thresholded to create binary image **I**.

specimens, but it is important not to disturb the boundary or surface of the object. For example, an object may naturally possess a highly tortuous or fractal boundary or surface that will not be well-preserved by morphological smoothing. In other cases, the extraneous objects/holes may be

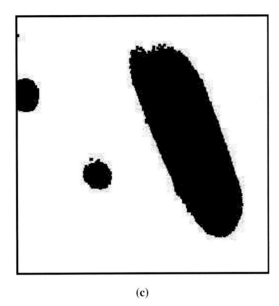

(c)

FIGURE 3.14 (Continued)   (c) DILATE(**I, B**).

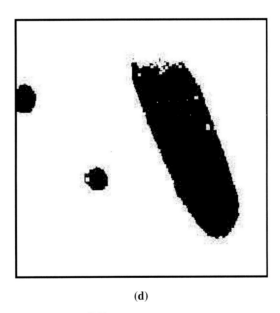

(d)

FIGURE 3.14 (Continued)   (d) ERODE(**I, B**).

of uniformly small enough size to be removed by morphological filtering without the need for blob coloring/removal. If the smoothing of object boundaries or surfaces is desirable or tolerable, then it may suffice to only use morphological filtering. In the general scenario, both methods should be considered.

3.6.2   APPLICATION TO BRIGHT-FIELD MICROGRAPHS OF PANCREAS ISLETS

The first application example illustrating the main themes given in this chapter is a very successful, generic, automated area analysis system that isolates an unknown object from its background and

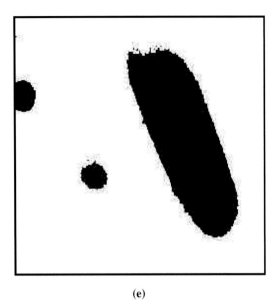

(e)

FIGURE 3.14 (Continued)   (e) CLOSE(**I, B**).

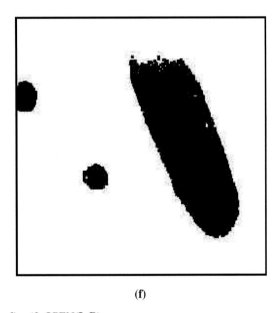

(f)

FIGURE 3.14 (Continued)   (f) OPEN(**I, B**).

measures the area of the object.[11] The algorithm has been applied quite successfully to determine the change in area in standard micrographic images of pancreatic islets of Langerhans. The system described affords significant improvement over earlier attempts,[12] which limited the analysis to the use of simple gray-level thresholding techniques.

The islet presents a difficult image to analyze because its surface tends to have large spatial variations in both gray level and light reflectance. Also, the boundary tends to be quite irregular, and there may be considerable variation in shape within a population. Therefore, standard object segmentation strategies often provide results that contain errors in object-background classification, or in the accurate identification of object boundary location. Errors of these types can render area

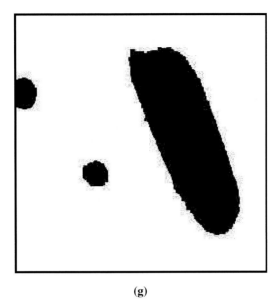

(g)

FIGURE 3.14 (Continued) (g) OPEN-CLOS(**I, B**).

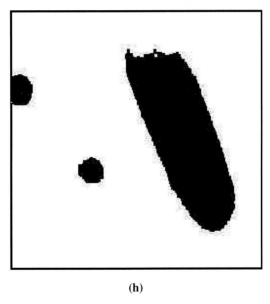

(h)

FIGURE 3.14 (Continued) (h) CLOS-OPEN (**I, B**).

measurement calculations highly inaccurate. Thus, approaches for correcting the segmentations are of interest.

The system is composed as follows: first, the image is segmented using simple gray-level thresholding. Classification errors inevitably occur. Then, misclassified regions and errors in the object estimates are corrected by blob coloring/removal and by morphological filtering. While the determination of an appropriate threshold is difficult and an ideal threshold is impossible because of the input image characteristics, the gray-level thresholding technique combined with object and shape-correcting devices produce measurements of area that are highly in accordance with accurate manual techniques.

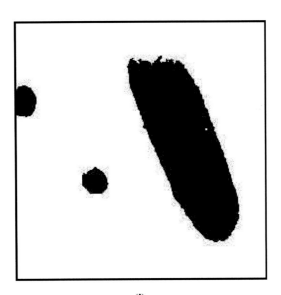

(i)

FIGURE 3.14 (Continued) (i) MAJORITY(**I, B**).

Figure 3.16a shows an example data image, a cross-sectional bright-field light micrograph of a rat pancreas islet. The islet presents a difficult image to analyze due to its complex shape, which somewhat resembles a cluster of grapes. Whereas a human operator could trace the boundary of the wound without difficulty, this kind of image represents a difficult subject for a machine vision system to analyze because of its complex shape, and the varying and overlapping gray-level distributions in both the object and the background regions. Thus, it is not easy even to subjectively identify an outline just by casual observation. Figure 3.16b illustrates a binary image obtained by the gray-level thresholding process. After the region correction and morphological operations are completed, the system gives the final output as the one illustrated in Figure 3.16c. It is worth noting the difference between the object boundary in Figure 3.16b and the smoothed boundary in Figure 3.16c, which is seen most clearly in the lower portion of the object. The smoothed boundary matches the actual shape quite well, as can be seen in Figure 3.16d, where the detected boundary is superimposed on the original image.

A second series is shown in Figure 3.17 on another islet micrograph obtained under the same circumstances. Figure 3.17c shows the output of this region correction algorithm and Figure 3.17d the detected boundary superimposed on the original image. Again, a high degree of accuracy was obtained.

An important application of this particular example is in the potential development of protocols for the successful cryopreservation of pancreas islets for curative transplantation into diabetic patients. In developing and assessing the viability of various freezing regimens (solutions, cooling rates, etc.), it is of interest to monitor the physical changes that the cells undergo during the freezing process. Often, crenulation of the cellular body or intrusion of ice crystals can be implicated in the death of the tissue. Hence, cryomicroscopy is an area of active research wherein it is of interest to rigorously quantify the shape and size changes in areas of cells and cell clusters undergoing freeze/thaw processes.[9,21,25,26] Different types of cells — ranging from yeast, human granulocytes,[27] and pancreas endocrine β-cells[28] to multicellular islets of Langerhans[29] — have been investigated using a variety of techniques.

In one algorithm that automatically computed the area of cells following a defined regimen of cooling and thawing, the segmentation of the digitized images into cells and background was accomplished by a more complex sequence of operations.[9] A modification of the Hough transform

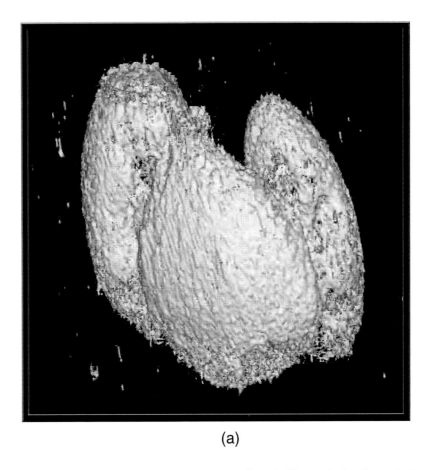

(a)

FIGURE 3.15   Processing of a 3-D image of a *Taraxacum officinale* (Compositae) pollen grain imaged with a laser-scanning confocal microscope. (a) Thresholded binary image containing considerable noise, (b) cleaned image using majority filter with a CUBE(125) window. A few extraneous objects remain which could be removed by blob coloring/removal. The resulting "clean" image is much more amenable for volumetric analysis of the specimen.

was applied as a means of evaluating all of the candidate object edges, in order to determine the one most likely of being the cell boundary.[5,25] When a series of images during a freezing sequence was analyzed, the Hough transform was continually updated so that in any frame of the image sequence, the program searched for an element having a shape identified for the cell in the immediately preceding frame. This step resulted in a reasonably good estimate of the cell outline in many cases. The result was a defined two-dimensional area, assumed to lie in a single plane. A center of gravity was identified for this area, from which the distance to all boundary pixel locations was measured. The average statistics of these distances were used to estimate cell irregularity, area, degree of circularity, etc. However, this approach is somewhat inexact and the measurement of area could be better accomplished by the more modern techniques described in the preceding.

3.6.3   FISH LABELED INTERPHASE CELLS

The application of digital image analysis algorithms to determine the area (size) of objects is demonstrated using images of FISH (Fluorescence *In Situ* Hybridization) labeled interphase cells. FISH is routinely used in clinical cytogenetics as a prenatal screening test to determine genetic

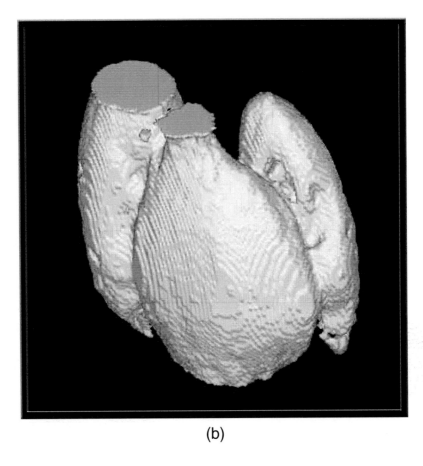

(b)

FIGURE 3.15 (Continued)   (b) cleaned image using majority filter with a CUBE(125) window.

aneuploidy. Aneuploidy screening via FISH, followed by classical cytogenetics, allows diagnosis of newborns with multiple congenital anomalies. Cells are typically counterstained with a nuclear stain such as DAPI (4′,6-diamidino-2-phenyl indole dihydrochloride) or PI (propidium iodide). Biotin- and digoxigenin-labeled DNA probes are used in conjunction with secondary reagents (e.g., avidin or antidigoxigenin conjugated to fluorescein or rhodamine) to detect the target chromosomes. Fluorescence imaging in conjunction with digital image analysis can be used to isolate cells, measure their areas, and count the number of chromosomes in each cell.

The techniques described here are demonstrated for this application. The digitized images are thresholded to obtain binary images of cells. A histogram of the pixel intensity data is initially generated. Because the images consist of cells at a uniform gray level that rest on a relatively uniform but contrasting background, the histogram is bimodal in nature. The larger peak corresponds to the background points, and the objects can be isolated by selecting a threshold value midway between the two peaks. The cell boundaries were smoothed using the CLOS-OPEN operation, where the window size varies with the size of the objects. This results in a smoothed cell boundary and consequently improved measurements of cell area. The cells are then uniquely identified using blob coloring. The number of pixels in each object is then counted to estimate its area in the image plane (which can be converted into true area as described later). In this example, a minimum size criterion is used to eliminate noise spikes and/or objects that fall below a minimum area threshold. In this way, the "BLACK" noise is removed from the image. As described earlier, the "WHITE"

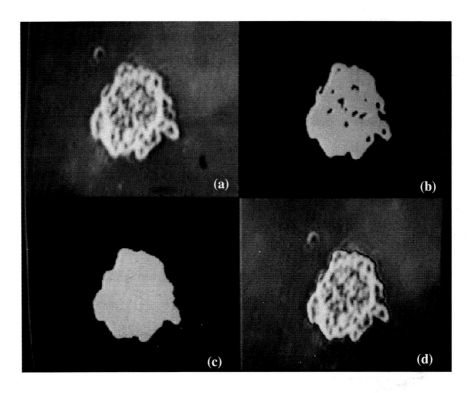

FIGURE 3.16   Example of micrograph image of rat pancreas islet at each processing step. (a) Original image, (b) gray-level thresholded image, (c) region-corrected and morphologically smoothed image, and (d) detected boundary superimposed on the original image.

noise is removed by inverting the image and applying the same blob coloring/removal procedure. Using a threshold, multiple objects are retained instead of a single object.

In the first example (Figure 3.18), the original image was thresholded at a value of 41 to obtain the binary image of objects, and was then smoothed using the CLOS-OPEN operation. Blob coloring was not utilized prior to filtering because the extraneous objects in the image were deemed to be of sufficiently small size to be handled by filtering. Moreover, the many small objects near the perimeters of the objects were considered to potentially be part of the main objects, and not necessarily to be removed. Filtering may incorporate such small objects into the main objects. Following filtering, the blob coloring procedure identified seven objects with (pixel-count) areas of 5793, 12480, 3239, 7453, 4813, 2918, and 3885 pixels, respectively. Each distinct object was displayed with a unique intensity (or "color"). Because these objects were deemed to be within a sufficient percentage of each other in size, blob removal was not utilized at this latter stage either. Hence, all objects were deemed to be of interest. The final panel of Figure 3.18 depicts the estimated boundaries ('0' – '1' transitions) overlaid on the original image.

In the second example (Figure 3.19), the threshold was determined at gray level 40, and the binarized image was smoothed using the CLOS-OPEN operation. The image was then blob-colored to identify four objects. Blob coloring and minor region removal was then used to remove both extraneous objects and extraneous holes. Initially, four objects were identified with areas of 6426, 69, 7644, and 2976 pixels, respectively. The smallest object was discarded because its area was a small fraction of any of the other objects. One of the remaining large objects was actually two large objects that were in contact, and initially identified as a single region. However, a **watershed algorithm** (described on p. 415 of Ref. 2), was applied to separate these two cells, as depicted in

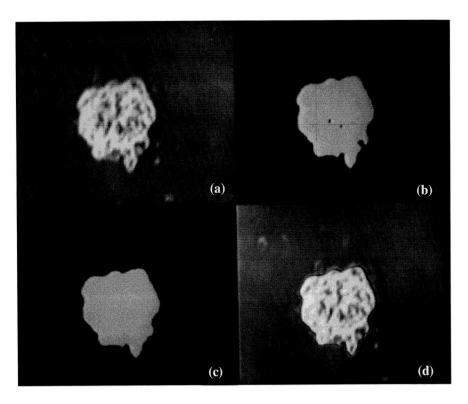

FIGURE 3.17   A micrograph of another pancreas islet. (a) Original image, (b) gray-level thresholded image, (c) region-corrected and smoothed image, and (d) detected boundary superimposed on the original image.

the final image, which shows the computed boundaries overlaid on the original. The watershed algorithm essentially operates by successively eroding a binary image until the various objects are separated, then selectively dilating the image while controlling the objects from re-merging. However, this description is somewhat simplistic and so the reader is referred to Ref. 2.

3.6.4 OBJECTIVE TESTING OF THE ALGORITHM

Naturally, it is of interest to determine whether the methods described here produce an accurate representation of true object occupancy and hence of object area or volume. In order to objectively assess the performance of the system described above, 650 photographic images of healing wounds were taken and processed through the stages of digitization, thresholding, blob coloring/removal, morphological smoothing, and image area measurement by pixel counting. Simultaneously, the same photographs were analyzed using a polar compensating radial planimeter, which represents the most sensitive of the traditional mechanical approaches to the problem. A statistical analysis of the two data sets (digital and mechanical) yielded an overall simple Pearson correlation of 0.9975 ($r = 0.0001$). This study was reported previously and so further details are not presented here.[31]

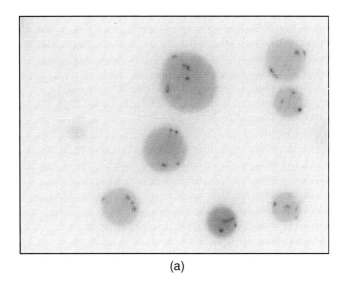

(a)

FIGURE 3.18 Example of processing FISH cells. (a) Original image; the negative is shown for better visual interpretation, since the image is otherwise dark and of low contrast. (b) Image histogram. Since the negative is being processed, brighter intensities are to the left. (c) Thresholded image using a threshold level 41. (d) Image cleaned using CLOS-OPEN. (e) Blob colored image. Different intensities denote distinct blob "colors." No blob was deemed to be too small to retain. (f) Original with computed boundaries overlaid.

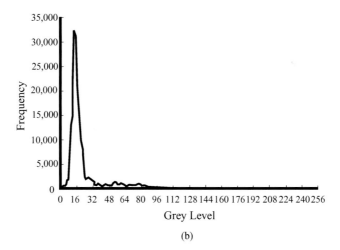

(b)

FIGURE 3.18 (Continued) (b) Image histogram. Since the negative is being processed, brighter intensities are to the left.

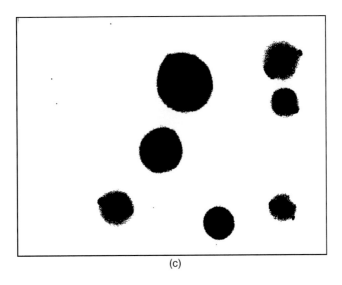

(c)

FIGURE 3.18 (Continued)   (c) Thresholded image using a threshold level 41.

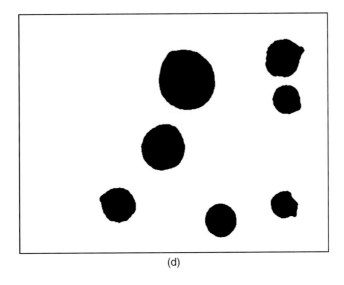

(d)

FIGURE 3.18 (Continued)   (d) Image cleaned using CLOS-OPEN.

3.7   SPECIMEN PREPARATION AND COMPUTATION OF TRUE OBJECT AREA/VOLUME OR CHANGE OF AREA/VOLUME

3.7.1   SPECIMEN PREPARATION

In biological work, a great deal of attention should be paid to the techniques applied to prepare a specimen for quantitative morphometric area analysis measurements. The reason for this is that, while there exist quite powerful digital image processing techniques for effecting corrections to

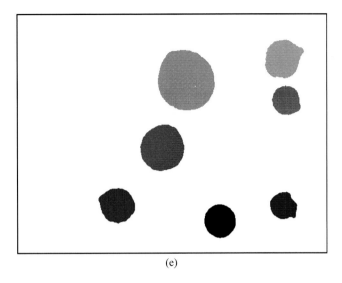

(e)

FIGURE 3.18 (Continued) (e) Blob colored image. Different intensities denote distinct blob "colors." No blob was deemed to be too small to retain.

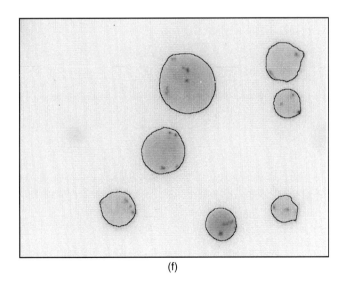

(f)

FIGURE 3.18 (Continued) (f) Original with computed boundaries overlaid.

errors made in the physical image process, certain errors, or errors of large magnitude, can be very difficult to correct. Particular care must be taken to keep any fixation, dehydration, and embedding procedures as constant and minimal as possible in order to avoid considerable volume and area changes that can occur, especially due to shrinkage. This is particularly important when it is desired to measure changes in area or volume following the application of some kind of regimen. A small difference in the histological technique can result in marked differences in the microscopic appearance of cells or tissue.

If the measurements are being done on tissue sections (a macroscopic application), then consideration must be given to the thickness of the sections and staining density variations in order to avoid errors caused by the Holmes effect. An ideal specimen is an infinitely thin section; however, in practice, of course, sections must have a finite thickness. This gives rise to two main problems. First, the area occupied by a given section may be overestimated, owing to contributions to the image from portions of the specimen lying outside the plane of focus. Second, the components being measured will have an imperfect contrast with the environment, particularly when small intra-cytoplasmic organelles are being imaged. For components such as these that do not contrast well with the surrounding medium, there will be a tendency to underestimate the area occupied on the surface of a section. Fortunately, these two effects tend to cancel each other out on many occasions, but it is essential to be aware of the problems involved and take steps to minimize and possibly correct any potential errors. A detailed account of these and associated problems is available in Aherne et al.[15] In general, sections thicker than 10 μm and measurements over a large number of the objects of interest are recommended.

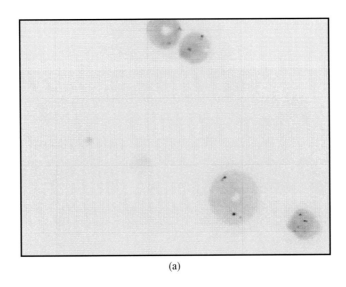

(a)

FIGURE 3.19 Example of processing FISH cells. (a) Original image (the negative is again shown); (b) image (negative) histogram; (c) thresholded image using a threshold level 40; (d) image smoothed using CLOS-OPEN; (e) removal of extraneous holes and blobs via blob coloring/removal algorithm, first applied to the image and then to its complement; (f) "colors" depicting the various distinct objects that remain in the image (note that two objects have been merged into one because of their proximity); and (g) original with computed boundaries overlaid. The merged objects were successfully separated using the **watershed algorithm**. (See text for details.)

Conversely, if isolated cells are the objects under consideration, then the use of smears or suspensions of isolated cells in order to measure projected area avoids the problem of partly sectioned organelles being excluded in the measurement. Particular attention must also be paid to the nature of the suspending media when measurements are being made on isolated cells. For example, the volume (or projected area) occupied by red blood cells suspended in plasma may be different than when suspended in a phosphate buffer.

Another point of interest for certain applications is the method of staining employed for the preparation of a cellular specimen. Staining is desirable in order to enhance the contrast between the background and the object under consideration. When using fluorescent dyes, double-labeling

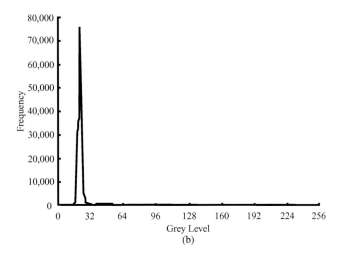

FIGURE 3.19 (Continued) (b) Image (negative) histogram.

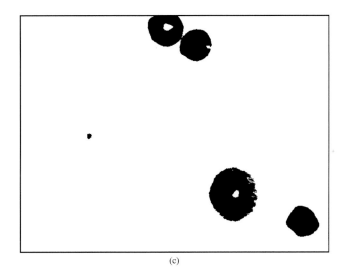

FIGURE 3.19 (Continued) (c) Thresholded image using a threshold level 40.

can significantly improve the selective demonstration and the contrast between two different objects. A variety of vital and specific fluorescent and nonfluorescent reagents are available for customized use and should be carefully chosen.[16,17]

Yet another important standardization problem for obtaining quantitative morphometric data is concerned with the calibration of the video camera in terms of optical density, scaling linearity, spectral sensitivity, and the data sampling rate.[18] Smooth curve dynamic range correction with a very minimum of run-out error over the full scaling range of the camera, together with a low spatial reading error, must be established. The sampling rate is one determinant of the image resolution because the resolution cannot be more than twice the pixel spacing distance. A general-purpose CCD video camera provides a relative, rather than an absolute scaling response, and neither the

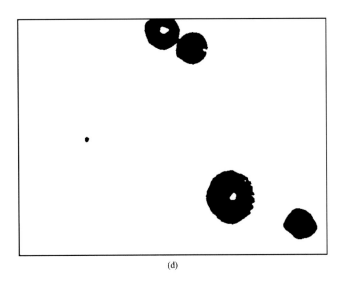

(d)

FIGURE 3.19 (Continued)   (d) Image smoothed using CLOS-OPEN.

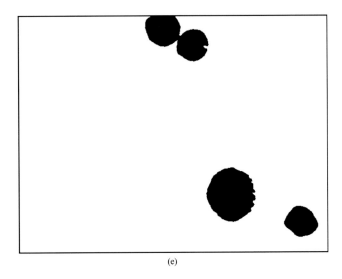

(e)

FIGURE 3.19 (Continued)   (e) Removal of extraneous holes and blobs via blob coloring/removal algorithm, first applied to the image and then to its complement.

camera scaling action nor the control facility is sufficiently drift-free to provide a completely repeatable and exacting readout. In microscopic applications using dyes, optical density gray-level intensity calibrations can be accomplished by first adjusting the transmission histogram peak of the blank digitized image, and then measuring the gray-level intensity of varying concentrations of the dye in a constant-geometry container (e.g., a hemocytometer). The least-squares fit of the normalized intensity data can then be determined. The normalized intensity, on a scale from 0 to 1, can then be used to represent the dynamic range of the imaging system.

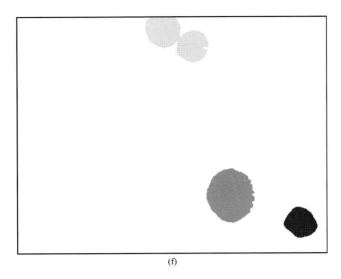

(f)

FIGURE 3.19 (Continued)   (f) "Colors" depicting the various distinct objects that remain in the image (note that two objects have been merged into one because of their proximity).

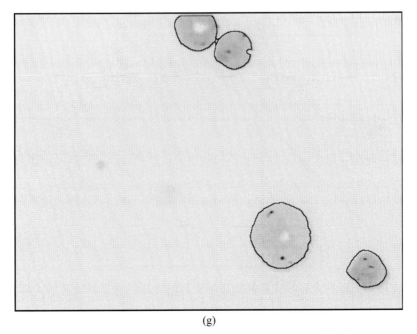

(g)

FIGURE 3.19 (Continued)   (g) Original with computed boundaries overlaid. The merged objects were successfully separated using the **watershed algorithm**.

3.7.2   AREA AND VOLUME MEASUREMENT

In the most common applications, it is of interest to measure or compute the true cross-sectional area, or the true volume, of an object or objects, rather than just the image-plane or "pixel" area/volume. The goal, then, is to transform the easily determined pixel area/volume into true area/volume, which requires that the general relationship between true area/volume and pixel area/volume, under the image projection process, be determined. This can be accomplished either by measuring the appropriate physical parameters of the imaging equipment directly, or alternately, by measuring objects of precisely known dimensions under the same imaging conditions as the objects of interest are to be studied. In either case, it is of use to understand the assumed relationship between image coordinates and true object coordinates.

Figure 3.20 depicts the general projective imaging geometry under a simple "pinhole-aperture" lens assumption, which is used here. Corrections to the model under a finite-aperture or shallow-depth of field model are beyond the current discussion as being too specific; but for 2-D specimens that are thin relative to the imaging scale, the assumption is highly accurate for most objectives of interest. For 3-D objects for which volumes are desired, additional work is required. In Figure 3.20, coordinates (X, Y, Z) denote points in true three-dimensional space, where the origin $(X, Y, Z) = (0, 0, 0)$ is taken to be the lens center. Thus the coordinate Z of an object feature is taken to be the distance of the object plane to the lens plane along the optical axis; the coordinates (X, Y) denote the orthogonal distances of the point from the optical axis. The image coordinates (x, y) denote points in the two-dimensional image plane, which is chosen to be parallel (congruent under projection) to the X-Y plane; the optical axis passes through the origins of both coordinate systems.

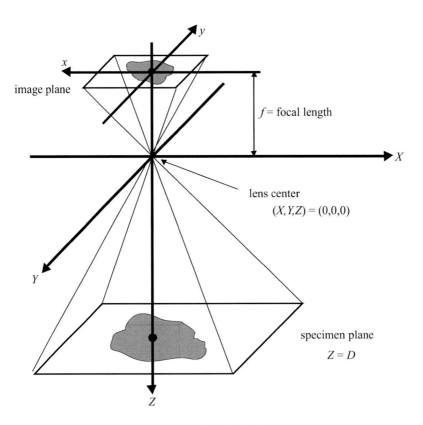

FIGURE 3.20 Projective imaging geometry under the pinhole lens assumption.

In the idealized imaging model used here, the lens is treated as a simple pinhole through which all light rays hitting the image plane pass. The image plane, which coincides with the photographic emulsion, or the CCD sensor array in a standard video camera, is a distance of one focal length f from the lens. Of course, in a complex multiple-lens system such as a microscope, the focal length parameter must be replaced by an estimate of the effective focal length, which may be difficult to determine, particularly if the microscope is interfaced with a video camera through an adapter. Moreover, when volumetric measurements are to be made, it may be necessary to associate different focal lengths with different sectional areas, which as a composite, comprise the volume. For each focusing position, the specimen plane is assumed to be a distance D from the lens plane; this can be determined via mechanical measurement in most instances, although again, in a complex microscope/adapter/camera system, determination of this quantity can be difficult. If multiple depth measurements are taken, viz., to obtain volumetric data, then D must be discovered for each depth plane, so that the aggregate areas may be combined into volumes. Thus, the process can be complex; indeed, in microscopic applications, it is usually preferred to make comparisons with an object of precisely known dimensions, as suggested below.

It is easily shown that under the simple projective geometry assumed here, a point $(X, Y, Z) = (X_0, Y_0, D)$ lying on the specimen (presumed perfectly flat) projects to the image-plane coordinates[5]

$$(x_0, y_0) = (f/D) (X_0, Y_0) \tag{3.33}$$

where the ratio f/D is the **magnification factor**, which varies with the distance D from the lens center to the specimen plane. Thus, for the distance D fixed or known, specimen coordinates are related to image coordinates by a simple scaling. It also follows that the true area A_{true} of the specimen of interest is related to the area A_{image} of its projected image by the (square of) the same scaling:

$$A_{\text{image}} = (f/D)^2 A_{\text{true}} \tag{3.34}$$

since the perspective projection is an affine transformation of the real-world coordinates. Thus, it remains to be able to measure the actual physical dimensions of points in the image plane (i.e., relate pixel areas to image plane areas). If a permanent photographic emulsion is used, this is easily accomplished by imaging the film negatives. Otherwise, a measurement of the true pixel dimensions must be made. Commercially available CCD camera manufacturers make available the physical dimensions of the sensor arrays, from which the dimensions of individual pixels may be deduced. Other types of cameras, such as vidicons, have overall physical sensor dimensions which are also specified. Estimates of the effective pixel dimensions can then be easily determined from the specifications of the frame-grabber apparatus (the sampling density).

If it is desired to image in the third dimension, viz., to compute object volumes, then ordinarily a sequence of depth images are combined and processed as a 3-D image. Suppose that M depth sections are taken, with associated distances $D_1, D_2, \ldots, D_M$, and hence, true depth differences $\delta_m = D_m - D_{m-1}$; $m = 1, \ldots, M$. If the true areas associated with the depth sections are (respectively) $A_1, A_2, \ldots, A_M$, then the overall true object volume may be estimated as

$$V_{\text{true}} = \sum_{m=1}^{M} A_m \delta_m \tag{3.35}$$

In many applications, the measurement of actual physical dimensions is not critical; it is often of interest instead to determine only the percent change in area, over time, of a specimen undergoing some type of regimen or evolving process such as healing. In this case, the determination of the desired quantities is vastly simplified (under the pinhole assumption) because the percent change in area remains unchanged with respect to the perspective transformation, and also with respect to image-to-pixel dimension changes. Thus,

$$\Delta A_{\text{true}}(\%) = \Delta A_{\text{image}}(\%) = \Delta A_{\text{pixel}}(\%) \qquad (3.36)$$

where $\Delta A(\%)$ represents percent change in area and A_{pixel} is the measured object area expressed as an integer number of pixels. Likewise, changes in volume will scale in the same way under the perspective (pinhole) model and, hence, percent volume changes can be computed at the pixel level.

Finally, the determination of the true area or volume of an object in a complex imaging situation, where the relative physical parameters may be difficult to measure, can be accomplished using objects of precisely known dimensions. Since the perspective projection equations of the test object and the specimen will be the same, then measurement of the test object's dimensions yields the direct relationship between the specimen's area A_{pixel} measured in pixels and the true area A_{true}. This is also true of the areas of depth planes in a volumetric 3-D object. In a microscope/adapter/camera system, calibrating the digital image in terms of pixels per micrometer can be accomplished with a stage micrometer. The spatial resolution of the camera and the orthogonal equality of the image can be determined by digitizing the image of an EM grid mounted on a standard microscope slide, and viewing the image. The camera is then rotated 90° on its long axis and the grid spacings are measured at right angles. When the limits of resolution are reached, the lines become deformed by a recognizable Moiré pattern.

3.8 APPLICATION OF SHAPE MEASUREMENT TECHNIQUES TO OTHER PROBLEMS OF BIOMEDICAL INTEREST

The preceding sections have used the example of video microscopy to illustrate the principles of semi-automated area and volume measurement, and associated image processing. This application is an excellent study because the images that are typically obtained conform nicely to the need for histogram bimodality, and hence for simple object extraction. However, the microscopic application also demonstrates the effective coupling of state-of-the-art digital microscopy with powerful digital image processing techniques. The capability to capture area and volume at the micro-dimension argues well for the synergism of these technologies.

A detailed description of the various other areas in the biomedical sciences in which quantitative size and shape parameters are applied as a means of evaluating a given state, and modifications following some physical intervention or pathology, is beyond the scope of this chapter. However, some representative areas are discussed below.

Anatomic pathology is a leading field for the application of computer-based size, shape, and pattern recognition because the numerical parameters of the object are the prime considerations for diagnostic purposes. Fully automated systems have thus far been of limited success because identification of the cell type within the scene under analysis is still a very complicated process and is best accomplished by an interactive operator-assisted process. However, once the objects have been identified, compilation of the morphometric data and statistical evaluations can be performed automatically. Tracing the specimens of interest is accomplished by any number of interactive peripherals; mouse, digitizing screens, and touch-sensitive screens are by far the most popular techniques. The most common measurement is the determination of nuclear and cytoplasmic diameters.[19]

Although the effective numerical shape factors aimed at recognition of an object presented in any orientation or angle are well-documented for the detection of man-made objects,[19] the application of shape factors for the interpretation of microscopic images have been of limited success. However, discrimination based on different perimeter/area relationships[21] that determine to what extent a shape approaches or differs from a circle have been widely used. Some investigators have used calibrated imaginary circles as graphics standards and have computed the ratio of the real perimeter measured from a cell to the standard as a measure of circularity.[19,22] Of course, this and other circularity factors that are based on perimeter/area relationships will shift whenever shape or size changes occur; they are also highly specific for a given shape. Semi-automated and automated methods using the "four-side connected region" concept for bone histiomorphometry have been developed by Beigbeder et al.[23] Their algorithm was designed to flood-fill simple geometric figures (polygons) with sharp, well-defined boundaries displayed on a video monitor. Simple gray-level thresholding has been used by Garrison et al.[24] for the determination of biomass and to compute the areas of developing colonies of chondriocytes during cartilage differentiation. It is likely that these, and other similar application domains, will benefit from the well-defined and more generic techniques developed in this chapter.

3.9   CONCLUSION

This chapter reviewed fundamental aspects of current microcomputer-based image area and volume measurement techniques and their biological applications. A rather generic machine vision system that has been successfully applied to the measurement of the areas and volumes of complex, biomedically generated shapes was demonstrated as a means of illustrating the basic techniques of object extraction and isolation, region correction and smoothing, and area/volume measurement. The example system has been demonstrated to perform at a high level of reliability with a minimum amount of intervention from a human operator. The system described requires online human input appropriate to the subject of interest, only as far as for determining an appropriate gray-level threshold and, possibly, for specifying the radius of a window or structuring element used in the morphological region smoothing operations. Experimental results obtained support the efficacy of this approach as well as its reliability.

Another merit of the techniques described and implemented for example purposes is that they can be implemented on a stand-alone workstation for the image processing task. In the current example system implementation, the program (not optimized) requires only a few seconds to process a single image; more time is spent in the human interaction. Still, the methods described here promise a very considerable reduction of manpower and man-hours for the simple yet important general task of measuring the areas and volumes of biological objects.

REFERENCES

1. Baker, D. G., Hwang, H. H., and Aggarwal, J. K., Detection and segmentation of objects in outdoor scenes: concrete bridges, *J. Opt. Soc. Am.*, A6, 938, 1989.
2. Serra, J., *Image Analysis and Mathematical Morphology*, Academic Press, New York, 1982.
3. Bovik, A. C., Clark, M., and Geisler, W. S., Multichannel texture analysis using localized spatial filters, *IEEE Trans. Pattern Anal. Machine Intell.*, 12, 55, 1990.
4. Gonzalez, R. C. and Woods, R. E., *Digital Image Processing*, Addison-Wesley, Reading, MA, 1993.
5. Ballard, D. H. and Brown, C. M., *Computer Vision*, Prentice-Hall, Englewood Cliffs, NJ, 1982.
6. Ballard, D. H. and Sklansky, J., A ladder-structured decision tree for recognizing tumors in chest radiographs, *IEEE Trans. Comput.*, 25, 503, 1976.
7. Chien, Y. P. and Fu, K. S., A decision function method for boundary detection, *Comput. Graph. Image Process.*, 3, 125, 1974.
8. Lester, J. M. Williams, H. A., Weintraub, B. A., and Brenner, J. F., Two graph searching techniques for boundary finding in white blood cell images, *Comput. Biol. Med.*, 8, 293, 1978.
9. Dietz, T. E., Davis, L. S., Diller, K. R., and Aggarwal, J. K., Computer recognition and analysis of freezing cells in noisy, cluttered images, *Cryobiology*, 19, 539, 1982.
10. Pavlidis, T., *Structural Pattern Recognition*, Springer-Verlag, New York, 1977.

11. Kim, N. H., Wysocki, A. B., Bovik, A. C., and Diller, K. R., A microcomputer-based vision system for area measurement, *Comput. Biol. Med.*, 17, 173, 1987.

12. Thompson, K. F. and Diller, K. R., Use of computer image analysis to quantify contraction of wound size in experimental burns, *J. Burn Care Rehab.*, 2, 307, 1981.

13. Lantuejoul, C. and Serra, J., M-filters, *Proc. IEEE Int'l Conf. Acoust., Speech, Signal Process.*, Paris, France, 1982, 2063.

14. Maragos, P. A., Tutorial on advances in morphological image processing and analysis, *Optical Engr.*, 26, 623, 1987.

15. Aherne, W. A. and Dunhill, M. S., *Morphometry,* Edward Arnold Publishers, London, 1982.

16. Haugland, R. P., *Molecular Probes: Handbook of Fluorescent Probes and Research Chemicals,* Molecular Probes, Eugene, OR, 1989.

17. Green, F. J., *The Sigma-Aldrich Handbook of Stains, Dyes and Indicators,* Aldrich Chemical, Milwaukee, WI, 1990.

18. Inoue, S., in *Video Microscopy,* Plenum Press, New York, 1986, 191.

19. Marchevesky, A. M., Gil, J., and Jeanty, H., Computerised interactive morphometry in pathology, *Hum. Pathol.*, 18, 320, 1987.

20. Horn, B. K. P., *Robot Vision,* MIT Press, Cambridge, MA, 1986.

21. Diller, K. R. and Aggarwal, S. J., Computer automated cell size and shape analysis in cryomicroscopy, *J. Microscopy*, 146, 209, 1987.

22. Steponkus, P. L., Role of plasma membrane in freezing injury and cold acclimation, *Annu. Rev. Plant Physiol.*, 35, 543, 1984.

23. Beigbeder, M., Chappard, D., Alexandre, L., Vico, L., Palle, S., and Riffat, G., Improved algorithms for automatic bone histomorphometry on a numerised image analysis system, *J. Microscopy*, 150, 151, 1988.

24. Garrison, J. C., Peterson, P., and Uyeki, E. M., Computer-based image analysis of cartilage differentiation in embryonic limb bud micromass cultures, *J. Microscopy*, 156, 353, 1989.

25. Dietz, T. E., Diller, K. R., and Aggarwal, J. K., Automated computer evaluation of time-varying cryomicroscopical images, *Cryobiology*, 21, 200, 1984.

26. Diller, K. E. and Knox, J. M., Automated computer analysis of cell size changes during cryomicroscope freezing: a biased trident convolution technique, *Cryo-Letters*, 4, 77, 1983.

27. Schwartz, G. J. and Diller, K. R., Osmotic response of individual cells during freezing. I. Experimental volume measurements, *Cryobiology*, 20, 61, 1983.

28. Aggarwal, S. J., Diller, K. R., and Davidson, I., Microscopic response of dog pancreas b cells to freezing and osmotic stresses, *Cryo-Letters*, 7, 218, 1986.

29. Macias-Garza, F., Diller, K. R., Bovik, A. C., Aggarwal, S. J., and Aggarwal, J. K., Obtaining a solid model from optical serial sections, *Pattern Recogn.*, 22, 577, 1989.

30. Bartels, K. S., *The Analysis of Biological Shape and Shape-Change from Multi-dimensional Image Sequences*, Ph.D. dissertation, The University of Texas at Austin, 1993.

31. Bovik, A. C., Aggarwal, S. J., Kim, N. H., and Diller, K. R., Quantitative area determination by image analysis, in *Image Analysis in Biology,* Häder, D.-P., Ed., CRC Press, Boca Raton, FL, 1991, 29.

32. Maragos, P. and Schafer, R.W., Morphological filters. Part II, *IEEE Trans. Acoust., Speech, Signal Process.*, 35, 1987.

33. Heijmans, H. J. A. M., Construction of self-dual morphological operators and modifications of the median, *IEEE Int. Conf. Image Process.*, Austin, TX, Nov. 13–16, 1994.

34. Heijmans, H. J. A. M., Mathematical morphology as a tool for shape description, in *Shape in Picture: Mathematical Description of Shape in Gray Level Images*, Ying-Lie, O. et al., Eds., Springer-Verlag, Berlin, 1994.

35. Pitas, I. and Venetsanopoulos, A. N., Morphological shape decomposition, *IEEE Trans. Pattern Anal. and Mach. Intell.*, 12, 38, 1990.

36. Serra, J. and Vincent, L., An overview of morphological filtering, *Circuits, Systems, and Signal Processing*, 11, 47, 1992.

37. Vincent, L., Morphological area openings and closing for grey-scale images, in *Shape in Picture: Mathematical Description of Shape in Grey-Level Images*, Ying-Lie, O., Ed., Springer-Verlag, Berlin, 1994, 197.

4 The Analysis of Shape Change in Multidimensional Image Sequences

Keith A. Bartels and Alan C. Bovik

CONTENTS

4.1 INTRODUCTION

A shared characteristic of all life forms, both single and multicellular, is the ability to move and change shape. The study of biological shape changes is, therefore, extremely important to biological and biomedical research. In this chapter, techniques for extracting shape-change parameters from biological image sequences are described. With current technologies, image sequences may be two-dimensional (2-D), such as those produced in standard video microscopy and X-ray cinematography, or three-dimensional (3-D), such as those produced by 3-D laser-scanning confocal microscopy (LSCM), magnetic resonance imaging (MRI), and other tomographic techniques. The image sequence contains information about the shape-change and motion of the imaged object. The nontrivial task of shape-change analysis is to extract the meaningful information from the image sequence.

In the best case, automated analysis would extract the shape-change and motion of the imaged specimen both globally and locally. The global information describes how the specimen has translated, rotated, or changed as a whole, whereas the local information describes the shape-change of individual regions within the specimen. For example, when studying a beating heart, it may be desirable to know the change in volume of the entire heart (global measurement), the change in volume of a single chamber (a more local measurement), or the change in position of a single point within the cardiac muscle (a very localized measurement). Image analysis that extracts shape-change at the local level is sufficient because global changes can be determined by integrating over the local changes.

Modern imaging techniques, such as those mentioned above, have made it possible to study biological shape-change through image sequences in both two and three spatial dimensions. MRI and other tomographic techniques are tools commonly used by physicians for making medical diagnoses and for surgical planning. The study of the progression of tumor growth and the study of visceral function are examples of shape-change analysis using these technologies. The confocal microscope, a relatively recent development in microscopy, is extremely important to the biologist because of its ability to study cellular and other microscopic motion in three spatial dimensions.

Along with a review of dynamic image analysis and shape modeling techniques, in this chapter, a technique for modeling the shape-change of biological specimens from multidimensional image sequences is described and demonstrated. The technique extracts the shape-change information at a very local level using no *a priori* knowledge of the specimen's deformations or of image correspondences. It is required, however, that the images of the sequence be sampled sufficiently fast that the specimen changes shape only a relatively small amount from one frame to the next. The multidimensional model is obtained by first segmenting the specimen in the initial image of the sequence with an appropriate technique and then "attaching" a material coordinate system to the specimen data. This gives a model for the specimen in the first image. To complete the model for the future time intervals, an energy functional defined over the space of possible deformation functions is minimized using the calculus of variations and a numerical finite difference technique. The energy functional is a linear combination of a shape-change energy term and a penalty function on the temporal brightness changes in the image data. A constant parameter is chosen that adjusts the trade-off between the shape-change energy and the brightness penalty. The shape-change analysis technique is applied to two- and three-dimensional image sequences of both real and synthetic data.

4.1.1 INITIAL DEFINITIONS

Before a discussion of image analysis techniques, it is important to have exact mathematical definitions and an understanding of the terms image and image sequence for the derivations in the following sections of this chapter. The term "specimen" will be used to describe the object being imaged. The N-dimensional image acquired at time t_i is denoted by the map $f_i: \mathcal{R} \to \mathcal{R}^N$, where $\mathcal{R}$ represents the set of real numbers. The function $f_i()$ assigns a real number, referred to as brightness, to each location in N-dimensional space. Although the value of $f_i()$ is referred to as brightness, it does not necessarily represent actual visual light intensity. It may represent acoustical intensities, as in the case of ultrasonic imaging, or the concentration of gamma-ray emitting isotopes, as in the case of positron emission tomography (PET). In any case, the function $f_i()$ gives the spatial distribution of some measured quantity and is usually displayed as a brightness image for visualization.

An image sequence of length L is defined as the collection $\{f_i(); i = 0, ..., L - 1\}$ where $t_0 \leq t_1 \leq ... t_{L-1}$. Figure 4.1 shows an example of three images from a 2-D image sequence. The specimen in this case is a *Euglena gracilis* that is imaged with a transmission laser-scanning microscope. The sequence was obtained by acquiring the 2-D image $f_0(x, y)$ at time $t_0 = 0$ s, image $f_1(x, y)$ at time $t_1 = 2.0$ s, and image $f_2(x, y)$ at time $t_2 = 4.0$ s. This image sequence has several important features that make it a good demonstrative example.

The specimen has obviously undergone a non-rigid transformation (i.e., more than a simple translation or rotation). A global shape-change is apparent; the specimen goes from very elliptical to a more circular shape within the three time frames. Along with the global change, local shape-changes are evident in the interior of the specimen. Given only the image sequence, the information

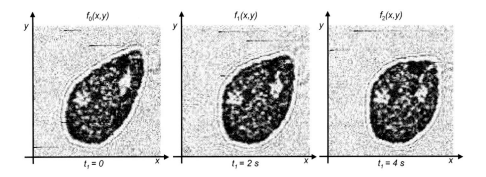

FIGURE 4.1   Sample 2-D image sequence of a *Euglena gracilis* imaged like a transmission laser-scanning microscope.

directly available is how the brightness of each position in space changes over time. Information on how the specimen actually changed shape is not directly available for quantitative analysis, although the human visual system can recognize some shape-changes. The shape-change analysis technique presented here will assign a unique material coordinate to each point on the specimen and compute the displacement of that material coordinate in the space for each image in the sequence. These concepts are discussed in detail in the following sections.

The images in the sequence of Figure 4.1 contain a great deal of noise, and two distinct types of noise are apparent in the images. Horizontal streaks can be seen in the images; these streaks were due to a fault in a power supply that caused the laser to intermittently switch off. The less coherent noise distributed throughout the entire image is due to the low intensity of the light received by the microscope's photodetector. The presence of noise will adversely effect the ability to isolate the specimen data within the image and the ability to extract shape-change information from the image sequence.

The images $f_i()$ have been introduced as functions with continuous domains and ranges. Images are most often captured and processed digitally, meaning the domain and range of $f_i()$ are actually countable (in fact, finite) sets. A more appropriate definition of the image is: $f_i: Z^N \rightarrow Z$, where $Z \subset \mathfrak{R}$ is the set of integers. However, as is often the case in digital signal and image processing, it will be assumed that the images have continuous domains and ranges in order to make use of tools from differential mathematics.

4.1.2 I<small>MPORTANT</small> C<small>ONCEPTS FROM</small> D<small>IFFERENTIAL</small> G<small>EOMETRY</small>

This section presents the concepts of differential geometry that are important to the mathematical development in this chapter. First, several definitions need to be made. A curve is defined to be the image of an infinitely differentiable (smooth) function $\alpha(u_1)$ that maps an interval $U \subset \mathfrak{R}$ into $\mathfrak{R}^N$. The dimension of a curve is said to be 1 since $\alpha(u_1)$ is parametrized by a single real number, or since U, the domain of $\alpha(u_1)$, is a one-dimensional set. A surface is defined to be the image of a smooth map $\alpha(u_1, u_2)$, mapping $U \subset \mathfrak{R}^2$ into $\mathfrak{R}^N$ where $N \geq 2$. Since U is a two-dimensional set, a surface is said to be two-dimensional. Similarly, a solid is defined to be the image of a smooth map $\alpha(u_1, u_2, u_3)$, mapping $U \subset \mathfrak{R}^3$ into $\mathfrak{R}^N$ where $N \geq 3$. A solid is a three-dimensional object. A solid is often called a 3-manifold, and surfaces and curves are sometimes referred to as 2-manifolds and 1-manifolds, respectively. A manifold is a generalization of the concept of a curve, surface, or solid to arbitrary dimension. Note that in the definition above, the curve was defined as the image of a set under a function α. Although this is the correct mathematical interpretation, it is common to call the function α itself a curve. The same is true for surfaces and solids.

The co-dimension of a k-manifold is defined as $(N - k)$ where N is the dimension of the space in which the manifold lies, sometimes called the ambient space. The co-dimension is the difference in the dimension of the manifold and the dimension of the ambient space. For example, a line curve has co-dimension 0, a plane curve has co-dimension 1, and a 3-space curve has co-dimension 2. A surface in the plane has co-dimension 0 and a surface in 3-space has co-dimension 1. A solid in 3-space has co-dimension 0. The concept of co-dimension is important to the mathematical development of the shape-change algorithm.

The following discussion will be specific to surfaces but it should be kept in mind that all concepts can be generalized to curves and solids by simply restricting or extending the dimension k.

Associated with every surface $\alpha(u_1, u_2)$ is its first fundamental form, known also as its metric tensor or simply its metric. The metric is given by the values

$$g_{ij} = \frac{\partial \alpha}{\partial u_i} \cdot \frac{\partial \alpha}{\partial u_j} \quad (i = 1, \ldots, k; \quad j = 1, \ldots, k) \tag{4.1}$$

where "$\cdot$" is the vector inner, or dot, product and $k = 2$ is the dimension of the surface. The metric is often written in the form of a $k \times k$ matrix as

$$g = \begin{bmatrix} E & F \\ F & G \end{bmatrix} \tag{4.2}$$

with

$$E = g_{11} = \alpha_{u_1} \cdot \alpha_{u_1}; \quad F = g_{12} = g_{21} = \alpha_{u_1} \cdot \alpha_{u_2}; \quad G = g_{22} = \alpha_{u_2} \cdot \alpha_{u_2} \tag{4.3}$$

and subscripts are used for partial differentiation with respect to the subscripted variable.

As its name implies, the metric g contains metrical information such as areas and angles on the surface. For example, the area of a region R on the surface is given by

$$A(R) = \iint_{\alpha^{-1}(R)} \left[\det(g) \right]^{1/2} du_1 du_2 \tag{4.4}$$

and the angle ϕ between coordinate curves on the surface is given by

$$\phi = \arccos\left(\frac{F}{\sqrt{EG}} \right) \tag{4.5}$$

A coordinate curve is a curve obtained by holding either u_1 or u_2 constant.

The notion of the metric generalizes to other dimensions. Using Eq. 4.1, the metric for a curve $\gamma(u_1)$ is given by

$$g = \left(\frac{d\gamma}{du_1} \right)^2 \tag{4.6}$$

Computation of arc length on a curve is analogous to that of area on a surface, and the formula for the length of the segment S of a curve is the same as Eq. 4.4 for area on a surface.

$$L(S) = \int_{\gamma^{-1}(S)} \left[\det(g) \right]^{1/2} du_1 \tag{4.7}$$

$$= \int_{\gamma^{-1}(S)} \left| \frac{d\gamma}{du_1} \right| du_1 \qquad (4.8)$$

Similarly, let $\zeta(u_1, u_2, u_3)$ be a solid, then g is a 3×3 matrix and the volume of a region C of the solid can be calculated via

$$V(C) = \int \iint_{\xi^{-1}(C)} \left[\det(g) \right]^{1/2} du_1 du_2 du_3 \qquad (4.9)$$

Such metrical computations can be performed once the deforming material coordinate system model is found using the shape-change analysis algorithm.

It is a fact of differential geometry that a curve, surface, or solid with co-dimension 0 is completely specified (up to a rigid motion in the ambient space) by its metric g. In other words, given its metric at all points in the domain, the intrinsic shape of a curve, surface, or solid with co-dimension 0 is known up to translations, rotations, and reflections in the ambient space. This fact is used in deriving the energy functional that will be minimized to find the material coordinate model.

4.1.3 RELATED WORK

Determining models of objects from image data is a problem that has been addressed since computer vision research was in its infancy. Early 3-D models were computed from surface maps acquired from stereo imaging, laser ranging, and algorithms such as shape-from-shading.[1] Modeling techniques used included membrane and thin-plate splines,[2-6] cubic B-splines,[7] spherical harmonic functions,[8] generalized cylinders,[8] and superquadrics.[9-11] With all of these techniques, a model is obtained for a single time instant and for only the object's surface. A time parameter can be added to these surface models in order to model motion in time. If it can be assumed that the object being imaged is rigid, the motion model can be obtained from simple rotations and translations of the original model. The difficulty, even with the rigid-body assumption, is that the object correspondences must be made between consecutive frames of the image sequence.

Non-rigid motion analysis has been addressed by many authors.[12-15] One of the first techniques developed was that of optical flow.[1,16,17] The power of this technique is that non-rigid motion vectors can be determined from a consecutive pair of 2-D image frames without any manual input of image correspondences. The technique presented in this chapter is closely related to the original optical flow idea. The new technique, however, is based on mathematical descriptions of the object's shape and is extended to be applicable in multiple dimensions and to multiple image frames.

4.2 MATHEMATICAL DEVELOPMENT

Before beginning the shape-change analysis, the specimen being analyzed must be isolated within the image space. This segmentation operation can often be performed by simple thresholding in situations of high signal-to-noise ratio. Such is often the case with fluorescent microscopy, for example, because information from the excitation wavelength can be removed using optical filters within the microscope. However, if the signal from the specimen is very weak, or there is information in the background, the image may contain substantial noise and processing other than simple thresholding will be necessary for segmentation.

Once the specimen is segmented, the domain of definition of the specimen is defined in the first image as the set

$$U = \{ \bar{u} : \bar{u} \text{ is in the segmented region of the image} \} \qquad (4.10)$$

The specimen data needs to be segmented in the first image of the image sequence only, because the domain in the following images will be defined by the image of U under the computed deformations.

4.2.1   DERIVATION FOR THE *N*-DIMENSIONAL CASE

The image sampled at time t_i is denoted $f_i: \mathfrak{R}^N \to \mathfrak{R}$, and it is assumed that $t_i - t_{i-1}$, which need not be constant, is small enough that shape changes result in displacements on the order of a few image pixels (voxels). There are several reasons for this assumption that will be pointed out below. The function $\bar{\alpha}_i : U \to \mathfrak{R}^N$ is a parametrization that describes the shape of the material coordinate system at time t_i. The desired deforming material coordinate model of the specimen's shape-change is represented by the collection $\{ \bar{\alpha}_i ; (i = 0, \ldots, L-1)\}$, where L is the number of images in the sequence.

An initial orthogonal material coordinate system is defined on the domain U by the identity function on U, $\bar{\alpha}_0(\bar{u}) = \bar{u}$. This gives an orthogonal parametrization and causes the initial coordinate system to exactly overlay the natural fixed coordinate system. The initial parametrization chosen could be any diffeomorphic function on U.[18]

A technique for finding the functions $\bar{\alpha}_i$ for $(i = 1, \ldots, L-1)$ to complete the shape-change model is now described. The deformation vector field $\bar{\Delta}_i : U \to \mathfrak{R}^N$ is defined for $i > 0$ such that $\bar{\alpha}_i = \bar{\alpha}_{i-1} + \bar{\Delta}_i$. Since $\bar{\alpha}_0()$ is known, finding the deformation functions $\bar{\Delta}_i$ for $i > 0$ is equivalent to finding the desired for the material coordinate model.

4.2.1.1   The Brightness Penalty Functional

It is reasonable to assume that the brightness of a particular material coordinate on the object will not vary greatly over time. It is reasonable also to assume that brightness change of a particular material coordinate is negligible from one image frame to the next. In other words, $f_i[\bar{\alpha}_i(\bar{u})] = f_{i-1}[\bar{\alpha}_{i-1}(\bar{u})]$; or in terms of the unknown deformation function $\bar{\Delta}_i$, $f_i[\bar{\alpha}_{i-1}(\bar{u}) + \bar{\Delta}_i(\bar{u})] = f_{i-1}[\bar{\alpha}_{i-1}(\bar{u})]$. The material-coordinate brightness penalty functional is defined by

$$P(\bar{\Delta}_i) = \int_U \left[f_i(\bar{\alpha}_{i-1} + \bar{\Delta}_i) - f_{i-1}(\bar{\alpha}_{i-1}) \right]^2 du_1 \ldots du_N \tag{4.11}$$

The deformation function $\bar{\Delta}_i$ that is sought will minimize this functional. Unfortunately, there may be many such functions that produce the same minimum to Eq. 4.11. The problem is therefore ill-posed.[19] Later in this section, additional constraints will be added that will make the problem well-posed.

Using Taylor's theorem for multidimensional functions,[20]

$$f_i(\bar{\alpha}_{i-1} + \bar{\Delta}_i) = f_i(\bar{\alpha}_{i-1}) + \bar{\Delta}_i \cdot \nabla f_i(\bar{\alpha}_{i-1}) + R_1(\bar{\Delta}_i, \bar{\alpha}_{i-1}) \tag{4.12}$$

where ∇ is the gradient operator and $R_1(\bar{\Delta}_i, \bar{\alpha}_{i-1})$ is the remainder term given by

$$R_1(\bar{\Delta}_i, \bar{\alpha}_{i-1}) = \sum_{j=1}^{N} \sum_{k=1}^{N} \int_0^1 (1-t) \frac{\partial^2 f}{\partial(^j x)\partial(^k x)} (\bar{\alpha}_{i-1} + t\bar{\Delta}_i)^j \Delta_i{}^k \Delta_i \, dt \tag{4.13}$$

where $(^1x, {}^2x, \ldots, {}^Nx)$ are coordinates in the N-dimensional image space and $(^1\Delta_i, {}^2\Delta_i, \ldots, {}^N\Delta_i)$ are the components of the N-dimensional deformation function Δ_i. When the specific cases of 2-D and 3-D are considered, $(^1x, {}^2x, \ldots, {}^Nx)$ and $(^1\Delta_i, {}^2\Delta_i, \ldots, {}^N\Delta_i)$ will be replaced by (x, y) and $(\varepsilon_i, \delta_i)$, respectively, for the 2-D case and (x, y, z) and $(\varepsilon_i, \delta_i, \gamma_i)$, respectively, for the 3-D case.

In order to write Eq. 4.11 so that f_i is evaluated only at the known value of $\overline{\alpha}_{i-1}$, Eq. 4.13, needs to be small enough that it can be neglected. Note that

$$R_1\left(\overline{\Delta}_i,\ \alpha_{i-1}\right) \le \sum_{j=1}^{N}\sum_{k=1}^{N}\left|{}^j\Delta_i{}^k\Delta_i\right|\int_0^1\left|\frac{\partial^2 f}{\partial\left({}^j x\right)\partial\left({}^k x\right)}\left(\overline{\alpha}_{i-1}+t\overline{\Delta}_i\right)\right|dt \tag{4.14}$$

The assumption that shape-changes between single time intervals are small ensures that the deformation functions ${}^j\Delta_i$ $(j = 1, \ldots, N)$ will be small. The integrand of Eq. 4.14 is ensured small if there are no large intensity changes (i.e., edges) between coordinate $\overline{\alpha}_{i-1}$ and $\overline{\alpha}_{i-1} + \overline{\Delta}_i$. Although such intensity changes will occur, they occur within only a small subset of the specimen domain U. In this case, the brightness penalty functional can be expressed as

$$P\left(\overline{\Delta}_i\right) = \int_U\left[f_i\left(\overline{\alpha}_{i-1}\right) - f_{i-1}\left(\overline{\alpha}_{i-1}\right) + \overline{\Delta}_i\cdot\nabla f_i\left(\overline{\alpha}_{i-1}\right)\right]^2 du_1\ldots du_N \tag{4.15}$$

4.2.1.2 The Shape-Change Smoothness Constraint

Minimizing Eq. 4.15 is an ill-posed problem since many minimizing functions $\overline{\Delta}_i$ can be found. However, such problems may be regularized by defining a regularization factor to limit the space of possible solutions.[21,22] The regularization factor imposed here will be a constraint on the allowable degree of shape-change.

The parametrization of the material coordinate system defines a manifold with co-dimension 0. A co-dimension of 0 allows the shape of the manifold to be completely described by its metric tensor. The metric tensor for the specimen at time t_i is g_{jk}^i. The shape-change of the specimen's parametrization between times t_i and t_{i-1} is $g_{jk}^i - g_{jk}^{i-1}$. A constraint to minimize this difference is

$$S\left(\overline{\Delta}_i\right) = \int_U\sum_{j=1}^{N}\sum_{k=1}^{N}\left(g_{jk}^i - g_{jk}^{i-1}\right)^2 du_1\ldots du_N \tag{4.16}$$

$S()$ is a function of $\overline{\Delta}_i$ alone, since

$$g_{jk}^j - g_{jk}^{i-1} = \frac{\partial\overline{\alpha}_i}{\partial u_j}\cdot\frac{\partial\overline{\alpha}_i}{\partial u_k} - \frac{\partial\overline{\alpha}_{i-1}}{\partial u_j}\cdot\frac{\partial\overline{\alpha}_i}{\partial u_k}$$

$$= \left(\frac{\partial\overline{\alpha}_{i-1}}{\partial u_j} + \frac{\partial\overline{\Delta}_i}{\partial u_j}\right)\cdot\left(\frac{\partial\overline{\alpha}_{i-1}}{\partial u_k} + \frac{\partial\overline{\Delta}_i}{\partial u_k}\right) - \frac{\partial\overline{\alpha}_{i-1}}{\partial u_j}\cdot\frac{\partial\overline{\alpha}_{i-1}}{\partial u_k} \tag{4.17}$$

$$= \frac{\partial\overline{\Delta}_i}{\partial u_j}\cdot\frac{\partial\overline{\alpha}_{i-1}}{\partial u_k} + \frac{\partial\overline{\alpha}_{i-1}}{\partial u_j}\cdot\frac{\partial\overline{\Delta}_i}{\partial u_k} + \frac{\partial\overline{\Delta}_i}{\partial u_j}\cdot\frac{\partial\overline{\Delta}_i}{\partial u_k}$$

A combined functional is formed as a linear combination of P and S:

$$E\left(\overline{\Delta}_i\right) = \lambda P\left(\overline{\Delta}_i\right) + S\left(\overline{\Delta}_i\right) \tag{4.18}$$

where $\lambda\in\mathfrak{R}$ is a smoothing parameter that gives a trade-off between fidelity to the data and shape-change smoothness. A low value of λ leads to little shape-change. Choosing an "optimum" value for λ is a difficult task since the choice is so highly data dependent.

The method of generalized cross validation[23] is probably the most widely used analytical method; however, the formulation of a strategy for λ is so highly dependent on the data characteristics (noise, etc.) that using a cross-validation technique is uncommon, and is not explored here. Equation 4.18 can be written in the form

$$E\left(\overline{\Delta}_i\right) = \int_U F\left(\overline{u}, \overline{\Delta}_i, \frac{\partial \overline{\Delta}_i}{\partial u_1}, \frac{\partial \overline{\Delta}_i}{\partial u_2}, \ldots \frac{\partial \overline{\Delta}_i}{\partial u_N}\right) \partial u_1 \, \partial u_2 \ldots \partial u_N \tag{4.19}$$

where

$$F\left(\overline{u}, \overline{\Delta}_i, \frac{\partial \overline{\Delta}_i}{\partial u_1}, \frac{\partial \overline{\Delta}_i}{\partial u_2}, \ldots, \frac{\partial \overline{\Delta}_i}{\partial u_N}\right) = \lambda\left[f_i\left(\overline{\alpha}_{i-1}\right) - f_{i-1}\left(\overline{\alpha}_{i-1}\right) + \overline{\Delta}_i \cdot \nabla f_i\left(\overline{\alpha}_{i-1}\right)\right]^2$$

$$+ \sum_{j=1}^{N} \sum_{k=1}^{N} \left(\frac{\partial \overline{\Delta}_i}{\partial u_j} \cdot \frac{\partial \overline{\alpha}_{i-1}}{\partial u_k} + \frac{\partial \alpha_{i-1}}{\partial u_j} \cdot \frac{\partial \overline{\Delta}_i}{\partial u_k} + \frac{\partial \overline{\Delta}_i}{\partial u_j} \cdot \frac{\partial \overline{\Delta}_i}{\partial u_k}\right)^2 \tag{4.20}$$

Minimizing Eq. 4.19 is a variational problem whose solution requires the resolution of N coupled nonlinear partial differential equations (the Euler equations).

Note that the shape change information at time t_i is calculated solely on information from time t_{i-1} and is independent of information from other times. Hence, the notation is simplified by eliminating subscripts from the $\overline{\Delta}_i$ and the $\overline{\alpha}_{i-1}$ variables. From here on, subscripts are used to indicate partial derivatives with respect to the subscripted variable. The subscript on $f_i()$, however, still indicates the image sampled at time t_i. With these changes, Eq. 4.19 becomes

$$\varepsilon\left(\overline{\Delta}_i\right) = \int_U F\left(\overline{u}, \overline{\Delta}, \overline{\Delta}_{u_1}, \overline{\Delta}_{u_2}, \ldots, \overline{\Delta}_{u_N}\right) du_1 \, du_2 \ldots du_N \tag{4.21}$$

4.2.2 THE TWO-DIMENSIONAL CASE

For the 2-D solution, the coordinate functions $x: U \to \mathfrak{R}$, $y: U \to \mathfrak{R}$, $\varepsilon: U \to \mathfrak{R}$, and $\delta: U \to \mathfrak{R}$ are defined such that $\overline{\alpha}\,(u_1, u_2) = [x(u_1, u_2), y(u_1, u_2)]$ and $\overline{\Delta}\,(u_1, u_2) = [\varepsilon(u_1, u_2), \delta(u_1, u_2)]$. Expanding $\overline{\alpha}$ and $\overline{\Delta}$ into their coordinate functions yields (from Eq. 4.16)

$$S\left(\overline{\Delta}\right) = \int\int_U \sum_{j=1}^{2} \sum_{k=1}^{2} \left(\varepsilon_{u_j} x_{u_k} + \delta_{u_j} y_{u_k} + \varepsilon_{u_k} x_{u_j} + \delta_{u_k} y_{u_j} + \varepsilon_{u_j}\varepsilon_{u_k} + \delta_{u_j}\delta_{u_k}\right)^2 du_1 du_2 \tag{4.22}$$

and Eq. 4.15 yields

$$P\left(\overline{\Delta}\right) = \int\int_U \left[f_i(x, y) - f_{i-1}(x, y) + \varepsilon \frac{\delta}{\delta x} f_i(x, y) + \delta \frac{\partial}{\partial y} f_i(x, y)\right]^2 du_1 du_2 \tag{4.23}$$

The function $F()$ given in Eq. 4.20 is rewritten, in terms of the coordinate functions and with the summations carried out, as:

$$F = \lambda\left[f_i(x, y) - f_{i-1}(x, y) + \varepsilon\left(f_i\right)_x(x, y) + \delta\left(f_i\right)_y(x, y)\right]^2$$

$$+ \left[2\varepsilon_{u_1} x_{u_1} + 2\delta_{u_1} y_{u_1} + \varepsilon_{u_1}^2 + \delta_{u_1}^2\right]^2$$

$$+ 2\left[\varepsilon_{u_1} x_{u_2} + \delta_{u_1} y_{u_2} + x_{u_1}\varepsilon_{u_2} + y_{u_1}\delta_{u_2} + \varepsilon_{u_1}\varepsilon_{u_2} + \delta_{u_1}\delta_{u_2}\right]^2 \tag{4.24}$$

$$+ \left[2\varepsilon_{u_2} x_{u_2} + 2\delta_{u_2} y_{u_2} + \varepsilon_{u_2}^2 + \delta_{u_2}^2\right]^2$$

The variational procedure to find functions ε and δ that minimize Eq. 4.19 requires solving the following Euler equations.[24]

$$F_\varepsilon - \frac{\partial}{\partial u_1} F_{\varepsilon_{u_1}} - \frac{\partial}{\partial u_2} F_{\varepsilon_{u_2}} = 0 \tag{4.25}$$

$$F_\delta - \frac{\partial}{\partial u_1} F_{\delta_{u_1}} - \frac{\partial}{\partial u_2} F_{\delta_{u_2}} = 0 \tag{4.26}$$

The above Euler equations assume a fixed boundary condition for ε and δ. Substituting Eq. 4.24 into Eq. 4.25 and Eq. 4.26, the following coupled pair of nonlinear, second-order, partial differential equations is obtained. From Eq. 4.25,

$$\frac{\lambda}{2}\left[(f_i)_x f_i - (f_i)_x f_{i-1} + \varepsilon(f_i)_x^2 + \delta(f_i)_x(f_i)_y\right]$$

$$- \varepsilon_{u_1 u_1}\left(3\varepsilon_{u_1}^2 + 6\varepsilon_{u_1} x_{u_1} + \varepsilon_{u_2}^2 + 2\varepsilon_{u_2} x_{u_2} + 2\delta_{u_1} y_{u_1} + \delta_{u_1}^2 + 2x_{u_1}^2 + x_{u_2}^2\right)$$

$$- \varepsilon_{u_2 u_2}\left(3\varepsilon_{u_2}^2 + 6\varepsilon_{u_2} x_{u_2} + \varepsilon_{u_1}^2 + 2\varepsilon_{u_1} x_{u_1} + 2\delta_{u_2} y_{u_2} + \delta_{u_2}^2 + 2x_{u_2}^2 + x_{u_1}^2\right)$$

$$- \varepsilon_{u_1 u_2}\left(4\varepsilon_{u_1}\varepsilon_{u_2} + 4\varepsilon_{u_1} x_{u_2} + 4\varepsilon_{u_2} x_{u_1} + 2\delta_{u_1} y_{u_2} + 2\delta_{u_2} y_{u_1} + 2\delta_{u_1}\delta_{u_2} + 2x_{u_1} x_{u_2}\right)$$

$$- \delta_{u_1 u_1}\left(2\varepsilon_{u_1}\delta_{u_1} + 2\varepsilon_{u_1} y_{u_1} + \varepsilon_{u_2}\delta_{u_2} + \varepsilon_{u_2} y_{u_2} + 2\delta_{u_1} x_{u_1} + \delta_{u_2} x_{u_2} + 2x_{u_1} y_{u_1} + x_{u_2} y_{u_2}\right)$$

$$- \delta_{u_2 u_2}\left(2\varepsilon_{u_2}\delta_{u_2} + 2\varepsilon_{u_2} y_{u_2} + \varepsilon_{u_1}\delta_{u_1} + \varepsilon_{u_1} y_{u_1} + 2\delta_{u_2} x_{u_2} + \delta_{u_1} x_{u_1} + 2x_{u_2} y_{u_2} + x_{u_1} y_{u_1}\right)$$

$$- \delta_{u_1 u_2}\left(\varepsilon_{u_1}\delta_{u_2} + \varepsilon_{u_1} y_{u_2} + \varepsilon_{u_2}\delta_{u_1} + \varepsilon_{u_2} y_{u_1} + \delta_{u_2} x_{u_1} + \delta_{u_1} x_{u_2} + x_{u_1} y_{u_2} + x_{u_2} y_{u_1}\right) \tag{4.27}$$

$$- x_{u_1 u_1}\left(3\varepsilon_{u_1}^2 + 4\varepsilon_{u_1} x_{u_1} + \varepsilon_{u_2}^2 + \varepsilon_{u_2} x_{u_2} + 2\delta_{u_1} y_{u_1} + \delta_{u_1}^2\right)$$

$$- y_{u_1 u_1}\left(2\varepsilon_{u_1}\delta_{u_1} + \varepsilon_{u_2}\delta_{u_2} + 2\delta_{u_1} x_{u_1} + \delta_{u_1} x_{u_2}\right)$$

$$- x_{u_2 u_2}\left(3\varepsilon_{u_2}^2 + 4\varepsilon_{u_2} x_{u_2} + \varepsilon_{u_1}^2 + \varepsilon_{u_1} x_{u_1} + 2\delta_{u_2} y_{u_2} + \delta_{u_2}^2\right)$$

$$- x_{u_1 u_2}\left(4\varepsilon_{u_1}\varepsilon_{u_2} + 3\varepsilon_{u_1} x_{u_2} + 3\varepsilon_{u_2} x_{u_1} + 2\delta_{u_1} y_{u_2} + 2\delta_{u_2} y_{u_1} + 2\delta_{u_1}\delta_{u_2}\right)$$

$$- y_{u_2 u_2}\left(2\varepsilon_{u_2}\delta_{u_2} + \varepsilon_{u_1}\delta_{u_1} + 2\delta_{u_2} x_{u_2} + \delta_{u_2} x_{u_1}\right)$$

$$- y_{u_1 u_2}\left(\varepsilon_{u_1}\delta_{u_2} + \varepsilon_{u_2}\delta_{u_1} + \delta_{u_1} x_{u_2} + \delta_{u_2} x_{u_1}\right) = 0,$$

and from Eq. 4.26,

$$\frac{\lambda}{2}\left[(f_i)_y f_i - (f_i)_y f_{i-1} + \varepsilon(f_i)_x (f_i)_y + \delta(f_i)_y^2\right]$$

$$-\delta_{u_1 u_1}\left(3\delta_{u_1}^2 + 6\delta_{u_1} y_{u_1} + \delta_{u_2}^2 + 2\delta_{u_2} y_{u_2} + 2\varepsilon_{u_1} x_{u_1} + \varepsilon_{u_1}^2 + 2y_{u_1}^2 + y_{u_2}^2\right)$$

$$-\delta_{u_2 u_2}\left(3\delta_{u_2}^2 + 6\delta_{u_2} y_{u_2} + \delta_{u_1}^2 + 2\varepsilon_{u_1} y_{u_1} + 2\varepsilon_{u_2} x_{u_2} + \varepsilon_{u_2}^2 + 2y_{u_2}^2 + y_{u_1}^2\right)$$

$$-\delta_{u_1 u_2}\left(4\delta_{u_1}\delta_{u_2} + 4\delta_{u_1} y_{u_2} + 4\delta_{u_2} y_{u_1} + 2\varepsilon_{u_1} x_{u_2} + 2\varepsilon_{u_2} x_{u_1} + 2\varepsilon_{u_1}\varepsilon_{u_2} + 2y_{u_1} y_{u_2}\right)$$

$$-\varepsilon_{u_1 u_1}\left(2\delta_{u_1}\varepsilon_{u_1} + 2\delta_{u_1} x_{u_1} + \delta_{u_2}\varepsilon_{u_2} + \delta_{u_2} x_{u_2} + 2\varepsilon_{u_1} y_{u_1} + \varepsilon_{u_2} y_{u_2} + 2y_{u_1} x_{u_1} + y_{u_2} x_{u_2}\right)$$

$$-\varepsilon_{u_2 u_2}\left(2\delta_{u_2}\varepsilon_{u_2} + 2\delta_{u_2} x_{u_2} + \delta_{u_1}\varepsilon_{u_1} + \delta_{u_1} x_{u_1} + 2\varepsilon_{u_2} y_{u_2} + \varepsilon_{u_1} y_{u_1} + 2y_{u_2} x_{u_2} + y_u x_u\right)$$

$$-\varepsilon_{u_1 u_2}\left(\delta_{u_1}\varepsilon_{u_2} + \delta_{u_1} x_{u_2} + \delta_{u_2}\varepsilon_{u_1} + \delta_{u_2} x_{u_1} + \varepsilon_{u_2} y_{u_1} + \varepsilon_{u_1} y_{u_2} + y_{u_1} x_{u_2} + y_{u_2} x_{u_1}\right) \tag{4.28}$$

$$-y_{u_1 u_1}\left(3\delta_{u_1}^2 + 4\delta_{u_1} y_{u_1} + \delta_{u_2}^2 + \delta_{u_2} y_{u_2} + 2\varepsilon_{u_1} x_{u_1} + \varepsilon_{u_1}^2\right)$$

$$-x_{u_1 u_1}\left(2\delta_{u_1}\varepsilon_{u_1} + \delta_{u_2}\varepsilon_{u_2} + 2\varepsilon_{u_1} y_{u_1} + \varepsilon_{u_2} y_{u_2}\right)$$

$$-y_{u_2 u_2}\left(3\delta_{u_2}^2 + 4\delta_{u_2} y_{u_2} + \delta_{u_1}^2 + \delta_{u_1} y_{u_1} + 2\varepsilon_{u_2} x_{u_2} + \varepsilon_{u_2}^2\right)$$

$$-y_{u_1 u_2}\left(4\delta_{u_1}\delta_{u_2} + 3\delta_{u_1} y_{u_2} + 3\delta_{u_2} y_{u_1} + 2\varepsilon_{u_1} x_{u_2} + 2\varepsilon_{u_2} x_{u_1} + 2\varepsilon_{u_1}\varepsilon_{u_2}\right)$$

$$-x_{u_2 u_2}\left(2\delta_{u_2}\varepsilon_{u_2} + \delta_{u_1}\varepsilon_{u_1} + 2\varepsilon_{u_2} y_{u_2} + \varepsilon_{u_1} y_{u_1}\right)$$

$$-x_{u_1 u_2}\left(\delta_{u_1}\varepsilon_{u_2} + \delta_{u_2}\varepsilon_{u_1} + \varepsilon_{u_1} y_{u_2} + \varepsilon_{u_2} y_{u_1}\right) = 0.$$

The functions $\varepsilon(u_1, u_2)$ and $\delta(u_1, u_2)$ are to be found from Eq. 4.27 and Eq. 4.28 above. All values besides ε and δ and their derivatives are known. The coefficients of the second-order derivatives of ε and δ are functions of the first-order derivatives of ε and δ, hence the nonlinearity. Equations 4.27 and 4.28 are simplified by collecting coefficients of ε, δ, and their second-order derivatives into single variables:

$$A_0\varepsilon + A_1\delta - A_2\varepsilon_{u_1 u_1} - A_3\varepsilon_{u_2 u_2} - A_4\varepsilon_{u_1 u_2} - A_5\delta_{u_1 u_1} - A_6\delta_{u_2 u_2} - A_7\delta_{u_1 u_2} = A_8 \tag{4.29}$$

$$B_0\varepsilon + B_1\delta - B_2\delta_{u_1 u_1} - B_3\delta_{u_2 u_2} - B_4\delta_{u_1 u_2} - B_5\varepsilon_{u_1 u_1} - B_6\varepsilon_{u_2 u_2} - B_7\varepsilon_{u_1 u_2} = B_8 \tag{4.30}$$

The numerical technique used to solve these Euler equations is developed later. In the next section, the Euler equations for the 3-D case are derived.

4.2.3 THE THREE-DIMENSIONAL CASE

For the 3-D case, $U \subset \mathcal{R}^3$, and the following coordinate functions are defined: $x: U \to \mathcal{R}$, $y: U \to \mathcal{R}$, and $z: U \to \mathcal{R}$ so that $\overline{\alpha}(u_1, u_2, u_3) = [x(u_1, u_2, u_3), y(u_1, u_2, u_3), z(u_1, u_2, u_3)]$. Analogously to the 2-D case, $\overline{\alpha}$ gives the (x, y, z) position of material coordinate (u_1, u_2, u_3) at the ith time sample.

The following components of the 3-D deformation function are defined: $\varepsilon: U \to \mathfrak{R}$, $\delta: U \to \mathfrak{R}$, and $\gamma: U \to \mathfrak{R}$ so that $\overline{\Delta}(u_1, u_2, u_3) = [\varepsilon(u_1, u_2, u_3), \delta(u_1, u_2, u_3), \gamma(u_1, u_2, u_3)]$. The 3-D shape-change constraint is obtained by letting $N = 3$ in Eq. 4.16:

$$S(\overline{\Delta}) = \iiint_U \sum_{j=1}^{3} \sum_{k=1}^{3} \left(\varepsilon_{u_j} x_{u_k} + \delta_{u_j} y_{u_k} + \gamma_{u_j} z_{u_k} + \varepsilon_{u_k} x_{u_j} + \delta_{u_k} y_{u_j} + \gamma_{u_k} x_{u_j} + \varepsilon_{u_j} x_{u_k} + \delta_{u_j} \delta_{u_k} + \gamma_{u_j} \gamma_{u_k} \right)^2$$

$$+ \, du_1 du_2 du_3$$

$$(4.31)$$

and the 3-D brightness penalty function is obtained by letting $N = 3$ in Eq. 4.15:

$$P(\varepsilon_i, \delta_i, \gamma_i) = \iiint_U \left[f_i(x, y, z) - f_{i-1}(x, y, z) + \varepsilon \frac{\partial}{\partial x} f_i(x, y, z) \right.$$

$$(4.32)$$

$$\left. + \delta \frac{\partial}{\partial y} f_i(x, y, z) + \gamma \frac{\partial}{\partial z} f_i(x, y, z) \right] du_1 du_2 du_3$$

For the 3-D case, the integrand $F()$ of Eq. 4.20 is

$$F = \gamma \left[f_i(x, y, z) - f_{i-1}(x, y, z) + \varepsilon(f_i)_x + \delta(f_i)_y + \gamma(f_i)_z \right]^2$$

$$+ 2 \left[\varepsilon_{u_1} x_{u_2} + \delta_{u_1} y_{u_2} + \gamma_{u_1} z_{u_2} + x_{u_1} \varepsilon_{u_2} + y_{u_1} \delta_{u_2} + \gamma_{u_2} z_{u_1} + \varepsilon_{u_1} \varepsilon_{u_2} + \delta_{u_1} \delta_{u_2} + \gamma_{u_1} \gamma_{u_2} \right]^2$$

$$+ 2 \left[\varepsilon_{u_1} 1 x_{u_3} + \delta_{u_1} y_{u_3} + \gamma_{u_1} z_{u_3} + x_{u_1} \varepsilon_{u_3} + y_{u_1} \delta_{u_3} + \gamma_{u_3} z_{u_1} + \varepsilon_{u_1} \varepsilon_{u_3} + \delta_{u_1} \delta_{u_3} + \gamma_{u_1} \gamma_{u_3} \right]^2$$

$$+ 2 \left[\varepsilon_{u_2} x_{u_3} + \delta_{u_2} y_{u_3} + \gamma_{u_2} z_{u_3} + x_{u_2} \varepsilon_{u_3} + y_{u_2} \delta_{u_3} + \gamma_{u_3} z_{u_2} + \varepsilon_{u_2} \varepsilon_{u_3} + \delta_{u_2} \delta_{u_3} + \gamma_{u_2} \gamma_{u_3} \right]^2 \quad (4.33)$$

$$+ \left[2\varepsilon_{u_1} x_{u_1} + 2\delta_{u_1} y_{u_1} + 2\gamma_{u_1} z_{u_1} + \left(\varepsilon_{u_1} \right)^2 + \left(\delta_{u_1} \right)^2 + \left(\gamma_{u_1} \right)^2 \right]^2$$

$$+ \left[2\varepsilon_{u_2} x_{u_2} + 2\delta_{u_2} y_{u_2} + 2\gamma_{u_2} z_{u_2} + \left(\varepsilon_{u_2} \right)^2 + \left(\delta_{u_2} \right)^2 + \left(\gamma_{u_2} \right)^2 \right]^2$$

$$+ \left[2\varepsilon_{u_3} x_{u_3} + 2\delta_{u_3} y_{u_3} + 2\gamma_{u_3} z_{u_3} + \left(\varepsilon_{u_3} \right)^2 + \left(\delta_{u_3} \right)^2 + \left(\gamma_{u_3} \right)^2 \right]^2$$

This time, the variational approach involves three coupled nonlinear partial differential equations:

$$F_\varepsilon - \frac{\partial}{\partial u_1} F_{\varepsilon_{u_1}} - \frac{\partial}{\partial u_2} F_{\varepsilon_{u_2}} - \frac{\partial}{\partial u_3} F_{\varepsilon_{u_3}} = 0 \qquad (4.34)$$

$$F_\delta - \frac{\partial}{\partial u_1} F_{\delta_{u_1}} - \frac{\partial}{\partial u_2} F_{\delta_{u_2}} - \frac{\partial}{\partial u_3} F_{\delta_{u_3}} = 0 \qquad (4.35)$$

$$F_\gamma - \frac{\partial}{\partial u_1} F_{\gamma_{u_1}} - \frac{\partial}{\partial u_2} F_{\gamma_{u_2}} - \frac{\partial}{\partial u_3} F_{\gamma_{u_3}} = 0 \qquad (4.36)$$

Performing the derivatives of Eq. 4.34 through 4.36 yields three very lengthy partial differential equations. These equations are given explicitly in Ref. 25.

4.3 FINITE DIFFERENCE SOLUTION OF THE EULER EQUATIONS

This section presents a finite difference solution to the 2-D and 3-D Euler equations, and applications to 2-D and 3-D examples are given.

4.3.1 SOLUTION OF THE 2-D EULER EQUATIONS

For numerical solution, the domain U of the specimen parametrization is discretized. Let i be the index in the u_1 direction and j be the index in the u_2 direction, so that discretized, $\varepsilon(u_1, u_2) \to \varepsilon^{i,j}$ and $\delta(u_1, u_2) \to \delta^{i,j}$. The second-order derivatives are replaced by their respective finite differences:

$$\varepsilon_{u_1 u_1} \to \varepsilon^{i-1,j} - 2\varepsilon^{i,j} + \varepsilon^{i+1,j}$$

$$\varepsilon_{u_2 u_2} \to \varepsilon^{i,j-1} - 2\varepsilon^{i,j} + \varepsilon^{i,j+1} \tag{4.37}$$

$$\varepsilon_{u_1 u_2} \to \varepsilon^{i-1,j-1} - \varepsilon^{i,j-1} - \varepsilon^{i-1,j} + \varepsilon^{i,j}$$

Analogous replacements are made for derivatives of δ.

Equations 4.29 and 4.30 are discretized by replacing the derivatives by the finite difference formula. Terms involving $\varepsilon^{i,j}$ and $\delta^{i,j}$ are factored out, and Eqs. 4.29 and 4.30 give

$$\varepsilon^{i,j}\left(A_0 + 2A_2 + 2A_3 - A_4\right) + \delta^{i,j}\left(A_1 + 2A_5 + 2A_6 - A_7\right) =$$
$$A_8 + A_2\left(\varepsilon^{i-1,j} + \varepsilon^{i+1,j}\right) + A_3\left(\varepsilon^{i,j-1} + \varepsilon^{i,j+1}\right) + A_4\left(\varepsilon^{i-1,j-1} - \varepsilon^{i,j-1} - \varepsilon^{i-1,j}\right) + \tag{4.38}$$
$$A_5\left(\delta^{i-1,j} + \delta^{i+1,j}\right) + A_6\left(\delta^{i,j-1} + \delta^{i,j+1}\right) + A_7\left(\delta^{i-1,j-1} - \delta^{i,j-1} - \delta^{i-1,j}\right)$$

$$\varepsilon^{i,j}\left(B_0 + 2B_5 + 2B_6 - B_7\right) + \delta^{i,j}\left(B_1 + 2B_2 + 2B_3 - B_4\right) =$$
$$B_8 + B_2\left(\varepsilon^{i-1,j} + \varepsilon^{i+1,j}\right) + B_3\left(\varepsilon^{i,j-1} + \varepsilon^{i,j+1}\right) + B_4\left(\varepsilon^{i-1,j-1} - \varepsilon^{i,j-1} - \varepsilon^{i-1,j}\right) + \tag{4.39}$$
$$B_5\left(\delta^{i-1,j} + \delta^{i+1,j}\right) + B_6\left(\delta^{i,j-1} + \delta^{i,j+1}\right) + B_7\left(\delta^{i-1,j-1} - \delta^{i,j-1} - \delta^{i-1,j}\right)$$

Equations 4.38 and 4.39 are simultaneous equations in the two unkowns, $\varepsilon^{i,j}$ and $\delta^{i,j}$. Let P_1 equal the right-hand side of Eq. 4.38, and P_2 equal the right-hand side of Eq. 4.39. Also let

$$M_1 = A_0 + 2A_2 + 2A_3 - A_4, \quad M_2 = A_1 + 2A_5 + 2A_6 - A_7 \tag{4.40}$$

$$M_3 = B_0 + 2B_5 + 2B_6 - B_7, \quad M_4 = B_1 + 2B_2 + 2B_3 - B_5 \tag{4.41}$$

The following system of equations is obtained:

$$\begin{bmatrix} M_1 & M_2 \\ M_3 & M_4 \end{bmatrix} \begin{bmatrix} \varepsilon^{i,j} \\ \delta^{i,j} \end{bmatrix} = \begin{bmatrix} P_1 \\ P_2 \end{bmatrix} \tag{4.42}$$

The finite difference algorithm implemented for solving the equations uses Jacobi iteration, successive under-relaxation of the nonlinear coefficients, and iterative decoupling of the two equations. The notation $\varepsilon_k^{i,j}$ and $\delta_k^{i,j}$ is introduced to denote the value of $\varepsilon^{i,j}$ and $\delta^{i,j}$ at iteration k. The following iteration scheme is used:

$$\varepsilon_k^{i,j} = \left[\frac{M_4 P_1 - M_2 P_2}{M_1 M_4 - M_2 M_3} \right]_{\varepsilon_{k-1}^{i,j}, \delta_{k-1}^{i,j}} \quad \text{and} \quad \delta_k^{i,j} = \left[\frac{M P_2 - M_3 P_1}{M_1 M_4 - M_2 M_3} \right]_{\varepsilon_{k-1}^{i,j}, \delta_{k-1}^{i,j}} \tag{4.43}$$

The iteration continues until

$$\left[\sup_{i,j} \left(\left| \varepsilon_k^{i,j} - \varepsilon_{-1}^{i,j} \right| \right) < \text{TOL} \right] \quad \text{and} \quad \left[\sup_{i,j} \left(\left| \delta_k^{i,j} - \delta_{-1}^{i,j} \right| \right) < \text{TOL} \right] \tag{4.44}$$

where TOL is a small tolerance parameter that determines the accuracy of the solution.

It was necessary to use iterative decoupling in the iteration process to obtain convergence for all data sets: ε is iterated without changing δ until convergence, and then δ is iterated until it converges. This is repeated until ε and δ do not change significantly. It is necessary to refrain from updating the nonlinear coefficients A_i, B_i for $i = 0, \ldots, 21$ in Eqs. 4.38 and 4.39 until the iteration converges. In other words, initial approximations of the nonlinear coefficients are made and Jacobi iteration is done until convergence. The nonlinear coefficients are then updated and iteration continues until convergence occurs again. This procedure is repeated until further updates of the coefficients do not change the solution significantly.

The iteration scheme described above may not converge for some input image sequences. For most image sequences, the algorithm may diverge after updating the nonlinear coefficients more than twice. The technique of successive under-relaxation can be used to eliminate problems such as this. With successive under-relaxation, the new values of ε and δ are not used to update the coefficients. Instead of $\varepsilon_k^{i,j}$ and $\delta_k^{i,j}$, the coefficients are updated with

$$\varepsilon_{\text{update}}^{i,j} = (1 - \omega) \varepsilon_{k-1}^{i,j} + \omega \left(\varepsilon_k^{i,j} \right) \quad \text{and} \quad \delta_{\text{update}}^{i,j} = (1 - \omega) \delta_{k-1}^{i,j} + \omega \left(\delta_k^{i,j} \right) \tag{4.45}$$

where $0 < \omega \leq 1$. The effect of this is to cause less of a "disturbance" to the iteration when updating the nonlinear coefficients. If $\omega = 1$, the update value of ε and δ is the full new value (the same as not using under-relaxation at all). If $\omega = 0$, the nonlinear coefficients never get updated past the first estimate.

4.3.1.1 Two-Dimensional Examples

A synthetic 2-D 64×64 image sequence, $f_i(x, y)$; $0 \leq x, y \leq 63$, representing a disk with a parabolic brightness distribution expanding in the y-direction and shrinking in the x-direction was created.

The sequence is given functionally by $f_i(x, y) = 255 - \left[\frac{(x-32)^2}{0.5 - 0.01i} \right] - \left[\frac{(y-32)^2}{0.5 + 0.01i} \right]$. All pixels with

values less than zero were set to zero and considered the background. Figure 4.2 shows sample images from the sequence for $i = 0$, 12, and 24. The domain of definition of the material coordinate system parametrization (see Eq. 4.10) is the region in the x-y plane covered by the object in image $f_0(x, y)$. This set is defined by $U = \{(x, y): f_0(x, y) > 0\}$.

Figure 4.3 shows the deforming material coordinate system model computed with the shape-change algorithm overlaid on the original images. Images sampled at $i = 0, 6, 12, 18$, and 24 are

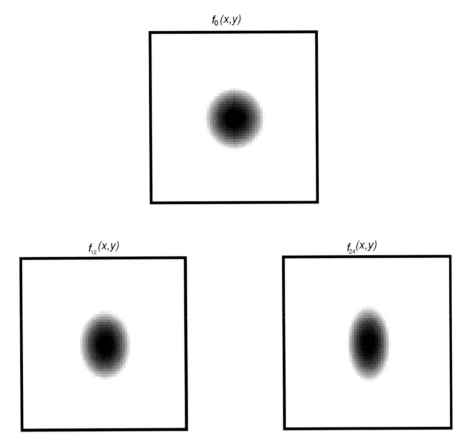

FIGURE 4.2   Sample images of the sequence for $i = 0$, 12, and 24.

shown. For this example, the smoothing parameter was $\lambda = 0.01$, the under-relaxation constant was $\omega = 1.0$ (no relaxation), and the iteration tolerance was 0.001. Next, the result of modeling the shape-change of a motile human white blood cell in 2-D is presented. The study of cell motility is extremely important in the biological sciences.[26-30] To date, no research has been reported for the modeling of both local and global shape-change of an entire cell. Most researchers have discussed tracking the global motion of one or more cells over time.

A live human white blood cell was imaged with a Zeiss laser-scanning confocal microscope in fluorescence mode. The confocal microscope is a relatively recent development in microscopy.[31-34] The white blood cell was stained with Acridine Orange and propidium iodide fluorescent stains. A total of seven frames were sampled at a constant rate of 30 s per frame. The white blood cell is about 15 μm across and was imaged during a relatively slow period of motility. Figure 4.4 shows the computed coordinate system model overlaid on the white blood cell data. For this model, $\lambda = 0.01$, $\omega = 0.01$, and the iteration tolerance was set to 0.01.

Although it is much more informative to view the images in an animation sequence, careful study of the images reveals that the material coordinate system has accurately tracked the shape-change undergone by the cell over a 3-min period. Also not recognizable from the still image representation is a global downward translation of the leukocyte. The algorithm is able to track this global "rigid" translation component of the cell motion as well as the local internal shape-changes of the cell.

Finally, an X-ray image sequence of a beating heart is analyzed. Figure 4.5 shows images $f_0(x, y)$ and $f_7(x, y)$ of the sequence. A radioactive stain is present in the blood so the motion of the interior wall of the left ventricle and aorta can be segmented. This image sequence is shown because it presents several difficulties for the shape-change analysis algorithm.

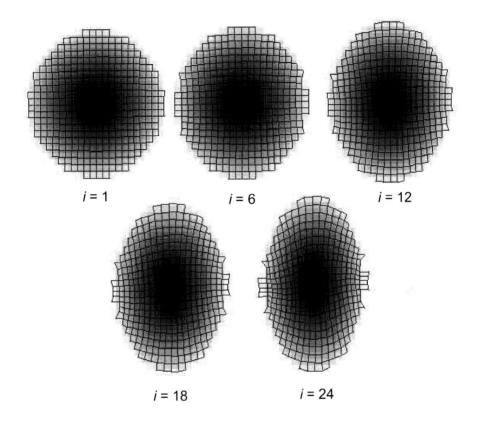

$i = 1$ $i = 6$ $i = 12$

$i = 18$ $i = 24$

FIGURE 4.3 Synthetic image sequence data with calculated deformed material coordinate system overlaid. Every sixth image of the entire sequence is shown. For this example, $\lambda = 0.01$, $\omega = 1.0$, and TOL $= 0.001$.

The sampling time between frames was large enough to cause problems for the algorithm. Another problem was that the image sequence contained two different classes of motion: the motion of the ventricular wall and the motion of the blood swirling within the ventricle and being ejected from the heart. The rapid and turbulent motion of the blood confused the algorithm since it could not differentiate the blood data from the ventricular data.

A spatio-temporal filtering scheme was used to ameliorate these problems. Gaussian low-pass filtering in both the spatial and temporal dimensions was performed on the image sequence. Spatial filtering smoothed out brightness variations due to imaging noise and the uneven distribution of radioactive stain. The temporal filtering smoothed swirling of the blood and other motions was too fast for the shape-change algorithm to handle. Temporal filtering allows modeling of the large-scale shape-changes in the sequence.

The information from the heart was segmented from the background information in the initial image with a simple thresholding procedure. The images of the sequence were reduced to 64×64 so that viewing the overlaid material coordinate system would be possible. With higher resolutions, the plot of the deforming parametrization was too dense and difficult to visualize. A 3-D Gaussian smoothing filter with a variance of 2 pixels in the spatial domain and 4 samples in the time domain

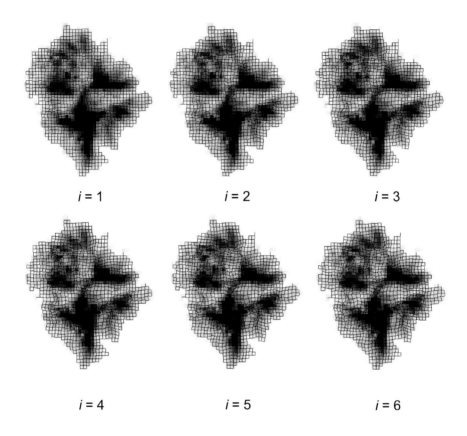

$$i = 1 \qquad\qquad i = 2 \qquad\qquad i = 3$$

$$i = 4 \qquad\qquad i = 5 \qquad\qquad i = 6$$

FIGURE 4.4 Output on a confocal microscope image sequence of a motile human white blood cell. ($\lambda = 0.01$).

was applied. Figure 4.6 shows the segmented, reduced, and smoothed images $f_0(x, y)$ and $f_7(x, y)$ from the sequence.

After smoothing, the high-velocity blood flow information was practically eliminated and the only motion visible in the sequence was that of the ventricular walls. The output of the algorithm on eight images of the sequence are shown in Figure 4.7. For this example, $\lambda = 0.001$, $\omega = 0.05$, and the iteration tolerance was set to 0.01. The algorithm tracked the contracting heart and expanding aorta well. There is an area, however, at the upper left side of the ventricle where the heart has "slipped away" from the original attachment of the material coordinate system. This area of the heart is undergoing shape-changes too quickly for the sample rate used. By sampling the images at a faster rate, this problem would not exist.

4.3.2 SOLUTION OF THE 3-D EULER EQUATIONS

For the 3-D numerical solution, the domain of the specimen parametrization is discretized and i, j, k are defined to be the indices into the u_1, u_2, u_3, directions, respectively. As in the 2-D case, the continuous second-order derivatives are replaced by finite differences:

$f_0(x,y)$ $f_7(x,y)$

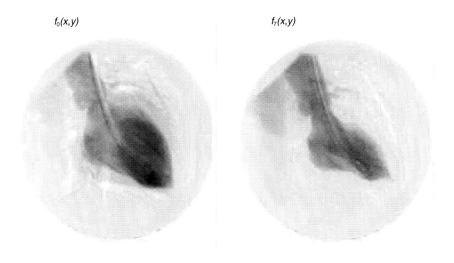

FIGURE 4.5   Images $f_0(x, y)$ and $f_7(x, y)$ of the heart sequence.

$f_0(x,y)$ $f_7(x,y)$

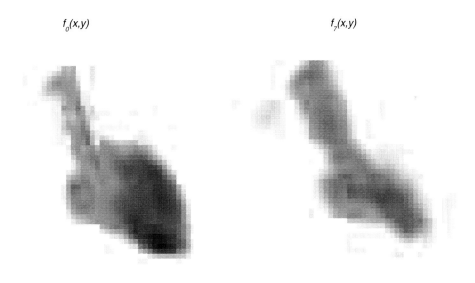

FIGURE 4.6   Segmented, reduced, and smoothed images $f_0(x, y)$ and $f_7(x, y)$ of the heart sequence.

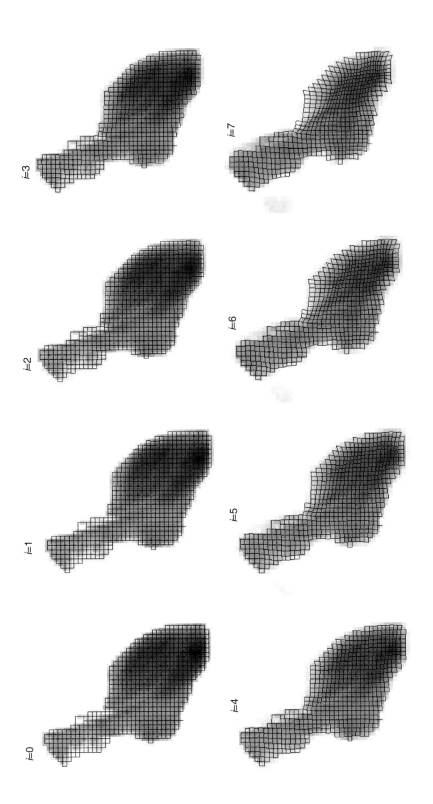

FIGURE 4.7 Algorithm results on a sequence of X-ray images of a heart with $\lambda = 0.001$. The image sequence was first filtered by a Gaussian smoothing filter in both space and time.

$$\varepsilon_{u_1} \to \varepsilon^{i-1,j,k} - 2\varepsilon^{i,j,k} + \varepsilon^{i+1,j,k}$$

$$\varepsilon_{u_2 u_2} \to \varepsilon^{i,j-1,k} - 2\varepsilon^{i,j,k} + \varepsilon^{i,j+1,k}$$

$$\varepsilon_{u_3 u_3} \to \varepsilon^{i,j,k-1} - 2\varepsilon^{i,j,k} + \varepsilon^{i,j,k+1}$$

$$\varepsilon_{u_1 u_2} \to \varepsilon^{i-1,j-1,k} - \varepsilon^{i,j-1,k} - \varepsilon^{i-1,j,k+1} + \varepsilon^{i,j,k} \qquad (4.46)$$

$$\varepsilon_{u_1 u_3} \to \varepsilon^{i-1,j,k-1} - \varepsilon'^{i,j,k-1} - \varepsilon^{i-1,j,k} + \varepsilon^{i,j,k}$$

$$\varepsilon_{u_2 u_3} \to \varepsilon^{i,j-1,k-1} - \varepsilon'^{i,j,k-1} - \varepsilon^{i,j-1,k} + \varepsilon^{i,j,k}$$

Analogous replacements are made for derivatives of $\delta(u_1, u_2, u_3)$ and $\gamma(u_1, u_2, u_3)$. After discretizing, terms involving $\varepsilon^{i,j,k}$, $\delta^{i,j,k}$ and $\gamma^{i,j,k}$ are factored out:

$$\begin{bmatrix} \varepsilon^{i,j,k} \\ \delta^{i,j,k} \\ \gamma^{i,j,k} \end{bmatrix} = \begin{bmatrix} M_1 & M_2 & M_3 \\ M_4 & M_5 & M_6 \\ M_7 & M_8 & M_9 \end{bmatrix}^{-1} \begin{bmatrix} P_1 \\ P_2 \\ P_3 \end{bmatrix} \qquad (4.47)$$

The elements in the above equation are too lengthy to present here explicitly. The explicit form of the equation is given in Ref. 25. Equation 4.47 leads straightforwardly to an iteration scheme identical to that described in Section 4.3.1 for the 2-D case. As in the 2-D case, Jacobi iterative decoupling was used for the three equations with successive under-relaxation of the nonlinear equations. Although much more computationally demanding, the solution of the 3-D Euler equations is exactly analogous to the 2-D case. In the following section, examples using both synthetic and real 3-D image sequences are presented.

4.3.3 Three-Dimensional Examples

Even more so than in the 2-D case, the synthetic images are extremely important in verifying that the model is meaningful for the 3-D case. Visual analysis of the results is very difficult to accomplish because an entire volume of data must be viewed through time. In essence, the 4-D data (three space dimensions and time) must be displayed in 3-D (two space dimensions of the computer screen and time). However, it is this difficulty that makes the shape-change analysis technique so valuable. The incredible amount of data within the four dimensions is very difficult to analyze manually in a quantitative manner.

A synthetic 3-D image sequence representing a ball with a parabolic brightness distribution that is expanding in the x- and y-directions and shrinking in the z-direction was created. Each image in the sequence contains 16×16 voxels. The sequence is given functionally by:

$$f_i(x, y, z) = 255 - \left[\frac{(x-8)^2}{0.08 + 0.005i}\right] - \left[\frac{(y-8)^2}{0.08 + 0.005i}\right] - \left[\frac{(z-8)^2}{0.12 - 0.005i}\right] \qquad (4.48)$$

A sequence of ten images was created by letting i range from 0 to 9. The 3-D images for $i = 0$ and $i = 9$ are shown in Figures 4.8 and 4.9. Laying out the images into 2-D slices along one of the coordinate axes is the only way to visualize the entire volume of brightnesses on a printed page.

The 3-D shape-change analysis algorithm was applied with $\lambda = 1$, $\omega = 0.1$, and TOL = 0.001. Figure 4.10 shows the deforming 3-D material coordinate system for $i = 0, 3, 6$, and 9. The results are shown as stereo-pairs that can be stereoscopically viewed either with a stereoscopic viewer or with practice by relaxing and focusing the eyes.

Until only recently, obtaining 3-D image sequences of biological systems undergoing shape-change was impossible. Tomographic 3-D imaging of moving specimens is still under development and is performed mostly in research labs. Acquiring 3-D image sequences with a laser-scanning confocal microscope is difficult because of the relatively slow image acquisition time. A specimen must be changing shape very slowly in order to capture meaningful images. Another problem is maintaining the viability of the specimen during the exposure to intense laser light.

Quality 3-D image sequences were obtained of islets of Langerhans subjected to osmotic stresses. A special stage was built for the microscope that allowed perfusion of various concentrations of the cryoprotectant dimethyl sulfoxide (DMSO) around a single islet. This setup allowed the islet shape-change to be somewhat under the control of the experimenter. The islet volume will decrease after the perfusion of a concentrated DMSO solution and then slowly increase as the osmotic forces equilibrate. The details of this experiment are given in Ref. 35.

In Ref. 35, a simple method for measuring the total volume of an imaged islet is given. The technique consists of segmenting the islet with a simple thresholding and median filtering technique, then counting the number of voxels in the segmented region. The perfusion experiment in Ref. 35 consisted of subjecting an islet to two step changes in DMSO concentration. At time 0 s, a 2 M DMSO solution is added and at 600 s, a 4 M DMSO solution is added. Figure 4.11 shows the volume as measured by the voxel counting technique versus time for this experiment. It has been shown in theory[36] and previous experiments[37] that step increases in solution molarity result in an immediate and rapid decrease in volume, followed by a slower increase back to the initial volume. This behavior is evident in Figure 4.11.

A 3-D image sequence of length 4 starting at time $t = 236$ min was extracted from this data set and modeled with the shape-change algorithm. The image corresponds to the volume samples marked with "*" in Figure 4.11. For this length-4 sequence, the islet volume increased in the first two intervals and then decreased.

Figure 4.12 shows sections down the z-axis of the islet for $i = 0$ (time $t = 236$ min) and $i = 2$ (time $t = 610$ min). The 3-D images consist of six 64×64 2-D images along the z-axis.

Figure 4.13 shows the result of 3-D shape-change modeling on the islet data with $\lambda = 0.01$, $\omega = 0.001$, and TOL = 0.001. The material coordinate model was subsampled by a factor of 3 in the x- and y-directions to reduce the complexity of the plots. The overall volume changes can be seen, as measured with the voxel counting routine described above, as well as the complex localized internal volume changes. Previous techniques for analyzing 3-D shape-change either measured only global properties, such as the voxel counting method, or measured shape and curvature changes on the surface of the specimen only. Obtaining such global measurements from the detailed model is straightforward.

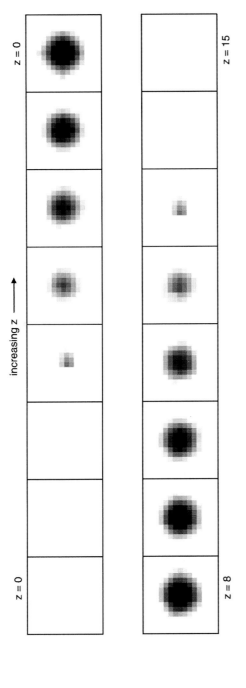

FIGURE 4.8 The 16 sections along the z-axis of $f_0(x, y, z)$.

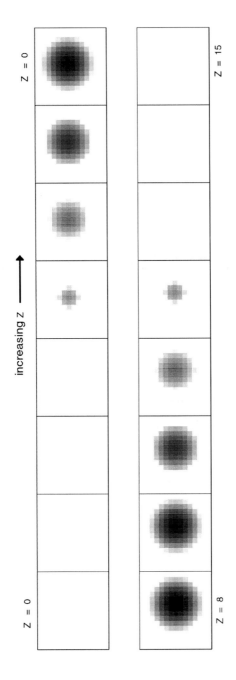

FIGURE 4.9 The 16 sections along the z-axis of $f_9(x, y, z)$.

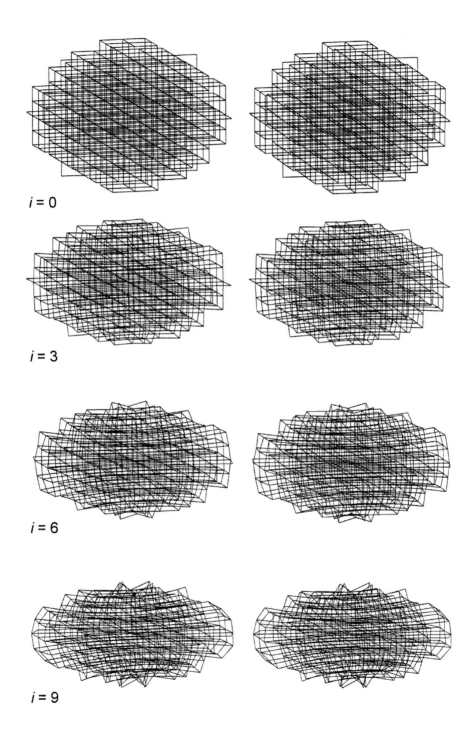

$i = 0$

$i = 3$

$i = 6$

$i = 9$

FIGURE 4.10   Stereo-pairs of the 3-D model at $i = 0$, 3, 6, and 9.

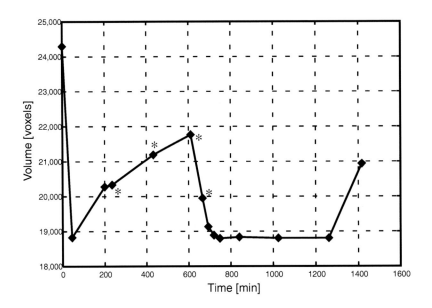

FIGURE 4.11   Islet volume vs. time for step changes in DMSO concentration. At $t = 0$, 2 M DMSO is added; at $t = 600$ s, 4 M DMSO is added. Images at points marked with "*" were used for shape-change modeling.

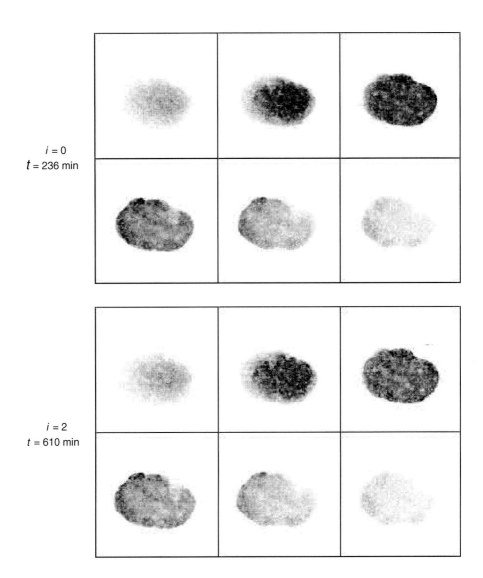

$i = 0$
$t = 236$ min

$i = 2$
$t = 610$ min

FIGURE 4.12 Sections down the z-axis of the islet for $i = 0$ (time $t = 236$ min) and $i = 2$ (time $t = 610$ min).

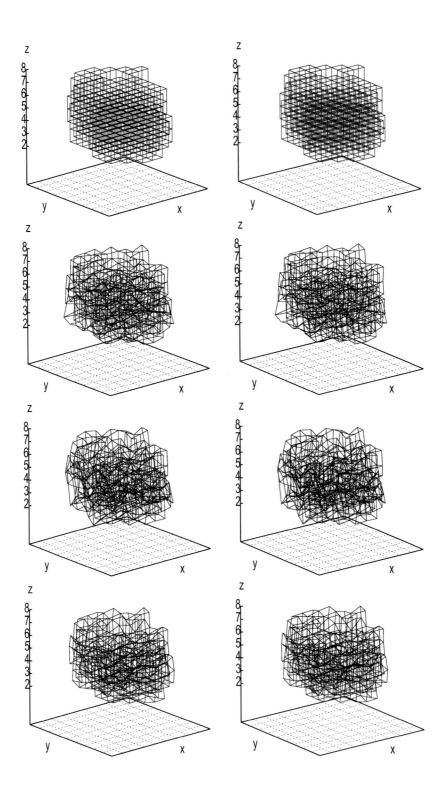

FIGURE 4.13   Stereo-pairs of 3-D model of pancreatic islet data at $i = 0$, 1, 2, and 3.

REFERENCES

1. Horn, B. K. P., *Robot Vision*, MIT Press, Cambridge, MA, 1986.
2. Grimson, W. E. L., A computer implementation of a theory of human stereo vision, *Phil. Trans. Roy. Soc. London B*, 298, 395, 1981.
3. Grimson, W. E. L., *From Images to Surfaces: A Computational Study of the Human Early Visual System*, MIT Press, Cambridge, MA, 1981.
4. Terzopoulos, D., Multilevel computation processes for visual surface reconstruction, *Comput. Vision, Graphics, Image Processing*, 24, 52, 1983.
5. Terzopoulos, D., Regularization of inverse visual problems involving discontinuities, *IEEE Trans. Pattern Anal. Machine Intell.*, 8, 413, 1986.
6. Terzopoulos, D., The computation of visible-surface representations, *IEEE Trans. Pattern Anal. Machine Intell.*, 10, 417, 1988.
7. Bartels, R. H., Beatty, J. C., and Barsky, B. A., *An Introduction to Splines for Use in Computer Graphics and Geometric Modeling*, Morgan Kaufmann, Los Altos, CA, 1987.
8. Ballard, D. H. and Brown, C. M., *Computer Vision*, Prentice-Hall, Englewood Cliffs, NJ, 1982.
9. Barr, A. H., Superquadrics and angle preserving transformations, *IEEE Comput. Graphics Applicat.*, 1, 11, 1981.
10. Solina, F. and Bajcsy, R., Recovery of parametric models from range images: the case for superquadrics with global deformations, *IEEE Trans. Pattern Anal. Machine Intell.*, 12, 131, 1990.
11. Terzopoulos, D. and Metaxas, D., Dynamic 3D models with local and global deformations: deformable superquadrics, *IEEE Trans. Pattern Anal. Machine Intell.*, 13, 703, 1991.
12. Leymarie, F. and Levine, M. D., Tracking deformable objects in the plane using an active contour model, *IEEE Trans. Pattern Anal. Machine Intell.*, 15, 7, 617, June 1993.
13. Moshfeghi, M., Ranganath, S., and Nawyn, K., Three dimensional elastic matching of volumes, *IEEE Trans. Image Processing*, 3, 7, 128, March 1994.
14. Price, J. L. and Denney, T., 3D displacement field reconstruction from planar tagged cardiac mr images, *Proc. IEEE Workshop on Biomedical Image Analysis*, Seattle, WA, June 1994.
15. Yeung, F., Levinson, S., Fu, D., and Parker, K., Feature-adaptive motion tracking of ultrasound image sequences using a deformable mesh, *IEEE Trans. Medical Imaging*, 17, 7, 945, December 1998.
16. Horn, B. K. P. and Schunck, B. G., Determining optical flow, *Artificial Intelligence*, 17, 185, 1981.
17. Denney, T. S. Jr. and Prince, J. L., Optimal brightness functions for optical flow, *IEEE Trans. Image Processing*, 3, 7, 178, March 1994.
18. Munkres, J. R., *Topology: A First Course*, Prentice-Hall, Englewood Cliffs, NJ, 1975.
19. Patne, L. E., *Improperly Posed Problems in Partial Differential Equations*, Society for Industrial and Applied Mathematics, Philadelphia, 1975.
20. Marsden, J. E. and Tromba, A. J., *Vector Calculus*, 2nd ed., W. H. Freeman and Company, San Francisco, 1981.
21. Tykhonov, A. N., Incorrectly posed problems and the method of regularization (English transl.), *Soviet Math.*, 4, 1963.
22. Tykhonov, A. N., On methods of solving incorrect problems, *Am. Math. Soc. Transl. Ser. 2*, 70, 222, 1968.
23. Wahba, G., *Spline Models for Observational Data*, Society for Industrial and Applied Mathematics, Philadelphia, 1990.
24. Weinstock, R., *Calculus of Variations with Applications to Physics and Engineering*, Dover, New York, 1974.
25. Bartels, K. A., Bovik, A. C., and Griffin, C. E., *The Computation of Biological Material Deformations from Image Sequences*, Technical Report UT-CVIS-TR-95-002, The University of Texas at Austin Center for Vision and Image Sciences, March 1995.
26. Murray, J., Vawter-Hugart, H., Voss, E., and Soll, D. R., Three-dimensional motility cycle in leukocytes, *Cell Motility and the Cytoskeleton*, 22, 211, 1992.
27. Noble, P. B. and Levine, M. D., *Computer-Assisted Analysis of Cell Locomotion and Chemotaxis*, CRC Press, Boca Raton, FL, 1986.
28. Segall, J. E., Quantification of motility and area changes of *Dictyostelium discoideum* amoebae in response to chemoattractants, *J. Muscle Res. Cell Motility*, 9, 481, 1988.

29. Soll, D. R., "DMS," a computer-assisted system for quantitating motility, the dynamics of cytoplasmic flow, and pseudopod formation: its application to *Dictyostelium* chemotaxis, *Cell Motility and the Cytoskeleton*, 10, 91, 1988.

30. Thurston, G., Spadinger, I., and Palcic, B., Computer automation in measurement and analysis of cell motility *in vitro*, *Cell Motility Factors*, Goldberg, I. D., Ed., Birkhäuser Verlag, Basel, Switzerland, 1991, 206.

31. Kino, G. S. and Xiao, G. Q., *Confocal Microscopy*, Academic Press, London, 1990, chap. 14, 361.

32. Wilson, T., *The Handbook of Biological Confocal Microscopy*, IMR Press, Madison, WI, 1989, chap. 11, 99.

33. Wilson, T., Ed., *Confocal Microscopy*, Academic Press, London, 1990.

34. Wilson, T. and Sheppard, C. J. R., *Theory and Practice of Scanning Optical Microscopy*, Academic Press, London, 1984.

35. Merchant, F. A., Aggarwal, S. J., Bartels, K. A., Diller, K. R., and Bovik, A. C., Analysis of volumetric changes in rat pancreatic islets under osmotic stress using laser scanning confocal microscopy, *ISA/IEEE 30th Annual Rocky Mountain Bioengineering Symposium*, San Antonio, TX, April 2–3, 1993.

36. Diller, K. R., Pegg, D. E., and Walcerz, D. B., A network thermodynamic model for analysis of water transport in a multicellular tissue during cryopreservation, *Biofluid Mechanics* 3, Schneck, D. J., Ed., New York University Press, New York, 1990, 55.

37. Schwartz, G. J. and Diller, K. R., Osmotic response of individual cells during freezing. I. Membrane permeability analysis, *Cryobiology*, 20, 542, 1983.

5 Digital Filters in Image Analysis

Roberto Marangoni, Giuliano Colombetti, and Paolo Gualtieri

CONTENTS

5.1 INTRODUCTION

The final result of several scientific experiments is often presented by an image. To be meaningful to the human observer, this image must often be subsequently elaborated; think, for example, of a satellite, microscope, or X-ray picture.

The very fast development in the field of electronics has made available, at relatively low cost, integrated systems that are capable of acquiring, storing, and analyzing images. These systems can easily be implemented on personal computers and thus are widely used in research laboratories.

If restricted to the biological sciences, one finds that image analysis techniques are mainly employed in the field of microscopy. Optical, electron, scanning, tunneling, and atomic force microscopic images are very often presented to the reader in a more or less elaborated form. Elaboration makes it possible to reduce the noise in the image, to enhance or even reveal details in the original picture, to correct optical aberration, to measure quantitatively morphological parameters (e.g., areas, dimensions, etc.), and to determine concentrations of chemical species (e.g., Ca-sensitive dyes). It is also possible to use similar techniques to analyze sequences of images; in other words, to follow the dynamic evolution of a system. It is thus possible to study motion parameters of a population of cells, automatically measuring their positions and velocities.

More recently, artificial intelligence and neural network techniques have contributed to the setup of expert systems capable of recognizing patterns present in a scene. Automatic pattern recognition is thus becoming possible, and further dramatic developments in this field are expected in the near future.

5.2   IMAGE FORMATION

In most cases, a biological image is obtained by means of a microscope. This chapter is limited to the study of optical images; in fact, other systems do not form an image *sensu strictu*. Not considered here are the physical principles that underlie image formation and their mathematical description. The interested reader is referred to the literature.[4,8,9] The knowledge of this may, however, be very useful in the development of image-restoring algorithms.

Independent of the system of image formation, the final result can be described by a light intensity function $f(x, y)$ where the dominion of the (x, y) points coincides with the image area and the value of f at a certain point is proportional to the brightness of the image at that point. On the basis of this definition, the function $f(x, y)$ represents a real image.

Assuming monochromatic images, this is not too restrictive an assumption. All the results to be discussed in the following sections are, in principle, also valid for color images, even if the situation in that case may be much more complicated. The acquisition of a color image may be useful, for example, to have a better discriminating capacity. From a technical viewpoint, a color-discriminating acquisition can be obtained in two different ways: either by using a commercial device or by acquiring three different monochromatic images taken with three different filters that form a base for colors (such as red – green – blue). It is also possible to digitize a monochromatic image and subsequently color it artificially in order to enhance certain features. These images are usually known as "false color images," since the colors visualized do not correspond to the real colors. A typical example is given by colored pictures taken from a meteorological satellite, where the different colors correspond to the different temperatures recorded. Further details are not considered in this chapter, the interested reader is encouraged to consult the appropriate literature (see, e.g., Refs. 4, 7, 15).

5.3   DIGITIZATION

Digitization is defined as the process by which the continuous function $f(x, y)$ defined above is transformed into a discrete function $l(m, n)$ defined over a set of discrete values m, n. This is a very important step because the use of a digital computer requires a discrete set of data to be processed. Digitization includes both spatial and amplitude values. Digitization in space is known as sampling: in amplitude, it is known as quantization.

5.3.1   SAMPLING AND QUANTIZATION

In order to convert $f(x, y)$ into $l(m, n)$, the following steps must be taken:

1. A usually square grid, formed by $N \times N$ squares, is superimposed onto the $f(x, y)$. Each square is identified by a couple of integers m, n with $0 < m < N$ and $0 < n < M$ (see Figure 5.1).
2. The average brightness is calculated over each square.
3. A function $g(m, n)$, having as value the average brightness previously calculated, is defined over each square. Its values are bound by two real numbers g_{min} and g_{max}.

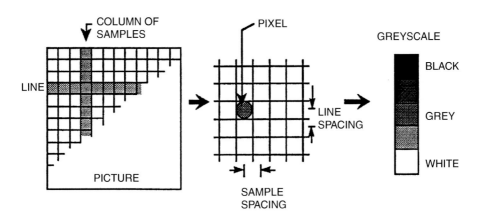

FIGURE 5.1   Sampling and quantization procedures (see text).

4. A set of discrete values is chosen equally spaced between 0 and an integer maximum L. These values are given by $(g(m, n) - g_{min})/g_{max} - g_{min}) \times L$ and are usually known as gray levels.
5. The discrete function $l(m, n)$ is defined over the same dominion as $g(m, n)$, with the values given in the previous point 4, and is the function being sought.

The discrete points of coordinates (m, n), elementary parts of the picture, are called pixels (from PICture ELements). The sampling and quantization described above are uniform, since a uniform square grid and equally spaced values are used. It is possible to use non-uniform sampling and quantization.[1,7] This is usually done for particular purposes. For example, a fine sampling obtained using a finely spaced grid for the central part of the image and a coarsely spaced grid elsewhere simulates the spatial sampling of the human retina better than a uniform sampling.[16,17] A nonlinear correspondence between brightness values and gray-level ranges may be useful to increase the visibility of poorly contrasted details (tapering).

5.3.2   SAMPLING AND QUANTIZATION TECHNIQUES

As seen in the previous paragraphs, a digital image is an image $f(x, y)$ discretized both spatially and in brightness. There are two major techniques that are currently used to digitize an image; they are known respectively as scan-in and scan-out (see Figure 5.2).[4]

In the scan-in procedure (Figure 5.2A), the scene is scanned by a light beam. A light sensor measures, one pixel at a time, the amount of transmitted light that gives the brightness value at that point. In the scan-out procedure (Figure 5.2B), the whole scene is illuminated at the same time and the sensor reads the image one point after the other.

From a formal viewpoint, the two procedures are not different, but the results obtained are different. In the scan-in procedure, the movement is mechanically driven, whereas in scan-out, it is electronically driven. The first system has the advantage of a good signal-to-noise ratio, the second of a very high scanning speed. It is also possible to use more refined methods that utilize both techniques.

There are many theoretical studies aimed at determining optimal procedures for sampling and quantization;[1,4,14] the average user has no need, however, to take care of such problems because commercial scanners and TV cameras are built in such a way as to optimize image digitization.

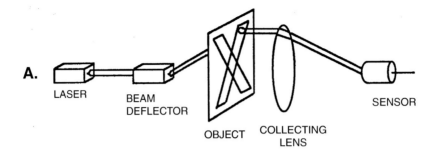

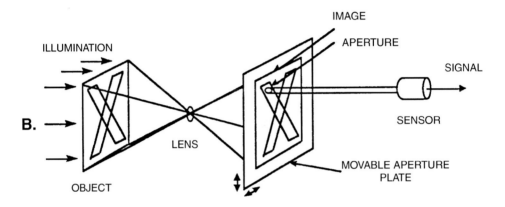

FIGURE 5.2   Scan-in (A) and scan-out (B) techniques (see text).

5.4   DIGITAL IMAGES

A digital image obtained according to the procedures described above can easily be represented by a square $N \times N$ matrix (a common value for N is 512), whose values are integer numbers proportional to the brightness values and ranging from 0 to an upper limit L (typically 255) in the so-called gray scale. In this representation it is possible to quantitatively analyze the image, for example, in terms of the histogram of the gray levels or of the spatial frequencies.

5.4.1   GRAY-LEVEL HISTOGRAM

A gray-level histogram is determined by counting all the pixels in an image that have a certain gray value and plotting on the abscissa the values of the gray levels and on the ordinate the frequencies of the different gray levels. From this histogram it is possible to obtain some information such as the average gray level, which gives an idea of the overall luminosity of the image. From the shape of the histogram it is possible to obtain some hints for further image processing. Assume, for example, that one is dealing with the microscopic image of cells on a strongly contrasted background. The gray-level histogram is bimodal, the background gray levels corresponding to the greater peak and the gray levels belonging to the second, smaller peak corresponding to the cell pixels. To determine automatically whether some pixels belong to a cell, it is sufficient to measure their values.

5.4.2 Spatial Frequency Content

Fourier analysis, as known from the theory of signal analysis, can also be applied to the study of digitized images. According to this theory, every image can be decomposed into a series of sinusoidal gratings with certain amplitudes and phases, the sum of which restores the original image. The spatial frequencies contained in an image are given by the reciprocal of the periods of these sinusoidal functions. It is thus possible to build a histogram having on the abscissa the different spatial frequencies and on the ordinate the frequency with which each spatial frequency is present in the image. The importance of this transformation lies in the fact that different spatial frequencies correspond to different regions of an image. The contours are the regions with higher spatial frequencies, the homogeneous portions of an image those with the lowest frequencies. It is thus possible to develop algorithms acting selectively on determined parts of an image.

5.5 IMAGE PROCESSING

It is usually necessary to process a digitized image in the following cases:

1. Image restoration: an image may be affected by factors such as optical system aberrations and noise.
2. Image enhancement: it is a common wish to enhance some details present in an image.
3. Information extraction: it is very useful to be able to measure automatically and in a quantitative manner some properties of the image. For example, it may be desirable to measure the surface of cells on a microscope slide, to count automatically the number of cells present in a field, to follow moving cells and determine their motion parameters, to measure the concentration of intracellular probes, etc.

The following sections briefly discuss these topics, giving some examples for each of them. Other chapters in this book deal in detail with specific applications.

5.5.1 Digital Image Transformations

Let $l(m, n)$ be the value of a randomly chosen pixel, and H_F the gray-level histogram of the original image, $l(m, n)$ and H_G their respective values after image processing, and T a generic transformation. An image can be processed in three different ways: point transformation, local transformation, and generic transformation. In the case of point transformation, a pixel is modified, taking into account only its actual value. A typical example of this is image thresholding: given a value x belonging to the gray-level interval, one obtains

$$l'(m, n) = \begin{cases} 0 & \text{if } l(m, n) < x \\ L & \text{elsewhere} \end{cases} \tag{5.1}$$

It should be noted that this transformation law is concerned only with pixel $l(m, n)$, and there is no consideration given to the other pixels belonging to the image. In the case of local transformation, a pixel is transformed, taking into account the values of the neighboring pixels. One can schematically represent the pixel and its neighbors, for example, with a 3×3 matrix, where the pixel to be modified is the central element in E.

A	B	C
D	E	F
G	H	I

A law such as $l'(m, n) = (A + B + C + D + E + F + G + H + I)/9$ is considered a local modification. Usually, a local transformation is a weighted average of the pixels of a small neighborhood; it is mathematically known as a convolution. When the transformation law is chosen, a matrix, usually called mask, is defined; its elements are the weights of the pixels. Let $K(s, t)$ be such a matrix. It can be shown that the correlation function between neighboring pixels becomes very small beyond the fifth neighbor.[4,7,14,15] It is thus sufficient to choose a 5×5 or 3×3 masking matrix. In order to save computing time, as small a mask as possible is usually chosen. The convolution between the mask and the image is then carried out; this means that $l(m, n)$ will be replaced by

$$l'(m, n) = l * K = \sum_i \sum_j l(i, j) K(m - i, n - j) \tag{5.2}$$

In order to visualize this operation, think of the mask as sliding over the image, superimposing each time to a 3×3 neighborhood of a point, and then compute the convolution products. It is desirable that the sum of the elements of the masking matrix be equal to 1, in order not to introduce artificial variations of the average brightness. This procedure is usually known as image filtering: filters may be used for different purposes, such as selection of determined bands of spatial frequencies, approximation, error dispersion, evaluation, etc. Most of the problems considered in what follows deal with appropriate filters.

In the case of a global transformation, the new pixel value depends on the values of all the pixels of the image. Such transformations are known as transforms and are widely used in signal analysis. The Fourier transform, which gives the frequency spectrum of an image, is probably the most well-known transform, but there are also other transforms that are useful to extract other types of information.[7,14,15] These transforms deal with the whole image, and usually require heavy calculations; they have, therefore, usually been replaced by digital filters, which are much simpler and faster. The situation might change in the near future since there are commercially available systems that can perform these transforms in real-time.

5.6   IMAGE RESTORING

The techniques of image restoring are used to improve image quality. They can be divided into two categories:

- Noise cleaning: this technique aims at improving the signal-to-noise ratio.
- Image restoring *sensu strictu*: it is used each time an image is systematically distorted and the distortion can be mathematically modeled. This is true, for example, for the spherical aberrations of a lens or for a motion blurred picture. In these cases it is possible to develop an algorithm that, by distorting the image in the opposite way, restores it to its original aspect.

5.6.1   "CLASSIC" FILTERS

We limit ourselves to the technique of noise cleaning. Among the different types of noise that affect an image at the different spatial frequencies, consider in particular the high-frequency noise, which causes loss of sharpness and alters the image contours. This section describes some low-pass filters that can reduce or even eliminate this type of noise.

In the examples shown, the low-pass filters are obtained by averaging the pixel value with the values of the pixels in its 3×3 neighborhood. These filters are easily obtained, but they have certain drawbacks. For example, the image contours become blurred since the high spatial frequencies are indiscriminately cut off; also, there is a broadening of the contours, so the image loses sharpness and appears a bit out of focus.

Figure 5.3A shows an image and Figure 5.3B shows the same image with superimposed Gaussian noise. The following are three examples for low-pass filters

$$A = \frac{1}{9}\begin{pmatrix} 1 & 1 & 1 \\ 1 & 1 & 1 \\ 1 & 1 & 1 \end{pmatrix} \quad B = \frac{1}{10}\begin{pmatrix} 1 & 1 & 1 \\ 1 & 2 & 1 \\ 1 & 1 & 1 \end{pmatrix} \quad C = \frac{1}{16}\begin{pmatrix} 1 & 2 & 1 \\ 2 & 4 & 2 \\ 1 & 2 & 1 \end{pmatrix} \quad (5.3)$$

Figures 5.4A and B show the results of the application of Filter A in Eq. 5.3 on the two images. It is possible to limit the broadening effect by choosing the pixels to be filtered by means of a previous thresholding such as

$$l'(m, n) = \begin{cases} l'(m, n) & \text{if } \left[8E - (A + B + C + D + F + G + H + I)\right] > \varepsilon \\ l(m, n) & \text{elsewhere} \end{cases} \quad (5.4)$$

A filter that does not bring about a broadening effect is a median filter. The median of a population of values is defined as the value that is greater than 50% of the population values and smaller than 50% of them. A median filter can, for example, be obtained by taking the values of a 3×3 neighborhood of a pixel, sorting them in ascending or descending order, and assigning to the original pixel the fifth value of the sorted sequence. Unlike the filters based on average values, this filter is relatively insensitive to noise; its main disadvantage is that being a nonlinear filter, it requires much longer computation times than the filters considered above. For the results, see Figures 5.5A and B.

(A) (B)

FIGURE 5.3 Original image (A) and superimposed Gaussian noise (B) (see text).

(A)   **(B)**

FIGURE 5.4   Result of convolution of both images in Figure 5.3 with filter *A* in Eq. 5.3 (see text).

(A)   **(B)**

FIGURE 5.5   Result of the application of the median filter on the images in Figure 5.3 (see text).

5.6.2   STATISTICAL ADAPTIVE FILTERS

Most of the noise-reducing filters presented above are based on low-pass filtering, and are therefore suitable for reducing the spot noise, which is mostly located at high spatial frequencies. Unfortunately, because contours and edges in an image are mainly high frequency, while using low-pass filters to reduce noise, one loses information about important features of an image (blurring effect). This happens because of the nonspecificity of the filters used, that simply cut off high spatial frequencies, no matter what the pixel can represent in an image.

A new generation of more sophisticated filters has been developed, which take into account the particular characteristics of some image features and thus perform a filtering operation without degrading the image. The simplest approach of this new generation is based on the statistical properties of the noise, and, in particular, on pixel variance analysis. The rationale of this method is that image areas (pixels or segments) containing structure and strong contrast between edges will have a higher variance than areas containing noise only. Therefore, in the statistical filtering process, the variance value of each image segment will determine whether noise reduction (low-pass filtering) is to be applied to that area. As a result of this process, noise is reduced and sharp edges are maintained in the image. In other words, by combining the search for high spatial frequencies and low variance values, one can design filters that can reduce noise without altering the object contours. If the statistical distribution of noise in an image is known, it is possible to design a filter based on this statistical distribution and to restore the original image.

This class of filters belongs to the more general class of adaptive filters, in which the coefficients of the used mask are not chosen independently of the image analyzed, but vary depending on its global and local characteristics.

A detailed discussion of this kind of filter goes beyond the scope of this chapter, also because it requires heavy statistical calculations. The interested reader is referred to the rich literature existing in the field.[18,19]

5.7   IMAGE ENHANCEMENT

Even if an image has a good signal-to-noise ratio and is not distorted, it may still need to be elaborated in order to be suitable for the required analysis. In general, some features may not be clear enough, either because of a low contrast or a wrong gray-scale choice. Some aspects of this problem are considered below.

5.7.1   EDGE SHARPENING

In order to selectively enhance contours, it is necessary to operate in an opposite way to that described in the previous section; in other words, high-pass filters are required.[5]

Some high-pass masks are defined below.

$$A = \begin{pmatrix} 0 & -1 & 0 \\ -1 & 5 & -1 \\ 0 & -1 & 0 \end{pmatrix} \quad B = \begin{pmatrix} -1 & -1 & -1 \\ -1 & 9 & -1 \\ -1 & -1 & -1 \end{pmatrix} \quad C = \begin{pmatrix} 1 & -2 & 1 \\ -2 & 5 & -2 \\ 1 & -2 & 1 \end{pmatrix} \quad (5.5)$$

As can be seen from the numerical coefficients used, they increase the brightness difference between a pixel and its neighbors. Figures 5.6A and B show the results of a high-pass filter.

(A)　　　　　　　　　　　　　　　　　　　　　(B)

FIGURE 5.6　Result of the application of filter *A* in Eq. 5.5 on the images in Figure 5.4 (see text).

5.7.2   CONTRAST OPTIMIZATION

Whenever an image has a low contrast, it is very difficult to recognize the finest details. The techniques that deal with this kind of problem are based on suitable manipulations of the gray-level histogram.

Contrast expansion is obtained by means of a linear expansion of a certain interval (between a maximum and a minimum determined by the user) of the gray-level histogram of the original image.

This is shown in Figure 5.7. It can be seen that this causes a loss of information in the region of higher or lower brightness. This transformation is mathematically represented by

$$l'(m, n) = \frac{l(m, n) - \min}{\max - \min} \cdot L \tag{5.6}$$

Figure 5.8 shows the result of contrast expansion on the sample image.

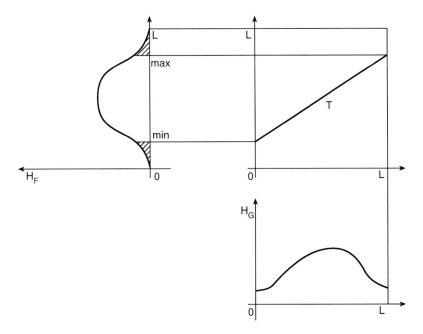

FIGURE 5.7   A representation of the transformation induced by contrast-expansion techniques on the gray-level histogram of an image (see text).

Another type of histogram manipulation is its equalization, so that it becomes rectangular. This transformation is more difficult to implement, but is much more useful because it maximizes image information.[7,15] Figure 5.9 shows the type of transformation that is brought about by histogram equalization. Mathematically, the transformation is

$$l'(m, n) = \frac{L^2}{N} \sum_{k=0}^{a} H_F(k) \quad \text{where} \quad a = l(m, n) \tag{5.7}$$

Histogram equalization requires time-consuming calculations, but it is worthwhile. Figure 5.10 has been obtained by equalizing the histogram of the sample image.

FIGURE 5.8 Contrast expansion on the image in Figure 5.4A (see text).

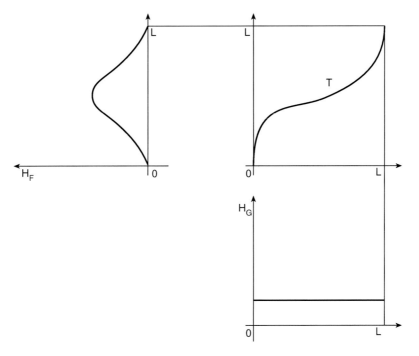

FIGURE 5.9 A representation of the transformation induced by a histogram-equalization technique on the gray-level histogram of an image (see text).

5.8 FEATURE EXTRACTION

For a long time, image analysis has dealt mainly with the problem of the development of edge detection algorithms. Nowadays, there are a certain number of good algorithms for this purpose, even if there is no definite conclusion on the type of algorithm or on the method to use.[2,3,10,13] The

FIGURE 5.10   Result of histogram-equalization techniques on the image of Figure 5.4A.

question of feature extraction has, however, gone far beyond this original purpose. The development of pattern recognition techniques, associative memories, and neural networks has made it possible to implement, even if at a still experimental level, expert systems capable of recognizing several complex patterns.[6] The development of relational semantics makes it possible to name an identified object, thus allowing computer recognition.[6] This discussion is restricted to the techniques of edge detection, referring the interested reader to the appropriate literature.[1,7]

The edge, defined as the spatial region where the object ends, is characterized by the region of strongest brightness variation. The function $f(x, y)$, the brightness of a real image, has in fact a discontinuity in the edge region. It has no meaning to look for a discontinuity of the discrete function $l(m, n)$; rather, one should look for a very high incremental ratio. The first researchers in the field moved along this line of thought, using linear filters such as derivatives or Laplacian.[3,11,12]

The three masks presented below are based on the methods of finite difference enhancement or Laplacian calculation:

$$A = \begin{pmatrix} 0 & -1 & 0 \\ -1 & 4 & -1 \\ 0 & -1 & 0 \end{pmatrix} \quad B = \begin{pmatrix} -1 & -1 & -1 \\ -1 & 8 & -1 \\ -1 & -1 & -1 \end{pmatrix} \quad C = \begin{pmatrix} 1 & -2 & 1 \\ -2 & 4 & -2 \\ 1 & -2 & 1 \end{pmatrix} \tag{5.8}$$

These masks are simple, but, unfortunately, are very sensitive to noise because they enhance the high spatial frequencies. The details of an image are strongly affected by such filters. One should compare Figure 5.11A with Figure 5.11B to see the result of the applications of such filters when strong noise is present. Because of this, it is recommended to smooth the image before the utilization of these edge detecting filters.

Nonlinear filters, which are much more robust to noise, have been developed in order to eliminate these drawbacks; they are known as the gradient and the Sobel filters. They are local filters, the mathematical expression of which is:

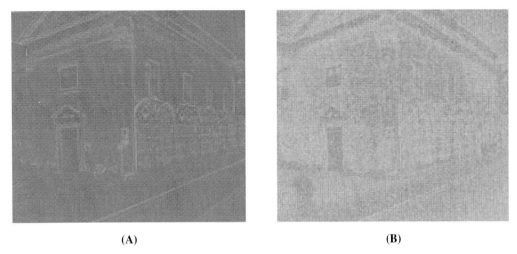

(A) (B)

FIGURE 5.11 Result of the application of filter A in Eq. 5.8 on the images in Figure 5.4 (see text).

$$l'(m, n) = |(A + B + C) - (G + H + I)| + |(A + D + G) - (C + F + H)| \tag{5.9}$$

$$l'(m, n) = \sqrt{[(C + 2F + I) - (A + 2D + G)]^2 + [(A + 2B + C) - (G + 2H + I)]^2} \tag{5.10}$$

for the Sobel one; the terms A, B, etc. refer to the 3×3 neighborhood defined in Section 5.5. These nonlinear filters are quite time-consuming, but also quite efficient. One should compare Figure 5.12, obtained applying these filters to sample images with Gaussian noise added, with Figures 5.11, where Gaussian masks have been employed.

(A) (B)

FIGURE 5.12 Result of the application of gradient (A, B) and Sobel (C, D) filters on the images in Figure 5.4 (see text).

(C)   (D)

FIGURE 5.12 (Continued)   Result of the application of gradient (*A, B*) and Sobel (*C, D*) filters on the images in Figure 5.4 (see text).

REFERENCES

1. Ballard, D. H. and Brown, C. M., *Computer Vision*, Prentice-Hall, Englewood Cliffs, NJ, 1984.
2. Burr, D. C. and Morrone, M. C., Feature detection in biological and artificial visual systems, in *Vision Coding and Efficiency*, Blakemore, C., Ed., Cambridge University Press, London, 1990, 56.
3. Canny, J. F., *Finding Edges and Lines in Images*, MIT Press, Cambridge, MA, 1983.
4. Castleman, K. R., *Digital Image Processing*, Prentice-Hall, Englewood Cliffs, NJ, 1979.
5. Davis, L. S., A survey of edge detection techniques, *Comput. Graph. Image Process.*, 4, 248, 1975.
6. Gonzales, R. C. and Thomason, M. G., *Syntactic Pattern Recognition*, Addison-Wesley, Reading, MA, 1978.
7. Gonzales, R. C. and Wintz, P., *Digital Image Processing*, Addison-Wesley, Reading, MA, 1987.
8. Gualtieri, P., Microscopia digitale, *Le Scienze*, 261, 52, 1990.
9. Gualtieri, P., Barsanti, L., and Coltelli, P., Computer processing of optical microscope image, *Micron Microsc. Acta*, 16, 154, 1985.
10. Hildreth, E. C., *Edge Detection*, A.I. Memo, MIT Press, Cambridge, MA, 1985, 858.
11. Marr, D., *Vision*, W. H. Freeman, San Francisco, 1979.
12. Marr, D. and Hildreth, E. C., Theory of edge detection, *Proc. R. Soc. London, Ser. B*, 207, 169, 1980.
13. Morrone, M. C. and Owen, R. A., Feature detection from local energy, *Pattern Recognition Lett.*, 6, 303, 1987.
14. Oppenheim, A. V. and Shafer, R. W., *Digital Signal Processing*, Prentice-Hall, Englewood Cliffs, NJ, 1975.
15. Pratt, W. K., *Digital Image Processing*, John Wiley & Sons, New York, 1977.
16. Sandini, G. and Tagliasco, V., An anthropomorphic retina-like structure for scene analysis, *Comput. Graph. Process.*, 14, 365, 1980.
17. Sandini, G., Il sistema visivo, in *Metodi di Analisi dei Sistemi Neurosensoriali*, Schmidt, R., Ed., Patron, Bologna, 1986, 34.
18. Haykin, S. S., *Adaptive Filter Theory*, Information and System Science Series, Prentice-Hall, Englewood Cliffs, NJ, 1995.
19. Farhang-Boroujeny, B., *Adaptive Filters: Theory and Applications*, John Wiley & Sons, New York, 1998.

6 Images on the Internet

Roberto Marangoni and Giuliano Colombetti

Contents

6.1   INTRODUCTION

The early systems used to share public information on the Internet (such as anonymous-FTP, Gopher, a.s.o.) had no graphic capabilities, so that the only way to exchange images on the Net was to download or upload files, without the possibility of displaying them until they had been transferred onto the local disk.

The World Wide Web, on the contrary, allows graphics, images, and multimedia objects to be displayed on screen directly during the connection to the remote node. This possibility has strongly facilitated the growth of the Web, and today it is probably the most diffused and used subset of the Internet.

At present, almost all institutions, organizations, and a large number of people who offer public information via the Web try to make it more attractive by using images and other multimedia objects such as audio and video files. Figure 6.1 shows an example of a Web page with some graphic elements.

This chapter offers a brief presentation of the image file formats compatible with the most diffused browsers and of the problems that one faces when adding graphical elements to Web pages.

6.2   NUMBER OF IMAGES IN A WEB PAGE

It is a common experience for the average user of the Internet that connection to a Web page has become more and more time-consuming, most of the time being spent in downloading graphical or pictorial elements. This is especially true for people using modem connections, but is also true for those using a LAN or a dedicated Internet connection.

Images, in fact, occupy a large amount of disk space and therefore require long downloading times. To make it easier and not boring for a visitor to connect to a certain site, it is necessary

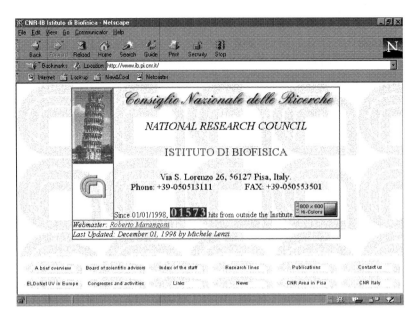

FIGURE 6.1   An example of a Web page with some different graphic elements: icons, pictures, animated GIFs, etc.

to keep to a minimum the amount of information (number of bytes) to be transferred and this is more easily done by keeping the number of images in a page reasonably low. This happens, of course, at the expense of page readability and appeal (people usually prefer images to written text) and it is therefore necessary to find a reasonable compromise between these two opposite requirements.

A first fact to keep in mind is that graphical effects do not necessarily require the use of images; the HTML language allows, in fact, one to set background and foreground colors for text, tables, and frames, so that one can obtain a "pictorial" scene in a page without inserting any image file, thus saving a lot of download time (see Figure 6.2, for an example). When it is necessary to insert images in a page, it is recommended to insert only a few images on the same page and to set their resolution and number in such a way that the Net surfer will be able to download the page in a reasonable amount of time.

6.3   HOW TO GET IMAGES

The simplest way to get an image to insert in a Web page is to copy already existing images from other sites. When original material is to be shown, there are several possible ways of generating digital images: scanners, digital cameras, and video cards able to digitize normal video frames.

This discussion is limited to the use of scanners, which are probably the most popular acquisition devices. Digital cameras and video cards usually come with detailed instructions on how to use them.

When using a scanner to acquire a picture, there are some parameters that must be determined by the user; in particular, the color representation and the scanning resolution. The first parameter is relatively easy to choose since it only depends on the type of picture scanned; it is advisable to use black-and-white (1 bit/pixel) to scan a text, gray levels (8 bits/pixel) for usual black-and-white photographs, and true color (24 bits/pixel) for usual color pictures. The choice of the second parameter (scanning resolution) is very important, in that it determines the size of the resulting file. One might think that the higher the resolution, the better the image quality obtained. This is

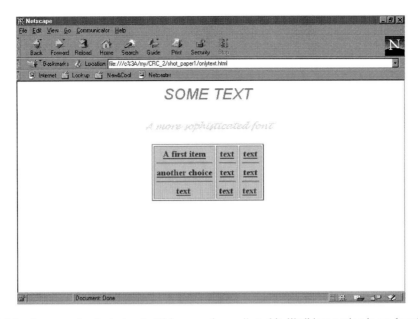

FIGURE 6.2 An example of a text-only Web page where a "graphic-like" impression is produced by means of different fonts and foreground and background colors.

true in principle, but the choice of too high a resolution can cause severe problems for the user. First, it is vain to try to display on-screen an image with a resolution higher than that of the screen itself: if one wants to keep resolution constant, the image will not fit into the screen; on the contrary, if one keeps the image size constant, many pixels will have to be discarded by the displaying program. Since the average Web user is only interested in seeing an image, it is better to keep the image to a dimension that fits on the screen and also to keep the image size in pixels as low as possible. The most common resolution for screens is at present 800 × 600. Keeping in mind that program windows take a certain amount of space and that a Web page may contain more than one image, a scanning resolution of 75 dpi (dots per inch) is often more than enough. Table 6.1 shows the file size resulting from the scanning of a 10 × 15 cm picture at different scanning resolutions; it is clear that to use resolutions higher than 150 dpi gives rise to files too large to be downloaded in reasonable times for the average Web surfer.[1]

TABLE 6.1
The Size of the (Uncompressed) File Obtained by Scanning a Standard Photographic Image (10 × 15 cm) at Different Resolution Levels

Resolution (DPI)	File Size (kb)
75	388
150	1550
300	6193
600	24,777
1200	99,066

6.4   TYPES OF IMAGES THAT CAN BE USED IN A WEB PAGE

The most popular browsers (i.e., the programs used to browse the Web pages, such as Internet Explorer, Netscape, Mosaic) support only a few types of graphic file formats, so that not any image file can be used to build an HTML file.

The image file formats supported by all browsers are:[1,2]

- GIF (Graphic Interchange Format)
- JPEG (Joint Photographic Expert Group)
- PNG (Portable Network Graphics)

Other formats can be accepted by some browsers, but not all. For example, the TIFF format can be read by some commercial browsers, but the most popular can only read the formats specified above. Figures 6.3 and 6.4 show typical examples of GIF and JPEG images.

FIGURE 6.3   A drawing digitized as a GIF file.

6.4.1   WHAT KIND OF FORMAT IS BETTER TO USE?

One might think that there is no relationship between the content of an image and the file format used to store it, but this is not true. Even if it is possible to store any image in any format, important factors to be considered are the color representation and the compression power characteristic of each format. The GIF format, for example, uses an 8-bit color map to represent the different colors, so that in a GIF image, it is not possible to have more than 256 different colors displayed at the same time, even if these colors can be chosen from a 32-bit true color space. This means that by storing a photographic picture in GIF format, the number of colors representable is greatly reduced and this may introduce some color artifacts that can alter the image perception.

The GIF format uses an LZW "lossless" compression (i.e., the decompression procedure can exactly restore the original image), whereas the JPEG format uses a "lossy" compression designed to preserve the perceptive characteristics of real images. A JPEG compression can thus greatly reduce the file size, keeping at the same time the perceptive characteristics almost constant.

Figure 6.5 shows the same image saved as JPEG with different levels of compression. It can be clearly seen that the image readability is globally preserved even at very high compression

FIGURE 6.4   A real scene photodigitized as a JPEG file.

factors that conserve 15% of the original information. Table 6.2 shows the size of the JPEG files as a function of the requested compression level. One should compare image readability and file size in order to appreciate the power of this procedure.

In summary, the GIF format is more suitable for graphic elements, pictures, icons, buttons, etc., whereas JPEG is better for photographic images.

6.5   ARRANGING MANY IMAGES IN A SINGLE WEB PAGE

The main problem in putting many images in a single Web page is that the reader has to endure a very long loading time because each image requires the transfer of a large amount of data. But there are many circumstances when it is necessary to offer the Net surfer a simultaneous presentation of many pictures. In these cases, the most commonly used strategy consists of presenting the pictures as "thumbnails." Thumbnails are very low-resolution images (see Figure 6.6), obtained from the originals by decreasing their size until they reach the dimensions of a stamp. By preparing a thumbnail page, in which each icon is linked to the original image, one gives the Net surfer the possibility to choose at a glance what image to load and then to save time in exploring the Web page.

6.6   BACKGROUNDS, ICONS, AND BUTTONS

In a Web page it is possible to use graphic elements that are not so heavy to download, but that can, nevertheless, produce a nicer look; background images, icons, and buttons are most frequently used for this purpose.

A background image is a GIF or JPEG file (the proper format can be chosen according to what is stated in Section 6.4.1), that fills the background of the page. It is usually a very small image, at a low resolution, that simulates some of the typical textures found in real materials such as wood, marble, velvet, a.s.o. The image is replicated horizontally and vertically until all the background is filled with it. If a suitable color match is chosen, the overall result of the application of a background image to a Web page is an increase in its readability and visual appeal (Figure 6.7).

FIGURE 6.5 A series of JPEG compressed images obtained from the same original image using different compression levels. The compression level of the JPEG format is expressed in terms of the quality of the resulting image: the greater the compression level, the lower the quality.

TABLE 6.2
The Size (in bytes) of the Image Files
Represented in Figure 6.5 and the Preserved
Quality of the JPEG Compression

Image in Figure 6.5	Quality	Size (bytes)
a	100%	160,702
b	50%	23,542
c	25%	15,997
d	15%	12,019
e	5%	7028
f	1%	4360

FIGURE 6.6   AltaVista Photo Finder™ is a search engine for pictures: all the pictures found are represented as thumbnails to give the reader synthetic information. By clicking on one of the thumbnails, the user is linked to the original picture.

(A)

(B)

FIGURE 6.7   (A) An example of emboss-filtered image and (B) its usage as background image in a Web page.

It is also possible to use photographic or realistic images as background, but this may be dangerous because of the contemporary presence of many objects on the scene that can confuse the reader. In the case where one has to use a real scene or a picture as the background image, it is possible to minimize the possible confusion between foreground and background images by altering the latter in some way.

One of the most common alterations of a background image consists of filtering it by means of an emboss filter. The resulting image suggests the original content, but cannot be confused with a real image.

Icons and buttons are other "light" graphic elements that can be inserted in a Web page. Usually, icons are used to suggest what a link is pointing to or what a document might contain. Buttons, on the other hand, are often used to better illustrate a bulletin list or to suggest, in an interactive program, what action the user can choose. Both icons and buttons are GIF files that can be easily created using the most diffused graphic or image programs (such as Adobe Photoshop and others), or by downloading them from public domain databases available on the network. Figure 6.8 shows one of the many Web pages from which icons can be downloaded.

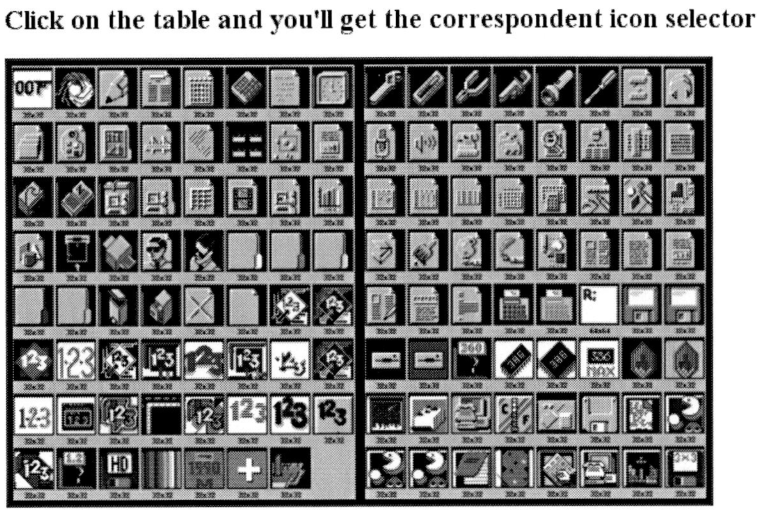

Click on the table and you'll get the correspondent icon selector

FIGURE 6.8 A small portion of the icons available through the IconBrowser (freely downloadable from the URL: http://www.cli.di.unipi.it/iconbrowser/icons.html).

6.7 IMAGES WITH A "MEANING"

Thumbnails are not the only situation in which a link is embedded into an image. Sometimes, it is very useful to have different links on the same image, arranged in such a way that by clicking different areas (or different pixels) within the image, the user can activate a different link.

In this situation, the image is used not only to display its content, but also like a map containing all the links of interest: the image, in this context, is called a "sensitive map." Links in a sensitive map are very useful when one has to insert data that has a geographic distribution or that is, in some way, related to a spatial localization. Figure 6.9 provides an example of a sensitive map: the EU has activated a network of UV-radiation measuring stations, the data of which are available on the Internet. Since the user may be interested to look at the data coming from each station, a sensitive map, in which all the different stations are indicated, is used. The Net surfer can choose a station by simply clicking on the corresponding point in the picture.

Also, interactive languages like Java or JavaScript generate graphic effects and can use images to interact with the user; this topic, however, is not considered here.

6.8 ANIMATED GIFS AND OTHER DYNAMIC GRAPHIC ELEMENTS

Modern browsers also support dynamic images and some video formats, so that it is possible to receive a television program directly on the Web page. The first format allowing animation on the Web was GIF. Animated GIFs can be used to store short animated icons or short trailers of a movie. As seen before, the GIF format is not so suitable for the representation of real scenes because of its color encoding and compression algorithm. This is also true for animated GIFs, which are, therefore, mostly used for small dynamic drawings (see Figure 6.10). The most commonly used formats for the representation of movies are AVI and QuickTime, which were the original standards for Windows and Apple computers, respectively. Today, there is no limitation of operating system and one can, for example, freely use QuickTime on a Windows-based computer.

Another widely used format is MPEG (Motion Picture Expert Group), which is very interesting because of the logic of its operation: the first image to be reproduced is a JPEG encoding of the original scene and, subsequently, only the pixels that are changed are displayed. The compression

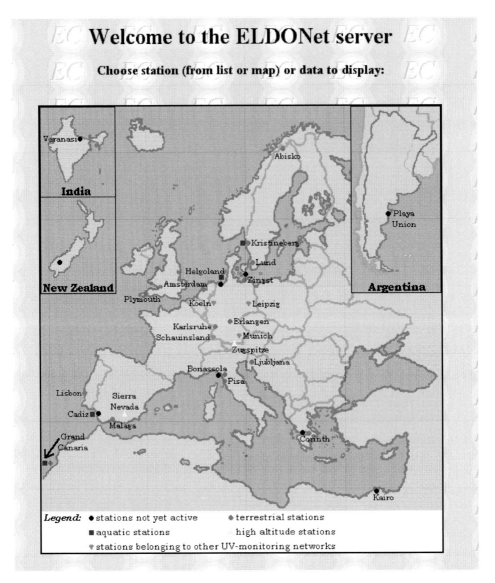

FIGURE 6.9 The sensitive map used in the ELDONet Web page. By clicking on the different stations, the user is linked to the data files of the corresponding location.

procedure is applied both to the images and to the difference between consecutive images. This allows one to obtain a significant increase in speed and a relatively limited need for memory.

Other formats (such as Real Video, Active Video or InterVU), which are usually adopted for synchronous transmission of images, have been developed by private companies for television networks and can be used, for example, in teleconferencing or in the broadcast of a TV station on the Net. All the above-mentioned formats also allow one to reproduce sound tracks, provided, of course, that the receiving computer has a sound card and loudspeakers or a headset.

In almost all cases, it is possible to freely download and install a proper plug-in in the browser used, in order to become able to read and represent one of those formats on screen. On the other hand, it is not easy to create animations or to transform into digital format a common "analogic" movie and it may often require the purchase of suitable hardware and software.

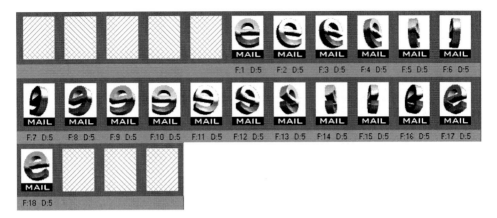

FIGURE 6.10 The different frames of a freely distributed animated GIF: the Web browser displays the successive frames giving rise to a motion impression (as in a movie projection).

ACKNOWLEDGMENT

The authors would like to thank Zoran Curto for suggestions in the preparation of some figures.

REFERENCES

1. Marangoni, R., Geddo, M., *Le Immagini Digitali*, Hoepli, Milano, Italy, 1996.
2. Murray, J. D. and Van Ryper, W., *Encyclopedia of Graphics File Formats*, O'Reilly and Associates, Sebastopol, 1994.

7 A Method for Conserving Perceptive Information in Digital Color Images

Primo Coltelli, Mauro Evangelisti, and Paolo Gualtieri

CONTENTS

7.1 INTRODUCTION

Didactic, archival, retrieval, and analysis applications in digital microscopy demand representation of color images of the same quality, either of microscope images or of their photographic reproduction. Moreover, textures in microscope images are colored textures. They may have either the same colors and different structural patterns, or different colors in the same structural patterns, or different colors and different structural patterns. Therefore, the texture description of an image should include both color and structural aspects.[1,2]

A major problem is to reduce the transmission and the processing time and the memory required to store the digital color images, which are usually $1000 \times 1000 \times 24$ bits (3 Mbytes), still maintaining a high information content, in order to save time and memory space for successive retrieval and analysis steps. Sophisticated compression algorithms, used to encode a color image in a new format, generate a quite satisfactory image characterized by a compression ratio (i.e., the ratio between the numbers of bytes of the original image and that of the final images), of about 30 (100 Kbytes). Dithering algorithms can be applied after compression algorithms as post processing techniques to improve overall image quality. Many compression algorithms are available; one of the most commonly used is JPEG.[3] With this algorithm, no differences are detectable between the original and the compressed image, at least at preliminary visual inspection. However, by analyzing the scene of the JPEG compressed color image, with algorithms that perform the partition between the background and the objects (segmentation) and/or extract features, one may notice that JPEG compression heavily modifies image frequencies, introduces new textures, and does not decrease the processing time. Different compression algorithms (e.g., those developed by the authors), based on the reduction of the number of colors to a characteristic set (i.e., the smallest number of colors satisfying some criteria or features), could be used to improve image clarity and decrease processing time. This chapter describes how to deal with the problems that can occur when working with digital microscope color images.

7.2   FORMULATION OF COLOR IMAGE THEORY

In computer graphics, colors can be defined as linear combinations of three numbers. These combinations form a three-dimensional color space, with defined metrics,[4] within which colors are specified. The RGB, the HBS, and human-oriented color space CIE*Luv*, which differ in the labeling of their three axes, are the principal color spaces used for graphics purposes.

The RGB color space is based on the common observation that most colors can be created by suitably mixing three primary lights. Hence, in this space, a color is defined by the relative proportions of the red, green, and blue display primaries required to produce it. This space can be considered a three-dimensional space with the origin at zero, and whose x-, y-, and z-axes are labeled by the Red, Green, and Blue display primaries, respectively. The value for each axis can lie between 0 and 1, and the entire color space is contained within a cube. Colors that plot at the origin would appear black. A gray scale would be represented by a vector from the origin along which Red, Blue, and Green values are the same. RGB color space is often chosen in computer graphics because of the three-gun hardware of commercial display and its low computation complexity.

In the case of the HBS color space, the three axes stand for Hue, Brightness, and Saturation, and the purpose of this color space is to provide users with a more intuitive means of mixing colors than the RGB color space. The HBS space is a distorted version of the RGB cube. Colors are specified by their hue angle, between 0° and 360°, (e.g., 120° for blue and 240° for green); by a brightness value located on the z-axis, which goes from black to white, through grays; and by a saturation value that is specified by moving from the hue plane toward the z-axis. The three-dimensional resulting space is a double cone. Intuitive terms used to describe colors, such as tint, shade, and tone, are easily represented in this space. However, a same change of saturation or brightness produces different results in different zones of the space.

The CIE*Luv* color space does not have an intuitive form, although it is perceptively uniform and computing the color difference between two color samples is straightforward. Any two couples of colors with equal Euclidean distance give the same visual perception of that distance. Because it is the human eye that judges the quality of the displayed digital microscope image, the best representation of the chromaticity of an image should be computed using the uniform color space (CIE*Luv*) that closely matches human perception.[5] The CIE LCH model, derived from the CIE*Luv* model, offers an easy-to-use representation in cylindrical coordinates without losing the important properties of CIE*Luv* model.

The devices for image acquisition and rendering (such as monitors and printers) represent the colors, depending on their hardware technology; such representations are far from the human visual model. In fact, the human perceptively uniform colorimetric space differs from the RGB space used by monitors and the CMYK space adopted by printers.

7.3   HARDWARE AND SOFTWARE FOR COLOR IMAGE
PROCESSING

Color images can be acquired directly from the microscope using a color video camera connected to a frame grabber plugged into a computer bus. An alternative method is to acquire slides of the microscope field by means of a table scanner. Both acquisition procedures declare 24 bits of color information per pixel in the RGB space.

The sampling step (600 × 600 dots per inch) is chosen to guarantee the best compromise between the need for storing image data of reasonable size (about 600 × 400 pixels) and the requirement of microscope resolution (0.5 μm). Sensitivity should be adjusted and input gamma corrected for the radiometric accuracy of the generated image.[6] For visualization, true color systems use 256 levels for each primary color (8 bits); therefore, more than 16 million colors can be displayed. The human eye is capable of recognizing 100 levels for each primary color; therefore, it can see a million of these different linear combinations.[7] Thus, it is possible to create and display many more colors than those recognizable by the human eye. However, the 24-bit frame buffer has some drawbacks, since image transmission by networks, image handling in window user interface environments, and image feature determinations have great complexity in the computational and transmission time and storage requirements.

For the handling of digital color images, the authors used a Windows-based system with a driver-direct color display (24 bits) as the hardware platform. User-friendly interfaces were assembled using libraries and objects implemented in C++.

7.4 COLOR IMAGE SEGMENTATION AND FEATURES EXTRACTION

7.4.1 INTRODUCTION

The characteristic colors of an image are those representing the set of colors with perceptive predominance in that image.

The algorithm implemented to extract characteristic colors operates on the CIE LCH space, which is uniform. To reduce computational costs, the method does not operate directly on the data within the CIE LCH three-dimensional space, but on their projections on each axis. The process consists of a recursive analysis of one-dimensional histograms.

The conversion between RGB and CIE LCH has been extensively described.[6] Briefly, it consists of:

- A linear transformation
- An algebraic transformation
- A change of coordinates

Cylindrical coordinates, which are named lightness (L^*), hue ($H°$), and chroma (C^*) in psychometry, are calculated as follows:

$$L^* = L^*$$

$$H° = \tan^{-1}\left(v^*/u^*\right) \tag{7.1}$$

$$C^* = \left(u^{*2} + v^{*2}\right)^{1/2}$$

These cylindrical coordinates correspond almost exactly to the physiological model of color vision. Therefore, they are more evident and they are preferred to the u* and v* variables.

Every horizontal section through this space (CIE LCH) defines a plane of constant lightness L. Every vertical plane intersecting the L-axis, has a constant hue. A cylindrical section concentric with the L-axis forms a surface of constant chroma, within which all the colors of an object have the same level of purity.

For an image expressed in cylindrical coordinates, one can calculate the three corresponding histograms for each of them. The algorithm operates by choosing one of these histograms and calculating its threshold value.

For example, by choosing the histogram relative to the H coordinate, one calculates four threshold values (H1, H2, H3, H4), and hence three intervals. For each couple of threshold values (i.e., for each interval), one calculates the histogram relative to the L and C coordinates of all the pixels whose third coordinate values belong to that interval. The procedure is repeated recursively until all the calculated histograms become unimodal. The results of this procedure are the characteristic colors, defined by triplets of intervals, each triplet determining a volume inside the colorimetric space:

$$L1 \leq L \leq L2$$

$$C1 \leq C \leq C2 \tag{7.2}$$

$$H1 \leq H \leq H2$$

Every characteristic color is represented by the color having coordinates equal to the weighted average of the coordinates forming the interval. This method begins subdivision along the Hue component because, due to the multimodal pattern of the relative histogram, it can provide more information with respect to the Chroma and Lightness coordinate histograms.

7.4.2 HISTOGRAM THRESHOLDING

Each pixel of an n-bit digital image is labeled by its line and column i and j values, respectively. The gray level at each point is $k(i, j)$; the associated histogram $H(k)$ measures the number of pixels with a gray level equal to k. The histogram is normalized to unity

$$\sum_{k=0}^{2^n-1} H(k) = 1 \tag{7.3}$$

and provides a good estimation of the probability distribution associated with the overall image-forming procedure. The distribution function $F(k)$ at point k is equal to the relative number of pixels with a gray level not greater than k,

$$F(k) = \sum_{l=0}^{k} H(l) \tag{7.4}$$

This function rises monotonically and is less than or equal to unity.[8] Techniques of histogram modification found throughout the literature[9-13] use the distribution function to calculate the desired transfer function. The F function can be used in segmentation and thresholding since no information is lost and, although $F(k)$ may be a piecewise constant function, it can be easily smoothed and simulated with much simpler mathematical models than those used for histogram simulation.

Were $F(k)$ a smooth curve, differential geometry would give its intrinsic characteristics. The arc length and the curvature torsion is not needed because one has a plane curve.[14,15] Now assume that F depends on the continuous variable x. Then, the curvature radius of F at $(x, F(x))$ can be given by:

$$\rho = \left(1 + \left(F'(x)\right)^2\right)^{3/2} \cdot \left(F''(x)\right)^{-1} \tag{7.5}$$

and the curvature is:

$$c = \rho^{-1} \tag{7.6}$$

Although F is not a smooth function, one can implement the calculation of the curvature using suitable approximations to the derivatives F' and F''. One can use

$$F'(k) = \frac{1}{2n}\left(F(k+n) - F(k-n)\right) \tag{7.7}$$

$$F''(k) = \frac{1}{2m}\left(F'(k+n) - F'(k-n)\right) \tag{7.8}$$

with n and m chosen according to the degree of smoothing of F.

These equations allow one to construct a function $C(k)$, also called the curvature of the distribution function, and on which one will rely heavily. $C(k)$ will be an oscillating and noisy function.

Approximation is done by a least-squares method using the Chebyshev basis.[16] The authors postulate an expansion of the kind

$$C(x) = \sum_{n=0}^{M} a_n T(x) \tag{7.9}$$

where $C(x)$ is known only for integer values of x, and M represents the order of the approximation; the coefficients a_n enter linearly and are chosen so as to minimize the sum of the squares of the residues:

$$\Omega^2 = \sum_{k=k_{min}}^{k_{max}} \left(C(k) - \sum_{n=0}^{M} a_n T_n(k) \right)^2 \tag{7.10}$$

This leads to a linear system of $M + 1$ equations with $M + 1$ unknowns that is easily solved by standard algorithms. The authors chose a Gaussian method with partial pivoting.[13]

The reasons for taking the Chebyshev basis are twofold:

1. It is well known that a polynomial approximation leads to matrices severely ill-conditioned, of the kind of the Hilber matrix,[11] where the biggest entry can be several orders of magnitude larger than the smallest. This leads to instability problems in the solution when parameters are slightly perturbed. The reason for such behavior is the non-orthogonality of the base functions (xk, $k = 0$, M);

2. The remarks in (1) do not apply when an orthogonal basis is chosen. This is the case for the Chebyshev basis, but there are other orthogonal bases one might choose. $C(k)$ may be a highly oscillating function with many zeros in the interval (k_{min}, k_{max}). A Chebyshev polynomial of order M has M real zeros in the $(-1, 1)$ interval. This basis allows one to correlate the degree of the fit to the behavior of the curvature data. Note that M will be the only parameter the user is asked to select.

Suppose that the curvature of the distribution function has been obtained, smoothed, and fitted; hence, it can be considered as a function of the continuous variable x. The critical points of this function are the zeros and the extrema. Because of its oscillating character, the zeros are almost always well-defined isolated points. Now one can relate the critical points of the curvature to its associated histogram.

If $C = 0$, the radius of curvature tends to infinity and the distribution function F presents an inflection point.[12,13] Whenever the curvature changes sign near zero — and this is the most typical case — the slope of the distribution function changes either more slowly or more rapidly. Whenever the associated histogram is modal, tops correspond to points like T and valleys to points like V.

Nevertheless, the curvature has other critical points that can be useful. First consider extrema (maxima or minima) that are surrounded by zeros (K point). Here, the distribution function really begins to grow. In the associated histogram, this corresponds to a leading edge of the mode. Whenever the sequence of critical points is ...V-K-T-K-V, (i.e., when there is only one extreme between two zeros), every T is surrounded by two K, which must be considered as the leading and final edges of a mode. Points K are not necessarily symmetrical with respect to the position of T. If asymmetrically placed, one can assume that two modes overlap and suitable algorithms might follow.

One can also have extrema that are not surrounded by zeros. In this case, the associated parts of the histogram cannot be analyzed in terms of modes.

Nevertheless, many pixels can be involved and segmentation is still needed. Whenever two extrema of the curvature are not separated by a zero, create a new zero between the extrema, for instance, equidistant from both. This new zero will be considered either as a valley-like point V or as a top-like point T. Alternatively, one can delete extrema if the oscillation of the curvature is considered small, or when there is reason to think that it was created by small irregularities on the histogram (noise).

In the sequence of critical points of the curvature, the series unit

$$min - zero - max - zero$$

repeats itself, even if one has to create new zeros or delete extrema. Valley-like points V will be chosen as thresholds.

The number of thresholds will be roughly equal to half the number of zeros of the curvature and so their number may not have to be specified at the beginning. This is the reason why this technique is referred to as a dynamic multiple thresholding technique.

7.4.3 Selecting a Characteristic Color Set

Step 1: Hue Component Thresholding

The first step is to calculate the hue component histogram and find the threshold values. For example, if the histogram shows two peaks, the algorithm finds three threshold values (H1, H2, H3). The LCH space is divided as in the following:

- The vertical plane defined by H1 and H2 values identifies the subspace M1.
- The vertical plane defined by H2 and H3 values identifies the subspace M2.

Two angular sections are defined in the color space (M1 and M2): the first is defined by the T1 and T2 hue values and the second is defined by the T2 and T3 hue values.

STEP 2: L AND C COMPONENTS THRESHOLDING

Consider the pixels of the subspace M1, and for them calculate the two one-dimensional histograms for the L and C components. The thresholding of the two histograms produces the values L1 and L2 for the lightness, and the values C1 and C2 for the chroma.

STEP 3: CIE LCH SPACE SUBDIVISION

By combining the results obtained by the previous space subdivision, one obtains volumes that identify color intervals.

STEP 4: CHARACTERISTIC COLOR CALCULATION

By calculating the weighted average of each resulting interval component by component, one defines the characteristic colors, which are therefore mere representations since colors defined in this way may not be present in the image.

STEP 5: IMAGE SEGMENTATION USING CHARACTERISTIC COLORS

The original image is segmented using the previously defined intervals, and giving to each segment the calculated characteristic color. The result is an image with a low number of colors (32), but with a high perceptive significance. The measurement of the perceptive difference between the original and the segmented image (difference represents the success of the operation) shows average differences of 2 to 3%, often concentrated along the edges of the image.

7.5 EXPERIMENTAL RESULTS

This section describes how features and/or textures can be extracted from an image and displayed using a limited set of colors without losing their biological significance. Selected examples are shown.

Figure 7.1a* shows the fluorescence image of the flagellate *Euglena gracilis*, characterized by the red emission of the chloroplasts and the yellow-green emission of its photoreceptor.[17] This color image has more than 2000 colors, which are reduced to eight by the algorithm (Figure 7.1b). The biological significance of the chloroplast and photoreceptor emission is still easily recognizable, despite the color number reduction.

The image in Figure 7.1c is extracted from the sequence of the digestive process of the raptorial feeder ciliate *Litonotus lamella*, which includes successive phases from the capture of the prey, the ciliate *Euplotes crassus*, to the formation of digestive vacuoles. The ciliate is stained with Acridine Orange, a fluorescent probe that has great affinity for the acid vesicle compartments.[18] Figure 7.1c, which possesses almost 5000 colors, shows the phase just before the capture of the prey: the starved *Litonotus* is characterized by orange acid vesicles localized in its neck. In Figure 7.1d, obtained by applying the algorithm on Figure 7.1c, the image is shown after color reduction (only five colors): the biological importance of the presence of acid vesicles in the neck of the ciliate is easily recognizable, even if the number of colors is dramatically reduced.

Figure 7.1e shows isolated *Euglena gracilis* photoreceptors that fluoresce in the yellow-green. This color image has more than 7000 colors. In Figure 7.1f, obtained by applying the algorithm to Figure 7.1e, the image is shown after color reduction (only four colors): the presence of the photoreceptors is easily recognizable despite the color number reduction.

---

* Color Figure 7.1 follows page 288.

7.6 CONCLUSIONS

In this chapter, the problem of representing the color aspect of image texture is analyzed. Characteristic colors (i.e., the smallest number of colors satisfying specific criteria) are proposed as a features set for representing the color aspect of visual texture. Algorithms for extracting features and the characteristic color set from biological images are analyzed and the results of their applications shown.

REFERENCES

1. Scharcansky, J., Hovis, J. K., and Shen, H. C., Representing the color aspect of texture images, *Patt. Recog. Lett.*, 15, 191, 1991.
2. Coltelli, P. and Gualtieri, P., Quantization and dithering algorithms applied to digital microscopy, *Micron*, 1994.
3. Murray, J. D. and VanRyper, W., *Encyclopedia of Graphics File Format*, O'Reilly & Associates, Sebastopol, 25, 309, 1994.
4. Apostol, T. M., *Calculus*, John Wiley & Sons, New York, 1969.
5. Travis, D., *Effective Colour Display: Theory and Practice*, Academic Press, San Diego, 1991.
6. Foley, J. D. and VanDam, A., *Computer Graphics*, Addison-Wesley, Reading, MA, 1990.
7. Pratt, W. K., *Digital Image Processing*, John Wiley & Sons, New York, 1978.
8. Papoulis, A., *Probability, Random Variables and Stochastic Processes*, McGraw-Hill, New York, 1965.
9. Boukharouba, S., Rebordao, J. M., and Wendel, P. L., An amplitude segmentation method based on the distribution function of an image, *Comput. Vision Graphics Image Process.*, 29, 47, 1985.
10. Bolc, L. and Kulpa, Z., Eds., *Digital Image Processing Systems*, Springer, Berlin/New York, 1981.
11. Hummel, R. A., Image enhancement by histogram transformation, *Comput. Vision Graphics Image Process.*, 6, 184, 1975.
12. Hall, E. H., Kruger, R. P., Dwyer, S. J., Hall, D. L., McLaren, R. W., and Lodwick G. S., A survey of preprocessing and feature extraction techniques for radiographic images, *IEEE Trans. Comput.*, C-20, 1032, 1971.
13. Scher A. and Rosenfeld A., Probability transforms of digital pictures, *Patt. Recog.*, 12, 457, 1980.
14. Lesieur, L. and Joulain, C., *Mathématiques, Algèbre et Géométrie*, Vol. 1, A. Collin-Collection U., Paris, 1974.
15. Piskounov, N. M., *Calcul Differentiel et Intégral*, Editions MIR, Moscow, 1976.
16. Abramowitz, M. and Stegun, I. A., *Handbook of Mathematical Tables*, Dover, New York, 1972.
17. Barsanti, L., Walne, P. L., Passarelli, V., and Gualtieri, P., *In vivo* photocycle of the *Euglena gracilis* photoreceptor, *Biophys. J.*, 72, 545, 1997.
18. Verni, F. and Gualtieri, P., Digestive process of the raptorial feeder ciliate *Litonotus lamella* visualized by fluorescence microscopy, *Micron*, 28, 447, 1997.

8 Single Photon Imaging

Tsuyoshi Hayakawa

CONTENTS

8.1 INTRODUCTION

Low light level imaging has been used in the life sciences under various conditions, including microscopes. The imaging method has become more effective in the analysis of biological objects, as image processing and laser technologies have widely and rapidly developed.

In the past, silicon intensified target (SIT) cameras and image intensifiers (II) were mainly used for imaging at low light levels. Although there are several practical problems in sensitivity, the development of dyes and labeling techniques with these image pickup devices enables one to visualize the behavior of biomolecules in solution. For example, the distribution of calcium ions in a single cell was detected by the combination of a newly developed dye (Fura-2) and ratio imaging,[1] and the behavior of actin filaments or DNA fragments in solution was also visualized and analyzed.[2,3] Thus, research using fluorescence microscopes requires cameras with higher sensitivity to analyze the biological phenomena.

On the other hand, a photon counting method using photomultiplier tubes has been widely used in detecting small amounts of luminescent biomolecules. However, information on the spatial distribution of those molecules is lost with this method since two-dimensional detection of biomolecules is required.

Radioisotopes have been extensively used as a highly sensitive method (e.g., autoradiography or electrophoresis). Alternative and more convenient methods to avoid the use of radioisotopes will probably be in high demand. This requirement can be more effectively achieved by two-dimensional photon counting.

An ultra-high-sensitivity TV camera,[4] whose main capability is single photon imaging, has been developed. In principle, the distribution of an extremely small amount of biomolecules can be measured by single photon imaging, which might be called "molecular counting imaging." Single photon imaging is useful in solving practical problems such as photobleaching, damage by illumination, and biological interference from reagents.

8.2 PRINCIPLES OF SINGLE PHOTON IMAGING

8.2.1 SAMPLING AND QUANTIZATION

As light intensity (photons/mm² s) decreases, light appears more and more localized in discrete areas on the photodetector. The degree of this problem depends on the temporal-spatial characteristics of the detector. Photon counting imaging has three main advantages:

1. Ultra-high sensitive detection
2. Excellent quantification
3. Acquisition of temporal-spatial information of a single photon

Figure 8.1 shows that there are two modes of photodetection, depending on the relationship between the light intensity and the temporal-spatial characteristics of the detector. In case (a) the superimposed signal (amplitude) of many photons is detected in order to measure the light intensity; while in case (b), each output signal corresponds to an individual photon which is counted. Therefore, these two photodetection modes can be called the "analog mode" and "counting mode," respectively. From the point of view mentioned above, photodetection can generally be achieved by counting photons if a photodetector has sufficient temporal-spatial characteristics.

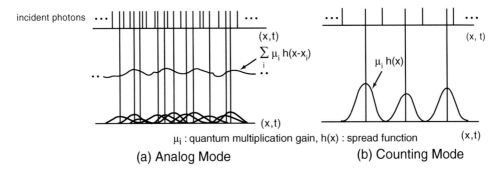

FIGURE 8.1   Analog mode and counting mode.

The counting mode is not possible in conventional low light level image pickup devices. In other words, no signal can be detected at ultra-low light levels, even when accumulated over a long time period.

8.2.2 SINGLE PHOTON IMAGING DEVICES

After dark adaptation, human eyes have a considerably high sensitivity. It is estimated that a few tens of photons can be detected by the naked eye. This could be called ultra-high sensitivity, but quantification of light is not possible.

Single photon imaging is achieved by a specially designed intensifier tube[5] (2-D photon counting tube) with a two-stage microchannel plate (MCP) (Figure 8.2).[6] An MCP consists of fiber bundles. The diameter of a single channel is about 10 μm. There is an electron multiplication function in each channel (Figure 8.3). When an optical image focuses on the photocathode of the photon counting tube, photons are converted to photoelectrons. This photoelectron image is then focused onto the MCP surface. In the two-stage MCP, the photoelectron image is amplified by a factor on the order of about 10^6, maintaining the positional information. Next, the electrons strike a phosphor screen. As a result of these processes, even a single photon can generate about 10^8 photons as the output of the 2-D photon counting tube. This high-density spot of a single photon can be easily detected by a successive image pickup device. Thus, single photon imaging is achieved.

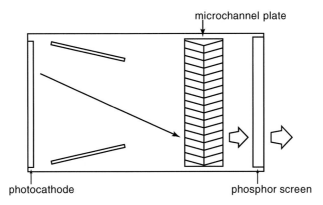

FIGURE 8.2 Operation of a 2-D photon counting tube. A single photon generates on the order of 10^8 photons on the phosphor screen.

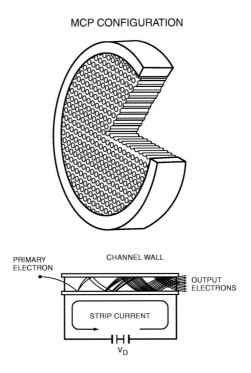

FIGURE 8.3 Structure of the microchannel plate and its operation.

Figure 8.4 shows several types of spectral sensitivity of the 2-D photon counting tube. Instead of using photoelectric spectral sensitivity S (A/W), it is more convenient to use quantum efficiency η (electron/photon) in the counting mode.

The relationship between S and η is given by:

$$\eta = \frac{hc}{\lambda} \cdot \frac{S}{e} \tag{8.1}$$

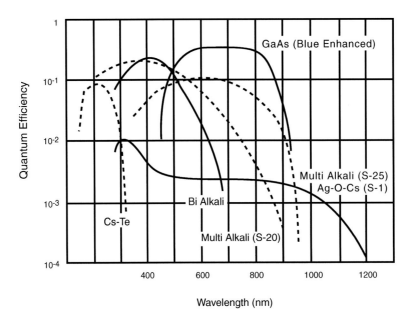

FIGURE 8.4   Quantum efficiency vs. wavelength for a 2-D photon counting tube.

$$\eta = \frac{1.24}{\lambda} \cdot S \tag{8.1a}$$

where h is Planck's constant (6.626×10^{-34} J·s), c is the velocity of light in vacuum (3×10^{8} m/s), λ is the wavelength (m in Eq. 8.1 and μm in Eq 8.1a), and e is the electron charge (1.6×10^{-19} Coulomb).

A cooled, charge coupled device (CCD) camera has been developed recently for the purpose of ultra-low light level imaging.[7] The photocathode of this device is basically the same as that of a silicon photodiode (Figure 8.5). In the operation of a cooled CCD camera, a signal of accumulated photoelectrons in each pixel is serially and slowly read out to form the image (ordinary readout of a frame takes a second to 1 hour). Up to now, as the readout noise is estimated to be greater than several electrons rms (root mean square value) per pixel, photon counting imaging is only partially possible.

Figure 8.6 shows the difference between a 2-D photon counting tube and a cooled CCD camera concerning the state of the photosignal readout.

8.3   PHOTON COUNTING CAMERA SYSTEM[4]

8.3.1   SYSTEM CONSTRUCTION AND CHARACTERISTICS

Figure 8.7 shows the schematic diagram of the photon counting camera system. The amplified image of photons on the phosphor screen of the 2-D photon counting tube is read out by a low-lag vidicon or CCD camera and stored within an image processor. The quality of images obtained depends on the spatial resolution as well as on the accuracy of the number of detected photons. As the spot of a single photon on the phosphor screen occupies about several to ten pixels in the frame memory, the centroid calculation of this spot improves both the spatial resolution (Figure 8.8) and the accuracy of the number of detected photons. In this system, the "analog mode" is also available by operating the MCP at a relatively low gain when the light intensity is sufficient.

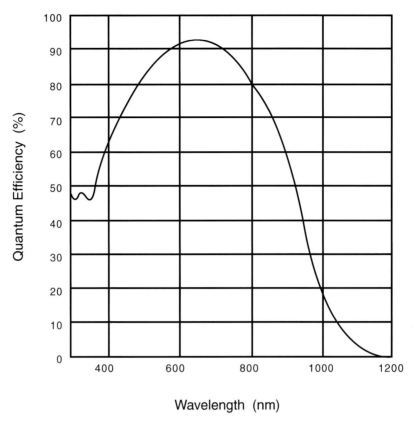

FIGURE 8.5 Quantum efficiency vs. wavelength (CCD).

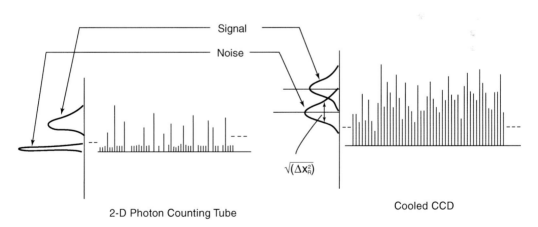

FIGURE 8.6 Schematic representation of photosignal readout. The output signal of a single photon in the 2-D photon counting tube is clearly distinct. On the other hand, in a cooled CCD, the possibility of photon counting imaging is strongly dependent on the readout noise $\sqrt{\langle \Delta x_R^2 \rangle}$.

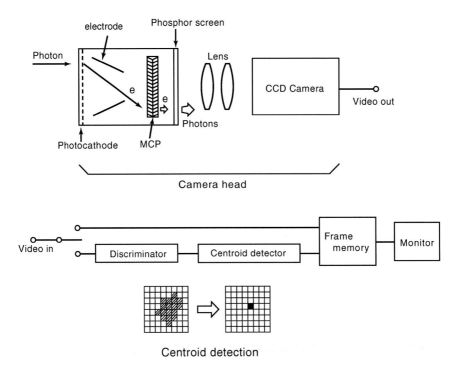

FIGURE 8.7   Schematic diagram of the photon counting camera.

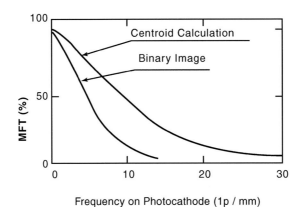

Frequency on Photocathode (1p / mm)

FIGURE 8.8   Modulation transfer function (MTF) of the photon counting camera.

The image quality is also determined by the signal-to-noise ratio. The signal-to-noise ratio *S/N* in the counting mode is given by

$$\frac{S}{N} = \sqrt{\frac{L\eta AT}{1 + \frac{2x_d + x_b}{L\eta}}} \tag{8.2}$$

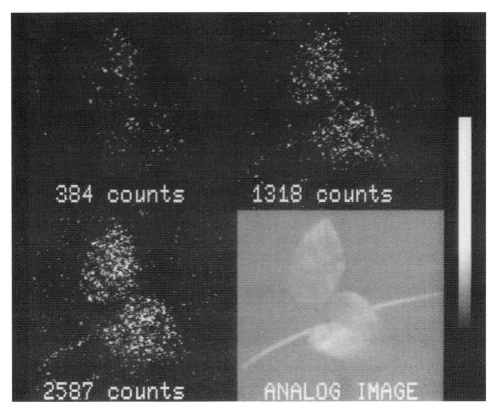

384 counts   1318 counts

2587 counts   ANALOG IMAGE

FIGURE 8.9 Examples of image quality vs. count number. (This photo was scanned from the first edition of this book.)

where L is the light intensity (photons/mm^2 s), η is the quantum efficiency of the photocathode (electron/photon), A is the signal detection area (mm^2), T is the signal integration time (s), x_d is the dark noise (counts/mm^2 s), and x_b is the background noise (counts/mm^2 s).

The dark noise of the photon counting camera is less than 20 counts/cm^2 s (with a bialkali photocathode at 25°C, or a multialkali photocathode at −20°C). An example of the dependence of the photon count number on the image quality is given in Figure 8.9, showing delayed fluorescence images of leaves. The lower right image was obtained in the "analog mode" by week illumination.

A cooled CCD camera system has recently been developed. As described above, the photon counting mode is not available in the strict sense of the word. However, if the readout noise $\sqrt{\langle \Delta x_R^2 \rangle}$ in Figure 8.6 is less than a few electrons, the counting mode becomes available. The pixel size of a cooled CCD camera is larger than 12×12 μm and, currently, the number of picture elements is 1000×1000. The signal-to-noise ratio S/N in 1 pixel is given by:

$$\frac{S}{N} = \sqrt{\frac{L\eta_{CCD}T}{1 + \dfrac{1}{L\eta_{CCD}}\left(\dfrac{2 \cdot \langle \Delta x_R^2 \rangle}{T} + 2x_d + x_b\right)}} \tag{8.3}$$

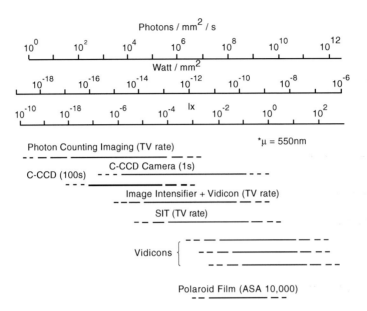

FIGURE 8.10   Method of image pickup vs. incident light intensity.

where L is the light intensity (photons/pixel s), η_{CCD} is the quantum efficiency of the photocathode (electron/photon), T is the integration time (s), $\sqrt{\langle \Delta x_R^2 \rangle}$ is the readout noise (electrons, rms), x_d is the dark noise (counts/s), and x_b is the background noise (counts/s).

A typical dark count per pixel is about 0.1 counts/s at $-30°C$. The readout noise (at several seconds per frame) is at least several electrons rms at present.

A suitable imaging method can be selected corresponding to the light intensity, as shown in Figure 8.10. The photon counting camera can detect a single photon image in real-time (TV rate). A cooled CCD camera will be available in the extremely low light range (on the order of 10^2 photons/mm² s) when the frame readout time is longer than several seconds.

8.3.2   APPLICATIONS OF PHOTON COUNTING CAMERAS

When single photon imaging is used under extremely low light levels, very often the dominant noise is the background noise x_b. For example, stray light, leak of excitation light, weak fluorescence emission from optical parts, and weak luminescence caused by impurities of the samples should be reduced in order to quantify the information of light. The fundamental capability of single photon imaging in biology is to measure the distribution and function of extremely small amounts of biomolecules. Its application has just started in the field of luminescence. There are some remarkable applications, such as two-dimensional detection of the biophotons[8,9] real-time visualization of oxyradical burst activities,[10] the monitoring of gene expression,[11,12] the distribution of ATP or glucose in tissues,[13] continuous observation of a calcium ion wave moving through a fertilized egg,[14] the imaging of the luminescence caused by a plant's defense response against fungal infection,[16] and single molecule imaging on silicon single-crystal wafers.[17]

Figure 8.11A shows the result of the visualization of an oxyradical burst from a single human neutrophil stimulated by opsonized zymosan in luminol-dependent chemiluminescence. Figure 8.11B provides a three-dimensional display of the two-dimensional distribution of the neutrophil luminescence.

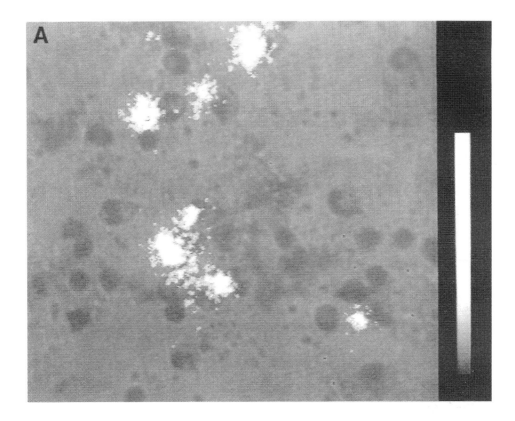

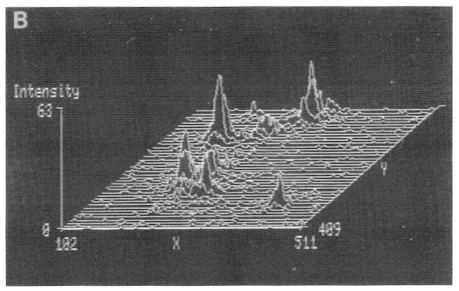

FIGURE 8.11   Image of neutrophil reaction (oxyradical burst) in luminol-dependent chemiluminescence. (From Suematsu, M., *Biochem. Biophys. Res. Commun.*, 155, 106, 1988. With permission. This photo was scanned from the first edition of this book.)

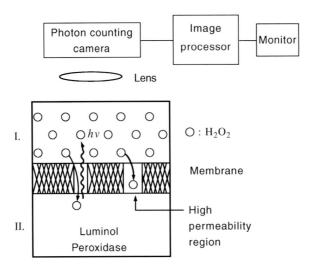

FIGURE 8.12   Experimental apparatus for imaging of permeants through a membrane.

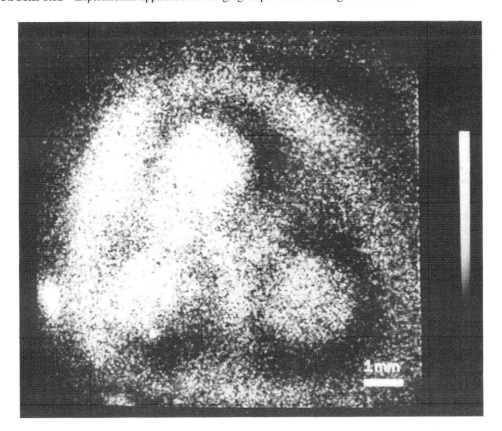

FIGURE 8.13   Two-dimensional distribution of permeant (hydrogen peroxide) through a polymethyl methacrylate film of nonuniform thickness. (From Hiramatsu, M., *J. Polym. Sci.*, 28, 133. Copyright 1990, John Wiley & Sons. With permission. This photo was scanned from the first edition of this book.)

   The two-dimensional distribution of a permeant moving through a membrane was measured in real-time using chemiluminescence.[15] Figure 8.12 shows the experimental system of the permeant measurement and Figure 8.13 shows one of the results.

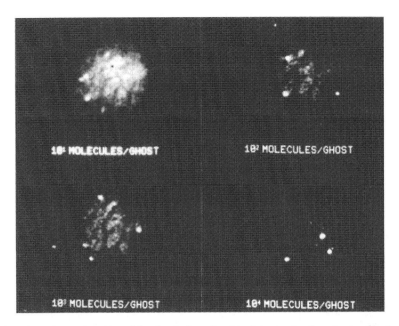

FIGURE 8.14 Experimental results of the simulation of molecular counting imaging by the photon counting camera (FITC-dextran in ghost of erythrocyte) objective 40x, excitation 100-W Hg lamp. (This photo was scanned from the first edition of this book.)

An example of a simulation experiment by fluorescence on molecular number counting imaging is shown in Figure 8.14. Currently, the limitation of molecular detection is determined not by the characteristics of the photon counting camera, but by the stray light (background noise). By reducing this stray light, one can detect a single molecule with its positional information.

The single photon imaging method has just been developed and is beginning use in video-intensified microscopy and other macroscopic imaging. It will be used more frequently in biology for the purpose of understanding biological phenomena *in vivo*. The transition from a "single photon" to a "single molecule," or from a "histological image" to a "function image," will be the most important application.

REFERENCES

1. Williams, D. A., Fogarty, K. E., Tsien, R. Y., and Fay, F. S., Calcium gradients in single smooth muscle cells revealed by the digital imaging microscope using Fura-2, *Nature*, 318, 558, 1985.
2. Yanagida, T., Nakase, M., Nishiyama, K., and Oosawa, F., Direct observation of motion of single F-actin filaments in the presence of myosin, *Nature*, 307, 58, 1984.
3. Matsumoto, S., Morikawa, K., and Yanagida, Y., Light microscopic structure of DNA in solution studied by the 4',6-diamidino-2-phenylindole staining method, *J. Mol. Biol.*, 152, 501, 1981.
4. Hayakawa, T., Kinoshita, K., Miyaki, S., Fujiwake, H., and Ohsuka, S., Ultra-low-light level camera for photon counting imaging, *Photochem. Photobiol.*, 43, 95, 1986.
5. Kinoshita, M., Kinoshita, K., Yamamoto, K., and Suzuki, Y., A two-dimensional photon counting tube, *Adv. Electron. Electron Phys.*, 64B, 323, 1985.
6. Matsuura, S., Umebayashi, S., Okumura, C., and Oba, K., Characteristics of the newly developed MCP and its assembly, *IEEE Trans. Nucl. Sci.*, NS-32, 1985.
7. Chandler, C. E., Bredthauer, R. A., Janesick, J. R., Westphal, J. A., and Gunn, J. E., Sub-electron noise charge coupled devices, *SPIE*, 1241, 238, 1990.
8. Scott, R. Q., Usa, M., and Inaba, H., Ultraweak emission imagery of mitosing soybeans, *Appl. Phys.*, B48, 183, 1989.

9. Ichimura, T., Hiramatsu, M., Hirai, N., and Hayakawa, T., Two-dimensional imaging of ultra-weak emission from intact soybean roots, *Photochem. Photobiol.*, 50, 283, 1989.

10. Suematsu, M., Oshio, C., Miura, S., and Tsuchiya, M., Real-time visualization of oxyradical burst from single neutrophil by using ultrasensitive video intensifier microscopy, *Biochem. Biophys. Res. Commun.*, 149, 1106, 1987.

11. Gallie, D. R., Lucas, W. J., and Walbot, V., Visualizing mRNA expression in plant protoplasts, *Plant Cell*, 1, 301, 1989.

12. Escher, A., O'Kane, D. J., Lee, J., and Szalay, A. A., Bacterial luciferase fusion protein is fully active as a monomer and higher sensitive *in vivo* to elevated temperature, *Proc. Natl. Acad. Sci.*, 86, 6528, 1989.

13. Mueller-Klieser, W., Walenta, S., Paschen, W., Kallinowski, F., and Vaupel, P., Metabolic imaging in microregions of tumors and normal tissues with bioluminescence and photon counting, *J. Natl. Cancer Inst.*, 80, 842, 1988.

14. Yoshimoto, Y., Iwamatsu, T., Hirano, K., and Hiramoto, Y., The wave pattern of free calcium release upon fertilization in medaka and sand dollar eggs, *Develop. Growth Diff.*, 28, 583, 1986.

15. Hiramatsu, M., Muraki, H., and Ito, T., A new method to characterize film inhomogeneities, *J. Polymer Sci., Part C: Polymer Lett.*, 28, 133, 1990.

16. Makino, T., Kato, K., Iyozumi, H., Honzawa, H., Tachiiri, Y., and Hiramatsu, M., Ultraweak luminescence generated by sweet potato and *Fusarium oxysporum* interactions associated with a defense response, *Photochem. Photobiol.*, 64, 953, 1996.

17. Ishikawa, M., Hirano, K., Hayakawa, T., Hosoi, S., and Brenner, S., Single-molecule detection by laser-induced fluorescence technique with a position-sensitive photon-counting apparatus, *Jpn. J. Appl. Phys.*, 33, 1571, 1994.

9 Image Analysis in Biochemistry

Donat-P. Häder

CONTENTS

9.1 INTRODUCTION

Preparative and analytical processes in biochemistry require the separation of proteins, DNA, RNA, and other biochemically important substances from cell homogenates. Separation of molecules can be performed based on different physical or chemical properties such as molecular weight, particle size or shape, electric charges, or behavior in different moieties. One separation procedure relies on the different retardations of substances by percolating them through a matrix of appropriate-sized beads or binding them to ligands on the matrix in a vertical column.[1,2] Recent developments in this field are HPLC,[3-5] FPLC,[6-8] and capillary electrophoresis,[9] which differ from simple column chromatography by using high or medium pressure, respectively, to force the substances to be separated through comparatively small columns.

9.2 GEL ELECTROPHORESIS

Gel electrophoresis is one of the most powerful tools in modern biochemistry to separate molecules that differ in their physical and chemical properties.[10-14] Samples obtained by solubilization from tissues, cells, or cellular organelles are separated from cellular debris and solubilized using biochemical methods, such as high salt (e.g., NaCl or urea) concentrations or detergents (e.g., Triton X-100), or physical methods such as grinding or ultrasound treatment.

9.2.1 ONE-DIMENSIONAL GEL ELECTROPHORESIS

The solubilized sample is applied to a gel matrix (e.g., agarose or polyacrylamide) in order to reduce diffusion at one end, and a dc voltage on the order of several hundred or thousand volts is applied across the gel by platinum electrodes to produce an electric field with (ideally) parallel field lines (Figure 9.1). The individual proteins (or DNA or other molecules) carry an electric charge

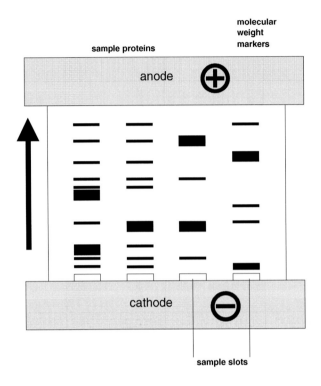

FIGURE 9.1 Gel electrophoresis of proteins in a gel using an external dc field. The proteins are applied in three sample slots, while the fourth is reserved for marker proteins of known molecular weight.

so that they are electrophoretically pulled by the external field across the gel matrix (native gel electrophoresis). Different sized molecules are retarded by the gel matrix to a different extent so that a separation (molecular sieving) is achieved. Alternatively, the proteins are treated with sodium dodecyl sulfate (SDS), which binds to the side groups of the protein and carries a negative charge. As a consequence, different molecules move different distances (R_f values) according to the charge they carry (SDS polyacrylamide gel electrophoresis = SDS PAGE).

Usually, the protein sample is applied to a stacking gel, which serves to sharpen the bands, from where it moves to a separation gel (Figure 9.2), which can have a constant mesh size in the polyacrylamide or form a gradient (e.g., 8 to 20% polyacrylamide).[15,16] For the separation of DNA, the externally applied electric field can be pulsed (pulsed field gel electrophoresis, PFG).[17]

A different technique is based on the fact that native proteins carry both positive and negative charges. Depending on the pH of the separation buffer, one of the charges exceeds the other, and at a specific pH both charges compensate each other, which is defined as the isoelectric point (IEP). At this point, the proteins become immobilized and are concentrated in the gel. This effect is employed by the technique of isoelectric focusing (IEF).[18-20] An IEF gel consists of a polyacrylamide or agarose matrix with 2 to 4% carrier ampholytes added. Carrier ampholytes are mixtures of amphoteric molecules with different pI values narrowly spaced. After the gel has set, two paper strip electrodes are placed under the two ends of the gel. The anode strip is imbibed with an acid solution and the cathode strip with an alkali. Upon application of an electric field, the ampholytes migrate through the gel in a direction depending on their net charge (Figure 9.3). In this way, a pH gradient is formed along the gel. Then the proteins are applied to the gel and the electric field is induced, so that the proteins migrate in the gel until they move to a pH that corresponds with

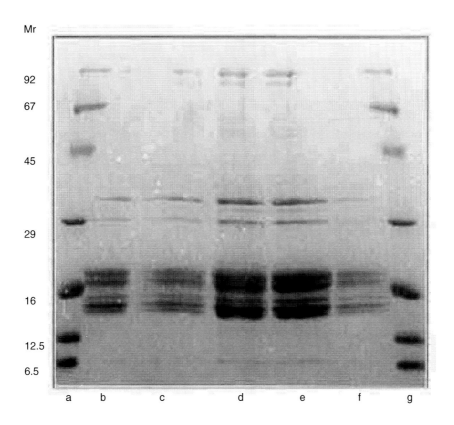

FIGURE 9.2 SDS PAGE of phycobilisome proteins from UV-irradiated cells (UV-A: 27.04 W m^{-2}, UV-B: 28.7 W m^{-2}). The proteins are denatured by SDS and carry a negative charge so that they move from the sample slots toward the anode (bottom). Equal amounts of protein (20 μg) were loaded onto the gel. Lanes a and g: molecular weight markers; lane b: phycobilisomes from control cells; lanes c to f: phycobilisomes from irradiated cells; lane c: 15 min; lane d: 30 min; lane e: 45 min; lane f: 60 min.

its isoelectric point and stop there (Figure 9.4). Thus, the individual proteins are separated according to their IEP.

A similar approach is used in isoelectric titration where the proteins are applied in a line to the center of the gel and move in the electric field toward the anode or the cathode, respectively. A pH gradient has been established previously perpendicular to the electric field (Figure 9.5). Calibration proteins have been added before the establishment of the pH gradient to indicate which region of the gel corresponds to which pH. Immunoelectrophoresis, affinity gel electrophoresis, agarose gel electrophoresis,[21-23] chromatofocusing, and other methods are also used for separation of the individual components in the protein mixture.

9.2.2 TWO-DIMENSIONAL SEPARATION

Especially for large numbers of proteins in a sample,[24,25] two-dimensional (2-D) techniques are being employed for better resolution.[26-34] Separation is a two-step process where IEF or non-equilibrium pH gradient electrophoresis (NEPHGE) are used in the first dimension[35-37] and usually SDS PAGE in the second direction perpendicular to the first (Figure 9.6).[38,39] Flat-bed gel electrophoresis[40,41] can be run horizontally or vertically and has a number of advantages, such as efficient cooling, small protein loss, and high resolution during separation in thin[42,43] or ultrathin gels.[44,45]

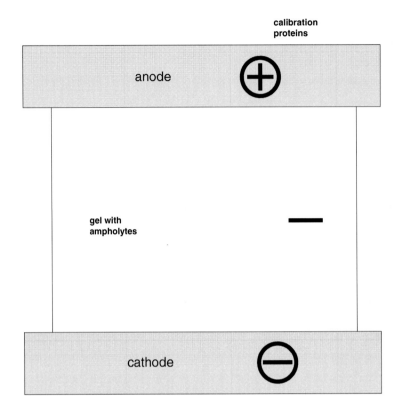

FIGURE 9.3 Prefocusing in an electric field establishes a pH gradient in the gel indicated by calibration proteins of known IEP.

9.2.3 STAINING AND VISUALIZATION

It is difficult to detect the location of individual proteins after separation in the gel, even when they carry a chromophoric group (chromoprotein). One possibility to visualize the spots is using fluorescence excitation in the ultraviolet or blue spectral region to take advantage of the autofluorescence of the proteins or the fluorescence of an added dye.[46,47] The alternative is to use one of a number of different stains, including Coomassie Brilliant Blue R-250,[48,49] silver,[50-52] Violet 17, India ink,[53] or ethidium bromide (a fluorescent label for DNA).[54,55] A third possibility is to radioactively label the substances and perform autoradiography by placing the gel after separation on a photographic film in the dark.[56-59]

The separated protein spots can be transferred (blotted) onto suitable carriers such as nitrocellulose or nylon membranes,[60-64] where they are stained or made visible by immunological techniques[65-67] or colloidal gold.[68,69]

9.3 DETECTION OF SEPARATED MOLECULES BY IMAGE ANALYSIS

To quantify and compare large numbers of protein spots in a gel, automatic analysis of the samples was developed.[70,71] One-dimensional (1-D) gels can be evaluated using a densitometer where the gel (or a photographic film) is placed on a stage which is moved across a light beam.[72] A photocell

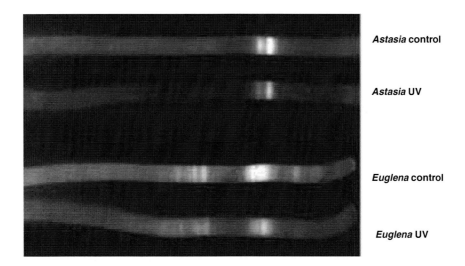

FIGURE 9.4 Isoelectric focusing of flagellar proteins from the flagellates *Euglena* and *Astasia* running in an electric field. The proteins move to a position where their isoelectric point coincides with the pH in the gel. Both flagellates contain α- and β-tubulins. In addition, *Euglena* carries a paraflagellar body, the putative photoreceptor which consists of four proteins, not present in *Astasia* flagella samples. UV radiation partially destroys the proteins.

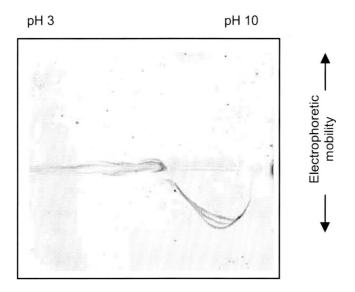

FIGURE 9.5 Isoelectric titration curves of proteins moving at different pHs in an electric field.

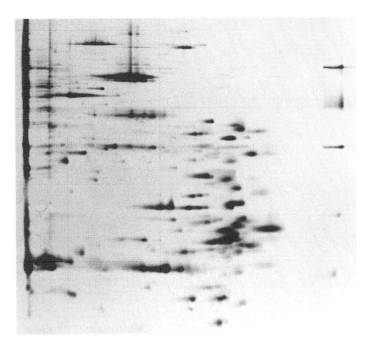

FIGURE 9.6   2-D gel electrophoresis using IEF in the first dimension and SDS PAGE in the second. (From Görg et al., *Electrophoresis*, 19, 1516, 1998. With permission.)

measures the attenuation when a stained protein spot passes the light beam. The signal is recorded on a chart strip recorder (Figure 9.7). The next step in automation is the digitization of the analog signal. Early attempts to achieve some kind of automation in protein spot analysis used the manual input into a computer using a digitization tablet.

Automatic image analysis of 2-D gels is basically performed by computer-controlled analysis as any other image.[73-76] These techniques are also used for cell, particle, or colony counting,[77] shape and size determination,[78] and tracking of motile microorganisms.[79-82]

The minimal requirements are a suitable scanner (CCD camera with a digitizer or a laser scanner), a PC computer, and a powerful software package. The graphic representation of the image should be based on high-resolution color video cards. Automatic spot detection and labeling should require minimal time (e.g., under 1 min for 500 spots). Also, the analysis of the key features such as position, area, and volume should be as fast as possible, preferably by running these algorithms in the background during image loading. The results of the analyses need to be documented graphically and numerically. Another important requirement is the development of a user-friendly surface to facilitate operation by untrained personnel to warrant reliable and efficient analysis.

An additional advantage of automatic image analysis of 2-D gels is that the data is available in electronic form after evaluation.[83-86] The data can then be used to develop a database that allows comparison between the results from different laboratories and previous experiments — provided the experimental procedures have been standardized.[87,88]

9.3.1 HARDWARE FOR DIGITIZING ELECTROPHORESIS GELS

Several hardware approaches have been developed to digitize the gel or a high-resolution photograph (or autoradiograph). A low-cost approach uses a video camera or a charge-coupled device (CCD) camera to record the image.[89-93] The advantage of video scanning is that the image is available in real-time. The disadvantages are the low spatial resolution and the small range of optical densities, limited to about 2.4 O.D. (optical density). Values above this point are cut off

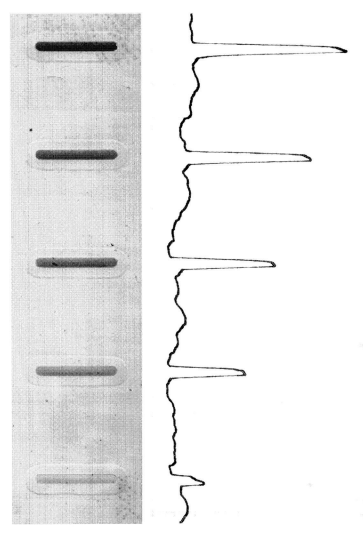

FIGURE 9.7 Slot blot of DNA samples on a nylon membrane and identified with primary antibodies against UV-induced thymine dimers. The secondary antibody is an alkaline phosphatase conjugated anti-mouse IgG, the substrate is BCIP/NBT which results in a blue stain (left), and densitometer scan to quantify the amount of thymine dimers (right).

even when the automatic gain control (AGC) of the camera is disabled.[94] In some applications, the low spatial resolution can be circumvented using a macro lens to record one small portion of the gel at a time. For low light level monitoring, special photon detection camera systems have been developed.[95]

After recording, the image is digitized (analog-to-digital [A/D] conversion) either in real-time using flash converters, or it may take up to a few minutes in sequential A/D conversion depending on the available hardware. The digitizer can be a stand-alone unit or a card within the host computer.[96-98] The spatial resolution of digitization is defined by the number of distinct image points (pixels) in the x- and y-directions. The sample density is given by its gray value defined for each pixel. The gray level is expressed by a digital value that ranges between black and white in a number of distinct steps that encode the absorption in a linear or logarithmic fashion. The gray-level resolution is limited by the number of bit planes during A/D conversion.

Higher resolution, especially of 2-D gels, can be achieved using a laser scanner. The measuring beam is produced from a well-focused laser beam. The gel is scanned by mechanically advancing it in x- and y-directions across the beam or by moving the laser beam using turning mirrors. The detector can be a photomultiplier that has the capability of measuring a wide range of optical densities. In these machines, the spatial resolution is determined by the step width of the stepper motor with which the gel or the beam is moved with respect to the photodetector. Resolutions of 5 μm have been achieved. One of the disadvantages of laser densitometers is the long time required for scanning a gel at high resolution, which can be on the order of several hours. Another suitable detector is a CCD array,[99] with a number (e.g., 2048) of parallel channels that measure the absorbance in adjacent pixels.[100]

In the Ultrascan (Amersham Pharmacia Biotech, Bromma, Sweden), the high-resolution scanning of gels or photographic negatives is performed with a laser scanner. The gel is placed on a backlit screen. The area to be analyzed is selected by motor-driven rulers. The resolution is defined for x- and y-directions, and the scanning is performed line by line with high-precision stepper motors, allowing for a spatial resolution of <40 μm. The attenuation of the laser beam by the sample is determined for each pixel in the range between 0 and 4 O.D. using a linear scaling curve.

After digitization, the pixel matrix is transferred to the host computer and stored in RAM (read-and-write memory) or on a storage medium (hard disk drive, floppy disk, magnetic tape) for analysis and feature extraction. Depending on the scanned area and the required resolution, the size of the data files can be up to 4 MByte.

The image of the gel can be displayed on a suitable screen, which can be the computer terminal or a separate display as a black and white binary or gray-level image. Pseudo-color representation can be used to better distinguish between different gray values. Overlay functions are used to draw indications on top of the image in different colors.

The digitized image stored in memory can be analyzed by the host computer on the basis of individual pixels.[101,102]

9.3.2 Software Approaches for Two-Dimensional Gel Electrophoresis

In several biological and medical applications it is useful and necessary to compare the spot patterns between two or more gels.[103] For example, the comparison of the protein patterns between the wild type and the mutant of an organism can identify the missing gene product. Likewise, this technique can be used to identify the induction of specific gene products during specific developmental stages or by external factors such as light or UV radiation.[104,105] The prerequisite for successful matching of corresponding spots in two parallel 2-D gels is a distance matching in both directions. This can be done by rubber banding techniques, in which one gel is stretched or compressed with respect to the other in one or two directions until a close alignment is found. This process may be complicated by nonlinear distortions between the gels.

Several search strategies have been developed to match corresponding spots, including hierarchical nearest-neighbor analysis, multidimensional classification, heuristic clustering, least-square fitting, and multivariance analysis.[106] Some software approaches need the manual input of the operator to start the search by matching "landmark" spots. Others attempt a fully automatic analysis. The main disadvantage of most programs developed thus far is the enormous time requirement, which can be on the order of hours or days.

A number of devoted computerized image analysis systems have been developed over the last few years to evaluate electrophoresis gels.[107] Computer-aided evaluation of one- and two-dimensional electrophoresis gels (CAEG) was developed soon after the introduction of the electrophoresis techniques themselves.[108,109]

Today, a number of commercial 2-D gel analysis packages are being marketed; for example, HERMeS II (Biolog Diffusion, Versailles, France), Gelsim (Biosoft, Cambridge, U.K.), Visage 1-D and 2-D gel software (Bio Image, Ann Arbor, MI), MCID (Imaging Research, Ontario, Canada),

PhastImage (Pharmacia LKB, Uppsala, Sweden), and GS365W (Hoefer Scientific Instruments, San Francisco, CA). Other programs include the Clip system,[110] the MasterScan (Scanalytics, Billerica, MA), the QUEST system,[111] and a system to identify mutant proteins from 2-D gels after video recording and digitization of the image. The performance of these packages varies widely, from mere digitization to a full analysis and spot matching. One of the programs (QUEST) is designed for the analysis of radiolabeled proteins. Two-dimensional Gaussian fitting is used for spot quantification and separation of overlapping spots. Spot matching is not restricted to two gels, but a whole set (called matchset), in which each gel is compared to every other gel. The authors claim that with an average of 2000 spots per gel, the system matches 97% correctly and that fewer than 1% are matched inconsistently.

Celis and coworkers[112] have used the PDQUEST software (Protein Databases, Inc.) running on a MASSCOMP CPV 5520 computer to analyze fluorescence images of 2-D gels of human embryonic lung MRC-5 fibroblast proteins. Editing the raw gel images with 1482 individual proteins took about 1.5 days. In the analyzed image, ellipses are drawn representing the area and position of the proteins in the original gel labeled and numbered for identification.[113,114] Spatial Data Analysis Laboratory at Virginia Tech developed a general image processing system that was used by Tyson and Haralick[115] to process 2-D gels of proteins from *Physarum polycephalum* photographed from the original gel and scanned with a laser scanner. This package is capable of adjusting for a nonhomogeneous background, can separate merged protein spots, and estimates the volume of the individual spots. Mathematical operations, such as adding or removing Gaussian components or curve-of-growth models to estimate peak volumes indicative of protein loading, are used to improve the analysis.

Automatic image analysis is also used to read and analyze DNA sequence gels.[116-118] Häder and Truß[119] used a UV-B emitting light source (transilluminator) to detect the fluorescence of ethidium bromide employed to stain nucleic acids. Automatic band detection algorithms have been developed to automatically read DNA restriction fragments in high-resolution images obtained by one-dimensional scanning of agarose gels.

The early computer program systems ran on large mainframe computers.[120,121] After the rapid development of both hardware and software, packages could be developed for the PC.[122,123] Commercial systems are available from Genomic Solutions (Ann Arbor, MI), Alpha Innotech Corp. (San Leandro, CA), and Wallac Oy (Turku, Finland). A gel documentation system has been developed by Ultra-Lum, Inc. (Paramount, CA). The following discussion is based on the recently developed Imagemaster 2D database (Amersham Pharmacia Biotech, Bromma, Sweden).

Gels are recorded and the files are opened by the software. Several files can be opened in parallel, which is important for the comparison between gels. **Spot detection** is the first step in gel electrophoresis analysis: a gray value threshold level is defined which separates background and spot areas.[124] The result of spot detection can be controlled since the gray image gel is overlaid with a blue binary image produced by the spot detection process. The first attempt may not produce an optimal result. This can be remedied by some user interaction. For this purpose, the image is enlarged using the zoom function. Changing the sensitivity factor, larger spots can be resolved into groups of smaller spots. Minimal spot areas can be defined to reject small artifactual spots. Larger spots can be split either by the software or manually.

The areas occupied by the protein spots are identified using one of several algorithms such as edge detection, chain coding,[125,126] or area filling.[127] The chain coding algorithm is explained in more detail in the chapter on real-time tracking of microorganisms by Häder and Lebert in this volume. The next step is the **quantification of the spots** in the current image.

In some 2-D gel calculations, the **area of the spots** is used to determine the amount of protein present. However, the protein loading depends on the absorption of the individual pixel. Since the absorption is an exponential function of the concentration, the Beer-Lambert law can be used to determine the amount of protein:

$$A(\lambda) = \log_{10}(I_0/I) = \varepsilon(\lambda)\, C\, l \qquad (9.1)$$

where $A(\lambda)$ is the absorbance at a wavelength λ, $\varepsilon(\lambda)$ is the molar extinction coefficient, I_0 and I are the incident and transmitted light intensities, respectively, C is the molar concentration (mol/l) of the absorbing species, and l is the optical path length.

Since the gray values of the measured pixels are obtained by digitization, filling in new values requires some interpolation. This can be done using the Lagrange equation:

$$P(x) = \frac{(x - x_2)(x - x_3)...(x - x_n)}{(x_1 - x_2)(x_1 - x_3)...(x_1 - x_n)} y_1 + \frac{(x - x_1)(x - x_3)...(x - x_n)}{(x_2 - x_1)(x_2 - x_3)...(x_2 - x_n)} y_2 + ...$$
$$+ \frac{(x - x_1)(x - x_2)...(x - x_{n-1})}{(x_n - x_1)(x_n - x_2)...(x_n - x_{n-1})} y_n$$

(9.2)

Interpolation is difficult when there are only few basepoints over long intervals because the precision decreases and the high degree interpolation polynomes tend to oscillate, especially at the end of the set interval, producing density artifacts.[128] For this reason, the volume calculation in 2-D gels is performed by cubic spline interpolations that use a third-degree polynomial:

$$f(x) = ax^3 + bx^2 + cx + d$$

(9.3)

During calculation, the algorithm determines the parameters a, b, c, and d for a curve fitted through the measured data points located at given x and y coordinates, making sure that the inclination and curvature of adjacent curve segments are identical at the data points. The cubic spline interpolation is derived from the commonly used one, modified for volume integration. The nodes for the cubic spline interpolation are of importance since they determine the efficiency and speed of fitting.

In order to determine these nodes, a preprocessing algorithm detects the optimal nodes by scanning along the density distribution before the spline interpolation commences. The cubic spline interpolation is not used when the scan deviates little from the baseline; it is only used when a sudden increase in absorption near a spot is detected. This dynamic adaptation to the conditions in the gel scan increase the calculation speed considerably. The cubic spline interpolation uses the following equations (for further details, see Press et al.[129]). Consider basepoints with coordinates at

$$y_i = y(x_i) \qquad \{i = 1, ..., N\}$$

(9.4)

with the corresponding abscissa values

$$x_i < x_{i+1}$$

(9.5)

In the ImageMaster 2D software, **background subtraction** is either done using the lowest value on the boundary or the average of the boundary pixels or may require some manual interaction from the user. For manual subtraction, it is advantageous to zoom into the image. A rectangle is defined with the mouse in an area with no spots. The average of the gray values is used as background value.

The final step in the analysis sequence is **matching the spots** in one gel (slave) with those in another (master or reference gel).[130] By this procedure, the software defines differences for each protein spot between gels. For this purpose, the slave gel file is loaded. While spot detection, spot measurement, and background subtraction have already been performed in the master gel, these steps need to be repeated for the slave gel. Both images are superimposed; the spots in the reference gel are shown in a red outline, those in the slave gel in blue. Unmatched spots in the master gel are shown in solid red and those in the slave gel in solid blue.

9.3.3 ANALYSIS OF ONE-DIMENSIONAL GELS

In principle, 1-D gels can be evaluated using the same software package developed to analyze 2-D gels.[131] Scanning a 1-D gel only along one scan line produces a gel scan that may be distorted by noise, both in the bands and the gaps. Treating the image as a 2-D window and integrating over the width of the bands smoothes out accidental flaws. However, the operator has to ensure that smaller bands fill the lane completely in order not to distort the analysis. This is facilitated using the rubber band technique. Further analysis of spot area and volume utilizes the same algorithms described above for 2-D gel analysis.

9.4 CONCLUSIONS

Image analysis is an appropriate tool to analyze and quantify sample spots (e.g., proteins, DNA, RNA) after separation in two dimensions by gel electrophoresis. Various systems are commercially available today for different tasks and are described in more detail in this chapter. Different separation techniques are used in the two directions to take advantage of the specific properties of the sample. Photographic films of gels can also be analyzed. All techniques for automatic analysis of two-dimensional electrophoresis gels by computer-aided image analysis start out with a 2-D scan of the gel using a laser scanner or a video snapshot. After digitizing the optical densities of the individual pixels, the image is stored in a computer. Individual sample spots are identified and quantified either in the whole field or in a zoomed portion. The found spots are numbered and shown in pseudo-color representation or as 3-D graphical output. A cubic spline algorithm, which automatically switches on and off depending on the necessity, is used to interpolate between individually measured points. Also, 1-D gels can be analyzed with the same software that allows averaging over a strip of adjustable width. The results can be documented on periphery devices (e.g., laser printer) for camera-ready output.

REFERENCES

1. Söderberg, L., Müller, R.-M., and Fägerstam, L., Chromatofocusing of desialylated human transferrin on Mono P, *Protides Biol. Fluids*, 30, 661, 1983.
2. Moreno-Lopez, J., Kristiansen, T., and Karsnäs, P., BVDV: affinity chromatography on *Crotalaria juncea* lectin, *J. Virol. Meth.*, 2, 293, 1981.
3. Cutler, P., Size exclusion chromatography, in *Molecular Biomethods Handbook*, Rapley, R. and Walker, J. M., Eds., Humana Press, Totowa, NJ, 1998, 451.
4. Neville, B., Reversed-phase HPLC, in *Molecular Biomethods Handbook*, Rapley, R. and Walker, J. M., Eds., Humana Press, Totowa, NJ, 1998, 479.
5. Welling, G. W., Groen, G., and Welling-Wester, S., Isolation of Sendai virus F protein by anion exchange high performance liquid chromatography in the presence of Triton X-100, *J. Chromatogr.*, 266, 629, 1983.
6. Markey, F., Rapid purification of deoxyribonuclease I using FPLC, *FEBS Lett.*, 167, 155, 1984.
7. Johns, E. H., Cooper, E. H., and Turner, R., Separation of cerebrospinal fluid by fast protein liquid chromatography (FPLC), *3rd Int. Symp. HPLC of Proteins, Peptides and Polynucleotides*, Monaco, 1983, Abstract 914.
8. Karsnäs, P., Moreno-Lopez, J., and Kristiansen, T., BVDV: purification of surface proteins in detergent-containing buffers by FPLC, *J. Chromatogr.*, 266, 643, 1983.
9. Begley, D. J., Free zone capillary electrophoresis, in *Molecular Biomethods Handbook*, Rapley, R. and Walker, J. M., Eds., Humana Press, Totowa, NJ, 1998, 425.

10. Lammel, B., A method for continuous preparative electrophoresis using a supporting medium, *Electrophoresis*, 2, 39, 1981.

11. Richards, P., Protein electrophoresis, in *Molecular Biomethods Handbook*, Rapley, R. and Walker, J. M., Eds., Humana Press, Totowa, NJ, 1998, 413.

12. Rabilloud, T., Hubert, M., and Tarroux, P., Procedures for two-dimensional electrophoretic analysis of nuclear proteins, *J. Chromatogr.*, 351, 77, 1986.

13. Andrews, A. T., *Electrophoresis. Theory, Techniques, and Biochemical and Clinical Applications*, Clarendon Press, Oxford, 1986.

14. Chrambach, A., Dunn, M. J., and Radola, B. J., Eds., *Advances in Electrophoresis*, Vol. 1, VCH Verlagsgesellschaft, Weinheim, 1989.

15. Shapiro, A. L., Vinuela, E., and Maizel, J. V., Molecular weight estimation of polypeptide chains by electrophoresis in SDS-polyacrylamide gels, *Biochem. Biophys. Res. Commun.*, 28, 815, 1967.

16. Righetti, P. G., On the pore size and shape of hydrophilic gels for electrophoretic analysis, in *Electrophoresis '81*, Allen, R. C. and Arnaud, P., Eds., W. de Gruyter, Berlin, 1981, 3.

17. Schwartz, D. C. and Cantor, C. R., Separation of yeast chromosome-sized DNAs by pulsed field gradient gel electrophoresis, *Cell*, 37, 67, 1984.

18. Patel, K., Dunn, M. J., Günther, S., Postel, W., and Görg, A., Dual-label autoradiographic analysis of human skin fibroblast and myoblast proteins by two-dimensional polyacrylamide gel electrophoresis using immobilised pH gradients in the first dimension, *Electrophoresis*, 9, 547, 1988.

19. Righetti, P. G., Morelli, A., Gelfi, C., and Westermeier, R., Direct recovery of proteins into a free-liquid phase after preparative isoelectric focusing in immobilized pH gradients, *J. Biochem. Biophys. Meth.*, 13, 151, 1986.

20. Morel, M. H. and Autran, J. C., Separation of durum wheat proteins by ultrathin-layer isoelectric focusing — A new tool for the characterization and quantification of low molecular weight glutenins, *Electrophoresis*, 11, 392, 1990.

21. Serwer, P., Electrophoresis of duplex deoxyribonucleic acid in multiple-concentration agarose gels: fractionation of molecules with molecular weights between 2×10^6 and 110×10^6, *Biochemistry*, 19, 3001, 1980.

22. Maniatis, T., Fritsch, E. F., and Sambrook, J., *Molecular Cloning, A Laboratory Manual*, Cold Spring Harbor Laboratory, New York, 1982.

23. McDonnel, M. W., Simon, M. N., and Studier, F. W., Analysis of restriction fragments of T7 DNA and determination of molecular weights by electrophoresis in neutral and alkaline gels, *J. Mol. Biol.*, 110, 119, 1977.

24. Scheele, G. A., Two-dimensional gel analysis of soluble proteins, *J. Biol. Chem.*, 250, 5375, 1975.

25. Roggero, P. and Pennazio, S., Two-dimensional polyacrylamide gel electrophoresis of extracellular soybean pathogenesis-related proteins using PhastSystem, *Electrophoresis*, 11, 86, 1990.

26. Celis, J. E. and Bravo, R., *Two Dimensional Gel Electrophoresis of Proteins*, Academic Press, New York, 1984.

27. Pahlic, M. and Tyson, J. J., Analysis of *Physarum* proteins throughout the cell cycle by two-dimensional PAGE, *Exp. Cell. Res.*, 161, 533, 1985.

28. Görg, A., Postel, W., Weser, J., Günther, S., Strahler, J. R., Hanash, S. M., and Somerlot, L., Horizontal two-dimensional electrophoresis with immobilized pH gradients in the first dimension in the presence of nonionic detergent, *Electrophoresis*, 8, 45, 1987.

29. Hanash, S. M., Strahler, J. R., Somerlot, L., Postel, W., and Görg, A., Two-dimensional electrophoresis with immobilized pH gradients in the first dimension: protein focusing as a function of time, *Electrophoresis*, 8, 229, 1987.

30. Strahler, J. R., Hanash, S. M., Somerlot, L., Weser, J., Postel, W., and Görg, A., High resolution two-dimensional polyacrylamide gel electrophoresis of basic myeloid polypeptides: use of immobilized pH gradients in the first dimension, *Electrophoresis*, 8, 165, 1987.

31. Görg, A., Postel, W., Weser, J., Günther, S., Strahler, J. R., Hanash, S. M., Somerlot, L., and Kuick, R., Approach to stationary two-dimensional pattern: influence of focusing time and immobiline/carrier ampholytes concentrations, *Electrophoresis*, 9, 37, 1988.

32. Görg, A., Postel, W., and Günther, S., The current state of two-dimensional electrophoresis with immobilized pH gradients, *Electrophoresis*, 9, 531, 1988.

33. Dunn, M. J. and Burghes, A. H. M., High resolution two-dimensional polyacrylamide gel electrophoresis. I. Methodological procedures, *Electrophoresis*, 4, 97, 1983.

34. Dunn, M. J. and Burghes, A. H. M., High resolution two-dimensional polyacrylamide gel electrophoresis. II. Analysis and applications, *Electrophoresis*, 4, 173, 1983.

35. O'Farrell, P. Z., Goodman, H. M., and O'Farrell, P. H., High resolution two-dimensional electrophoresis of basic as well as acidic proteins, *Cell*, 12, 1133, 1977.

36. O'Farrell, P. H. and O'Farrell, P. Z., Two dimensional polyacrylamide gel electrophoretic fractionation, in *Methods Cell Biol.*, Vol. XVI, Stein, G., Stein, J., and Kleinsmith, L. J., Eds., 1977, 407.

37. Mattila, K., Pirttil, T., and Frey, H., Horizontal two-dimensional electrophoresis of cerebrospinal fluid proteins with immobilized pH gradients in the first dimension, *Electrophoresis*, 11, 91, 1990.

38. O'Farrell, P. H., High resolution two-dimensional electrophoresis of proteins, *J. Biol. Chem.*, 250, 4007, 1975.

39. Tarroux, P., Analysis of protein patterns during differentiation using 2-D electrophoresis and computer multidimensional classification, *Electrophoresis*, 4, 63, 1983.

40. Dunn, M. J. and Patel, K., Immobilized pH gradients for the first dimension of 2-D page, in *Electrophoresis '86*, Dunn, M. J., Ed., VCH, Weinheim, 1986, 574.

41. Görg, A., Postel, W., Günther, S., and Weser, J., Electrophoretic methods in horizontal systems, in *Electrophoresis '86*, Dunn, M. J., Ed., VCH, Weinheim, 1986, 435.

42. Dunn, M. J., Burghes, A. H. M., Patel, K., Witkowski, J. A., and Dubowitz, V., Analysis of genetic neuromuscular diseases by 2D-page: a system for comparisons, in *Electrophoresis '84*, Neuhoff, V., Ed., VCH, Weinheim, 1984, 281.

43. Burghes, A. H. M., Dunn, M. J., and Dubowitz, V., Enhancement of resolution in two-dimensional gel electrophoresis and simultaneous resolution of acidic and basic proteins, *Electrophoresis*, 3, 354, 1982.

44. Görg, A., Postel, W., and Westermeier, R., Ultrathin-layer horizontal electrophoresis, isoelectric focusing, and protein mapping in polyacrylamide gels on cellophane, in *Electrophoresis '79*, Radola, B. J., Ed., Walter de Gruyter, Berlin, 1980, 67.

45. Görg, A., Postel, W., and Westermeier, R., SDS electrophoresis of legume seed proteins in horizontal ultrathin-layer pore gradient gels, *Z. Lebensm. Unters.-Forsch.*, 174, 282, 1982.

46. Jackson, P., Urwin, V. E., and Mackay, C. D., Rapid imaging, using a cooled charge-coupled-device, of fluorescent two-dimensional polyacrylamide gels produced by labeling proteins in the first-dimensional isoelectric focusing gel with the fluorophore 2-methoxy-2,4-diphenyl-3(2H)furanone, *Electrophoresis*, 9, 330, 1988.

47. Rodriguez, L. V., Gersten, D. M., Ramagli, L. S., and Johnston, D. A., Towards stoichiometric silver staining of proteins resolved in complex two-dimensional electrophoresis gels: real-time analysis of pattern development, *Electrophoresis*, 14, 628, 1993.

48. Neuhoff, V., Arold, N., Taube, D., and Erhardt, W., Improved staining of proteins in polyacrylamide gels including isoelectric focusing gels with clear background at nanogram sensitivity using Coomassie Brilliant Blue G-250 and R-250, *Electrophoresis*, 9, 255, 1988.

49. Neuhoff, V., Stamm, R., Pardowitz, I., Arold, N., Ehrhardt, W., and Taube, D., Essential problems in quantification of proteins following colloidal staining with Coomassie Brilliant Blue dyes in polyacrylamide gels, and their solution, *Electrophoresis*, 11, 101, 1990.

50. Merril, C. R., Goldman, D., Sedman, S. A., and Ebert, M. H., Ultrasensitive stain for proteins in polyacrylamide gels show regional variation in cerebrospinal fluid proteins, *Science*, 211, 1437, 1981.

51. Sammons, D. W., Adams, L. D., and Nishizawa, E. E., Ultrasensitive silver-based color staining of polypeptides in polyacrylamide gels, *Electrophoresis*, 2, 135, 1981.

52. Hoffman, E. P., Brown, R. H., and Kunkel, L. M., Dystrophin: the protein product of the Duchenne muscular dystrophy locus, *Cell*, 51, 919, 1987.

53. Aebersold, R. H., Pipes, G., Hood, L. H., and Kent, S. B. H., Amino-terminal and internal sequence determination of microgram amounts of proteins separated by isoelectric focusing in immobilized pH gradients, *Electrophoresis*, 9, 520, 1988.

54. Laskey, R. A. and Mills, A. D., Quantitative film detection of ^{3}H and ^{14}C in polyacrylamide gels by fluorography, *Eur. J. Biochem.*, 56, 335, 1975.

55. Gray, A. J., Beecher, D. E., and Olson, M. V., Computer-based image analysis of one-dimensional electrophoretic gels used for the separation of DNA restriction fragments, *Nucl. Acids Res.*, 12, 473, 1984.

56. Johnston, R. F., Pickett, S. C., and Barker, D. L., Autoradiography using storage phosphor technology, *Electrophoresis*, 11, 355, 1990.

57. Patel, A. C. and Matthewson, S. R., Radiolabeling of peptides and proteins, in *Molecular Biomethods Handbook*, Rapley, R. and Walker, J. M., Eds., Humana Press, Totowa, NJ, 1998, 401.

58. Bateman, J. E., Quantitative autoradiographic imaging using gas-counter technology, *Electrophoresis*, 11, 367, 1990.

59. Sjöberg, S., Carlsson, P., Enerbäck, S., and Bjursell, G., A compact, flexible and cheap system for acquiring sequence data from autoradiograms with a digitizer and transferring it to an arbitrary host computer, *CABIOS*, 5, 41, 1989.

60. Southern, E. M., Detection of specific sequences among DNA fragments separated by gel electrophoresis, *J. Mol. Biol.*, 98, 503, 1975.

61. Beisiegel, U., Protein blotting, *Electrophoresis*, 7, 1, 1986.

62. Burnette, W. N., "Western blotting": electrophoretic transfer of proteins from sodium dodecyl sulfate-polyacrylamide gels to unmodified nitrocellulose and radiographic detection with antibody and radio-iodinated protein A, *Anal. Biochem.*, 137, 195, 1981.

63. Shewry, P. R. and Fido, R. J., Protein blotting: principles and applications, in *Molecular Biomethods Handbook*, Rapley, R. and Walker, J. M., Eds., Humana Press, Totowa, NJ, 1998, 435.

64. Scherberich, J. E., Fischer, P., Bigalke, A., Stangl, P., Wolf, G. B., Haimerl, M., and Schoeppe, W., Routine diagnosis with PhastSystem compared to conventional electrophoresis: automated sodium dodecyl sulfate-polyacrylamide gel electrophoresis, silver staining and Western blotting of urinary proteins, *Electrophoresis*, 10, 58, 1989.

65. Wolff, J. M., Pfeifle, J., Hollmann, M., and Anderer, A., Immunodetection of nitrocellulose-adhesive proteins at the nanogram level after trinitrophenly modification, *Anal. Biochem.*, 147, 396, 1985.

66. Heegaard, N. H. H., Hagerup, M., Thomsen, A. C., and Heegaard, P. M. H., Concanavalin A crossed affinity immunoelectrophoresis and image analysis for semiquantitative evaluation of microheterogeneity profiles of human serum transferrin from alcoholics and normal individuals, *Electrophoresis*, 10, 836, 1989.

67. Macnamara, E. M. and Whicher, J. T., Electrophoresis and densitometry of serum and urine in the investigation and significance of monoclonal immunoglobulins, *Electrophoresis*, 11, 376, 1990.

68. Jones, A. and Moeremans, M., Colloidal gold for the detection of proteins on blots and immunoblots, in *Methods in Molecular Biology*, Vol. 3, Walker, J. M., Ed., The Humana Press, Clifton, New Jersey, 1988, 441.

69. Surek, B. and Latzko, E., Visualization of antigenic proteins blotted onto nitrocellulose using the immuno-gold-staining(IGS)-method, *Biochem. Biophys. Commun.*, 121, 284, 1984.

70. Dunn, M. J., Paper symposium — Quantitative evaluation and densitometry, *Electrophoresis*, 11, 353, 1990.

71. Horgan, G. W. and Glasbey, C. A., Uses of digital image analysis in electrophoresis, *Electrophoresis*, 16, 298, 1995.

72. Pardowitz, I., Ehrhardt, W., and Neuhoff, V., Quantitative densitometry from the point of view of information theory, *Electrophoresis*, 11, 400, 1990.

73. Shipton, H. W., The microprocessor, a new tool for the biosciences, *Annu. Rev. Biophys. Bioeng.*, 8, 269, 1979.

74. Serra, J., Digitalization, *Mikroskopie Suppl.*, 37, 109, 1980.

75. Kemnitz, H.-D. and Hougardy, H. P., Design of an interface for digital image storage, *Mikroskopie*, 7, 415, 1980.

76. Skarnulis, A. J., A computer system for on-line image capture and analysis, *J. Microscopy*, 127, 39, 1982.

77. Erhardt, R., Reinhardt, E. R., Schlipf, W., and Bloss, W. H., FAZYTAN — A system for fast automated cell segmentation, cell image analysis and feature extraction based on TV-image pickup and parallel processing, *Anal. Quant. Cytol. J.*, 2, 25, 1980.

78. Preston, K., Jr. and Dekker, A., Differentiation of cells in abnormal human liver tissue by computer image processing, *Anal. Quant. Cytol. J.*, 2, 1, 1980.

79. Desai, V. and Reimer, L., Digital image recording and processing using an Apple II microcomputer, *Scanning*, 7, 185, 1985.

80. Häder, D.-P. and Lebert, M., Real-time computer-controlled tracking of motile microorganisms, *Photochem. Photobiol.*, 42, 509, 1985.

81. Jaffe, M. J., Wakefield, A. H., Telewski, F., Gulley, E., and Biro, R., Computer-assisted image analysis of plant growth, thigmomorphogenesis, and gravitropism, *Plant Physiol.*, 77, 722, 1985.

82. Bryan, S. R., Woodward, W. S., Griffis, D. P., and Linton, R. W., A microcomputer based digital imaging system for ion microanalysis, *J. Microscopy*, 138, 15, 1985.

83. Garrels, J. and Franza, B. R., Jr., Transformation-sensitive and growth-related changes of protein synthesis in REF52 cells, *J. Biol. Chem.*, 264, 5299, 1989.

84. Garrels, J. and Franza, B. R., Jr., The REF52 protein database, *J. Biol. Chem.*, 264, 5283, 1989.

85. Yang, R.-C., Tsuji, A., and Suzuki, Y., Two-dimensional electrophoresis aided by personal computer analysis for screening of mutant proteins in inherited diseases, *Electrophoresis*, 10, 785, 1989.

86. Neidhardt, F. C., Report of workshop on cellular protein databases derived from two-dimensional polyacrylamide gel electrophoresis, *Electrophoresis*, 10, 73, 1989.

87. Neidhardt, F. C., Appleby, D. B., Sankar, P., Hutton, M. E., and Phillips, T. A., Genomically linked cellular protein databases derived from two-dimensional polyacrylamide gel electrophoresis, *Electrophoresis*, 10, 116, 1989.

88. Gill, P. and Werrett, D. J., Interpretation of DNA profiles using a computerized database, *Electrophoresis*, 11, 444, 1990.

89. Santarén, J. F., Towards establishing a protein database of *Drosophila*, *Electrophoresis*, 11, 254, 1990.

90. Mancini, P., Benassi, A., Valli, G., and Donato, L., Minimum computer system for videodensitometry and image analysis, *Med. Biol. Eng. Comput.*, 16, 542, 1978.

91. Jansson, P. A., Grim, L. B., Elias, J. G., Bagley, E. A., and Lonberg-Holm, K. K., Implementation and application of a method to quantitate 2-D gel electrophoresis patterns, *Electrophoresis*, 4, 82, 1983.

92. Freeman, S. E., Larcom, L. L., and Thompson, B. D., Electrophoretic separation of nucleic acids — Evaluation by video and photograpic densitometry, *Electrophoresis*, 11, 425, 1990.

93. Boniszewski, Z. A. M., Comley, J. S., Hughes, B., and Read, C. A., The use of charge-coupled devices in the quantitative evaluation of images, on photographic film or membranes, obtained following electrophoretic separation of DNA fragments, *Electrophoresis*, 11, 432, 1990.

94. Häder, D.-P., Computer-assisted image analysis in biological sciences, *Ind. Acad. Sci. (Plant Sci.)*, 98, 227, 1988.

95. Wiedemann, G., Müller-Klieser, W., Walenta, S., Schleinkofer, L., and Wood, W. G., Low light level *in vitro* monitoring of cellular and antigen-antibody reactions using a photon detection camera system — New perspectives for clinical diagnostics and research, *Klin. Wochenschr.*, 68, 33, 1990.

96. Danielsson, P. E. and Kruse, B., PICAP II — A second generation picture processing system, *Mikroskopie Suppl.*, 37, 425, 1980.

97. Steinbach, T., Unland, F., and Müller, K.-M., Kostengünstiges Mikroprozessorsystem zur Ergänzung eines Quantimet 720-Bildanalysegerätes, *Microsc. Acta*, 86, 139, 1982.

98. Allen, R. D. and Allen, N. S., Video-enhanced microscopy with a computer frage memory, *J. Microscopy*, 129, 3, 1983.

99. Toda, T., Fujita, T., and Ohashi, M., A method of microcomputer-aided two-dimensional densitometry: an apparatus equipped with a charge-coupled device camera, and an algorithm of microcomputer programming, *Electrophoresis*, 5, 42, 1984.

100. Elder, J. K., Green, D. K., and Southern, E. M., Automatic reading of DNA sequencing gel autoradiographs using a large format digital scanner, *Nucl. Acids Res.*, 14, 417, 1986.

101. Russ, J. C. and Russ, J. C., Image processing in a general purpose microcomputer, *J. Microscopy*, 135, 89, 1984.

102. Mayfield, C., A simple computer-based video image analysis system and potential applications to microbiology, *J. Microbiol. Meth.*, 3, 61, 1984.

103. Nugues, P. M., Two-dimensional electrophoresis image interpretation, *IEEE Trans. Biomed. Eng.*, 40, 760, 1993.

104. Lemkin, P. F. and Lipkin, L. E., 2-D electrophoresis gel data base analysis: aspects of data structures and search strategies in GELLAB, *Electrophoresis*, 4, 71, 1983.

105. Appel, R., Hochstrasser, D., Roch, C., Funk, M., Muller, A. F., and Pellegrini, C., Automatic classification of two-dimensional gel electrophoresis pictures by heuristic clustering analysis: a step toward machine learning, *Electrophoresis*, 9, 136, 1988.

106. Vohradsky, J. and Panek, J., Quantitative analysis of gel electrophoretograms by image analysis and least squares modeling, *Electrophoresis*, 14, 601, 1993.

107. Häder, D.-P., Analysis of two-dimensional electrophoresis gels, in *Image Analysis in Biology*, Häder, D.-P., Ed., CRC Press, Boca Raton, FL, 1992, 87.

108. Ford-Holevinski, T. S., Agranoff, B. W., and Radin, N. S., An inexpensive, microcomputer-based, video densitometer for quantitating thin-layer chromatographic spots, *Analyt. Biochem.*, 132, 132, 1983.

109. Miller, M. J., Olson, A. D., and Thorgeirsson, S. S., Computer analysis of two-dimensional gels: automatic matching, *Electrophoresis*, 5, 297, 1984.

110. Potter, D. J., A review of the clip system for the quantitative analysis of 2-dimensional electrophoresis gels, *Electrophoresis*, 11, 415, 1990.

111. Garrels, J., The QUEST system for quantitative analysis of two-dimensional gels, *J. Biol. Chem.*, 264, 5269, 1989.

112. Celis, J. E., Ratz, G. P., Madsen, P., Gesser, B., Lauridsen, J. B., Hansen, K. P., Kwee, S., Rasmussen, H. H., Nielsen, H. V., Crüger, D., Basse, B., Leffers, H., Honore, B., Moller, O., and Celis, A., Computerized, comprehensive databases of cellular and secreted proteins from normal human embryonic lung MRC-5 fibroblasts: identification of transformation and/or proliferation sensitive proteins, *Electrophoresis*, 10, 76, 1989.

113. Tarroux, P., Vicens, P., and Rabilloud, T., HERMES: a second generation approach to the automatic analysis of two-dimensional electrophoresis gels. Part V. Data analysis, *Electrophoresis*, 8, 187, 1987.

114. Krauss, M. R., Collins, P. J., and Blose, S. H., Computer-analysed 2-D electrophoresis, *Nature*, 337, 669, 1989.

115. Tyson, J. J. and Haralick, R. H., Computer analysis of two-dimensional gels by a general image processing system, *Electrophoresis*, 7, 107, 1986.

116. Mount, D. W. and Conrad, B., Improved programs for DNA and protein sequence analysis on the IBM personal computer and other standard computer systems, *Nucl. Acids Res.*, 14, 443, 1986.

117. Elder, J. K., Maximum entropy image reconstruction of DNA sequencing gel autoradiographs, *Electrophoresis*, 11, 440, 1990.

118. Khashoggi, A. and Lichtenstein, C., SequAlign: a computer program that displays DNA sequence alignments as a compact 'bar-code' graph, *Plant Mol. Biol.*, 23, 639, 1993.

119. Häder, D.-P. and Truß, M., High resolution scanning of absorbing and fluorescent electrophoresis gels using video image analysis, *CABIOS*, 3, 339, 1987.

120. Anderson, L., Identification of mitochondrial proteins and some of their precursors in two-dimensional electrophoretic maps of human cells, *Proc. Natl. Acad. Sci. U.S.A.*, 78, 2407, 1981.

121. Celis, J. E., Ratz, G. P., and Celis, A., Secreted proteins from normal and SV40 transformed human MRC-5 fibroblasts: toward establishing a database of human secreted proteins, *Leukemia*, 1, 707, 1987.

122. Klerk, H. and Jespers, A., Gelanal, a personal computer-program to compare protein patterns on 2-dimensional polyacrylamide gels, *Electrophoresis*, 11, 420, 1990.

123. Czerniak, B., Herz, F., Wersto, R. P., Alster, P., Puszkin, E., Schwarz, E., and Koss, L. G., Quantitation of oncogene products by computer-assisted image analysis and flow cytometry, *J. Histochem. Cytochem.*, 38, 463, 1990.

124. Patton, W. F. and Tempst, P., Enhancing spot detection and reducing noise from digitized electrophoretic gel images using area processing filters, *Electrophoresis*, 14, 650, 1993.

125. Freeman, H., Computer processing of line-drawing images, *Comput. Surveys*, 6, 57, 1974.

126. Freeman, H., Analysis and manipulation of lineal map data, in *Map Data Processing*, Freeman, H. and Pieroni, G. G., Eds., Academic Press, New York, 1980, 151.

127. Häder, D.-P., Automatic area calculation by microcomputer-controlled video analysis, *EDV Med. Biol.*, 18, 33, 1987.

128. Bronstein, I. N. and Semendjajew, K. A., *Taschenbuch der Mathematik*, Verlag Harri Deutsch, Frankfurt, 1981.

129. Press, W. H., Flannery, B. P., Teukolsky, S. A., and Vetterling, W. T., *Numerical Recipes*, Cambridge University Press, London, 1988.

130. Rohde, K. and Bork, P., A fast, sensitive pattern-matching approach for protein sequences, *CABIOS*, 9, 183, 1993.

131. Tietz, D., Analysis of one-dimensional gels and two-dimensional Serwer-type gels on the basis of the extended Ogston model using personal computers, *Electrophoresis*, 12, 28, 1991.

132. Görg, A., Boguth, G., Obermaier, C., and Weiss, W., Two-dimensional electrophoresis of proteins in an immobolized pH 4-12 gradient, *Electrophoresis*, 19, 1516, 1998.

Microscopy and Ultramicroscopy

10 Image Restoration in Two-Dimensional Microscopy

Luigi Bedini, Anna Tonazzini, and Paolo Gualtieri

CONTENTS

10.1 INTRODUCTION

Some of the most recent and important findings in cell molecular biology have been obtained not only by means of biochemistry and molecular biology techniques, but also by staggering improvements in light microscopy.[1,2] Developments in this field are many: implementation of laser beam and stage scanning systems incorporating confocal optics; new electro-optical detectors with great sensitivity, linearity, and dynamic range; the possibility of two-dimensional (2-D) fast image enhancement, reconstruction, analysis, and three-dimensional (3-D) display; and application of luminescence techniques that provide the possibility to investigate the chemical and molecular details of life processes, thanks to their sensitivity, specificity, and potentiality for yielding spatial information.[3-6]

The purpose of digital image restoration is to provide images with a higher spatial resolving power with respect to the original degraded images, by increasing sharpness and clarity, thus enhancing fine structure.

In confocal microscopy, techniques for 2-D (two-dimensional) image restoration proved to greatly ameliorate the quality of the acquired optical section. Conversely, wide-field microscopy has very low depth discrimination, so that the optical section of 3-D object could be recovered only by means of 3-D deconvolution. Nevertheless, a 3-D deconvolution is computationally very demanding, and requires *a priori* information about the imaging system, which is often not available. Fortunately, as shown by results reported in this chapter, techniques for 2-D image restoration can also be successfully applied in processing data acquired in wide-field microscopy.

For this reason, 2-D image restoration is commonly considered a preliminary step for recognition, classification, and 3-D reconstruction, although methods for recovering 3-D objects directly from the observed data have been proposed, for example, for fluorescent micrographs,[7] for wide-field and confocal microscopy,[8,9] and for other fields such as single photon emission computer tomography (SPECT).[10]

Two-dimensional restoration algorithms[11,12] have been extensively applied to conventional wide-field fluorescence microscopy. Gerace et al.[13] used an edge-preserving restoration algorithm in a probabilistic framework; Carrington, Fogarty, and Fay[14] applied a method based on regularization; Agard and co-workers[15] used the Jansson-van Cittert method; whereas Holmes[16,17] used the maximum likelihood estimation method with an expectation-maximization (EM) algorithm.

The aim of this chapter is to introduce the reader to the most recent general principles of 2-D image restoration. The authors surveyed the latest methodologies, starting from regularization, and focus on the edge-preserving image restoration methods they have developed. The problem of the joint estimation of the model parameters, both of the degradation model and of the *a priori* image model, is also addressed. The principles of the analyzed methods are outlined with computer simulations and recent results on 2-D fluorescence images.

10.2 FORMULATION OF THE IMAGE RESTORATION PROBLEM

10.2.1 Data Generation Model

In microscopy, a 2-D image g(x, y) can be expressed as:

$$g(x, y) = f(x, y) \cdot h(x, y) + n(x, y) \qquad (10.1)$$

where f(x, y) represents the true chromophore(s) concentration, $*$ denotes the 2-D convolution operator, h(x, y) is the point spread function (PSF) of the system, which can be expressed as the Airy diffraction pattern, and n(x,y) is the measurement noise. Refer to Ref. 4 for a complete survey of the physical description of a microscope system. In discrete form, Eq. 10.1 becomes:[18]

$$\mathbf{g} = \mathbf{H}\,\mathbf{f} + \mathbf{n} \qquad (10.2)$$

where $\mathbf{g}$, $\mathbf{f}$, and $\mathbf{n}$ are the vectors of size N^2 of the lexicographic notation for the image, the object, and the noise, respectively, and H is the matrix associated with the samples of the PSF. It is common to assume that the noise is white and Gaussian with zero mean and known variance σ^2. Indeed, at the light levels usually reached in microscopy, the Gaussian model is a reasonable approximation to the more accurate Poisson model, which must be instead applied in the case of very low numbers of photons per pixel (e.g., ten).[16]

The inverse problem of image restoration consists of estimating $\mathbf{f}$ from $\mathbf{g}$ and H. This can be done by imposing data consistency, that is, by searching for an $\mathbf{f}$ such that Hf is not too far from $\mathbf{g}$. However, this is always an ill-posed problem, in the sense of Hadamard, in that the existence, uniqueness, and stability of the solution are never guaranteed.[19] Although a continuous generalized inverse for H always exists,[20] so that the problem becomes well-posed, this pseudo-inverse is generally ill-conditioned, and finite (although very small) errors in the data may be highly amplified in the solutions.

The means to overcome ill-posedness and ill-conditioning is regularization,[21] that is any technique that defines a well-posed and well-conditioned restriction of the original problem by exploiting extra information.

Although one can develop the theory of regularization in the deterministic context, the authors prefer to introduce it in the framework of the Bayesian techniques, which give a more general and comprehensive point of view on image restoration and to which the various regularization approaches can be brought back in particular cases.

10.2.2 BAYESIAN APPROACH AND MAP ESTIMATION

In the Bayesian setting, the original image $\mathbf{f}$ to be recovered, the data $\mathbf{g}$, and the noise $\mathbf{n}$ are regarded as $N \times N$ matrices (or N^2 vectors, in the lexicographic notation) of random variables associated to the pixel intensities. Regularization can be achieved by expressing the *a priori* knowledge of the problem in terms of appropriate probability distributions. By virtue of the Bayes rule, the so-called prior probability, which expresses the extra information on the solution, is combined with the likelihood function, derived from the measurements and from the data model, and this produces the posterior probability:

$$P(\mathbf{f}|\mathbf{g}) = \frac{P(\mathbf{g}|\mathbf{f})P(\mathbf{f})}{P(\mathbf{g})} \tag{10.3}$$

This can be maximized to give the Maximum *a posteriori* (MAP) estimate, or used to derive other estimates.[22,23]

Since the noise process is assumed to have components that are independent, white, and Gaussian, with zero mean and variance σ^2, the likelihood function is given by:

$$P(\mathbf{g}|\mathbf{f}) = \left(2\pi\sigma^2\right)^{-N^2/2} \exp\left[-\frac{1}{2\sigma^2}\|\mathbf{g} - \mathbf{H}\mathbf{f}\|^2\right] \tag{10.4}$$

Ignoring the constant term $P(\mathbf{g})$, and considering the function in Eq. 10.4, Eq. 10.3 now becomes:

$$P(\mathbf{f}|\mathbf{g}) = \left(2\pi\sigma^2\right)^{-N^2/2} \exp\left[-\frac{1}{2\sigma^2}\|\mathbf{g} - \mathbf{H}\mathbf{f}\|^2\right]P(\mathbf{f}) \tag{10.5}$$

The key role is thus played by the expression chosen for $P(\mathbf{f})$, that is, the *a priori* probabilistic model for the desired solution. Often, $P(\mathbf{f})$ has an exponential form, that is, $P(\mathbf{f}) = \frac{1}{Z}\exp[-\lambda U(\mathbf{f})]$, where $U(\mathbf{f})$ is called either stabilizer or prior energy, and Z is the normalizing constant, or partition function. $U(\mathbf{f})$ represents a global or local measure of some desired property of the solution. On the basis of physical considerations (e.g., coherence of matter), this property is often smoothness.

Maximizing the posterior probability (Eq. 10.5) is equivalent to minimizing the posterior energy (or cost function):

$$E(\mathbf{f}) = \|\mathbf{g} - \mathbf{H}\mathbf{f}\|^2 + \lambda U(\mathbf{f}) \tag{10.6}$$

where parameter λ includes also $2\sigma^2$. The squared norm term is here called residual and the set of its minima defines the generally infinite feasible solutions of the problem.

In the following sections, the authors consider in detail the most common choices for P($\mathbf{f}$), or equivalently for U($\mathbf{f}$), and show how they lead to various image restoration methods.

As a limit case, recall that if P($\mathbf{f}$) is assumed constant, the MAP estimation reduces to the Maximum Likelihood (ML) criterion, that is, to the minimization of the residual in this case. Since no regularization is enforced, when the matrix H is singular, this technique will produce a solution strongly dependent on the starting point. As a remedy to this drawback, pseudo-inversion or generalized inversion techniques have been proposed. They amount to selecting, among all the feasible solutions, the one of minimum energy.[20] An alternative approach is based on the iterative Expectation-Maximization (EM) algorithm,[24] that was first proposed for ML estimation in emission tomography,[10] and in fluorescence microscopy with Poisson data.[16,17] EM is also suitable for the treatment of Gaussian data in image restoration/reconstruction.[25,26] As well as the pseudo-inverse solution, EM solutions are often very unstable, and special stopping criteria must be adopted to avoid too many artifacts in the images.[27]

10.3 GLOBAL SMOOTHNESS CONSTRAINT

10.3.1 STANDARD REGULARIZATION

In standard regularization,[19] the stabilizer U($\mathbf{f}$) is linear and quadratic, related to linear combinations of derivatives of the solution, and enforces global smoothness on the solution, whose order depends on the order of the derivatives. It has been proven that this is equivalent to restricting the solution space to generalized splines.[28,29] In this case, U($\mathbf{f}$) = $\mathbf{f}^{\mathrm{T}}$A$\mathbf{f}$, where A is a symmetric, positive semidefinite matrix. Provided that the null spaces of A and H^{T}H intersect only in the null vector, the posterior energy (Eq. 10.6) has a unique solution, in the form:

$$\mathbf{f}(\lambda) = (\mathbf{H}^{\mathrm{T}}\mathbf{H} + \lambda\mathbf{A})^{-1}\mathbf{H}^{\mathrm{T}}\mathbf{g} \tag{10.7}$$

which depends on the particular value chosen for the regularization parameter λ.

The simplest and most used forms for the stabilizer are the following:

$$U(\mathbf{f}) = \|\mathbf{f}\|^2 \tag{10.8}$$

$$U(\mathbf{f}) = \|\nabla\mathbf{f}\|^2 \tag{10.9}$$

When used in Eq. 10.6, the stabilizer (Eq. 10.8) prevents the solution from having a large energy, while the stabilizer (Eq. 10.9) prevents the solution gradient from having a large energy.

10.3.2 MAXIMUM ENTROPY

In maximum entropy methods, the stabilizer or prior energy U($\mathbf{f}$) has the following form:

$$U(\mathbf{f}) = \Sigma f_i \log f_i \tag{10.10}$$

This stabilizer has two indisputably appealing properties. First, it forces the solution to be always positive. Second, it yields the most uniform solution consistent with the data, ensuring that the image features result from the data and are not artifacts.[30-33]

10.3.3   ALGORITHMS FOR CONVEX MINIMIZATION

In the cases considered above, the cost function is convex. This means that standard descent algorithms, such as the steepest descent or the conjugate gradient, can be used to find the unique minimum.[34] Since the dimension of the space where the optimization is performed is the same as the image size (typically 256×256 pixels or more), the cost for implementing these techniques is very high, especially for the entropy stabilizer, which makes the cost function highly nonquadratic.

Nevertheless, when the stabilizer is quadratic, as in the case of standard regularization, an alternative to descent algorithms can be the solution of the linear system of Eq. 10.7 by inversion of the non-singular matrix $H^T H + \lambda A$. Although these matrices have a formidable size, (i.e., $N^2 \times N^2$), in most cases they are block-Toeplitz and thus can be made diagonal by discrete FFT when approximated as block-circulant matrices.[18,35] An explicit expression of $f(\lambda)$ can thus be found in terms of the eigenvalues and the eigenvectors of the matrices $H^T H$ and A.

Also, neural networks could be a powerful tool for solving convex, even non-quadratic, optimization problems. Because of the high connectivity, this is typical of neural systems, and the convergence speed of a stable continuous system in reaching an equilibrium state, which is the minimum of an associated Liapunov function. Bedini and Tonazzini[36] suggested using the Hopfield nonlinear neural network model[37,38] to effectively solve the maximum entropy problem for the restoration of blurred and noisy images.

10.4   LOCAL SMOOTHNESS CONSTRAINT

The stabilizers (Eqs. 10.8 and 10.9) introduce a global smoothness constraint that acts on the whole image. Indeed, although formally expressed as sums of local functions, stabilizers do not allow any spatial control, so that the smoothness propagates throughout the image domain. The smoothness constraint has its validity because microscope images usually vary smoothly. However, the smoothness assumption is not valid for all regions of the image. Object boundaries, occlusions, and textures can cause discontinuities in image intensity, so that any microscope image can be assumed piecewise smooth (i.e., composed of a number of connected regions where smoothness is verified), separated by boundaries where the smoothness constraint has no physical meaning. Reconstructing such an image using a global smoothness constraint leads to a solution that is oversmoothed across the boundaries. In image analysis, boundary detection and location are very important tasks; therefore, it is clear that a reconstructed image with incorrect boundary information is unacceptable for most applications. To restore images and preserve their boundaries (edge-preserving image restoration), constraints that act locally on the image to be reconstructed must be imposed.

10.4.1   MARKOV RANDOM FIELD MODELS AND GIBBS PRIORS

This section defines a class of probabilistic image models that are suitable for describing the behavior of both the intensity and the discontinuity fields (i.e., the class of Markov Random Field (MRF) models on finite lattices).[39]

One begins by characterizing the intensity attribute. In this case, the lattice will be the $N \times N$ pixel grid, and the site set S will be any arrangement of the grid; for example, $S = \{(i, j), i = 1, 2, ..., N; j = 1, 2, ..., N\}$.

For each site r, one can define a neighborhood G_r such that if s and t are two distinct sites, their neighborhoods, G_s and G_t, respectively, have the following properties:

$$s \notin G_s$$

$$s \in G_t \Leftrightarrow t \in G_s$$

Associated with the neighborhood system, there is a set C of cliques. A clique C is either a single site, or a subset of sites such that any two distinct sites in C are neighbors of each other. The cliques are thus uniquely determined by the neighborhood system chosen.

The vector of pixel intensities $\mathbf{f}$ will then be regarded as a family of random variables associated with S, $\mathbf{f} = \{f_s, s \in S\}$, whose values lie on a common discrete or continuous set Λ. If Ω is the set of all the possible configurations (images)

$$\Omega = \{f_s: f_s \in \Lambda, s \in S\} \tag{10.11}$$

then $\mathbf{f}$ is an MRF with respect to the pair (S, G) if and only if

$$P(\mathbf{f}) > 0 \; \forall \mathbf{f} \in \Omega \tag{10.12a}$$

$$P\left(f_s \middle| f_t, \forall t \neq s\right) = P\left(f_s \middle| f_t, \forall t \in G_s\right) \qquad \forall s \in S, \forall \mathbf{f} \in \Omega \tag{10.12b}$$

where $P(\mathbf{f})$ is the probability distribution of $\mathbf{f}$. Hence, the conditional probability of the element in s, given all the other elements, depends only on the values of the elements of its neighborhood.

By virtue of the Clifford-Hammersley theorem, the joint probability of an MRF has the form of a Gibbs distribution

$$P(\mathbf{f}) = \frac{1}{Z} \exp\left[-U(\mathbf{f})\right] \tag{10.13a}$$

$$U(\mathbf{f}) = \sum_{c \in C} V_c(\mathbf{f}) \tag{10.13b}$$

where Z is the normalizing constant, $U(\mathbf{f})$ is called the energy function, and the potentials $V_c(\mathbf{f})$ are functions supported on the cliques of the field. This important result allows the joint probability of an MRF to be directly derived by specifying the potentials instead of the conditional probabilities. This makes it very easy to model and constrain the local behavior of an MRF with a specified neighborhood and clique systems. In particular, this makes it possible to design prior energies for images that permit the smoothness constraint to be locally broken or relaxed where discontinuities are likely to occur. This can be accomplished by augmenting the MRF model, through the introduction of an explicit, auxiliary line process, or using particular non-quadratic stabilizers that are able to preserve discontinuities without treating extra variables.

10.4.2 Implicit Treatment of the Discontinuities

Consider the stabilizer $U(\mathbf{f})$ of Eq. 10.9. In the MRF formalizm, it can be rewritten as

$$U(\mathbf{f}) = \sum_{c \in C} \left[D_c(\mathbf{f})\right]^2 \tag{10.14}$$

where $D_c(\mathbf{f})$ are finite-difference approximations of the first-order partial derivatives of $\mathbf{f}$, and C are the sets of pixels involved in the computation.[40] Each potential is thus a quadratic function of the partial derivatives. This function has the desirable properties of being positive, even, finite in zero, and increasing with the magnitude of its argument. Nevertheless, it increasingly penalizes high differences between neighboring pixels, and this prevents the discontinuities from being recovered. This can be avoided by replacing the quadratic function with a non-quadratic function ϕ (also called neighbor interaction function, see Blake and Zisserman[41]), which retains the good properties mentioned above, but allows sharp transitions between distinct regions to be preserved. The general form of the prior energy in this case is

$$U(\mathbf{f}) = \sum_{c \in C} \alpha \phi \left(\frac{D_c(\mathbf{f})}{\Delta} \right) \tag{10.15}$$

where α and Δ are positive parameters. Various forms for ϕ have been proposed in the literature. In particular, the following non-convex functions:[40-42]

$$\phi_1(t) = \min\left(1, \ t^2\right) \tag{10.16}$$

$$\phi_2(t) = \frac{t^2}{t^2 + 1} \tag{10.17}$$

$$\phi_3(t) = \frac{|t|}{|t| + 1} \tag{10.18}$$

share the two properties that their limit at infinity is finite, and that $\phi\left(\sqrt{t}\right)$ is concave. When used in Eq. 10.15, these functions encourage neighboring pixels to have similar values if the derivatives are lower than Δ. Beyond this value, a further increase in the derivatives is allowed, with a relatively small increase in the penalty. The differences between neighboring pixels within smooth regions are thus penalized without excessively penalizing the larger differences occurring at the boundaries between different regions of the image.

As a major difference among these three functions, it must be noted that, while the first function, called truncated parabola, presents a clean, sharp threshold, so that only discontinuities higher than Δ are admitted, the other two functions allow a more graduated reconstruction of discontinuities of different amplitude. For this intuitive argument, and also for theoretical reasons not reported here (see Geman and Reynolds[40] for details and proof), the truncated parabola (Eq. 10.16) is said to implicitly refer to binary discontinuities, while functions 10.17 and 10.18 are said to implicitly refer to graded discontinuities.

Other authors proposed non-quadratic but convex functions:[43,44]

$$\phi_4(t) = |t| \tag{10.19}$$

$$\phi_5(t) = \log \cosh(t) \tag{10.20}$$

whose behavior at infinity is linear; discontinuities are thus allowed, but excessive intensity jumps are penalized. The effect of these functions is thus a compromise between a parabola and an asymptotically finite stabilizer as those shown in Eqs. 10.16 through 10.18.

All the functions shown above can take discontinuities into account without introducing extra variables. However, no significant available constraints can be enforced on the geometry of the image boundaries. Conversely, the discontinuities associated with edge contours are often connected and thin, and introducing this information into the problem would greatly improve the quality of the reconstructed images. Some preliminary attempts have been made to design neighbor interaction functions that are able to enforce at least simple constraints,[45-47] but the problem is still open. A more manageable and comprehensive way of treating geometrically constrained discontinuities is to consider them as explicit unknowns of the problem, as shown in Section 10.4.3.

10.4.3 EXPLICIT TREATMENT OF THE DISCONTINUITIES

In this approach, the original image is regarded as a pair of interacting MRFs, (**f**, **l**), where **f** is the matrix of the pixel intensities, and **l** is a new binary field that marks the presence or absence of the discontinuities. **f** is called the intensity process, and **l** is called the line process.

Typically, the line elements are considered as localized in a rectangular interpixel grid and are distinguished into vertical and horizontal elements. Under this assumption, **l** will be given by $\mathbf{l}_h$ and $\mathbf{l}_v$, which are the $(N-1) \times N$ and $N \times (N-1)$ random matrices associated with the horizontal and vertical line elements, respectively. The values assumed by the elements of $\mathbf{l}_h$ and $\mathbf{l}_v$ will be denoted by $h_{i,j}$ and $v_{i,j}$, respectively.

The set of sites for the global field (**f**, **l**) will be given by the union of the intensity and line sites. In this case, the neighborhood system G and the related clique system C must be defined on the mixed set of sites, allowing adjacent pixels and lines to be neighbors. Thus, the prior distribution for (**f**, **l**) is:

$$P(\mathbf{f}, \mathbf{l}) = \frac{1}{Z} \exp\left[-U(\mathbf{f}, \mathbf{l})\right] \qquad (10.21a)$$

where

$$U(\mathbf{f}, \mathbf{l}) = \sum_{c \in C} V_c(\mathbf{f}, \mathbf{l}) \qquad (10.21b)$$

is the prior energy function.

The potentials can be defined on homogeneous cliques (made of intensity sites alone or line sites alone), or mixed cliques (made of intensity **and** line sites). The general form (Eq. 10.21b) thus admits the following decomposition:

$$U(\mathbf{f}, \mathbf{l}) = U_1(\mathbf{f}) + U_2(\mathbf{l}) + U_3(\mathbf{f}, \mathbf{l}) \qquad (10.21c)$$

where $U_1(\mathbf{f})$ models the local constraints of the intensity process, $U_3(\mathbf{f}, \mathbf{l})$ enforces the dependence between pixel intensities and line element configurations, and $U_2(\mathbf{l})$ represents the mutual relationships among neighboring line elements.

Besides making the introduction of constraints on the line geometry simple and direct, this approach also allows cooperative processing; for example, simultaneous reconstruction and edge detection can be performed.

Now focus on the class of piecewise smooth images, with connected and thin binary discontinuities. For this class, the neighborhood system in Figure 10.1, which is of first order with respect to the intensity sites, is considered sufficient. The mixed cliques allow us to enforce a local first-order smoothness constraint, according to the following expression for $U_3(\mathbf{f}, \mathbf{l})$:

$$U_3(\mathbf{f}, \mathbf{l}) = \sum_{i,j} \lambda\left(f_{i,j} - f_{i,j+1}\right)^2\left(1 - v_{i,j}\right) + \alpha v_{i,j}$$
$$+ \sum_{i,j} \lambda\left(f_{i,j} - f_{i+1,j}\right)^2\left(1 - h_{i,j}\right) + \alpha h_{i,j} \qquad (10.22)$$

where λ and α are positive parameters and $\sqrt{\alpha/\lambda}$ represents a threshold on the gradient, above which a discontinuity is likely to be created. In other words, the term $U_3(\mathbf{f}, \mathbf{l})$ in the prior encourages

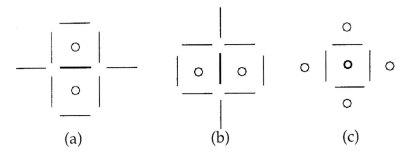

FIGURE 10.1 First-order neighborhood system for: (a) horizontal line element; (b) vertical line element; (c) intensity element.

solutions that have discontinuities where the horizontal or vertical gradient is higher than the threshold and are smoothly varying elsewhere. Considered individually, λ is a regularization parameter that promotes smoothing in the absence of discontinuities, and α represents the cost of creating a discontinuity, so as to prevent the creation of too many discontinuities.

The term $U_2(\mathbf{l})$ should reflect prior expectations concerning the structure of discontinuities; for example; it is known that lines are generally thin and connected. With reference to the line configurations derived from the neighborhood system of Figure 10.1, a simple analytical form for $U_2(\mathbf{l})$ is:

$$
\begin{aligned}
U_2(\mathbf{l}) = &\gamma_1 \sum_{i,j} h_{i,j} h_{i+1,j} + \gamma_2 \sum_{i,j} v_{i,j} v_{i,j+1} + \beta_1 \sum_{i,j} h_{i,j} h_{i,j+1} \\
&+ \beta_2 \sum_{i,j} h_{i,j} v_{i,j} + \beta_3 \sum_{i,j} h_{i,j} v_{i+1,j} + \beta_4 \sum_{i,j} v_{i,j} v_{i+1,j} \\
&+ \beta_5 \sum_{i,j} v_{i,j} h_{i,j+1} + \beta_6 \sum_{i,j} v_{i+1,j} h_{i,j+1} + \varepsilon_1 \sum_{i,j} v_{i,j} v_{i+1,j} h_{i,j+1} \\
&+ \varepsilon_2 \sum_{i,j} h_{i,j} v_{i,j} v_{i+1,j} + \varepsilon_3 \sum_{i,j} h_{i,j} h_{i,j+1} v_{i+1,j} + \varepsilon_4 \sum_{i,j} h_{i,j} h_{i,j+1} v_{i,j} \\
&+ \kappa \sum_{i,j} h_{i,j} h_{i,j+1} v_{i,j} v_{i+1,j}
\end{aligned}
\tag{10.23}
$$

where the first and second terms penalize the formation of adjacent parallel lines (double lines), the next six terms favor the formation of continuous lines, and the last five terms penalize the formation of branches and crosses. The values of the parameters β_i should thus be negative and can be chosen to give different probabilities to straight lines and turns.

10.5 ALGORITHMS FOR NON-CONVEX OPTIMIZATION

10.5.1 Metropolis Algorithm and Gibbs Sampler

Almost all the MRF models described in Section 10.4 for treating image discontinuities lead to posterior energies that are non-convex and depend on a very large number of variables. This, at least in principle, would require the adoption of stochastic relaxation algorithms to escape from local minima and find the global minimum.

One of them is the well-known Metropolis algorithm,[48] which simulates the evolution of a multivariate physical system at its thermal equilibrium.

Consider $\mathbf{x}$ to represent either the only intensity process or the coupled intensity-line process. Given a probability distribution in the form $P(\mathbf{x}) = \exp\{-E(\mathbf{x})\}$, the elements of $\mathbf{x}$ are visited in any order and a random value is generated for each of them. Let $\mathbf{x}^*$ be the state of $\mathbf{x}$ when an update for the value of an element is proposed, and let $\mathbf{x}^k$ be the previous state. The value $\Delta E = E(\mathbf{x}^*) - E(\mathbf{x}^k)$ is then calculated; if it is negative (i.e., if the update lessens the total energy of the system), the update is accepted and $\mathbf{x}^{k+1} = \mathbf{x}^*$. If the energy change is positive, the update of the element is accepted with probability $\exp\{-\Delta E\}$. In practice, a random number τ, uniformly distributed in $(0, 1)$, is generated and, if $\tau < \exp\{-\Delta E\}$, the update is accepted; otherwise, the proposed update is refused and $\mathbf{x}^{k+1} = \mathbf{x}^k$. After a sufficiently large number of updates, the system reaches the "thermal equilibrium," and the successive updates of $\mathbf{x}$ are distributed in accordance to $P(\mathbf{x})$.

Geman and Geman[39] showed another technique, called Gibbs sampler, for drawing a sample from a probability distribution. They proved that if $P(\mathbf{x})$ is a Gibbs distribution, a random sample distributed as $P(\mathbf{x})$ can be drawn by successively updating each element on the basis of its local conditional probability, which only depends on the state of its neighborhood. This means that the algorithm can be easily parallelized.

10.5.2 SIMULATED ANNEALING

In the Metropolis algorithm or in the Gibbs sampler, transitions with increasing energy are allowed with non-zero probability. One can control the generation of these transitions and the amplitude of the energy jumps by introducing a parameter, namely the "temperature" T, that regulates the peaking of the posterior. By allowing the temperature to decrease, the Metropolis algorithm or the Gibbs sampler generate a nonuniform Markov chain whose stationary state converges to the uniform distribution over the modes of the energy function. This is the principle underlying the Simulated Annealing (SA) minimization algorithm.[49]

One can modify the posterior distribution $P(\mathbf{x})$ as follows:

$$P(\mathbf{x}, T) = \exp\left(-\frac{E(\mathbf{x})}{T}\right) \qquad (10.24)$$

One observes that if the temperature is high, the distribution is practically flat over its domain and all the changes proposed by a Metropolis algorithm are accepted. When the temperature goes to zero, it can be proven that Eq. 10.24 becomes uniform on the set of the global energy minima and zero elsewhere. If one starts from high temperatures and reaches the thermal equilibrium for several, slowly decreasing values of the temperature, one is guaranteed to reach a global energy minimum. Asymptotic convergence criteria for simulated annealing can be found in the literature,[39,50] even if they lead to computational requirements that cannot be fulfilled by any feasible procedure. However, practical criteria to reach a good solution can be established and validated experimentally. The most efficient rules have been proposed by Kirkpatrick et al.,[49] and Aarts and van Laarhoven (see Aarts and Korst[50]).

Despite their good performances, these algorithms are computationally very demanding, due to the size of the problems treated and the number of iterations required. Two main strategies have been adopted to reduce the complexity of the problem and/or the execution time. In the first, the problem is transformed or approximated to allow the use of totally or partially deterministic algorithms.[22,51] In the second, parallel algorithms are studied, for use on general-purpose or dedicated parallel architectures.[52]

10.5.3 MIXED ANNEALING

For the case of an explicit line process, Marroquin proposed an alternative procedure to simulated annealing (mixed annealing), where stochastic steps are alternated with deterministic steps that support almost all the computational load of the minimization.[22]

Since the posterior energy E($\mathbf{f}$, $\mathbf{l}$) is a convex function of $\mathbf{f}$ for any fixed configuration of the line process, the search for the global minimum can be restricted to the set of configurations ($\mathbf{f}$*($\mathbf{l}$), $\mathbf{l}$), where $\mathbf{f}$*($\mathbf{l}$) is the minimizer of the posterior energy with the line process fixed at $\mathbf{l}$ (optimal conditional estimate). The non-convex posterior energy restricted to this set is now a Gibbs energy, with the same neighborhood system as that of the prior. E($\mathbf{f}$*($\mathbf{l}$), $\mathbf{l}$) can be thus minimized by an annealing scheme where the random samples can be drawn by a Gibbs sampler. Bedini et al.[53] described this annealing procedure as follows:

1. An initial temperature, T_o, and a cooling schedule are chosen; the number N_O of iterations spent at temperature T_o and an initial guess $\mathbf{l}_o$ for the line process are given.
2. For each temperature T_k and each number of iterations N_k, starting from the line configuration $\mathbf{l}_k$, the Gibbs sampler draws a line process sample $\mathbf{l}_{k+1}$, with distribution

$$P\left(\mathbf{f}^*(\mathbf{l}), \mathbf{l}\right) \propto \exp\left(-\frac{E\left(\mathbf{f}^*(\mathbf{l}), \mathbf{l}\right)}{T_k}\right) \qquad (10.25)$$

3. New values T_{k+1} and N_{k+1} are selected, and the algorithm repeats steps 2 and 3 until the stop criterion is satisfied.

Note that, at each update of a single line element by the Gibbs sampler, a new optimal conditional estimate $\mathbf{f}$* should be evaluated. For this reason, the algorithm is still very expensive, but its most expensive part is the deterministic one. Due to the small size of the neighborhoods, and because the line process is binary, the cost of drawing a sample from the line process is low.

The authors have previously shown[51] that the availability of an analogue Hopfield-type neural network would permit the computation of $\mathbf{f}$*($\mathbf{l}$) in almost real-time, as the stable state of an electrical circuit. Conversely, the stochastic sampling with respect to $\mathbf{l}$ can be accomplished by means of a low-cost binary Gibbs sampler. It is well known that simulated annealing with the Gibbs sampler or the Metropolis algorithm represents the running modality of the Boltzmann Machine, which is a stochastic neural network whose units evolve until they stabilize at the minimizer of an internal energy.[50] This internal energy can be made to coincide with the conditional prior energy of the lines given the intensity, so that it can be used, at fixed temperature, to produce samples of the line process.[54] This Boltzmann Machine can be realized by means of a grid of processors working in parallel.

10.5.4 GRADUATED NON-CONVEXITY

Blake and Zisserman[41] derived a fully deterministic algorithm to minimize the non-convex posterior energy for implicit line treatment, when the neighbor interaction function ϕ is of the type in Eq. 10.16. Blake proved its lower complexity when compared with stochastic relaxation techniques.[55]

This algorithm is called Graduated Non-Convexity (GNC). It is based on a sequence of approximations $E^p(\mathbf{f})$ of the posterior energy, depending on a real parameter, $p \in [0, p^*]$, and such that $E^0(\mathbf{f}) \equiv E(\mathbf{f})$ and $E^{p^*}(\mathbf{f})$ is a convex function. Gradient descent algorithms are then applied to minimize the modified posterior energies, for decreasing values of p, starting from $p = p^*$. The starting point of each minimization is the minimizer found for the previous value of p. If E^{p^*} is already a good approximation of E, then the algorithm can reach from the first iteration a point that is close to the desired global minimizer. As E^p approaches E, this estimate is refined, so that a good approximation of the global minimizer can be obtained.

Assuming that

$$E(\mathbf{f}) = \|\mathbf{g} - \mathbf{Hf}\|^2 + \sum_{i,j} \phi\left(f_{i,j} - f_{i,j+1}\right) + \phi\left(f_{i,j} - f_{i+1,j}\right) \tag{10.26}$$

Blake and Zisserman constructed a series of approximations for the neighbor interaction function:[41]

$$\phi(t) = \min(\alpha, \lambda t^2) \tag{10.27}$$

Bedini et al.[56] derived a series of approximations for a neighbor interaction function of the type in Eq. 10.18, rewritten as:

$$\phi(t) = \frac{\lambda|t|}{\dfrac{\lambda}{\alpha}|t| + 1} \tag{10.28}$$

Besides GNC, many other deterministic algorithms have been proposed to find good sub-optimal solutions for the case of implicit discontinuities treatment. In their mean field annealing approach, Geiger and Girosi[45] provided a parametric family of cost functions, converging to the same cost function as Blake and Zisserman's, and approximated the global minimum by iteratively minimizing the cost functions through the solution of deterministic equations.

Iterated conditional modes (ICMs) is a deterministic algorithm proposed by Besag[57] for approximating the MAP estimate. ICM iteratively computes the maximum of the posterior probability of each image element, conditioned on the values assumed by all other elements at the previous iteration.

Geman and Yang[58] proposed a linear algebraic method to implement regularization with implicit discontinuities. By introducing suitable auxiliary variables, the posterior distribution becomes Gaussian in the intensity variables, with a block circulant covariance matrix. A simulated annealing algorithm with simultaneous updating of all the pixels can thus be designed using FFT techniques.

Generalized Expectation-Maximization (GEM) algorithms have also been proposed to solve the MAP problem that arises when edge-preserving priors are introduced to stabilize the solutions.[10,25,26]

10.6   BLIND UNSUPERVISED IMAGE RESTORATION

All the restoration methods described in Sections 10.3 through 10.5 have been developed assuming that the parameters appearing in the image model are known in advance. This assumption is not actually valid, in that usually one does not have enough information on the true image to satisfactorily determine them. When an MRF model for the image is adopted, these parameters are called hyperparameters or Gibbs parameters. Estimating the hyperparameters is a critical issue in image restoration methods, as they considerably affect the solutions obtained.

Similarly, in many image processing applications, including digital microscopy, the parameters of the imaging system are not known or it is cumbersome and/or expensive to measure them. These are typically the coefficients of the PSF or blur mask and the noise variance, and are called degradation parameters.

Early works on image model hyperparameter estimation focused on the regularization or smoothness parameter λ in standard regularization. This non-negative parameter determines the compromise, in the solution, between data consistency, expressed by the residual, and regularity, enforced by the stabilizer or prior. It is easy to verify that, by modifying its value, very different solutions can be obtained, ranging from the ultrarough least-squares solution, when $\lambda = 0$, to the ultrasmooth solution, when λ goes to $+\infty$. Moreover, experiments showed that the best value for λ is very sensitive to the stabilizer adopted, to the image structure, and to the amount of noise affecting the data. Intuitively, λ should be large if the data set is heavily corrupted by noise, and small otherwise. Regarding the convex cost function as the Lagrangian associated with a constrained minimization problem, and the parameter as the Lagrange multiplier, the necessary conditions for the minimum also specify an equation to be satisfied by the Lagrange multiplier.[59] In many cases, however, these equations are nonlinear. Bedini et al.[60] proposed a method based on the primal-dual theory for convex optimization[61] to estimate simultaneously both the solution and the Lagrange multiplier, via the solution of a simpler dual problem. The method has been derived for the restoration of blurred, noisy images, and for different kinds of stabilizers such as cross-entropy and energy.

When the stabilizer is quadratic, statistical methods have been proposed to estimate the smoothness parameter from the data. The most popular among them are the Chi-squared, the equivalent degrees of freedom, and the generalized cross-validation methods.[62-65] The computation of these estimates entails the iterative solution of nonlinear equations, and the computation of the solution $\mathbf{f}(\lambda)$ of Eq. 10.7 at each iteration. Nevertheless, since the computations of the eigenvalues and the eigenvectors of the matrices in Eq. 10.7 can be accomplished only once and off-line, the cost of these estimates is relatively low. Moreover, by solving the equations for λ, both the estimation of the regularization parameter and the optimal reconstruction for $\mathbf{f}$ are simultaneously obtained.

With concern to blur identification, early works were based on maximum likelihood (ML) and generalized cross-validation (GCV) methods using stochastic linear ARMA (autoregressive moving average) models for the distorted image.[66-70] Another approach, which has been widely experimented, consists of forcing positivity and/or size constraints on both the image and the blur, alternatively and iteratively estimated through inverse filtering, Wiener filtering, simulated annealing, or projections onto constraint sets.[71-74] Positivity and size constraints are obviously useful in reducing the number of admissible solutions for the blind restoration problem, but they are by no means enough. Other constraints, such as the knowledge of the type of blur (e.g., motion blur, etc.) or the global smoothness of the blur itself are too restrictive or force the estimate toward uniform blurs. On the other hand, the choice of a good image model, which allows for a satisfactory estimation of the true image, is crucial for the success of joint image restoration and blur estimation. Furthermore, assuming that the image is piecewise smooth, most of the information about the blur is expected to be located across the discontinuity edges. Although global smoothness for the image has been used in ML- and GCV-based methods, it is well known that this constraint destroys sharp intensity transitions in the image, and then it cannot perform well for blur identification. Thus, it can be expected that further improvement can be achieved via the incorporation of the piecewise smoothness of the image, through models that allow for a reliable edge location. As an example, the piecewise smoothness of the image is coupled to the piecewise smoothness of the blur and incorporated in space-adaptive regularization to recover uniform blurs.[75] Following some previous promising results of blur identification based on the EM algorithm, Zhang[76] proposed to augment this approach with MRF image models, where the MRF is given by a continuous-valued intensity process, coupled with a binary line process, and the MRF model hyperparameters are assumed to be known in advance.

More recently, fully data-driven approaches have been designed that allow for the joint estimation of the restored image, the image model hyperparameters, and the degradation parameters. The most natural setting to formalize these methods is again the Bayesian framework.

Assign $\mathbf{w}$ to the vector of the model hyperparameters, for example, those appearing in the Gibbsian prior given in Eq. 10.21 through 10.23. Assuming that the degradation operator is a blur, matrix H in the data formation model of Eq. 10.2 or, equivalently, in the likelihood function of Eq.

10.4 is a block Toeplitz matrix, whose elements derive, according to a known rule, from a usually small size mask d. One can rewrite H as H(d) and call θ the set of observation parameters d and σ^2. Unless specified, in the following, **x** stands for both the intensity process alone and the intensity process plus the line process.

By explicitly introducing the dependence of all the functions defined thus far from their own parameters [e.g., U(**x**) + U(**x**|**w**), P(**g**|**x**) = P(**g**|**x**, θ)], the Bayesian point of view to reconstruct the image, estimate the "best" Gibbs parameters, and accomplish the degradation process identification for a given problem is to assume a prior distribution for **w** and θ, and then simultaneously compute **x**, **w**, and θ by maximizing some joint distribution of all of them. Assuming a uniform prior for **w** and θ, and considering the distribution P(**x**, **g**|**w**, θ), the problem can be formulated as:

$$\max_{\mathbf{x, w}, \theta} P(\mathbf{x}, \mathbf{g}|\mathbf{w}, \theta) \qquad (10.29)$$

The joint maximization (Eq. 10.29) is a very difficult task. Nevertheless, given the separability of the observation and the model parameters, the considered distribution can be rewritten as:

$$P(\mathbf{x}, \mathbf{g}|\mathbf{w}, \theta) = P(\mathbf{x}|\mathbf{g}, \mathbf{w}, \theta) \, P(\mathbf{g}|\mathbf{w}, \theta) = P(\mathbf{g}|\mathbf{x}, \theta) \, P(\mathbf{x}|\mathbf{w}) \qquad (10.30)$$

This allows for the adoption of the following sub-optimal iterative procedure, proposed by Besag[57] and Lakshmanan and Derin,[77] among others:

$$\mathbf{x}^{(k)} = \arg \max_{\mathbf{x}} P(\mathbf{x}|\mathbf{g}, \mathbf{w}^{(k)}, \theta^{(k)}) \qquad (10.31a)$$

$$\theta^{(k+1)} = \arg \max_{\theta} P(\mathbf{g}|\mathbf{x}^{(k)}, \theta) \qquad (10.31b)$$

$$\mathbf{w}^{(k+1)} = \arg \max_{\mathbf{w}} P(\mathbf{x}^{(k)}|\mathbf{w}) \qquad (10.31c)$$

Starting from an initial guess $\mathbf{w}^{(0)}$, $\theta^{(0)}$, and iterating steps 10.31a, 10.31b, and 10.31c, a sequence $(\mathbf{x}^{(k)}, \mathbf{w}^{(k)}, \theta^{(k)})$ that converges to a local maximum of P(**x**, **g**|**w**, θ) can be obtained. In this sense, the problem in Eq. 10.31 is smaller than the problem of Eq. 10.29. Nevertheless, if $(\mathbf{x}^*, \mathbf{w}^*, \theta^*)$ is the solution of Eq. 10.31, then **x*** is the MAP estimate of **x** based on **g**, **w***, and θ^*; θ^* is the ML estimate of θ based on the likelihood function computed in **x***; and **w*** is the ML estimate of **w** based on the prior computed in **x***.[77] The same considerations hold at each stage of the iterative procedure in Eq. 10.31. Thus, the final solution $(\mathbf{x}^*, \mathbf{w}^*, \theta^*)$ is adaptively obtained by the iterative execution of a MAP estimation for **x** and ML estimations for **w** and θ in turn.

For the MAP estimate, depending on the form of the prior and on the explicit or implicit presence of a line process, one of the algorithms described in Section 10.5 can be adopted.

Problem 10.31b, (i.e., the estimation of the degradation parameters), does not depend on the line field eventually present and can be reduced to the two following computational steps:

$$d^{(k+1)} = \arg \min \left\| \mathbf{g} - H(d)\mathbf{x}^{(k)} \right\|^2 \qquad (10.32a)$$

$$\left(\sigma^2 \right)^{(k+1)} = \frac{\left\| \mathbf{g} \, H\left(d^{(k+1)}\right)\mathbf{x}^{(k)} \right\|^2}{N^2} \qquad (10.32b)$$

which are derived by taking the negative logarithm of the likelihood function and dropping the constant terms. In particular, the closed form solution for σ^2 is obtained by putting to zero the first derivative with respect to σ of the log-likelihood. Step 10.32a formulates the estimation of the blur mask as a least squares problem. By setting to zero the gradient, the original minimization problem reduces to the solution of a usually small size linear system.

Thus, the main problem in solving Eq. 10.31 is to find an efficient algorithm to compute the ML estimate of $\mathbf{w}$. Indeed, the difficulty is related to the presence in the prior distribution of the normalizing constant, which depends not linearly on $\mathbf{w}$.

In the coding method, $\mathbf{w}$ is estimated by maximizing the conditional likelihood:[78,79]

$$\prod_{i \in M} P\left(x_i^k \big| x_j^k, j \in G_i, \mathbf{w}\right) \tag{10.33}$$

where M, called "coding," is a set of sites that does not contain any pair of neighbors in the MRF sense. For example, considering the intensity process alone as a first-order MRF, at least two disjoint codings can be defined, corresponding to a checkerboard partition of the sites. Maximizing the conditional likelihoods related to different codings gives different estimates of the parameter vector. The final estimate can be obtained by averaging these estimates. In this method, the effort in computing the normalizing constants is reduced, since, for each distribution in Eq. 10.33, the summation required to compute the normalizing constant is made on the values of x_i, and not on all the possible configurations for $\mathbf{x}$.

An extension of the coding methods is the maximum pseudo-likelihood estimation (MPL),[57,77] in which the product in Eq. 10.33 is extended over all sites. Although this is not a true likelihood, in many cases of parameter estimation, MPL is a good approximation of the ML estimate, and it is also consistent, in the sense that it converges in probability to the ML estimate when the size of the image increases. Alternatively, a mean field approximation for the original distribution can be adopted by neglecting statistical fluctuations of the field and assuming that each single variable depends only of the mean values of the others.[76,80] More recently, Higdon et al.[81] and Descombes et al.[82] applied Monte Carlo Markov Chain (MCMC) techniques to compute, via time averages, the ratio between the partition functions taken at two different values of the hyperparameter set, with application to PET tomography and image segmentation, respectively.

Nevertheless, most works proposing MPL and mean field approximations for the estimation of the MRF parameters are in the context of image segmentation, where the image pixels can assume only a few values. Moreover, these works do not consider, at least not explicitly, the line process. In the authors' opinion, the introduction of the discontinuities into the MRF model considerably complicates the problem of parameter estimation, in that intensity and line elements are highly correlated, thus preventing the pseudo-likelihood from being a good approximation of the original prior distribution and still very expensive to compute. On the other hand, the correlation between intensities and lines can be exploited to greatly simplify the ML hyperparameters estimation step, by adopting different approximations based on the assumption that the line process alone can retain a good deal of information about the hyperparameters that best model the whole image.

In the case of a coupled MRF, the hyperparameter estimation step (Eq. 10.31c) entails maximizing the prior distribution of the image field computed in the current image estimate ($\mathbf{f}^{(k)}$, $\mathbf{l}^{(k)}$). By reformulating this ML estimation problem as the one that minimizes the negative log-prior, this function can easily be shown to be convex with respect to $\mathbf{w}$, so that a possible criterion to get its minimum is to look for the zero of its gradient, that is, by solving the following system of equations:

$$\frac{\partial}{\partial w_r} \left\{ -\log P\left(\mathbf{f}^{(k)}, \mathbf{l}^{(k)} \big| \mathbf{w}\right) = \frac{\partial U\left(\mathbf{f}^{(k)}, \mathbf{l}^{(k)} \big| \mathbf{w}\right)}{\partial w_r} + \frac{1}{Z(\mathbf{w})} \frac{\partial Z(\mathbf{w})}{\partial w_r} \right.$$

$$= V_r\left(\mathbf{f}^{(k)}, \mathbf{l}^{(k)}\right) - E_{\mathbf{w}}\left[V_r(\mathbf{f}, \mathbf{l})\right] = 0 \tag{10.34}$$

where the prior energy is in the form of a scalar product between the hyperparameter vector $\mathbf{w}$ and the vector of the clique potential functions V_r $(\mathbf{f}, \mathbf{l})$. The expectations in Eq. 10.34 are computed with respect to the prior distribution $P(\mathbf{f}, \mathbf{l}|\mathbf{w})$, and $Z(\mathbf{w})$ is the related partition function, which depends on $\mathbf{w}$. Hence, a gradient descent technique could be applied to update the parameters vector $\mathbf{w}$ according to the following iterative scheme:

$$w_r^{t+1} = w_r^t - \eta \frac{\partial}{\partial w_r}\left\{-\log P\left(\mathbf{f}^{(k)}, \mathbf{l}^{(k)}|\mathbf{w}\right)\right\}\bigg|\mathbf{w}^t$$

$$= w_r^t + \eta\left\{E_{w^t}\left[V_r(\mathbf{f}, \mathbf{l})\right] - V_r\left(\mathbf{f}^{(k)}, \mathbf{l}^{(k)}\right)\right\}$$

(10.35)

where η is a positive, small parameter, to be suitably chosen to ensure convergence. Starting with $\mathbf{w}^{(0)} = \mathbf{w}^{(k)}$, at the end of the iterative procedure 10.35, one would obtain the new estimate $\mathbf{w}^{(k+1)}$ of the parameters, according to Eq. 10.31c. The physical meaning of this updating rule is to look for those parameters that make the various potentials, computed in the MRF realization $(\mathbf{f}^{(k)}, \mathbf{l}^{(k)})$, equal to their averages when $(\mathbf{f}, \mathbf{l})$ is let free to assume all the possible configurations. In other words, the parameter updating rule aims to determine the probability under which the frequencies of occurrence of the various local configurations in $(\mathbf{f}^{(k)}, \mathbf{l}^{(k)})$ are equal to their statistical expectations.

Nevertheless, the computational charge of the iterative scheme in Eq. 10.35 as it stands is insurmountable, due to the computation of the expectations, or equivalently of the partition function, which requires summation over binary variables and integration over continuous variables. In previous papers,[54,83] the authors have shown that the explicit presence in the MRF model of a line process enables the use of a well-known result of the statistical mechanical theory (i.e., the saddle point approximation), to feasibly approximate the partition function. It was shown that this leads to base the estimation on the conditional prior $P(\mathbf{l}^{(k)}|\mathbf{f}^{(k)},\mathbf{w})$ rather than on the joint prior $P(\mathbf{f}^{(k)}, \mathbf{l}^{(k)}|\mathbf{w})$. The hyperparameter updating rule of Eq. 10.35 then becomes:

$$w_r^{t+1} = w_r^t - \eta \frac{\partial}{\partial w_r}\left\{-\log P\left(\mathbf{f}^{(k)}, \mathbf{l}^{(k)}|\mathbf{w}\right)\right\}\bigg|\mathbf{w}^t$$

$$= w_r^t + \eta\left\{E_{w^t}\left[V_r(f, l)\right] - V_r\left(\mathbf{f}^{(k)}, \mathbf{l}^{(k)}\right)\right\}$$

(10.36)

The meaning for the parameter updating rule is now the search for that prior probability under which the potentials computed in $(\mathbf{f}^{(k)}, \mathbf{l}^{(k)})$ are equal to the corresponding potentials computed in $(\mathbf{f}^{(k)}, \bar{\mathbf{l}})$, where $\bar{\mathbf{l}}$ is the expected value of the line process for the given intensity $\mathbf{f}^{(k)}$. The physical point of view is that the fluctuations of $\mathbf{l}$, with respect to sample $\mathbf{l}^{(k)}$ and conditioned on some relevant value for $\mathbf{f}$, are assumed to give almost the same information about the parameters as the fluctuations of both fields around $(\mathbf{f}^{(k)}, \mathbf{l}^{(k)})$. It is straightforward to observe that, since $\mathbf{f}^{(k)}$ is already available from step 10.31a, the computation of the expectations now requires summation over binary variables only.

If the same assumption to keep $\mathbf{f}$ always clamped to $\mathbf{f}^{(k)}$ is made in the context of a Monte Carlo algorithm to update the hyperparameters, similar performance in terms of computation saving can be obtained when Monte Carlo Markov Chain techniques are applied to approximate, via time averages, the ratio between partition functions.[84]

In both cases, the time averages can be computed by means of a low-cost binary Gibbs sampler. As already noted, a Boltzmann Machine can be used for performing stochastic sampling. Moreover, it is immediate to recognize in the updating rule of Eq. 10.36 the learning algorithm associated with this network.[54]

The implementation of the entire procedure for the unsupervised, blind restoration of images requires the iterative application in cascade of MAP and ML estimators according to the scheme of Eq. 10.31. In deriving the final formulations of these estimators, the authors exploited the presence in the image model of an explicit binary line process and made use of decompositions and mathematically grounded approximations, such as the saddle point approximation, with the aim of reducing the computational complexity of the original problem.

10.7   RESULTS AND DISCUSSION

The performances of the various methods reviewed in this chapter for the restoration of microscope images have been tested both on synthetic and real images, assuming different amounts of degradation. The methods can be grouped into two classes: the class of methods that exploit global smoothness constraints, such as standard regularization and maximum entropy, and the class of edge-preserving methods that exploit stabilizers addressing discontinuities. In particular, consideration is given to stabilizers for the implicit treatment of non-interacting discontinuities, and stabilizers for the explicit treatment of self-interacting discontinuities. For comparison, reconstructions produced by the ML method, which does not employ any *a priori* information, are also considered. Some preliminary results obtained by means of blind restoration techniques are also reported.

10.7.1   RESTORATION OF SYNTHETIC IMAGES

In a first set of experiments, the authors considered as original image the synthetic 32×32 image shown in Figure 10.2a, which is similar to that used by Holmes and Liu.[17] This image was first degraded by means of convolution with a 7×7 mask, which approximates the PSF corresponding to an objective lens of NA = 1.2 in a wide-field microscope, and then adding a white Gaussian noise of zero mean and variance $\sigma^2 = 49$. The degraded image is shown in Figure 10.2b. Figures 10.2d and 10.2e show the reconstructions obtained by maximum entropy and standard regularization, respectively. In both cases, a conjugate gradient algorithm was employed to minimize the cost functions, and the best value of the regularization parameter λ was selected empirically, by trial and error, resulting in $\lambda = 0.16$ for standard regularization, and $\lambda = 0.64$ for maximum entropy. As expected, both standard regularization and maximum entropy are quite efficient in removing the noise, but they leave the blur almost unchanged (with a slight superiority of maximum entropy with respect to standard regularization). On the other hand, ML is very sensitive to the noise present on the data, producing a reconstruction that is more noisy than the original degraded image.

The same degraded image of Figure 10.2b was then restored using edge-preserving stabilizers. More precisely, both implicit and explicit discontinuities were considered. The first case adopted the posterior energy of Eq. 10.26, and chose the truncated parabola of Eq. 10.27 as neighbor interaction function ϕ. Indeed, it could be argued that, due to the sharp discontinuities present in the original image, better results can be obtained using a stabilizer that presents a clean threshold. The cost function that arises in this case is non-convex. To minimize it, both the GNC algorithm (Figure 10.2f) and simulated annealing (Figure 10.2g) with same parameters $\lambda = 0.16$ and $\alpha = 20$ were used. One immediately appreciates the excellent reconstruction produced by simulated annealing, which proves the superiority of stabilizers that account for discontinuities, when compared with stabilizers that enforce global smoothness. When using the GNC for minimizing the same edge-preserving posterior energy, the computational costs are severely reduced, but the reconstruction can be highly unsatisfactory (see Figure 10.2f). The rationale for this behavior lies in the fact that the GNC does not ensure convergence to the global minimum — especially when blur is present — because, in that case, it can be difficult or even impossible to find a first convex approximation. However, in other experiments, a good convex approximation, and consequently good local minima, can be found through GNC when no blur or moderate blur affects the degraded image.

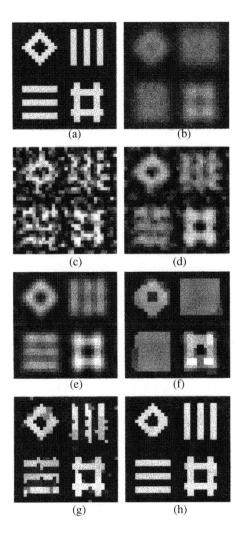

FIGURE 10.2 Restoration of a synthetic image: (a) original 32×32 image; (b) image blurred by a 7×7 convolution mask, plus Gaussian noise with $\sigma^2 = 49$; (c) ML reconstruction; (d) maximum entropy reconstruction ($\lambda = 0.64$); (e) standard regularization reconstruction ($\lambda = 0.16$); (f) reconstruction obtained using the neighbor interaction function of Eq. 10.27, and GNC ($\lambda = 0.16$, $\alpha = 20$); (g) reconstruction obtained using the neighbor interaction function of Eq. 10.27 and simulated annealing ($\lambda = 0.16$, $\alpha = 20$); and (h) reconstruction obtained using stabilizer of Eqs. 10.22 and 10.23 and simulated annealing ($\lambda = 0.16$, $\alpha = 40$, $\beta_1 = \beta_4 = -20$, $\gamma_1 = \gamma_2 = 20$).

A second case considered as stabilizer, the prior energy for explicit line treatment given by the sum of Eq. 10.22 and 10.23. For simplicity sake, only the vertical and horizontal line continuation constraint, as well as the double line penalization constraint were enforced, so that we dropped from the prior potentials which are weighted by parameters β_i, ε_i, and κ. The reconstruction shown in Figure 10.2h was obtained through simulated annealing, with parameters $\lambda = 0.16$, $\alpha = 40$, $\beta_1 = \beta_4 = -20$; $\gamma_1 = \gamma_2 = 20$. The quality of this reconstruction is better than that of Figure 10.2g. This is essentially due to the exploitation of two important features of the edges in an image: their connectivity and thinness. As a counterpart, with this method the discontinuities are considered as extra variables of the problem, so that the complexity of simulated annealing increases. Nevertheless, it must be said that the same result would have been obtained through the mixed-annealing algorithm, with a saving in computational time of about 20%. Also in these cases, the best values for the parameters were chosen empirically, by trial and error.

As a conclusive consideration, it is noted that the synthetic image considered so far, despite its small size, can be considered a good test to highlight the most salient features and mutual differences of the various models and algorithms reviewed in this chapter. Indeed, that image presented very fine structures, which were almost completely destroyed by the superimposed blur and noise. From this point of view, that image was a "difficult" image to restore. The use of a larger image with the same resolution characteristics, would have essentially been a matter of computational cost. For $128\infty128$ images, it was found that an IBM Risc 6000/380 uses about 4 minutes to run the GNC algorithm, about 1 hour to run algorithms based on simulated annealing, and about 10 minutes to obtain the same results of SA with mixed annealing. Thus, unless specialized parallel hardware is available, it can be concluded that the GNC algorithm is the best trade-off between performance and computational cost when a moderate blur affects the image, as often happens in practical contexts.

Thus far, the restoration algorithms employed assumed the PSF of the imaging system to be known *a priori*. Nevertheless, as already said, in many practical cases, the PSF is not known, at least not exactly, or it is difficult to measure. In these situations, it is necessary or more practical to make use of blind image restoration techniques for the joint estimation of the restored image and the PSF coefficients directly from the data. Section 10.6 we developed an iterative procedure that allows for a fully data-driven, blind, and unsupervised image restoration, where both the PSF and all the model hyperparameters are estimated along with the image. The performance of this procedure was analyzed by testing it on the same synthetic image of Figure 10.2b. For computational reasons, the authors restricted themselves to the sub-case of blind restoration, assuming the model hyperparameters to be known. This experiment allows for a qualitative comparison with the results produced by classic image restoration techniques with known blur, and the goodness of the estimated PSF can also be quantitatively measured. As already highlighted, joint image restoration and blur identification is an extremely difficult task, and its success is mainly related to the choice of a proper model for the image. For this reason, the image model adopted was the one given by the prior energy of Eq. 10.21 to 10.23, with the same hyperparameters as employed for the supervised restoration case. After running the procedure, the estimated image obtained was slightly worse than that shown in Figure 10.2h; and the blur estimate differed from the true one (exactly known in this case) for a mean square error (MSE) of about 0.004. From a computational point of view, this procedure amounts to a classic image restoration procedure, augmented, at each iteration, by a step of blur estimation. Since this step is performed by means of a conjugate gradient algorithm whose dimension corresponds to the size of the blur mask, and then usually very low, the computational cost of the entire procedure is slightly higher than that of supervised restoration with known blur.

10.7.2 Restoration of Real Images

In another set of experiments, the authors applied the same restoration methods used for the synthetic images to the real image (Figure 10.3b) of a filamentous autofluorescent photosynthetic alga (Figure 10.3a). This image is a 128×128 portion of a 640×512 image, acquired by a wide-field microscope.[85] As model for the PSF of the microscope, the above-described 7×7 mask was used. Some moderate noise was introduced by the acquisition process. Figure 10.3c shows the result of the ML method, Figure 10.3d shows the result of the maximum entropy method ($\lambda = 1.44$), while Figure 10.3e shows the result of the restoration performed by means of standard regularization ($\lambda = 0.01$). For these three techniques, the same considerations made for the synthetic images remain valid. However, some differences can be observed when edge-preserving restoration methods are used. Indeed, it must be pointed out that the synthetic images restored above were precisely piecewise constant, so that they fit at best the models for local smooth functions expressed by the various stabilizers adopted. On the other hand, this real image, as well as most microscope images,

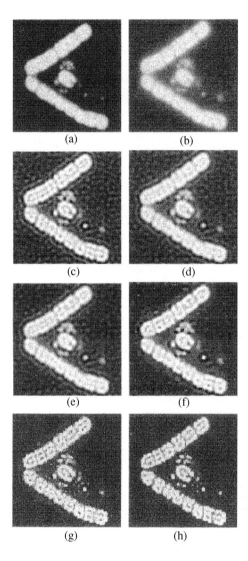

FIGURE 10.3 Restoration of a fluorescence image of filamentous photosynthetic alga: (a) photographic black and white image; (b) video camera image; (c) ML reconstruction; (d) maximum entropy reconstruction ($\lambda = 1.44$); (e) standard regularization reconstruction ($\lambda = 0.01$); (f) reconstruction using the neighbor interaction function of Eq. 10.28, and GNC ($\lambda = 0.64$, $\alpha = 20$); (g) reconstruction using the neighbor interaction function of Eq. 10.28, and simulated annealing ($\lambda = 0.64$, $\alpha = 20$); (h) reconstruction using stabilizer of Eqs. 10.22 and 10.23 and simulated annealing ($\lambda = 0.64$, $\alpha = 40$, $\beta_1 = \beta_4 = -20$, $\beta_2 = \beta_3 = \beta_5 = \beta_6 = -4$, $\gamma_1 = \gamma_2 = 20$).

cannot exactly be modeled as a piecewise constant function because it presents regions where the intensity varies slowly, but is not constant. A better model for a microscope image could be obtained using stabilizers that are functions of the intensity derivatives of order two or more.[40] Another way to ameliorate the model for microscope images is to consider graded discontinuities instead of binary discontinuities. This can be accomplished using a prior energy for implicit line treatment where the neighbor interaction function is given by Eq. 10.28. Indeed, experiments have shown[56] that with this neighbor interaction function, the planar regions of the images are quite well-reconstructed, even if first-order derivatives are considered. For this reason, the neighbor interaction function (Eq. 10.27) was not applied to this image because it presents a too pronounced threshold. Thus, Figures 10.3f and 10.3g show the results of GNC and simulated annealing, applied to the posterior energy of Eq. 10.26 when the neighbor interaction function is that of Eq. 10.28, with parameters $\lambda = 0.64$ and $\alpha = 20$. Again, simulated annealing performs better than GNC because a non-negligible blur operator was introduced in the posterior energy. Finally, Figure 10.3h shows the result of simulated annealing when the prior energy is given by Eqs. 10.21 to 10.23, with parameters $\lambda = 0.64$, $\alpha = 40$, $\beta_1 = \beta_4 = -20$, $\beta_2 = \beta_3 = \beta_5 = \beta_6 = -4$, and $\gamma_1 = \gamma_2 = 20$.

As general remarks, it is observed that all the algorithms performed as deblurring filters. This is highlighted in the results from the good recovering and localization of the parallel-arranged thylakoid membranes in the cell periphery, not visible in the original image.[86] However, as for the synthetic images, the methods that address the discontinuities are advantageous with respect to classic regularization with global smoothness. Nevertheless, from Figure 10.3h, it is possible to observe that only little improvement in the quality of the reconstruction can be obtained when explicit, binary, and constrained lines are considered instead of implicit, unconstrained, but graded lines. In other words, the result is excellent, but not significantly better than those produced by a much simpler stabilizer as that of Eq. 10.28.

The successive, last set of experiments regards the recovering of real images when the blur operator is not known. Two different real images were considered, as shown in Figures 10.4a and 10.5a, respectively. The first is a 96×160 image drawn from the sequence of digitized fluorescence images showing the variation of the emission of the *Euglena gracilis* photoreceptor under a 365-nm excitation beam.[87] In this case, the emission of the photoreceptor is almost not detectable because it is at the beginning of the photocycle. The second image is a 200×200, highly defocused and noisy image of an isolated *Euglena gracilis* photoreceptor, badly acquired by means of a transmission electron microscope. Since the blur masks that affected the two images were unknown, the blind restoration procedure, already tested in the synthetic case, was used. As already highlighted, the success of blind restoration depends on the choice of a good model for the image to be recovered. For these images, it was found that a model that addresses explicit and constrained lines is necessary for obtaining satisfactory reconstructions. Thus, consideration was given to the prior of Eqs. 10.21 to 10.23 where the hyperparameters were chosen heuristically. To reduce the computational costs, the mixed-annealing algorithm was employed instead of simulated annealing.

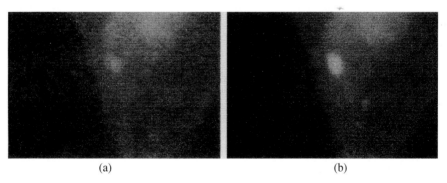

(a)   (b)

FIGURE 10.4 Blind restoration of the emission of a *Euglena gracilis* photoreceptor: (a) 96×160 digitized image of the fluorescent photoreceptor under the 365-nm excitation beam (in this case, the photoreceptor is barely detectable); (b) blind restoration result obtained by setting $\alpha = 100$, $\lambda = 2$, $\beta_1 = \beta_4 = -50$, $\beta_2 = \beta_3 = \beta_5 = \beta_6 = -50$, and $\gamma_1 = \gamma_2 = 50$, and assuming a 7×7 size for the unknown blur mask (this reconstruction shows a photoreceptor with a detectable emission).

Figure 10.4b shows the result of the restoration of the image of Figure 10.4a, obtained by setting $\alpha = 100$, $\lambda = 2$, $\beta_1 = \beta_4 = -50$, $\beta_2 = \beta_3 = \beta_5 = \beta_6 = -50$, and $\gamma_1 = \gamma_2 = 50$, and assuming a 7×7 size for the unknown blur mask. Figure 10.5b shows the result of the restoration of the image of Figure 10.5a, obtained by setting $\alpha = 200$, $\lambda = 1$, $\beta_1 = \beta_4 = -100$, $\beta_2 = \beta_3 = \beta_5 = \beta_6 = -20$, and

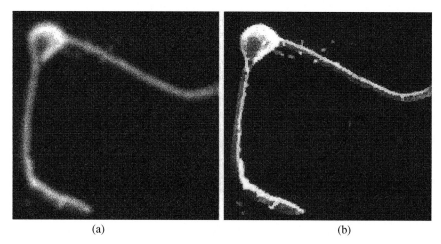

(a)   (b)

FIGURE 10.5 Blind restoration of the anatomical characteristics of an isolated *Euglena gracilis* photore-ceptor: (a) 200×200 highly defocused and noisy image, badly acquired by means of a transmission electron microscope; (b) result of the blind restoration obtained by setting $\alpha = 200$, $\lambda = 1$, $\beta_1 = \beta_4 = -100$, $\beta_2 = \beta_3 = \beta_5 = \beta_6 = -20$, and $\gamma_1 = \gamma_2 = 100$, and assuming a 7×7 size for the unknown blur mask (this reconstruction shows a photoreceptor with very sharply defined anatomical structures).

$\gamma_1 = \gamma_2 = 100$, and assuming again a 7×7 size for the unknown blur mask. In both cases, the blind restoration procedure generates images of good quality, that is, a photoreceptor with a detectable emission (Figure 10.4b), and a photoreceptor with very sharply defined anatomical structures (Figure 10.5b).

10.8 CONCLUSIONS

Procedures for software image restoration can be useful for studying biological events observed by means of microscopy techniques. This chapter has described most of the algorithms for image restoration, with particular emphasis on the edge-preserving procedures, and introduced recent techniques of blind restoration. Some comparative results of the application of these procedures to synthetic and real microscopy 2-D images were also provided.

REFERENCES

1. Preston, T. M., King, C. A., and Hyams, J. S., *The Cytoskeleton and Cell Motility*, Blackie, London, 1990, 1.
2. Gualtieri, P., Microspectrophotometry of photoreceptor pigment in flagellated algae, *Crit. Rev. Plant Sci.*, 9, 475, 1991.
3. Jovin, T. M. and Arndt-Jovin, D. J., Luminescence digital imaging microscopy, *Annu. Rev. Biophys. Biophys. Chem.*, 18, 271, 1989.
4. Gualtieri, P., Molecular biology in living cells by means of digital optical microscopy, *Micron Microscop. Acta*, 23, 239, 1992.
5. Taylor, D. L. and Wang, Y., Eds., *Fluorescence Microscopy of Living Cells in Culture, Part B*, Academic Press, San Diego, 1989.
6. Pluta, M., *Advanced Light Microscopy*, 3, Elsevier, Warszawa, 1993.
7. Agard, D. A., Optical sectioning microscopy: cellular architecture in three dimensions, *Annu., Rev. Biophys. Bioeng.*, 13, 191, 1984.
8. Carrington, W., Fogarty, K. E., Lifschitz, L., and Fay, F. S., 3D imaging on confocal and wide-field microscopes, in *Handbook of Biological Confocal Microscopy*, Pawley, J. B., Ed., Plenum Press, New York, 1989, 151.

9. Conchello, J. and Hansen, E., Enhanced 3-D reconstruction from confocal scanning microscope images. I. Deterministic and maximum likelihood reconstruction, *Appl. Optics*, 29, 3795, 1990.

10. Hebert, T. and Leahy, R., A generalized EM algorithm for 3-D Bayesian reconstruction from Poisson data using Gibbs priors, *IEEE Trans. Med. Imag.*, 8, 194, 1989.

11. Katsaggelos, A. K., *Digital Image Restoration*, Springer-Verlag, Berlin, 1991.

12. Sezan, M. H., *Selected Papers on Digital Image Restoration*, SPIE Milestone Series, MS-47, SPIE Optical Engineering Press, Washington, 1992.

13. Gerace, I., Bedini, L., Tonazzini, A., and Gualtieri, P., Edge-preserving restoration of low-light-level microscope images, *Micron*, 26, 195, 1995.

14. Carrington, W. A., Fogarty, K. E., and Fay, F. S., 3-D Fluorescence imaging of single cells using image restoration, in *Noninvasive Techniques in Cell Biology*, Wiley-Liss, New York, 1990.

15. Agard, D. A., Hiraoka, Y., Shaw, P., and Sedat, J. W., Fluorescence microscopy in three dimensions, *Meth. Cell Biol.*, 30, 353, 1989.

16. Holmes, T. J., Maximum-likelihood restoration adapted for noncoherent optical imaging, *J. Opt. Soc. Am.*, 5, 666, 1988.

17. Holmes, T. J. and Liu, Y.-H., Image restoration for 2-D and 3-D fluorescence microscopy, in *Visualization in Biomedical Microscopies*, Kriete, A., Ed., VCH, New York, 1992, 283.

18. Andrews, H. C. and Hunt, B. R., *Digital Image Restoration*, Prentice-Hall, Englewood Cliffs, 1977.

19. Tikhonov, A. N. and Arsenin, V. Y., *Solutions of Ill-Posed Problems*, John Wiley & Sons, Washington, D.C., 1977.

20. Pratt, W. K., *Digital Image Processing*, Wiley, New York, 1978.

21. Bertero, M., Poggio, T., and Torre, V., Ill-posed problems in early vision, *IEEE Proc.*, 76, 869, 1988.

22. Marroquin, J. L., *Probabilistic Solution of Inverse Problems*, Ph.D. thesis, MIT T. R. 1985, 860.

23. Marroquin, J. L., Mitter, S., and Poggio, T., Probabilistic solution of ill-posed problems in computational vision, *J. Am. Stat. Assoc.*, 82, 76, 1987.

24. Dempster, A. P., Laird, N. M., and Rubin, D. B., Maximum likelihood from incomplete data via the EM algorithm, *J. Roy. Stat. Soc., B*, 39, 1, 1977.

25. Bedini, L., Salerno, E., and Tonazzini, A., Edge-preserving tomographic reconstruction from Gaussian data using a Gibbs prior and a generalized expectation-maximization algorithm, *Int. J. of Imaging Syst. Technol.*, 5, 231, 1994.

26. Hebert, T. and Lu, K., Expectation-maximization algorithms, null spaces, and MAP image restoration, *IEEE Trans. Image Process.*, 4, 1084, 1995.

27. Veklerov, E. and Llacer, J., Stopping rule for the MLE algorithm based on statistical hypothesis testing, *IEEE Trans. Med. Imag.*, 6, 313, 1987.

28. Reinsch, C. H., Smoothing by spline functions, *Numer. Math.*, 10, 177, 1967.

29. Poggio, T., Torre, V., and Koch, C., Computational vision and regularization theory, *Nature*, 317, 314, 1985.

30. Jaynes, E. T., On the rationale of maximum-entropy methods, *IEEE Proc.*, 70, 939, 1982.

31. Burch, S. F., Gull, S. F., and Skilling, J., Image restoration by a powerful maximum entropy method, *Computer Vision, Graphics, and Image Processing*, 23, 113, 1983.

32. Gull, S. and Skilling, J., Maximum entropy method in image processing, *IEEE Proc., Part F.*, 131, 646, 1984.

33. Frieden, B. R., Dice, entropy, and likelihood, *IEEE Proc.*, 73, 1764, 1985.

34. Scales, L. E., *Introduction to Non-linear Optimization*, Macmillan, New York, 1985.

35. Hunt, B. R., The application of constrained least squares estimation to image restoration by digital computer, *IEEE Trans. on Computers*, 22, 805, 1973.

36. Bedini, L. and Tonazzini, A., Neural networks use in maximum entropy image restoration, *Image and Vision Computing*, 8, 108, 1990.

37. Hopfield, J. J., Neurons with graded response have collective computational properties like those of two-state neurons, *Proc. Natl. Acad. Sci. U.S.A.*, 81, 3088, 1984.

38. Hopfield, J. J. and Tank, D. W., Computing with neural circuits: a model, *Science*, 233, 625, 1986.

39. Geman, S. and Geman, D., Stochastic relaxation, Gibbs distributions, and the Bayesian restoration of images, *IEEE Trans. Patt. Anal. Machine Intell.*, 6, 721, 1984.

40. Geman, D. and Reynolds, G., Constrained restoration and the recovery of discontinuities, *IEEE Trans. Patt. Anal. Machine Intell.*, 14, 367, 1992.

41. Blake, A. and Zisserman, A., *Visual Reconstruction*, MIT Press, Cambridge, MA, 1987.
42. Geman, S. and McClure, D. E., Bayesian image analysis: an application to single photon emission tomography, *Proc. Am. Stat. Assoc., Stat. Comp. Sect.*, 12, 1985.
43. Besag, J., Towards Bayesian image analysis, *J. Appl. Statistics*, 16, 395, 1989.
44. Green, P. J., Bayesian reconstructions from emission tomography data using a modified EM algorithm, *IEEE Trans. Med. Imag.*, 9, 84, 1990.
45. Geiger, D. and Girosi, F., Parallel and deterministic algorithms for MRFs: surface reconstruction, *IEEE Trans. Patt. Anal. Machine Intell.*, 13, 401, 1991.
46. Bedini, L., Gerace, I., and Tonazzini, A., A deterministic algorithm for reconstructing images with interacting discontinuities, *CVGIP: Graphical Models and Image Processing*, 56, 109, 1994.
47. Bedini, L., Gerace, I., and Tonazzini, A., Sigmoidal approximations for self-interacting line processes in edge-preserving image restoration, *Pattern Recog. Lett.*, 16, 1011, 1995.
48. Metropolis, N., Rosenbluth, A. W., Rosenbluth, M. N., and Teller, E., Equations of state calculations by fast computing machines, *J. Chem. Phys.*, 21, 1087, 1953.
49. Kirkpatrick, S., Gellatt, C. D., and Vecchi, M. P., Optimisation by simulated annealing, *Science*, 220, 671, 1983.
50. Aarts, E. and Korst, J., *Simulated Annealing and Boltzmann Machines: A Stochastic Approach to Combinatorial Optimization and Neural Computing*, Wiley, Chichester, 1989.
51. Bedini, L. and Tonazzini, A., Image restoration preserving discontinuities: the Bayesian approach and neural networks, *Image and Vision Computing*, 10, 108, 1992.
52. Jeng, F., Woods, J. W., and Rastogi, S., Compound Gauss-Markov Random Fields for parallel image processing, in *Markov Random Fields*, Chellappa, R. and Jain, A., Eds., Boston Academic Press, Boston, 11, 1993, 11.
53. Bedini, L., Benvenuti, L., Salerno, E., and Tonazzini, A., A mixed-annealing algorithm for edge preserving image reconstruction using a limited number of projections, *Signal Processing*, 32, 397, 1993.
54. Bedini, L., Tonazzini, A., and Minutoli, S., A Neural Architecture for Simultaneous MAP Image Restoration and ML Estimation of Edge-Preserving Gibbs Priors, Technical Report B4-48, IEI-CNR, 1996.
55. Blake, A., Comparison of the efficiency of deterministic and stochastic algorithms for visual reconstruction, *IEEE Trans. Patt. Anal. Machine Intell.*, 11, 2, 1989.
56. Bedini, L., Gerace, I., and Tonazzini, A., A GNC algorithm for constrained image reconstruction with continuous-valued line processes, *Pattern Recog. Lett.*, 15, 907, 1994.
57. Besag, J., On the statistical analysis of dirty pictures, *J. Roy. Statist. Soc. B*, 48, 259, 1986.
58. Geman, D. and Yang, C., Nonlinear image recovery with half-quadratic regularization, *IEEE Trans. Image Process.*, 4, 932, 1995.
59. Luenberger, D. G., *Optimization by Vector Space Methods*, John Wiley & Sons, New York, 1969.
60. Bedini, L., Fantini, E., and Tonazzini, A., A dual approach to regularization in image restoration, *Pattern Recog. Lett.*, 12, 687, 1991.
61. Luenberger, D. G., *Linear and Nonlinear Programming*, 2nd ed., Addison-Wesley, Reading, MA, 1984.
62. Kay, J. W., On the choice of regularization parameter in image restoration, in *Springer Lecture Notes in Computer Science*, 1988, 301, 587.
63. Thompson, A. M., Brown, J. C., Kay, J. W., and Titterington, D. M., A study of methods of choosing the smoothing parameter in image restoration by regularization, *IEEE Trans. Patt. Anal. Mach. Intell.*, 13, 326, 1991.
64. Hall, P. and Titterington, D. M., Common structure of techniques for choosing smoothing parameters in regression problems, *J. Royal Stat. Soc. B*, 49, 184, 1987.
65. Golub, G. H., Heath, M., and Wahba, G., Generalized cross-validation as a method for choosing a good ridge parameter, *Technometrics*, 21, 215, 1979.
66. Tekalp, A. M. and Kaufman, H., On statistical identification of a class of linear space-invariant blurs using nonminimum-phase arma models, *IEEE Trans. Acoust. Speech Signal Process.*, 38, 1360, 1988.
67. Lagendijk, R. L., Tekalp, A. M., and Biemond, J., Maximum likelihood image and blur identification: a unifying approach, *Opt. Eng.*, 29, 422, 1990.
68. Lagendijk, R. L., Biemond, J., and Boekee, D. E., Identification and restoration of noisy blurred images using the expectation-maximization algorithm, *IEEE ASSP*, 38, 1180, 1990.

69. Katsaggelos, A. K. and Lay, K. T., Maximum likelihood blur identification and image restoration using the EM algorithm, *IEEE Trans. Signal Process.*, 39, 729, 1991.

70. Reeves, S. I. and Mersereau, R. M., Blur identification by the method of generalized cross-validation, *IEEE Trans. Image Processing*, 1, 301, 1992.

71. Ayers, G. R. and Dainty, J. G., Iterative blind deconvolution method and its applications, *Opt. Lett.*, 13, 547, 1988.

72. Davey, B. L. K., Lane, R. G., and Bates, R. H. T., Blind deconvolution of noisy complex-valued images, *Optics Commun.*, 69, 353, 1989.

73. McCallum, B. C., Blind deconvolution by simulated annealing, *Optics Commun.*, 75, 101, 1990.

74. Yang, Y., Galatsanos, N. P., and Stark, H., Projection-based blind deconvolution, *J. Opt. Soc. Am., A*, 11, 2401, 1994.

75. You, Y. and Kaveh, M., A regularization approach to joint blur identification and image restoration, *IEEE Trans. Image Process.*, 5, 416, 1996.

76. Zhang, J., The mean field theory in EM procedures for blind Markov Random Field image restoration, *IEEE Trans. Image Process.*, 2, 27, 1993.

77. Lakshmanan, S. and Derin, H., Simultaneous parameter estimation and segmentation of Gibbs random fields using simulated annealing, *IEEE Trans. Patt. Anal. Machine Intell.*, 11, 799, 1989.

78. Besag, J., Spatial interaction and the statistical analysis of lattice systems (with discussion), *J. Roy. Stat. Soc., B*, 36, 192, 1974.

79. Cross, G. R. and Jain, A. K., Markov random field texture models, *IEEE Trans. Patt. Anal. Machine Intell.*, 5, 25, 1983.

80. Zhang, J., The mean field theory in EM procedures for Markov Random Fields, *IEEE Trans. Signal Process.*, SP-40, 2570, 1992.

81. Higdon, D. M., Johnson, V. E., Turkington, T. G., Bowsher, J. E., Gilland, D. R., and Jaszczak, R. J., Fully Bayesian Estimation of Gibbs Hyperparameters for Emission Computed Tomography Data, Technical Report 96-21, Institute of Statistics and Decision Sciences, Duke University, 1995.

82. Descombes, X., Morris, R., Zerubia, J., and Berthod, M., Maximum likelihood estimation of Markov Random Field parameters using Markov Chain Monte Carlo algorithms, in *Lecture Notes in Computer Science*, Pelillo, M. and Hancock, E. R., Eds., 133, 1223, 1997.

83. Tonazzini, A. and Bedini, L., Using intensity edges to improve parameter estimation in blind image restoration, *SPIE's Int. Symp. Optical Science, Engineering, and Instrumentation, Bayesian Inference for Inverse Problems (SD99)*, 1998.

84. Tonazzini, A., Bedini, L., and Minutoli, S., Joint MAP image restoration and ML parameter estimation using MRF models with explicit lines, *Proc. IASTED Int. Conf. Signal and Image Processing (SIP'97)*, 1997, 215.

85. Gualtieri, P. and Coltelli, P., An image-processing system (IPS100) applied to microscopy, *Comp. Methods Prog. Biomed.*, 36, 15, 1991.

86. Carr, N. G. and Whitton, B. A., Eds., *The Biology of Cyanobacteria*, Blackwell, London, 1982.

87. Barsanti, L., Passarelli, V., Walne, P. L., and Gualtieri, P. *In vivo* photocycle of the *Euglena gracilis* photoreceptor, *Biophys. J.*, 72, 545, 1997.

11 Image Cytometry

Jean Paul Rigaut

CONTENTS

11.1   INTRODUCTION

The term "cytometry" defines the quantitative image analysis of cells and tissues with the help of a computerized system. Using conventional microscopy, the specimens (smears, imprints, and tissue sections) are observed in two-dimensional (2-D) projection. It is proposed that the name "Confocal Cytometry" be given to the quantitative analysis of three-dimensional (3-D) images of cells and tissues, obtained by confocal microscopy.[1,2]

Confocal microscopy was invented in 1955 by Marvin Minsky,[3,4] who was to become famous in another domain — artificial intelligence. Interestingly, his motivation, far from being unrelated to the latter when building his confocal microscope, arose from a superbly quixotic search for a way to study the 3-D arrangement of nerve cells in the brain. Minsky may therefore be considered the father of confocal cytometry, although many years would pass before his dream of obtaining 3-D microscopical images of the brain would actually become progressively true.

An independent inventor, Mojmir Pétrañ[5,6] from Pilzen, developed a confocal microscope of a similar type, also to observe brain cells.[7] Several experimental laser instruments followed.[8-11] The first commercial laser confocal microscope became available in 1988 (MRC 500, Bio-Rad, U.K.).

After a long technical evolution, confocal microscopy now represents an easy way to obtain 3-D images from thin, independent, optical "sections" deep inside unsectioned cells or tissues. Therefore, confocal imaging allows straightforward use of the full 3-D information present in a tissue block. This offers a revolutionary approach to quantitative image analysis in cell biology. Up until recent years, access to the third dimension, in terms of quantitative image analysis, relied mostly on probabilistic stereological estimations from limited sets of physical sections.[12,13] Stereological methods are not devoid of severe limitations.[14] The only solutions for the 3-D imaging of thick preparations of cells or tissues were either tedious reconstructions from serial microtome-obtained sections, or computationally intensive reconstructions from serial optical "sections," using deconvolution methods to eliminate the optical contributions from above and under each "section."[15]

Most works published up to now on confocal microscopy have either dealt with theoretical and instrumental aspects or have been satisfied by obtaining strikingly sharp images in 2-D confocal planes on cells or tissues. Images obtained in confocal mode allow 3-D imaging, including reconstruction and quantitation. Although some authors understood early[16] that 3-D images could be produced by stacking up consecutive confocal images, relatively few results of quantitative biological applications on such images have been presented.

11.2   CONFOCAL MICROSCOPY

11.2.1   BASIC KNOWLEDGE

11.2.1.1   Principles of Confocal Microscopy

The reader is referred to leading articles in the field for detailed information on the fundamental principles of confocal microscopy.[17-28]

In classical optical microscopy, the depth of field is defined by the thickness of the part of the viewed specimen for which the image appears in focus. However, all the out-of-focus information participates in the imaging process and, therefore, no real resolving power exists in the direction of the optical axis.

In scanning optical microscopy, the specimen is scanned by a light beam focused to a diffraction-limited spot by a high numerical aperture (NA) objective. This scanning can be achieved by many simultaneous light beams, using a Nipkow disk (scanning reflected light microscope), or by a single laser beam (confocal scanning laser microscope). In both systems, the confocal effect (i.e., the elimination of out-of-focus light contributions), is achieved using a pinhole aperture on the imaging plane in front of the detector. With laser instruments, confocal images can be obtained only by reflectance or by fluorescence.

11.2.1.2   Confocal Microscopes

In the first type of confocal scanning optical microscopy, the scanning reflected light microscope, invented by Minsky[3,4] and independently, later, by Pétrañ,[5,6] the reflected or emitted fluorescence light passes back in "tandem," after beam-splitting, through holes, in a Nipkow disk diametrically opposed to those through which the illuminating light had passed.[29] A non-tandem type, in which both illuminating and emitted beams pass through the same Nipkow disk holes, has been described.[30] The first biological confocal images were published in 1967/1968.[6,7]

The confocal scanning laser microscope derives from the marriage between the confocal scanning optical microscopy principle and scanning laser microscopy.[31-33] The first biological images obtained by laser were published in 1972/1973.[33,34] Laser scanning may be brought about either by translating the specimen under a fixed beam (scanned stage) or by moving the beam on a fixed specimen (scanned beam). The latter system is usually achieved by rotating or vibrating mirrors, although other techniques using acousto-optic deflectors or objective lens scanning[35,36] are under experimentation. Scanned beam ("off-axis") instruments, first developed in 1971,[32,37] are used in all present commercial laser confocal instruments. An important advantage is that the scanning module can be easily adapted to any conventional microscope. Scanned stage ("on-axis"), first developed in 1977,[38] has the obvious advantages of minimizing optical aberrations and offering a field of view that is less limited. Scanned stage suffers, however, from a slow scan rate due to the speed limitation of present motorized high-precision scanning stages.[39]

The confocal laser, because of its higher $\{x, y\}$ and z resolution and its ability to use much smaller light intensities, is preferable to tandem systems for the preparation of 3-D images. Confocal imaging can be achieved by reflectance or by fluorescence and only with epi-illumination, with the exception of stage-scanning instruments.[17] Several commercial laser confocal instruments allow simultaneous imaging at two different emission wavelengths,[40-43] a technique that produces less photobleaching than when using two consecutive imagings[44] and eliminates the risk of a misalignment produced by the change of emission filter. The emission spectra always overlap to some extent, but it has been shown that the resulting "cross-talk" can be corrected.[40] Ratio imaging can be used to evaluate spectral changes induced by ion-binding and fluorescence resonance energy transfer.[25]

An additional non-confocal image for the scanned transmitted light can be obtained by implementing a fiber-optic guide collecting the light from below the microscope condenser to a second detector. Many other imaging modes, such as phase-contrast, differential phase contrast, differential amplitude contrast, dark field, polarization, optical-beam-induced contrast, and confocal interference contrast, can also be used.[28]

11.2.1.3   Lasers for Confocal Microscopy

Among the many types of continuous wave lasers,[45] it seems that, until now, only gas lasers (argon ion, krypton ion, helium-cadmium, and helium-neon) have been used. Dye lasers have a broad tuning range. Most solid-state lasers operate in the near-infrared. The tunable solid-state titanium-sapphire argon ion-pumped laser is replacing infrared dye lasers. Solid-state diode-pumped lasers, which can operate in the red at 527 nm (yttrium lithium fluoride) or at 532 nm (neodymium-yttrium aluminum garnet), are an attractive alternative to helium-neon lasers. A pulsed dye laser has been used in the recently developed two-photon and multiphoton techniques.[46]

Low-powered, air-cooled argon ion lasers are by far the most commonly used, as their two main excitation lines, at 488 and 514 nm, are very convenient for many markers and especially for three of the most common ones, fluorescein (with 488 nm) and Rhodamine and Texas Red (with 514 nm). Among their many other, less intense excitation wavelengths (between 458 and 529 nm), the 458-nm line has been used in confocal microcopy for Chromomycin A321 and for Lucifer Yellow.[9,10,47]

Krypton ion lasers have numerous wavelengths (between 338 and 799 nm). The strongest lines are in the red range. Lines at 476,[48] 531,[49] and 647[48,50] nm have been used. Helium-cadmium lasers also have numerous wavelengths (between 325 and 888 nm), but only those at 325 and 442 nm are strong enough to be used in practice, with the exception of additional lines in the green and the red with "white-light" He-Cd lasers. The line at 442 nm has been used.[50] Helium-neon lasers are mostly used for red excitation. Among their numerous wavelengths (between 543 and 1523 nm), the strong 633-nm line has been used in laser microscopy[17] and in tandem confocal microscopy,[51] and a weak line at 543 nm (wavelength close to the classic mercury lamp green line at 546 nm) has been used in confocal microscopy.[52]

Imaging at UV wavelengths (using a helium-cadmium or a krypton ion laser, or the pumped rare-gas "excimer" laser) is to be considered with extreme care, as presently available objectives are not achromatic under UV light[53] and this may produce substantial image deformations (Amos, B., personal communication, 1990); new types of objectives have been proposed (e.g., Zeiss, Germany).

11.2.1.4 Ongoing Developments

New instruments and techniques are being developed. Most of them concentrate on increasing the image acquisition speed of confocal laser systems. Cell scanning rates approaching those achieved by flow cytometry should be possible in the near future.[50] Novel scanning systems are being tried, such as a fast-rotating polygon[54] or high-resolution acousto-optical deflectors.[55-57] The polygon has been used to scan in the x-direction while the specimen was translated in the y-direction,[58] but this is not easy to achieve.[50] Acousto-optical deflectors allow imaging at standard video rates.[57] They have, however, been reported to introduce small image distortions.[59] They usually require that one of the scan axes still use a conventional mirror and the partially de-scanned fluorescence emission be imaged through a slit (instead of a circular pinhole) or by CCD detectors, due to the fact that the fluorescence emission cannot be sent back through the acousto-optic modulator, which is wavelength specific. A new instrument that allows the use of two acousto-optical deflectors and achieves confocality by raster-scanning the aperture of an image dissector tube has recently been presented.[57]

New microscope objectives that include a correction for spherochromatism in the near-UV range are being developed.[50] "Super-resolution"[60] can be achieved by deconvoluting images obtained with an array of detectors.[61] Preliminary results have been presented on particle trapping with a confocal microscope, the trap manipulation being done either by moving the beam or by moving in $\{x, y, z\}$ a special high-NA objective.[62]

A technique based on nonlinear molecular excitation by the simultaneous absorption of two photons allows an intrinsic 3-D resolution and the elimination of out-of-focus bleaching and of UV chromatic aberrations.[46] The simultaneous absorption of two photons from a stream of strongly focused femtosecond pulses of red light, generated by a colliding pulse mode-locked dye laser, stimulates fluorescence from fluorophores having single-photon absorption in the UV.

Finally, confocal Raman microspectroscopy[63] should allow studies of compositional and structural information inside cells without the use of fixatives or stains.

11.2.2 CYTOLOGICAL AND HISTOLOGICAL PREPARATION METHODS

11.2.2.1 Reflectance

Gold immunolabeling,[64,65] silver-intensified[66] or not,[64,65] has been used for reflectance with confocal laser microscopy. It was discovered that very small colloidal gold particles (1 nm) attached to RNA can be visualized[67,68] and quantified.[68] This technique has been achieved in transformed breast MCF7 cells in culture.

11.2.2.2 Fluorescence

Immunofluorescence is by far the most common preparation technique used for confocal laser microscopy.[69,70] The quantitation of markers imaged simultaneously at different wavelengths is possible. Confocal "sectioning" eliminates the problem encountered in conventional microscopy of underestimating particle counts because of projection effects.

It is beyond this chapter's scope to cite all the fluorochromes that might be useful in confocal microscopy. The reader is referred to two major reviews[65,71] and to Table 11.1 for a non-exhaustive list.

Specific dyes (e.g., for DNA, cytoskeletal microfilament proteins, mitochondria) and fluorescinated monoclonal antibodies can be used. In pathology, markers for DNA,[72,73] oncogenes,[25] BrdU,[74] and many antigens, such as Ki67 (or MIB-1, preferable to Ki-67 when paraffin is used)[75] or c-erbB2, PS2, p53, are especially interesting.

Specific DNA probes for *in situ* hybridization[76-78] have been used in fluorescence. A quantitative evaluation of the hybridized sequence can be envisaged.[79] Sequences distant from less than 100 Kb can be visualized separately in interphase nuclei.[80] *In situ* hybridization is increasingly used to locate specific chromosome regions, and there is no doubt that multiple DNA probes will become an important way to search for chromosomal abnormalities in cancers.

Markers of DNA, which are the most routinely used fluorochromes, are still in big demand. The most common ones are pararosanilin,[64] quinacrine,[27] propidium iodide,[74,81] acridine orange,[82,83] ethidium bromide, mithramycin,[64,82,84] and chromomycin A3.[85] The Hoechs No. 33342 and DAPI stains require UV excitation, which presents major difficulties with confocal laser instruments. The same problem limits the use of Indo-1 for Ca^{2+} imaging and quantitation, which is more easily achieved with Rhod-2 or Fluo-3.

The use of particular fluorochromes makes observation *in vivo* possible in some cases, for example, in the study of living embryos,[86] molecule and drug microkinetics[87,88] and intracellular pH, flavoprotein (FAD),[89] and Ca^{2+}.[90] Voltage-sensitive styryl dyes open the way to the imaging of the activity of single cells in a functioning brain or tissue slice preparation.[8] Experiments on living plant cells and tissues require a treatment at pH 10 to permeate the cell membrane,[91] before being able to use most fluorescent dyes.

For dual-wavelength imaging, a convenient method consists of combining fluorescein and Texas Red. A double fluorochrome conjugate with a large Stokes shift, FR-1 (covalently linked molecules of fluorescein and rhodamine), has recently been developed for single-wavelength double-labeling applications.[65] Auto-fluorescence can also be used with suitable biological materials (e.g., chloroplasts).[19,49]

Photobleaching is a major problem in conventional fluorescence imaging[25,92] and this is also true in confocal laser imaging.[25,71,93] When using fixed specimens, anti-bleaching agents can be used.[71,94] Many antioxidants, such as *p*-phenylenediamine (PPD), diazobicyclo-octane (DABCO), propyl-gallate, hydroquinone, crocetin, and etretinate, can be used, but they are not devoid of problems, as the first four can have pharmacological effects, the first one is dangerous to manipulate (carcinogenic), and the last two (carotenoids) are fluorescent.

11.2.2.3 Penetration

Analyzing cells or tissues by confocal microscopy requires good penetration of the fluochrome into the specimen. This requires a permeation treatment during the histological preparation when preparing very thick 3-D images. Saponin, for example, can be added to the fixation solution.[40-43] Good results are now being obtained in paraffin-embedded material.[40-43] Staining can be made on cryostat slices or *en-bloc* after paraformaldehyde fixation. Rigaut and Vassy[95] have shown that there is no limitation of the penetration of chromomycin A3 up to at least z = 300 μm.

TABLE 11.1
Main Fluorochromes

Fluorochrome	Excitation (nm)	Emission Peak (nm)	
Tracers and Receptors			
Cascade Blue hydrazide	376, 389	423	Immunocytological fluorochromes, conjugates, and lectins
AMCA	350	450	
Coumarin 138	365	460	
Bodipy phallicidine	505	512	
Mono-Polysaccharides and Various Proteins			
Nitrobenzoxadiazole (NBD)	468	520	
Fluorescein-ITC (FITC)	490	520	Immunocytological fluorochromes, conjugates, and lectins
Acridine Orange	490	530	
Eosin-5-ITC	524	548	Immunocytological fluorochromes, conjugates, and lectins
Erythrosin-5-ITC	535	558	
Tetramethyl-rhodamine (TMRA)	549	570	Immunocytological fluorochromes, conjugates, and lectins
Tetramethyl-rhodamine (TRITC)	541	572	
Phycoerythrin-B (PE-B) 545	545	576	Immunocytological fluorochromes, conjugates, and lectins
Phycoerythrin-R (PE-R)	495, 545	578	Immunocytological fluorochromes, conjugates, and lectins
Dansyl chloride	340	578	Immunocytological fluorochromes, conjugates, and lectins
Lissamine rhodamine B	567	590	
Rhodamine X-ITC (XRITC)	578	604	Immunocytological fluorochromes, conjugates, and lectins
Texas Red sulfonyl chloride	596	620	
Allophycocyanin	620	660	Immunocytological fluorochromes, conjugates, and lectins
Indopentamethine-cyanines (CY5)	630	670	
Nuclear			
DAPI	372	456	DNA, AT, Q
Hoechst 33258	365	465	DNA, AT, Q
ACMA	430	474	DNA, AT, Q
DiOC1(3)	482	510	DNA, RNA
Acriflavine-Feulgen	455	515	DNA
Acridine Orange	490	530	Single/double
Acridine Orange	640	DNA, D, Y	
Thiazole Orange (TO)	509	533	DNA, RNA
Adriamycine	480	555	DNA, GC
Chromomycin A3	450	570	DNA, GC
Mithramycin	395	570	DNA, GC
Pyronin Y	540	570	RNA
Ethidium bromide	545	610	DNA, RNA
Propidium iodide	530	615	DNA, RNA
7-Aminoactinomycin-D	523	647	DNA, GC

TABLE 11.1 (CONTINUED)
Main Fluorochromes

Fluorochrome	Excitation (nm)	Emission Peak (nm)	
	Labeling of Vital Probes		
Cascade Blue hydrazide	376, 389	423	Covalent
Diphenylhexatriene (DPH)	351	430	Hydrophobe
Phalloidin-fluorescein	490	520	F-actin
Nile Red	450-500	525	Neutral lipids
Thiazole Orange (TO)	509	533	Reticulocytes
Rhodamine 123	505-511	534	Mitochondria
Lucifer Yellow CH	430	535	Neurons
DASMI	429	557	Mitochondria
Pyronin Y	545	580	Mitochondria
Phalloidin-rhodamine	540	580	F-actin
Nile Blue A	515-560	605	Neutral lipids
	pH Probes		
DCDHB	340	500, 580 (basic)	pH
FD	490	515 (acid)	pH
BCECF	500	530, 620 (basic)	pH
SNAFL-2	485, 514	546 (acid)	pH
SNARF-1	518, 548	587 (basic)	pH
	Ca^{2+} Probes		
Indo-1	331	410	Ca indicator
Quin-2	339	492	Ca indicator
Fura-2	335, 362	505	Ca indicator
Fura-3	506	526	Ca indicator
Rhod-2	553	576	Ca indicator

Note: The ranges are due to determinations of wavelengths made in various solvents.[66-67] For a more extensive list of fluorescent probes, see, for example, *The Handbook of Fluorescent Probes and Research Chemicals* (Haugland, 1992) and also Brown (1996).

11.3   THREE-DIMENSIONAL IMAGES

11.3.1   THREE-DIMENSIONAL IMAGING

11.3.1.1   The Imaging Process

The axial resolution in confocal microscopy is limited by the imaging process, that is, the convolution of the original image with the transfer function of the instrument.[17,18,38,96-106] The maximum theoretical z resolution (Rayleigh criterion) is in laser confocal systems between 650 and 750 nm, depending on instrumental and imaging conditions.[16,17,48,84] Experimental values are found between 700 and 750 nm in reflectance imaging[16,17,48] and between 800 and 900 nm in fluorescence imaging[19,64,83,84,104] under usual optimal conditions (planapochromat objective of 1.40 NA, immersion oil of 1.515 refraction index, excitation wavelength of 488 or 514 nm, emission wavelength above 520 nm, small detector pinhole opening). Microscope objectives of high quality and high NA (1.3 or 1.4) must be used to ensure optimal confocal "sectioning" effects.[103,107,108]

The $\{x, y\}$ and z resolutions will decrease if the detection pinhole is opened wider to increase the image intensity. The z resolution, however, is not too sensitive to a modest increase in the pinhole aperture.[64,102] The MRC-600 (Bio-Rad, U.K.) instrument, which uses "quasi-infinity optics" by folding the optical path by means of mirrors, has an adjustable pinhole, allowing the imaging of weakly fluorescent specimens; other instruments are following this trend.

The maximum theoretical $\{x, y\}$ resolution, increased by a factor of approximately 1.4 as compared to conventional microscopy, is between 159 and 185 nm, depending on instrumental and imaging conditions.[17,84] Experimental values are found between 190 and 250 nm.[16,17,48,83,104]

11.3.1.2 Construction of Three-Dimensional Images

A three-dimensional (3-D) image is usually obtained by stacking up $\{x, y\}$ images from consecutive confocal planes.[16] It is also possible to assemble consecutive $\{x, z\}$ images.[16,109] Scanning in the z-direction can be achieved by a motor drive on the microscope's focus knob, or by a piezoelectric translator between the microscope stage and the specimen.[102,107] The x, y calibration can be modified by the microscope objective, but also by the laser scanning step ("electronic zoom").

The problem of determining the optimal laser power is not trivial, as all undesirable effects increase when the intensity is increased to approach the saturation of the relevant fluorochrome.[71]

Rigaut and Vassy have obtained images of up to 300 μm deep at high magnification in normal and cancerous livers and esophagus.[95,110] Other workers have obtained images of up to 200 μm deep at low magnification in the brain.[9] High NA objectives have a limited working distance. Lower NA values can be used when high resolution is not essential.[85,111,112]

11.3.1.3 Undesirable Effects Encountered in Three-Dimensional Imaging

11.3.1.3.1 Photobleaching and attenuation of the fluorescence intensity in depth

Even under ideal conditions, an attenuation of the detected fluorescence intensity with depth is observed in 3-D images. There is no decrease of the intensity of the excitatory beam up to at least 60 μm deep, as demonstrated by the fact that the bleaching half-life does not depend on the depth,[95] whereas it does depend of course on the laser intensity.[95] This is at variance with an *a priori* belief that the opposite may be true.[64,85]

Therefore, the main limiting factors are the attenuation of the intensity with depth of the emission beam as it passes through the tissue, and the attenuation of the intensity by photobleaching ("bleaching"). The former can be modeled by a negative exponential function.[95] Contrary to what seems to be a common belief,[113] this is not true for the latter.[95]

The chemistry of bleaching is complex.[71,92,93] Attempts at deriving analytical equations for bleaching are dependent on limiting assumptions; for example, bleaching from the triplet electronic energy level state only and no return from this state to the ground-level singlet state.[93] Even a sum of three exponential functions, derived from theoretical considerations,[93] does not allow a good fit of experimental curves of bleaching versus time.[95] Rigaut and Vassy[95] have obtained a very good empirical fit with a modified log-logistic function.

The intensity of the excitatory beam at the in-focus voxel is enormously higher than with conventional fluorescence,[85] but the total energy absorbed by the specimen may nevertheless be lower with confocal microscopy, since only a very small fraction of the field is illuminated at any given time. An important reduction of bleaching rates can be obtained by scanning fast enough to reduce the dwelling time to approximately 0.1 μs/pixel.[114] The dwelling time is 2.5 μs/pixel under our conditions, for one scan; the use of averaging for noise reduction (ten scans per image) leads, in practice, to a value of 25 μs/pixel. The use of improved imaging and scanning techniques should help in obtaining high-resolution 3-D images with shorter scanning dwell times and, therefore, less bleaching.

High excitation levels may produce images of reduced contrast.[85] This does not seem to be due to increased bleaching in excitation-saturated regions,[28] but only to saturation effects per se.[85] A spatial heterogeneity in bleaching rate constants has been reported in cytology.[115] This was not observed by Rigaut and Vassy[95] under their experimental conditions.

When acquiring consecutive, stacked-up confocal images, the effects of bleaching are more important in the last planes than in the first ones and it is therefore preferable, when preparing a 3-D image, to stack up confocal images from in-depth to the surface. Indeed, contrary to some reports,[85,116,117] bleaching is not limited to the confocal plane being viewed. Experimental demonstration of this fact[95] is in agreement with theoretical notions.[93]

11.3.1.3.2 Aliasing

The ideal spacing (z) between confocal planes depends on the imaging conditions. According to the Nyquist theorem,[118] the $\{x, y\}$ and z sampling frequencies required for an accurate digital representation should be more than twice the spatial resolution of the optical system, lest the image be disastrously deformed (stripes). This rule, even more crucial when deconvolution is envisaged, leads to $\{x, y\}$ and z scanning intervals of approximately 90 nm, 90 nm, and 350 nm, respectively.[72]

11.3.1.3.3 Refractive index problems

Mismatches in the refractive index can cause elongation or flattening of 3-D histological images, depending on the embedding medium. With an aged Mowiol (a common medium), a spherical particle seems artificially flattened in z,[119] whereas water embedding produces an elongated ellipsoid.[120,121]

11.3.1.3.4 Other undesirable effects

Photodestruction of biological structures (distinct from photobleaching) can occur, saturating with increasing laser power. It appears that the damage can be reduced by intermittent illumination, using higher intensities for shorter times.[71] Rayleigh scattering can produce unwanted signals, due either to excitation light passing through the dichroic mirror and the barrier filter, or to imperfect monochromaticity of the excitation beam (this is often the case when using a simple interference filter to select one line from a multiline laser).[71]

Raman scattering contributes fluorescence-like signals, for example, due to the characteristic Raman band of water (producing an emission peak at 584 nm for an excitation at 488 nm); additional wavelengths closer to the excitation may appear at high concentrations of protein or embedding media.[71] Rayleigh and Raman scattering, being proportional to the laser power, do not saturate, as opposed to specific fluorescence.[71] Therefore, excessive laser power decreases the image contrast.

Unwanted autofluorescence may arise from endogenous fluorochromes. Flavins and flavoproteins, which absorb at 488 nm and emit in the same region as fluorescein,[71] are the main source of such problems. Reduced pyridine nucleotides and lipofuscin absorb UV light and are difficult to saturate.[71] The predominant source of noise in confocal images is photon noise.[9] A random noise component due to the use of barrier filters for dual-channel imaging can also be observed.

11.3.1.4 Visualization of Three-Dimensional Images

A 3-D image can of course be displayed as a series of consecutive $\{x, y\}$ or $\{x, z\}$ 2-D images, but everyone's dream is a real 3-D visualization. Real-time holograms from digital data are not quite feasible yet. Creating images with a pseudo-relief effect is at present the only available solution.

The most popular way consists of preparing stereoscopic pairs ("stereo-pairs"), by photographic means,[122] or by computing orthographic projections from arbitrary viewing directions.[118] Stereo-microphotography has mainly been used with tandem scanning instruments.[29,123] With laser confocal systems, digital methods are particularly easy to use. The most common one consists of assigning the maximum values of the pixel intensities encountered along symmetric oblique "viewing" lines in the 3-D image to the pixels in the corresponding images of the stereo-pair.[84,124,125] The intensity may also be assumed to be inversely proportional to the distance from the viewer. An enhancement of the stereoscopic perception can be obtained by preprocessing each confocal image by a modified Wallis transformation or a gradient operator.[126] One person in ten is unable to perceive depth from stereo-pairs.[127] Stereoprojection, now used in some congresses, seems more easily accepted by most people than printed stereo-pairs. It requires two ordinary slide projectors with additional polarizing jelly on their optical path, a special metallized screen, and polarizing viewing glasses. The image fusion is achieved by the projectors. Stereoscopic animation can also be used.[125]

A ray-tracing algorithm[128] has been used to reconstruct a "cuberille" model.[129] A simulated fluorescence emission process has also been proposed.[130] A 3-D effect can be obtained by a rotation, with an adequate speed, of the full 3-D image on the computer graphics screen (e.g., with the Thru-View software, Bio-Rad, U.K.).

Several types of algorithms can be used to visualize 3-D binary images containing segmented biological objects. The need for standards has been stressed.[131] State-of-the-art methods include contour polygonation and triangulation between contour points,[132] or a voxel representation[133] with surfacing, smoothing,[134] and shading.[134,135] At least four commercial software packages are based on such methods, Visilog-3D (Noesis, France), 3D-GMR (Apollo, U.S.), VoxelView (U.S.) and the Huygens system with the Imaris system (Bitplane, Switzerland). The latter is probably the most complete system, offering friendly procedures for 3-D data manipulation (including deconvolution).[136]

11.3.1.5   Storage of Three-Dimensional Images

The storage requirements of 3-D images are huge. For example, the images routinely prepared at present, which are composed of $512 \times 512 \times 64$ voxels, require 16 MBytes of storage space each. Standard computer hard disks will never have enough space for a complete experiment, which will usually require a large number of images (especially for quantitation, where sampling becomes crucial). The solution is to use an archive medium. The possibilities at present are either a large Winchester disk or removable media.[137] Among the latter, removable and erasable magnetic cartridge tapes (40 to 200 MBytes) are the cheapest. Non-erasable laser optical disks (WORM) have a very large capacity (2 GBytes). Erasable laser optical disks (600 MBytes) are commercially available. Data compression procedures, such as run-length encoding and octree representations,[138] can be used. A vector representation can be used after data reduction.[47]

11.3.2   ANALYSIS OF THREE-DIMENSIONAL IMAGES

11.3.2.1   Introduction

Image analysis includes image enhancement, segmentation of relevant objects, and their measurement. Segmentation is facilitated by image enhancement. The latter can be achieved by non-specific means,[139] but it is always preferable to start by restoring the image, that is, by applying methods based on the knowledge of the distortions due to the imaging process.

The image processing of large 3-D images requires a powerful workstation interfaced with the host computer, which then remains free for image evaluation and acquisition. Rigaut et al.'s image analysis network, for example, is composed of an MRC-600 (Bio-Rad, U.K.) controlled through an Ethernet to an Alpha UNIX (D.E.C., U.S.) and two image analyzers: a Pentium (U.S.), and an IMCO-1000 (Kontron, Germany).

11.3.2.2   Image Restoration

11.3.2.2.1 Correction for the attenuation of intensity in depth

Correcting mathematically for depth- and bleaching-related attenuation of intensity is useful when preparing images for 3-D visualization and before the automated segmentation of 3-D images, where the use of a constant gray-level threshold at all depths will facilitate discrimination operations.[95] Correcting becomes compulsory before any quantitative analysis based on gray levels, such as chromatin texture analysis of nuclei and quantitation of nuclear DNA. A correction method using a proportionally increasing factor has been described,[64] but it yields imperfect results because the attenuation of fluorescence intensity in depth is not linear.[95] Rigaut and Vassy[95] have recently proposed a more satisfactory mathematical correction method, based on a formula derived from the log-logistic equation (Figure 11.1).

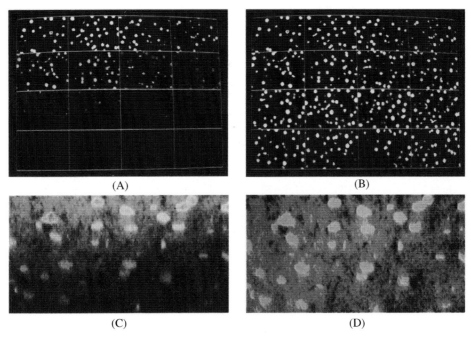

(A)　　　　　　　　　　　　　　　(B)

(C)　　　　　　　　　　　　　　　(D)

FIGURE 11.1 Bio-Rad MRC 600 confocal laser microscope, chromomycin A3 stain,[95] and laser band at 458 nm,[21] as for all other figures. DEC Alpha computer (Digital, U.S.). Normal adult rat liver. (A) Successive $\{x, y\}$ confocal sections through a 3-D image, every $\{z\} = 10$ μm, up to $\{Z\} = 160$ μm thick. The attenuation of the fluorescence intensity in depth is obvious; (B) same, after mathematical correction for photobleaching and attenuation of fluorescence emission in depth;[95] (C) $\{x, z\}$ confocal planes ($\{z\}$ and $\{Z\}$ as above); (D) same, corrected. Note, in $\{z\}$, a degree of flattening of the nuclei, as expected with aged Mowiol.[119]

11.3.2.2.2 Deconvolution

As seen above, the axial resolution in confocal microscopy is limited by the imaging process. As the modulation transfer function of the instrument can be evaluated,[83,105,140] deconvolution can be achieved,[61] producing an important gain in resolution (Figure 11.2).[136] "Super-resolution"[60] can be achieved using an array of detectors, allowing the use of the out-of-focus information in deconvolution schemes.[61]

11.3.2.2.3 Other restoration methods

Enhancement by geostatistical filters has been reported.[141] This allows both a certain degree of signal restoration and an interpolation of voxels onto regular lattices. As the predominant source of noise in confocal images is photon noise,[9] better signal-to-noise ratios are obtained by averaging several image inputs and by increasing the laser power if it is too far from saturation. The "cross-talk" and the noise component observed when using dual-channel imaging may be eliminated.[40]

11.3.2.3 Image Segmentation

Automated image segmentation, although difficult in most cases in biology,[14] is facilitated in confocal laser microscopy by the fact that the intensity of scattered light is decreased by several orders of magnitude, producing highly contrasted images.

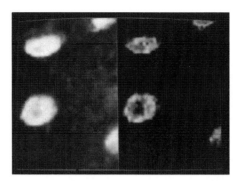

FIGURE 11.2   Deconvolution by inverse system,[61,178] using a DEC Alpha computer (Digital, U.S.). (A) Epithelial breast carcinoma nuclei in a $\{z\}$ confocal section; (B) after deconvolution, the details of the chromatin become visible.

Image segmentation can of course be used separately on each 2-D confocal plane of a 3-D image.[82,83,129] It makes more sense, however, to use direct 3-D algorithms, notably for particle de-agglomeration and for noise reduction.[130] It must be noted that the signal strength is dependent on the orientation of the relevant objects with respect to the confocal plane.[105]

With confocal-obtained images, a major difficulty in 3-D processing arises from the anisotropic character of the voxels, due to the lower resolution in $\{z\}$ than in $\{x, y\}$. Interpolation algorithms must be used to create artificial planes between the actual ones, so as to equalize the $\{x, y\}$ and $\{z\}$ resolutions.

As the cubic digital grid is ill-suited for the estimation of connectivity in space, the lattice system must be transformed into another one. The tetradecahedron (also called cuboctohedron), described by Archimedes and discussed by Kepler,[142] was first used as a lattice kernel in the early 1980s[143-146] for 3-D image analysis, including skeletonization.[143,146]

A renewed interest in 3-D image analysis, due to confocal microscopy, has spurred the development of theoretical derivations for the optimal generalization of mathematical morphology[147,148] to 3-D.[141,146] The tetradecahedron (based on the center of the unit cube and the centers of its edges) and also the rhombododecahedron (based on the center and the vertices of the unit cube) can be used as lattice kernels and structuring elements.[147] Using a transformed lattice, the distance transformation is easily implemented with a sequential algorithm, allowing the use of watershed-derived methods[149] and of a fast skeletonization procedure.[130,149] Skeletonization is impossible in 3-D by the thinning method commonly used in 2-D.[149]

Rigaut et al. have presented the first results of direct 3-D segmentation by gray-tone mathematical morphology on the tetradecahedral lattice (Visilog-3D, Noesis, France). Rat liver cells were segmented in 3-D images obtained with an MRC-60041-42. Some earlier results were obtained with Gauss, Laplace, Sobel, and median filters and with binary dilation and erosion, but with no modification of the original lattice (cubic filters were used, but discarding corner voxels improved the results for median filtering).[130]

11.3.2.4   Measurements

For any quantitation of discrete objects, a guard-frame must be used[150] to avoid the frame-edge bias related to particles of different sizes. Measurements may require some theoretical developments concerning, for example, the estimation of surface areas, curvatures, etc., from binarized representations using, either a voxel or a surface-triangulated model.

The estimation of particle volume by the Cavalieri principle[13] is straightforward (the number of 1-voxels is an unbiased estimator). The estimation of its variance is more complex.[151] The surface area can be estimated by the average number of chords, over the major directions of the lattice, with an appropriate weighting. The shell method for the correction of the frame-edge bias in the estimation of the Euler-Poincar characteristic has been adapted to the tetradecahedral lattice.[152] Rigaut et al.[73] have developed a method for the estimation of nuclear DNA amounts.

Spatial statistics[153-155] and marked point process theory[156,157] can be used to quantitate histological architectural patterns.[158] Graph theory, which has been used to study tissue architecture in 2-D,[159] can also be applied to 3-D data.

11.4 APPLICATIONS

11.4.1 THREE-DIMENSIONAL RECONSTRUCTION

Obtaining thick 3-D images is required for studies by confocal cytometry. Biological objects must be segmented and binarized before being measured. They can be reconstructed in 3-D surfaced and shaded mode and then used as masks for intensity measurements on the original 3-D gray-tone image.[136]

Constructing thick 3-D images in histology has been achieved at low magnification for buccal epithelial cells,[97] neurons,[9,10] and embryos,[27,28,85] and at high magnification for actin,[18,64] tubulin,[85] cytokeratin,[40,160] vimentin,[40,160] chloroplasts,[49] chromosomes,[18,85,139] chromatin,[18,84] and sets of nuclei from normal and pathological tissues.[161-164] Ray-tracing (Figure 11.3) 3-D rendering and reconstruction into a cuberille model[128] has been reported for murine trophoblast giant-cell chromatin.[129]

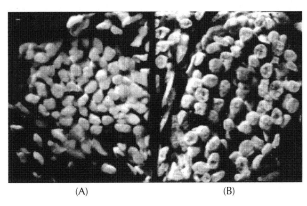

(A)   (B)

FIGURE 11.3   Ray tracing[128] 3-D visualization of a human prostate tissue[163] (100 mm thick), using a Pentium (U.S.) computer for all calculations. Prostate epithelium: (A) dysplasia; (B) carcinoma.

Images of nuclei in 3-D were obtained, for example, in thick 3-D images of nuclei from hepatocytes (Figure 11.4), dysplastic and carcinoma prostate,[163] esophagus cancer *in situ*,[73] chromosomes from *Chironomus thummi*,[81,83] *Crepis capillaris*,[139,165] and the nuclear lamina of a cell derived from the mouse embryocarcinoma line P19.[130]

11.4.2 QUANTITATION

Few quantitative applications have been published up to now in confocal microscopy, whereas it is increasingly difficult to keep track of the number of theoretical papers and of application papers dealing with qualitative aspects.

Rigaut's team has reported quantitative studies on DNA in normal tissues and in tumors of the liver and of the esophagus.[73,166] For the evaluation of DNA histograms, confocal cytometry has the advantage, as compared to other methods, that DNA profiles can be obtained from different regions inside a tissue block. It has been shown, for example, that the hepatocytic tetraploid:diploid DNA ratio is higher in pericentrolobular zones than in periportal zones of the liver.[166] Another advantage

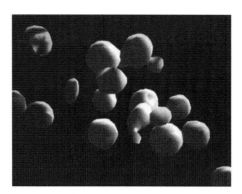

FIGURE 11.4 Normal rat hepatocyte nuclei shown by 3-D rendering, using a DEC Alpha computer (Digital, U.S.) and a Visilog software (Noesis, France).

is the possibility of analyzing different cellular types separately. This eliminates the contamination of DNA profiles by irrelevant cells[73] and allows an estimation of the diploid reference value from stromal cells, or from epithelial cells in a normal zone neighboring a tumor, or from mesenchymal cells in the liver (Figure 11.5).[73]

Rigaut et al. have also made confocal cytometry studies on the normal fetal liver skeleton[40-43] and the modeling of cytoskeleton in transformed MCF7 cells.[167] We have also dealt with DNA[161,162] and chromatin 3-D texture in the prostate.[163,164]

A confocal 3-D cytometric evaluation of architectural patterns observed in normal and pathological tissues is possible using spatial statistics[153-155] and stochastic modeling by unmarked and marked point processes,[156,157,166,168] applied to the 3-D distributions of particle centroids. Rigaut et al.[156,157] have studied the architecture of the liver, using the centroids of hepatocytic and mesenchymal nuclei. The numerical density[150] and the spatial arrangement[158,166,168] of osteocyte lacunae have also been studied. Various quantities have been estimated, including numerical density,[150] K and L functions,[157,158,166,168] pair correlation function, product density function, distribution of nearest neighbor distances, spheric contact distribution, mark covariance function,[157] orientation distribution function, and Schmidt projection.[156] Another team obtained interesting information from confocal 3-D visualization of AgNOR particles.[169]

An analysis of nuclear anaphases from *Crepis capillaris* showed several non-random arrangements of chromosomal material inside the nuclei.[64] Imaging of intact isolated islets of Langerhans has been described.[170]

Chromosome banding offers very interesting possibilities in multi-labeling[171] in confocal microscopy. There is, however, a competition with recent spectral methods. *In situ* hybridization was used to construct mappings of centromeric and telomeric regions in the interphase nucleus,[77,78,172] of human genomic fragments,[80,173] and of cellular RNA.[77] A statistical approach showed that chromosome 1 centromeres were not positioned randomly inside interphase blood cell nuclei.[78]

Transmitter-identified axon terminals were quantitated automatically.[11] The number of microtubules was evaluated in NGF-activated PC12 cells.[174] The levels of nuclear SV40 large-T antigen in cultures of immortalized mouse skin fibroblasts were evaluated in correlation with the increase of the cell density.[175] The phagocytosis of fluorescent microspheres ingested by murine macrophages was studied quantitatively.[176]

Intracellular Ca^{2+} amounts and pH have been evaluated *in vivo* by confocal laser cytometry.[90] An attempt to use voltage-sensitive dyes to quantitate voltage-dependent changes in neuronal tissues was unsuccessful (although the dye could be seen) because of a low signal-to-noise ratio.[8]

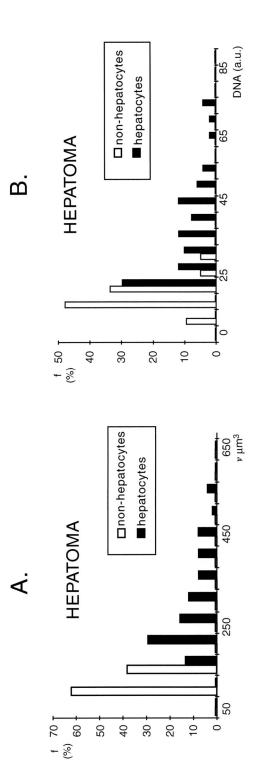

FIGURE 11.5 Histogram of nuclear volume (A) and DNA content (B) from a human cancerous hepatoma. Automatic image analysis by DEC Alpha computer (Digital, U.S.). Computer measurements were made on 200 hepatocytic nuclei and 50 from stromal nuclei, corrected mathematically for fluorescence attenuation in depth.[95] Aneuploidy is obvious.

11.5   THE FUTURE

The microscope itself is under constant evolution, for example, with two- and multi-photon laser microscopy,[46] together with Bio-Rad (U.K.), Zeiss (Germany), and Bitplane[136] software. Ever faster instruments[57,58,136] allow super-resolution.[60,61,136,177] Optical trapping can be integrated in a confocal microscope.[62] Confocal Raman microscopy has allowed the study of living cells in chromosomes.[63] Fast computers and software now allow 3-D deconvolution of large arrays.[136,178] New fluorochromes appear constantly.[65,69,71] *In situ* hybridization[79,80] and multidimensional image[40,106] rendering are used commonly now in confocal microscopy. Reflectance laser microscopy[67,68] can lend itself to quantitation and is a new, promising method. There is an unending future for confocal microscopy and confocal cytometry. Table 11.2 lists a number of Web sites for image analysis.

TABLE 11.2
Some Web Sites for Image Analysis

http://rsb.info.nih.gov/nih-image	NIH: image Web site
http://rsb.info.nih.gov/nih-image/confocal.html	NIH: image processing
http://www.soils.agri.umn.edu/infoserv/lists/nih-image/	NIH: image archives
http://corn.eng.buffalo.edu/www/ConfocalList/readme.html	3-D confocal animations
http://www.cyto.purdue.edu/hmarchive/Cytometry/	3-D confocal animations
http://cotf.edu/ETE/strange.html	NIH: measurements
ftp://www.zippy.nimh.nih.gov/pub/nih-image	NIH: FTP site

REFERENCES

1. Rigaut, J. P., What's new from the field: a new technology is born — confocal image cytometry, *Anal. Cell. Pathol.*, 3, 137, 1991.
2. Rigaut, J. P. and Serra, J., Editorial (Guest Eds.), Image analysis, Special issue, *J. Microscopy*, 156, 1, 1989.
3. Minsky, M., Microscopy Apparatus, U.S. Patent No. 3013467, Dec. 19, 1961 (filed Nov. 7, 1957).
4. Minsky, M., Memoir on inventing the confocal scanning microscope, *Scanning*, 10, 128, 1988.
5. Pétrañ, M. and Hadravsky, M., Czechoslovakian Patents No. 128936 and 128937, 1968 (filed Dec. 5, 1966).
6. Pétrañ, M., Hadravsky, M., Egger, M. D., and Galambos, R., Tandem-scanning reflected light microscope, *J. Opt. Soc. Am.*, 58, 661, 1968.
7. Egger, M. D. and Pétrañ, M., New reflected light microscope for viewing unstained brain and ganglian cells, *Science*, 157, 305, 1987.
8. Fine, A., Amos, W. B., Durbin, R. M., and McNaughton, P. A., Confocal microscopy: applications in neurobiology, *T. I. N. S.*, 11, 346, 1988.
9. Carlsson, K., Wallén, P., and Brodin, L., Three-dimensional imaging of neurons by confocal fluorescence microscopy, *J. Microscopy*, 155, 15, 1989.
10. Wallen, P., Brodin, L., Carlsson, K., Liljeborg, A., Mossberg, K., Ericsson, M., Grillner, S., Hskfelt, T., and Ohta, Y., Confocal laser scanning microscopy utilized for 3-D imaging of fluorescent neurons in the central nervous system, *Eur. J. Cell Biol.*, 48/S25, 43, 1989.
11. Mossberg, K. and Ulfhake, B., Automated quantification of transmitter identified axon terminals from 3D-data recorded with a confocal scanning laser microscope, *Roy. Microsc. Soc. Proc.*, 25, 76, 1990.
12. Weibel, E. R., *Stereological Methods*, Academic Press, London, 1979.
13. Gundersen, H. J. G., Stereology of arbitrary particles. A review of unbiased number and size estimators and the presentation of some new ones, in memory of William R. Thomson, *J. Microscopy*, 143, 3, 1986.
14. Rigaut, J. P., Image analysis in histology — hope, disillusion, and hope again, *Acta Stereol.*, 8, 3, 1989.

15. Agard, D. A. and Sedat, J. W., Three-dimensional architecture of a polytene nucleus, *Nature*, 302, 676, 1983.
16. Wijnaendts van Resandt, R. W., Marsman, H. J. B., Kaplan, R., Davoust, J., Stelzer, E. H. K., and Stricker, R., Optical fluorescence microscopy in three dimensions: microtomoscopy, *J. Microscopy*, 138, 29, 1985.
17. Brakenhoff, G. J., Blom, P., and Barends, P., Confocal scanning light microscopy with high aperture immersion lenses, *J. Microscopy*, 117, 219, 1979.
18. Brakenhoff, G. J., van der Voort, H. T. M., van Spronsen, E. A., and Nanninga, N., Three-dimensional imaging in fluorescence by confocal scanning microscopy, *J. Microscopy,* 153, 151, 1989.
19. Brakenhoff, G. J., van Spronsen, E. A., van der Voort, H. T. M., and Nanninga, N., Three-dimensional confocal fluorescence microscopy, in *Methods in Cell Biology*, Vol. 30, *Fluorescence Microscopy of Living Cells in Culture. Part B. Quantitative Fluorescence Microscopy — Imaging and Spectroscopy*, Taylor, D. L. and Wang, Y. L., Eds., Academic Press, San Diego, 1989, chap. 14, 379.
20. Shotton, D., Electronic light microscopy, in *Techniques in Modern Biomedical Microscopy*, Shotton, D., Ed., Wiley-Liss, London, 1993.
21. Rigaut, J. P., Carvajal-Gonzalez, S., and Vassy, J., Three-dimensional image cytometry, in *Visualization in Biomedical Microscopies — 3-D Imaging and Computer Applications*, Kriete, A., Ed., VCH, Weinheim, 1992, 205.
22. Rigaut, J. P. and Vassy, J., Confocal microscopy and three-dimensional imaging, in *Compendium on the Computerized Cytology and Histology Laboratory*, Wied, G. L., Bartels, P. H., Rosenthal, D. L., and Schenck, U., Eds., Tutorials of Cytology, Chicago, 1994, 194.
23. Wilson, T. and Sheppard, C. J. R., *Theory and Practice of Scanning Optical Microscopy*, Academic Press, New York, 1984.
24. Wilke, V., Optical scanning microscopy: the laser scan microscope, *Scanning*, 7, 88, 1985.
25. Jovin, T. M. and Arndt-Jovin, D. J., Luminescence digital imaging microscopy, *Annu. Rev. Biophys. Biophys. Chem.*, 18, 271, 1989.
26. Pawley, J., Ed., *The Handbook of Biological Confocal Microscopy*, I. M. R. Press, Madison, WI, 1989.
27. Shuman, H., Murray, J. M., and DiLullo, C., Confocal microscopy: an overview, *Biotechniques*, 7, 154, 1989.
28. Shotton, D. M., Confocal scanning optical microscopy and its applications for biological specimens, *J. Cell Sci.*, 94, 175, 1989.
29. Boyde, A., Stereoscopic images in confocal (tandem scanning) microscopy, *Science*, 230, 1270, 1985.
30. Xiao, G. Q. and Kino, G. S., A real-time confocal scanning optical microscope, *Proc. S. P. I. E.*, 809, 107, 1987.
31. Davidovits, P. and Egger, M. D., Scanning laser microscope, *Nature*, 223, 831, 1969.
32. Davidovits, P. and Egger, M. D., Scanning laser microscope for biological investigations, *Appl. Opt.*, 10, 1615, 1971.
33. Slomba, A. F., Wasserman, D. E., Kaufman, G. I., and Nester, J. F., A laser flying spot scanner for use in automated fluorescence antibody instrumentation, *J. Assoc. Adv. Med. Instrum.*, 6, 230, 1972.
34. Davidovits, P. and Egger, M. D., Photomicrography of corneal endothelial cells *in vivo*, *Nature*, 244, 366, 1973.
35. Hamilton, D. K. and Wilson, T., Scanning optical microscopy by objective lens scanning, *J. Phys. E. (Sci. Instr.)*, 19, 52, 1986.
36. Reimer, L., Egelkamp, S., and Verst, M., Lock-in technique for depth-profiling and magneto-optical Kerr effect imaging in scanning optical microscopy, *Scanning*, 9, 17, 1987.
37. Phelan, R. J. and DeMeo, N. L., Rapid scanning microscope for light probing and infrared mapping, *Appl. Opt.*, 10, 858, 1971.
38. Sheppard, C. J. R. and Choudhury, A., Image formation in the scanning microscope, *Optica Acta*, 24, 1051, 1977.
39. Marsman, A. J. B., Stricker, R., Wijnaendts van Resandt, R. W., and Brakenhoff, G. J., Mechanical scan system for biological applications, *Rev. Sci. Instrum.*, 54, 1047, 1983.
40. Vassy, J., Rigaut, J. P., Briane, D., and Kraemer, M., Confocal microscopy immunofluorescence localization of desmin and other intermediate filament proteins in fetal rat liver, *Hepatology*, 17, 293, 1993.

41. Vassy, J., Beil, M., Irinopoulou, T., and Rigaut, J. P., Quantitative image analysis of cytokeratin filament distribution during fetal rat liver development, *Hepatology*, 23, 630, 1996.

42. Vassy, J., Irinopoulou, T., Beil, M., and Rigaut, J. P., Spatial distribution of cytoskeleton intermediate filaments during fetal hepatocyte differentiation, *Microsc. Res. Tech.*, Special Issue: *Histology of the Fetal Liver*, 39, 436, 1997.

43. Vassy, J., Kraemer, M., Briane, D., and Rigaut, J. P., Expression of desmin in fetal rat liver. Immunofluorescence visualization by confocal microscopy, in *Cells of the Hepatic Sinusoid*, Vol. 4, Knook, D. L. and Wisse, E., Eds., The Kupffer Cell Foundation, Leiden, 1993, 592, 158.

44. Mossberg, K. and Ericsson, M., Detection of doubly stained fluorescent specimens using confocal microscopy, *J. Microsc*opy, 158, 215, 1990.

45. Gratton, E. and vandeVen, M. J., Laser sources for confocal microscopy, in *The Handbook of Biological Confocal Microscopy*, Pawley, J., Ed., I. M. R. Press, Madison, WI, 1989, 47.

46. Denk, W., Strickler, J. H., and Webb, W. W., Two-photon laser scanning fluorescence microscopy, *Science*, 248, 73, 1990.

47. Carlsson, K. and Liljeborg, A., A confocal laser microscope scanner for the digital recording of optical serial sections, *J. Microscopy*, 153, 171, 1989.

48. Brakenhoff, G. J., van der Voort, H. T. M., van Spronsen, E. A., and Nanninga, N., 3-Dimensional imaging of biological structures by high resolution confocal scanning laser microscopy, *Scanning Microsc.*, 2, 33, 1988.

49. van Spronsen, E. A., Sarafis, V., Brakenhoff, G. J., van der Voort, H. T. M., and Nanninga, N., Three-dimensional structure of living chloroplasts as visualized by confocal scanning laser microscopy, *Protoplasma*, 148, 8, 1989.

50. Shack, R. V., Bartels, P. H., Buchroeder, R. A., Shoemaker, R. L., Hillman, D. W., and Vukobratovich, D., Design for a fast fluorescence scanning microscope, *Anal. Quant. Cytol. Histol.*, 9, 509, 1987.

51. Bianco, P. and Boyde, A., Alkaline phosphatase cytochemistry in confocal scanning light microscopy for imaging the bone marrow stroma, *Bas. Appl. Histochem.*, 33, 17, 1989.

52. Kapitza, H. G., Advanced fluorescence methods in confocal laser scan microscopy, *Roy. Microsc. Soc. Proc.*, 25, 55, 1990.

53. Keller, H. E., Objective lenses for confocal microscopy, in *The Handbook of Biological Confocal Microscopy*, Pawley, J., Ed., I. M. R. Press, Madison, WI, 1989, 69.

54. Shack, R. V., Bell, B., Hillman, D., Landesman, A., Shoemaker, R. L., Vukobratovich, D., and Bartels, P. H., Ultrafast laser scanner microscope: first performance tests, in *Proc. Int. Workshop Phys. Engin. Med. Imaging*, Nalcioglu, O., Ed., Pacific Grove, CA, 1982, 49.

55. Suzuki, T. and Horikawa, Y., Development of a real-time scanning laser microscope for biological use, *Appl. Opt.*, 25, 4115, 1986.

56. Draaijer, A. and Houpt, P. M., A standard video-rate confocal laser-scanning reflection and fluorescence microscope, *Scanning*, 10, 139, 1988.

57. Goldstein, S., Hubin, T., Rosenthal, S., and Washburn, C., A confocal video-rate laser-beam scanning reflected-light microscope with no moving parts, *J. Microscopy*, 157, 29, 1990.

58. Bartels, P. H., Buchroeder, R. A., Hillman, D. W., Jonas, J., Kessler, D., Shoemaker, R. L., Shack, R. V., Towner, D., and Vukobratovich, D., Ultrafast laser scanner microscope design and construction, *Anal. Quant. Cytol.*, 3, 55, 1981.

59. Horikawa, Y., Yamamoto, M., and Dosaka, S., Laser scanning microscope: differential phase images, *J. Microscopy*, 148, 1, 1987.

60. Bertero, M., Brianzi, P., and Pike, E. R., Super-resolution in confocal scanning microscopy, *Inverse Problems*, 3, 195, 1987.

61. Bertero, M., Boccacci, P., Brakenhoff, G. J., Malfanti, F., and van der Voort, H. T. M., Three-dimensional image restoration and super-resolution in fluorescence confocal microscopy, *J. Microscopy*, 157, 3, 1990.

62. Visscher, K. and Brakenhoff, G. J., Single beam optical trapping integrated in a confocal microscope for biological applications, *Cytometry*, 4, 17, 1990.

63. Puppels, G. J., de Mul, F. F. M., Otto, C., Greve, J., Robert-Nicoud, M., Arndt-Jovin, D. J., and Jovin, T. M., Studying single living cells and chromosomes by confocal Raman microspectroscopy, *Nature*, 347, 301, 1990.

64. Brakenhoff, G. J., van der Voort, H. T. M., Baarslag, M. W., Mans, B., Oud, J. L., Zwart, R., and van Driel, R., Visualization and analysis techniques for three dimensional information acquired by confocal microscopy, *Scanning Microsc.*, 2, 1831, 1988.

65. Haugland, R. P., Molecular Probes: Handbook of Fluorescent Probes and Research Chemicals, *Molecular Probes Inc.*, Eugene, OR, 1992.

66. van den Pol, A. N., Neuronal imaging with colloidal gold, *J. Microscopy*, 155, 27, 1989.

67. Linares-Cruz, G., Rigaut, J. P., Vassy, J., de Oliveira, T. C., de Crémoux, P., Olofsson, B., and Calvo, F., Reflectance *in situ* hybridisation (RISH). Detection, by confocal reflectance laser microscopy, of gold-labelled riboprobes in breast cancer cell lines and histological specimens, *J. Microscopy*, 173, 27, 1994.

68. Linares-Cruz, G., Millot, G., de Crémoux, P., Vassy, J., Olofsson, G., Rigaut, J. P., and Calvo, F., Combined analyses of multiple in situ hybridization and BrdU staining using reflectance and immunofluorescence confocal microscopy, *J. Histochem.*, 27, 15, 1995.

69. Brown, S., Mieux comprendre les fluorochromes pour la microscopie confocale, in *INSERM, Formation Permanente: La Microscopie Confocale,* I.F.R. "Cellules Epithéliales," C. H. U. Xavier Bichat, Paris, 1996, 26.

70. Rigaut, J. P., de Oliveira, T. C., and Vassy, J., Microscopie confocale, in *Cytométrie par Fluorescence. Apports Comparatifs des Techniques de Flux, Image et Confocale*, Métézeau, P., Ratinaud, M. H., and Carayon, P., Eds., Techniques en…, Editions INSERM, Paris, 1995, 97.

71. Tsien, R. Y. and Waggoner, A., Fluorophores for confocal microscopy: photophysics and photochemistry, in *The Handbook of Biological Confocal Microscopy*, Pawley, J., Ed., I. M. R. Press, Madison, WI, 1989, 153.

72. Rigaut, J. P., Vassy, J., Herlin, P., Duigou, F., Masson, E., Mandard, A. M., Calard, P., and Foucrier, J., DNA cytometry by confocal scanning laser microscopy in thick tissue blocks — Methodology and preliminary results in histopathology, *Trans. Roy. Microsc. Soc.*, 1, 385, 1990.

73. Rigaut, J. P., Vassy, J., Herlin, P., Duigou, F., Masson, E., Briane, D., Foucrier, J., Carvajal-Gonzalez, S., Downs, A. M., and Mandard, A. M., Three-dimensional DNA image cytometry by confocal scanning laser microscopy in thick tissue blocks, *Cytometry*, 12, 511, 1991.

74. Arndt-Jovin, D. J., Robert-Nicoud, M., and Jovin, T. M., Probing DNA structure and function with a multi-wavelength fluorescence confocal laser microscope, *J. Microscopy*, 157, 61, 1990.

75. Verheijen, R., Kuijpers, H. J. H., van Driel, R., Beck, J. L. M., van Dierendonck, J. H., Brakenhoff, G. J., and Ramaekers, F. C. S., Ki-67 detects a nuclear matrix-associated proliferation-related antigen. II. Localization in mitotic cells and association with chromosomes, *J. Cell Sci.*, 92, 531, 1989.

76. Harders, J., Lukacs, N., Robert-Nicoud, M., Jovin, T. M., and Riesner, D., Imaging of viroids in nuclei from tomato leaf tissue by *in situ* hybridization and confocal laser scanning microscopy, *EMBO J.*, 8, 3941, 1989.

77. van Dekken, H., van Spronsen, E., Bauman, J., Jonker, R., and Visser, J., Three dimensional analysis of intranuclear DNA and cellular RNA distribution by confocal microscopy after fluorescent *in situ* hybridization, *Cytometry*, 2, 14, 1988.

78. van Dekken, H., van Rotterdam, A., Jonker, R., van der Voort, H. T. M., Brakenhoff, G. J., and Bauman, J. G. J., Confocal microscopy as a tool for the study of the intranuclear topography of chromosomes, *J. Microscopy*, 158, 207, 1990.

79. du Manoir, S. and Brugal, G., Is fluorescent DNA-DNA *in situ* hybridization a quantitative approach?, *Anal. Cell. Pathol.*, 1, 325, 1989.

80. Lawrence, J. B., An *in situ* hybridization approach to interphase gene mapping and chromatin organization, *Cytometry*, 4, 13, 1990.

81. Takamatsu, T. and Fujita, S., Microscopic tomography by laser scanning microscopy and its three-dimensional reconstruction, *J. Microscopy*, 149, 167, 1988.

82. Robert-Nicoud, M., Arndt-Jovin, D. J., Schormann, T., and Jovin, T. M., 3-D imaging of cells and tissues using confocal laser scanning microscopy and digital processing, *Eur. J. Cell Biol.*, 48, 49, 1989.

83. Schormann, T. and Jovin, T. M., Optical sectioning with a fluorescence confocal SLM: procedures for determination of the 2-D digital modulation transfer function and for 3-D reconstruction by tesselation, *J. Microscopy*, 158, 153, 1990.

84. Brakenhoff, G. J., van der Voort, H. T. M., van Spronsen, E. A., Linnemans, W. A. M., and Nanninga, N., Three-dimensional chromatin distribution in neuroblastoma nuclei shown by confocal scanning laser microscopy, *Nature*, 317, 748, 1985.

85. White, J. G., Amos, W. B., and Fordham, M., An evaluation of confocal versus conventional imaging of biological structures by fluorescence light microscopy, *J. Cell Biol.*, 105, 41, 1987.

86. Hyman, A. A. and White, J. G., Determination of cell division axes in the early embryogenesis of *Caenorhabditis elegans*, *J. Cell Biol.*, 105, 2123, 1987.

87. Adler, J. and Cheema, M., Drug microkinetics using confocal (laser scanning) microscopy, *Eur. Microsc. Anal.*, 5, 15, 1990.

88. Entwistle, A., Hoffmann, H. H., Noble, M., and Stroobant, P., Measurement of the diffusion time of small molecules into cell aggregates using confocal microscopy, *Roy. Microsc. Soc. Proc.*, 25, S67, 1990.

89. Weinlich, M. and Acker, H., Flavoprotein-fluorescence imaging for metabolic studies in multicellular spheroids by means of confocal scanning laser microscopy, *J. Microscopy*, 160, RP1, 1990.

90. Hernandez-Cruz, A., Sala, F., and Adams, P. R., Subcellular dynamics of [Ca^{2+}] monitored with laser scanned confocal microscopy in a single voltage-clamped vertebrate neuron, *Biophys. J.*, 55, 216, 1989.

91. van der Valk, H. C. P. M., Blaas, J., van Eck, J. W., and Verhoeven, H. A., Vital DNA-staining of agarose-embedded protoplasts and cell suspensions of *Nicotiana plumbaginifolia*, *Plant Cell Rep.*, 7, 489, 1988.

92. Linden, S. M. and Neckers, D. C., Bleaching studies of Rose Bengal onium salts, *J. Am. Chem. Soc.*, 110, 1257, 1988.

93. Wells, K. S., Sandison, D. R., Strickler, J., and Webb, W. W., Quantitative fluorescence imaging with laser scanning confocal microscopy, in *The Handbook of Biological Confocal Microscopy*, Pawley, J., Ed., I. M. R. Press, Madison, WI, 1989, 23.

94. Johnson, G. D., Davidson, R. S., McNamee, K. C., Russell, G., Goodwin, D., and Holborow, E. J., Fading of immuno-fluorescence during microscopy: a study of the phenomenon and its remedy, *J. Immunol. Methods*, 55, 231, 1982.

95. Rigaut, J. P. and Vassy, J., High resolution three-dimensional images from confocal scanning laser microscopy. Quantitative study and mathematical correction of the effects due to bleaching and fluorescence attenuation in depth, *Anal. Quant. Cytol. Histol.*, 13, 223, 1991.

96. Sheppard, C. J. R. and Wilson, T., Depth of field in the scanning microscope, *Optics Lett.*, 3, 115, 1978.

97. Sheppard, C. J. R., Confocal microscopy of thick structures, *Eur. J. Cell Biol.*, 48/S25, 33, 1989.

98. Sheppard, C. J. R., Axial resolution of confocal fluorescence microscopy, *J. Microscopy*, 154, 237, 1989.

99. Sheppard, C. J. R. and Cogswell, C. J., Three-dimensional image formation in confocal microscopy, *J. Microscopy*, 159, 179, 1990.

100. Wilson, T., Three-dimensional imaging in confocal systems, *J. Microscopy*, 153, 161, 1989.

101. Wilson, T., Optical sectioning in confocal fluorescent microscopes, *J. Microscopy*, 154, 143, 1989.

102. Wilson, T. and Carlini, A. R., Three-dimensional imaging in confocal imaging systems with finite sized detectors, *J. Microscopy*, 149, 51, 1988.

103. Wilson, T. and Carlini, A. R., The effect of aberrations on the axial response of confocal systems, *J. Microscopy*, 154, 243, 1989.

104. van der Voort, H. T. M., Brakenhoff, G. J., and Janssen, G. C. A. M., Determination of the 3-dimensional optical properties of a confocal scanning laser microscope, *Optik*, 78, 48, 1988.

105. van der Voort, H. T. M. and Brakenhoff, G. J., 3-D image formation in high-aperture fluorescence confocal microscopy: a numerical analysis, *J. Microscopy*, 158, 43, 1990.

106. Cheng, P. C., Lin, T. H., Wu, W. H., and Wu, J. L., *Multidimensional Microscopy*, Springer Verlag, Berlin, 1994.

107. Wilson, T. and Hamilton, D. K., Dynamic focusing in the confocal scanning microscope, *J. Microscopy*, 128, 139, 1982.

108. Cogswell, C. J., Sheppard, C. J. R., Moss, M. C., and Howard, C. V., A method for evaluating microscope objectives to optimize performance of confocal systems, *J. Microscopy*, 158, 177, 1990.

109. Stelzer, E. H. K. and Wijnaendts van Resandt, R. W., Applications of fluorescence microscopy in three dimensions: microtomoscopy, *Proc. S.P.I.E.*, 602, 63, 1986.178.

110. Rigaut, J. P., Imagerie Confocale Multidimensionnelle En Pathologie Cellulaire, Coll. organisé par l'Académie des Sciences, Instrumentation Physique en Biologie et en Médecine, Paris, 1994, Technologie et Documentation, 1995, 109.

111. White, J. G. and Amos, W. B., Confocal microscopy comes of age, *Nature*, 328, 1, 1987.

112. Carlsson, K. and Aslund, N., Confocal imaging for 3-D digital microscopy, *Appl. Opt.*, 26, 3232, 1987.

113. Koppel, D. E., Carlson, C., and Smilowitz, H., Analysis of heterogeneous fluorescence photobleaching by video kinetics imaging: the methods of cumulants, *J. Microscopy*, 155, 199, 1989.

114. Webb, J. P., McColgin, W. C., Peterson, O. G., Stockman, D. L., and Eberly, J. H., Intersystem crossing rate and triplet state lifetime for a lasing dye, *J. Chem. Phys.*, 53, 4227, 1970.

115. Benson, D. M., Bryan, J., Plant, A. L., Gotto, A. M., Jr., and Smith, L. C., Digital imaging fluorescence microscopy: spatial heterogeneity of photobleaching rate constants in individual cells, *J. Cell Biol.*, 100, 1309, 1985.

116. Amos, W. B., Results obtained with a sensitive confocal scanning system designed for epifluorescence, *Cell Motil. Cytoskel.*, 10, 54, 1988.

117. van Oostveldt, P. and Bauwens, S., Quantitative fluorescence in confocal microscopy. The effect of the detector pinhole aperture on the re-absorption and inner filter phenomena, *J. Microscopy*, 158, 121, 1990.

118. Castleman, K., *Digital Image Processing*, Prentice Hall, Englewood Cliffs, NJ, 1979, 368.

119. Downs, A. M., Vassy, J., and Rigaut, J. P., Refractive index mismatches can cause elongation or flattening of 3D histological images, depending on the embedding medium, *7th. Eur. Congr. Stereol.*, Amsterdam, 1998.

120. Carlsson, K., The influence of specimen refractive index, detector signal integration, and non-uniform scan speed on the imaging properties in confocal microscopy, *J. Microsopy*, 163, 167, 1991.

121. Visser, T. D., Oud, J. L., and Brakenhoff, G. J., Refracting index and axial distance measurements in 3D microscopy, *Optik*, 90, 17, 1992.

122. Böhme, N., Neue Methoden der Mikro-Stereoskopie und ihre mikrophotographische Auswertung, *Photo-Techn. Wirtsch.*, 6, 218, 1953.

123. Boyde, A., Direct recording of stereoscopic pairs obtained directly from disk-scanning confocal light microscopes, in *The Handbook of Biological Confocal Microscopy*, Pawley, J., Ed., I.M.R. Press, Madison, WI, 1989, 147.

124. Cox, I. J. and Sheppard, C. J. R., Digital image processing of confocal images, *Image Vision Comput.*, 1, 52, 1983.

125. Carlsson, K., Danielsson, P. E., Lenz, R., Liljeborg, A., Majlöf, L., and Aslund, N., Three-dimensional microscopy using a confocal laser scanning microscope, *Optics Lett.*, 10, 53, 1985.

126. Schormann, T., Robert-Nicoud, M., and Jovin, T. M., Improved stereovisualization method for confocal laser scanning microscopy, *Eur. J. Cell Biol.*, 48/S25, 53, 1989.

127. Richards, W., Stereopsis and stereoblindness, *Exp. Brain Res.*, 10, 380, 1970.

128. Herman, G. T., Reynolds, R. A., and Udupa, J. K., Computer techniques for the representation of three-dimensional data on a two-dimensional display, *Proc. S.P.I.E.*, 367, 3, 1982.

129. Montag, M., Spring, H., Trendelenburg, M. F., and Kriete, A., Methodical aspects of 3-D reconstruction of chromatin architecture in mouse trophoblast giant nuclei, *J. Microscopy*, 158, 225, 1990.

130. van der Voort, H. T. M., Brakenhoff, G. J., and Baarslag, M. W., Three-dimensional visualization methods for confocal microscopy, *J. Microscopy*, 153, 123, 1989.

131. Huijmans, D. P., Integration of software computer-aided three-dimensional reconstruction from parallel serial sections; a plea for standards, *Eur. J. Cell Biol.*, 48/S25, 57, 1989.

132. Keppel, E., Approximating complex surfaces by triangulation of contour lines, *I. B. M. J. Res.*, 19, 2, 1975.

133. Goldwasser, S. M. and Reynolds, R. A., Real-time display and manipulation of 3-D medical objects: the voxel processor architecture, *Comput. Vision Graph. Image Proc.*, 39, 1, 1987.

134. Gouraud, H., Continuous shading of curved surfaces, *IEEE Comput.*, C-20, 623, 1971.

135. Phong, B. T., Illumination for computer generated pictures, *Commun. A.C.M.*, 18, 311, 1975.

136. Bitplane AG, *Scientific Solutions*, Switzerland, 1999.

137. Carrington, W. A., Fogarty, K. E., Lifschitz, L., and Fay, F. S., Three-dimensional imaging on confocal and wide-field microscopes, in *The Handbook of Biological Confocal Microscopy*, Pawley, J., Ed., I.M.R. Press, Madison, WI, 1989, 137.

138. Smith, J., Jongenelen, J., Lamers, W. H., Los, J. A., and Strackee, J., Octree representation for 3D image analysis, *Eur. J. Cell Biol.*, 48/S25, 97, 1989.

139. Houtsmuller, A. B., Oud, J. L., van der Voort, H. T. M., Baarslag, M. W., Krol, J. J., Mosterd, B., Mans, A., Brakenhoff, G. J., and Nanninga, N., Image processing techniques for 3-D chromosome analysis, *J. Microscopy*, 158, 235, 1990.

140. Shaw, P. J. and Rawlins, D. J., Measurements of the point spread function and its use in deconvolution of confocal microscope images, *Roy. Microsc. Soc. Proc.*, 25, S31, 1990.

141. Conan, V., Howard, V., Jeulin, D., Renard, D., and Cummins, P., Improvement of 3D confocal microscope images by geostatistical filters, *Roy. Microsc. Soc. Proc.*, 25, S48, 1990.

142. Kepler, J., *Omnia Opera*, Vol. 5, Harmonices Mundi, 1619.

143. Preston, K., Jr., The crossing number of a three-dimensional dodecamino, *J. Combin. Info. Sys. Sci.*, 5, 281, 1980.

144. Preston, K., Jr. and Duff, M. J. B., *Modern Cellular Automata — Theory and Applications*, Plenum Press, New York, 1984, 49.

145. Sternberg, S., Cellular computers and biomedical image processing, in *Lecture Notes in Medical Informatics*, Vol. 17, *Biomedical Images and Computers*, Sklansky, J. and Bisconte, J. C., Eds., Springer-Verlag, Berlin, 1980, 274.

146. Hafford, K. J. and Preston, K., Jr., Three-dimensional skeletonization of elongated solids, *Comput. Vision Graph. Image Proc.*, 27, 78, 1984.

147. Serra, J., *Image Analysis and Mathematical Morphology*, Academic Press, London, 1982.

148. Serra, J., Ed., *Image Analysis and Mathematical Morphology*, Vol. 2, *Theoretical Advances*, Academic Press, London, 1988.

149. Meyer, F., 3D mathematical morphology, *Roy. Microsc. Soc. Proc.*, 25, S47, 1990.

150. Howard, V., Reid, S., Baddeley, A., and Boyde, A., Unbiased estimation of particle density in the tandem scanning reflected light microscope, *J. Microscopy*, 138, 203, 1985.

151. Mattfeldt, T., Volume estimation of biological objects by systematic sections, *J. Math. Biol.*, 25, 685, 1987.

152. Bhanu Prasad, P., Jernot, J. P., Lantuéjoul, C., and Chermant, J. L., Use of the shell correction method for quantitation of three dimensional digitized images, *Roy. Microsc. Soc. Proc.*, 25, S49, 1990.

153. Ripley, B. D., *Spatial Statistics*, John Wiley & Sons, New York, 1981.

154. Diggle, P. J., *Statistical Analysis of Spatial Point Patterns*, Academic Press, London, 1983.

155. Fisher, N. I., Lewis, T., and Embleton, B. J. J., *Statistical Analysis of Spherical Data*, Cambridge University Press, Cambridge, 1987.

156. König, D., Blackett, N., Clem, C. J., Downs, A. M., and Rigaut, J. P., Orientation distribution for particle aggregates in 3-D space based on point processes and laser scanning confocal microscopy, *Acta Stereol.*, 8, 213, 1989.

157. König, D., Carvajal-Gonzalez, S., Downs, A. M., Vassy, J., and Rigaut, J. P., Modelling and analysis of 3-D arrangements of particles by point processes with examples of application to biological data obtained by confocal scanning light microscopy, *J. Microscopy*, 161, 405, 1991.

158. Baddeley, A. J., Howard, C. V., Boyde, A., and Reid, S., Three-dimensional analysis of the spatial distribution of particles using the tandem-scanning reflected light microscope, *Acta Stereol.*, 6/S2, 87, 1987.

159. Kayser, K., Stute, H., Bubenzer, J., and Paul, J., Combined morphometrical and syntactic structure analysis as tools for histomorphological insight into human lung carcinoma growth, *Anal. Cell. Pathol.*, 2, 167, 1990.

160. Vassy, J., Rigaut, J. P., Hill, A. M., and Foucrier, J., Analysis by confocal scanning laser microscopy imaging of the spatial distribution of intermediate filaments in fetal and adult rat liver cells, *J. Microscopy*, 157, 91, 1990.

161. Irinopoulou, T., Vassy, J., and Rigaut, J. P., Application of confocal scanning laser microscopy for three-dimensional DNA image cytometry of prostatic lesions, *Anal. Quant. Cytol. Histol.*, 20, 351, 1998.

162. Irinopoulou, T., Vassy, J., Beil, M., Nicolopoulou-Stamati, P., and Rigaut, J. P., Three-dimensional DNA image cytometry by confocal scanning laser microscopy in thick tissue blocks of prostatic lesions, *Cytometry*, 27, 99, 1997.

163. Beil, M., Irinopoulou, T., Vassy, J., and Rigaut, J. P., Application of confocal scanning laser microscopy for an automated nuclear grading of prostate lesions in three dimensions, *J. Microscopy*, 183, 231, 1996.

164. Beil, M., Irinopoulou, T., Vassy, J., and Rigaut, J. P., Chromatin texture analysis in three-dimensional images from confocal scanning laser microscopy, *Anal. Quant. Cytol. Histol.*, 17, 323, 1995.

165. Oud, J. L., Mans, A., Brakenhoff, G. J., van der Voort, H. T. M., van Spronsen, E. A., and Nanninga, N., Three-dimensional arrangement of *Crepis capillaris* in mitotic prophase and anaphase as studied by confocal scanning laser microscopy, *J. Cell Sci.*, 92, 329, 1989.

166. Carvajal-Gonzalez, S., König, D., Downs, A. M., Nguyen, Q., Vassy, J., and Rigaut, J. P., Analysis of histological architecture by point process modelling and spatial statistics applied to three-dimensional images from laser scanning confocal microscopy, *Acta Stereol.*, 8, 407, 1989.

167. Portet, S., Vassy, J., Beil, M., Millot, G., Hebbache, A., Rigaut, J. P., and Schoëvaërt, D., Quantitative analysis of cytokeratin network topology in the MCF7 cell line, *Cytometry*, 35, 1999, in press.

168. Carvajal-Gonzalez, S., Rigaut, J. P., Vassy, J., and König, D., Three-dimensional architecture analysis in cellular pathology by confocal microscopy and point process modelling, *Trans. Roy. Microsc. Soc.*, 1, 301, 1990.

169. Ploton, D., Gilbert, N., Ménager, M., Kaplan, H., and Adnet, J. J., Three-dimensional colocalization of nucleolar argyphilic components and DNA in cell nuclei by confocal microscopy, *J. Histochem. Cytochem.*, 42, 137, 1994.

170. Brelje, T. C., Scharp, D. W., and Sørenson, R. L., Three-dimensional imaging of intact isolated islets of Langerhans with confocal microscopy, *Diabetes*, 38, 808, 1989.

171. Leeman, T., Walt, H., Emmerich, P., and Anliker, M., Computer-aided 3-D localization of chromosomes or parts thereof within interphase nuclei of human cells, *Roy. Microsc. Soc. Proc.*, 25, S76, 1990.

172. Bartholdi, M. F., Higher order structure in the cell nucleus, *Cytometry*, S4, 43, 1990.

173. Lichter, P., Boyle, A., Ferguson, M., Ballard, G., and Ward, D. C., High resolution mapping of DNA probes and detection of chromosome aberrations by non-isotopic *in situ* hybridization, *Cytometry*, S4, 13, 1990.

174. Stevens, J. K., Trogadis, J., and Leatio, C., The application of 3D volume investigation methods to serial confocal and serial EM data: distribution of microtubules in PC12 cells, *Roy. Microsc. Soc. Proc.*, 25, S70, 1990.

175. Entwistle, A. and Noble, M., The semi-quantitative measurement of nuclear SV40 large-T antigen employing confocal microscopy, *Roy. Microsc. Soc. Proc.*, 25, S75, 1990.

176. Hook, G. R. and Odeyale, C. O., Confocal scanning fluorescence microscopy: a new method for phagocytosis research, *J. Leukocyte Biol.*, 45, 277, 1989.

177. Cox, I. J., Sheppard, C. J. R., and Wilson, T., Super-resolution by confocal fluorescent microscopy, *Optik*, 60, 391, 1982.

178. Rigaut, J. P. and Vassy, J., Mathematical models for the enhancement of three-dimensional images from laser scanning confocal fluorescent microscopy, *Anal. Quant. Cytol. Histol.*, 12, 210, 1990.

12 Digital Light Microscopy Techniques for the Study of Living Cytoplasm

Dieter G. Weiss, Vladimir P. Tychinsky, Walter Steffen, and Axel Budde

CONTENTS

12.1 INTRODUCTION

The cytoplasm of eukaryotic cells is highly structured. It consists of the fluid phase (cytosol) and the cytoskeleton made up from several types of proteinaceous filaments. Embedded in this so-called "cytomatrix," which is highly viscous in animal cells, are various kinds of membrane-bounded organelles ranging from 50-nm vesicles to the mitochondria, which are up to several micrometers long. Eukaryotic cells require ATP to move actively and redistribute the organelles. The cytoskeletal filaments serve as tracks, as part of the force-generating mechanism, and they organize the distribution of organelles to their places of destination.

All organelles are structurally well-characterized in the fixed and dehydrated state by electron microscopy. The study of most of their details in the living cell was, however, impossible for almost all but the largest organelles due to the limited resolution of light microscopy. When using visible light, this limit is around 200 nm, thus making the observation of most organelles such as secretory vesicles, synaptic vesicles, peroxisomes, small lysosomes, and essentially all cytoskeletal filaments, such as actin filaments (6 nm diameter), intermediate filaments (10 nm), and microtubules (25 nm), impossible. Considerable progress in the study of cytoplasmic structures and their dynamics had, therefore, to await new techniques capable of visualizing these structures, identifying their chemical nature, and quantitatively analyzing their dynamic behavior.

Until the beginning of the 1980s, the power of light microscopy was limited by the properties of the human eye. In the meantime, the application of suitable electronic devices in light microscopy made it not only possible to obtain images at much lower light levels as required for human vision, but also to digitize and process such microscope images in "real-time" (i.e., at video rates and at high spatial resolution). This led to dramatic improvements in resolution, brightness, and contrast of microscopic images. This "electronic revolution in light microscopy"[1-3] provided new digital techniques suitable to study living cells in many instances where previously only dehydrated material could be studied electron-microscopically.

The new quality of light microscopy emerges, if one observes the specimen — instead of with the human eye — with a video camera connected to state-of-the-art analog and digital video processing equipment, usually working in real-time. Video microscopy is thus much more than just adding a camera and monitor to the microscope to share the images with a larger audience. New electronic devices other than video cameras, such as high-sensitivity charge-coupled device (CCD) cameras and scanning light detector systems have been added to microscopes. The three fields

- *Video-enhanced contrast microscopy* for highest resolution work
- *Video-intensified microscopy* for low light applications
- *Electronic scanning microscopy* for confocal microscopy and 3-D imaging

differ in the type of device generating the electronic image, but all three use basically the same type of analog and digital image processing routines. While all three techniques are generally defined as *electronic light microscopy*, the first two techniques are called *video microscopy*.

Four digital light microscopy techniques especially well-suited for the study of living cytoplasm are discussed in this chapter. Most of these techniques are based on video microscopy, which depends on analog and digital image processing, but they all require, in addition, extensive digital image processing. The term "digital light microscopy" is used here to emphasize the power of digital procedures to visualize specific parameters of the intracellular dynamics, especially movements and dynamic changes of molecules.

Quantitative data for different morphological and biochemical parameters can be recorded and calculated from digitized images in real-time (i.e., at video rate) and encoded in the form of gray-shaded or pseudo-color images. These can be recorded continuously, yielding video films of the intracellular events. Digital microscopy also allows multiparametric studies that yield a wealth of information that was inaccessible in the past. A synopsis of some of the present video microscopic techniques is given in Table 12.1.

Considerable progress has also been made in increasing the resolution in the vertical or z-axis (optical sectioning). While classical differential interference contrast (DIC) microscopy allows resolutions down to 300 nm, this can be increased by video enhancement (VEC-DIC) to 150 nm or better.[4,5] Also, the digitized images can be subjected to advanced 3-D reconstruction algorithms

TABLE 12.1
Parameters That Can Be Measured by Digital Light Microscopy in Living Cells or Tissue Slices

Morphology-based parameters:

1. Visualization of cells and cellular processes in thick tissue slices (IR VEC-DIC, anaxial illumination)[10,37,38]
2. Number, position, size, and shape parameters of free cells (e.g., video microinterferometry)[151]
3. Number, position, total area, size, and shape of organelles (organelle-specific vital dyes):[10,44] nucleus,[44,152] mitochondria,[83] endoplasmic reticulum,[16,19,81] Golgi apparatus,[82] secretory vesicles,[88,92] lysosomes,[126] endosomes[44]
4. Motion analysis of organelles and microtubules:[49,50,52] velocity, fluctuation of velocity, straightness of path, rhythmicity, directionality, pauses, contribution of Brownian motion
5. Cytoplasmic viscosity (Brownian motion) (requires microinjection)[84,126]
6. Elastic properties of organelles, microtubules, membranes, and cells[129-131]
7. Local motile activity in cells (root-mean-square average of changes of pixel brightness values)[153]
8. Determination of dry matter in free cells (video microinterferometry):[151] total dry matter, dry matter distribution, pattern of flow of dry matter
9. Measurement of forces in the microscopic and molecular domain ("magnetodrome," microscopy in controlled magnetic fields[154]), laser tweezers[29,131]

Biochemical parameters:

1. Thickness profile of cells (optical pathlength)[31,138,139]
2. Distribution and amount of antigens (requires microinjection or fixation): fluorescence-labeled antibodies,[43] gold-labeled antibodies (also for receptor labeling and receptor or antigen transport studies)[26,89,90]
3. Concentration of substances (monochromatic illumination):[77] concentration of free intracellular Ca^{2+}, Mg^{2+}, Zn^{2+}, K^+, Na^+, Cl^- [31-33,142]; intracellular H^+-concentration (pH-value);[31,33] enzyme reactions[146] endogenous fluorescent or absorbing compounds[75]
4. Membrane potential (plasma membrane and endomembranes)[155,156]
5. Metabolites can be measured by coupled bioluminescence assays; "metabolic imaging" (requires frozen sections or microinjection):[93,94] lactate, glucose, ATP
6. Intracellular metabolites and messengers can be photoactivated[157]
7. Intracellular diffusion[144]
8. Transfer of fluorescence-labeled compounds between neighboring cells[143,145]
9. Dynamic behavior and cooperativity of enzyme clusters (dynamic phase microscopy)[64,135]
10. Measurement of the activation of gene expression associated with a given promoter (by promoter-controlled expression of the luciferase gene or the GFP moiety)[87,147-149]
11. Molecular distances (FRET)[150]

Note: The references given are to be understood as examples only.

that process full grey-tone images.[5] In the case of fluorescence microscopy, the depth resolution is relatively poor (a few micrometers). This can be improved to a resolution of about 0.7 μm by confocal light microscopy.[6,7] Although similar image processing routines are used in confocal microscopy, this technique will not be covered in detail (*see* Chapter 11).

12.2   TECHNIQUES AND EQUIPMENT REQUIRED

12.2.1   VIDEO-ENHANCED CONTRAST (VEC) MICROSCOPY

A dramatic improvement of image quality is obtained with *video-enhanced contrast (VEC) microscopy*. By increasing contrast and magnification, it is possible to extend the limits of both true resolution and visualization and to visualize and analyze in the living state positions and movements of biological objects smaller than 1/10 of the limit of resolution of conventional light microscopy.

All membrane-bounded organelles and part of the cytoskeleton can be imaged and their motion and assembly can be studied.[8]

VEC microscopy,[4,9-11] and especially the type introduced by R.D. Allen (AVEC microscopy),[11-14] should be applied when finest details are to be visualized with bright-field, differential interference contrast (DIC), or polarization optics. The required procedures are described by Allen et al. and Weiss et al.[10,11,14,15] A schematic representation of the equipment needed is shown in Figure 12.1. As a result, resolution of objects of about 150 nm and visualization of even smaller biological objects such as endoplasmic reticulum[16-21] (Figures 12.2 and 12.3b), vesicles[22,23] (Figures 12.3a and 12.3c) actin bundles[18,19,20,24] (Figure 12.3a) and microtubules[3,8,13,22] (Figures 12.3e to 12.3h), down to 20 nm are achieved with visible light in situations where the limit of resolution of conventional microscopy was about 250 nm.[4,10,25] Individual colloidal gold particles, which are used as tags for proteins, can directly be visualized in the 5- to 10-nm range[26] — in theory, even down to 1 to 3 nm.[25,27]

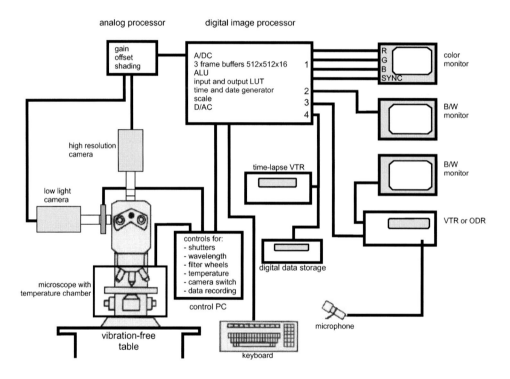

FIGURE 12.1 Digital light microscopy setup showing the two-stage image processor and the periphery required. Color monitor, raw image monitor, and time-lapse recorder are not essential, but are very useful. VTR, video tape recorder; ODR, optical disk recorder or other high-capacity digital storage medium; 1, pseudo-color processed image; 2, raw image for focusing; 3,4, processed image B/W; A/DC, analog-to-digital converter; ALU, arithmetic logic unit; LUT, look up table; D/AC, digital to analog converter.

In VEC microscopy, the resolution is increased by a factor of almost 2 and the visualization of small objects by a factor of 10, provided that optimal optics are used. This is due to the fact that optics with the highest working numerical aperture can be used at full aperture settings, that the resulting excessive image brightness due to straylight can be suppressed electronically by an offset voltage, and that for electronic imaging devices, Rayleigh's criterion of the limit of resolution is replaced by the more advantageous Sparrow criterion.[4,10] According to Rayleigh's criterion, two

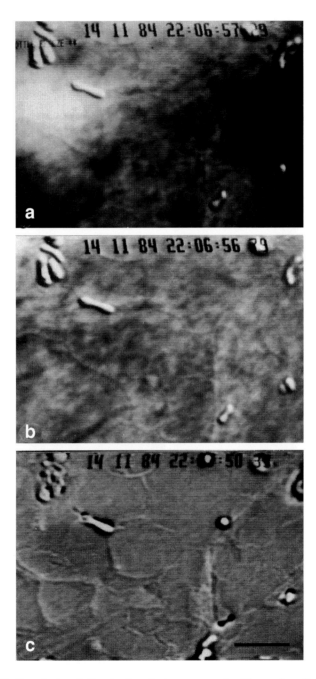

FIGURE 12.2   Visualization of subresolution structures by image processing. The specimen is a plant epithelial cell. (a) Analog contrast enhancement of the microscopic image brings about more image detail but also enhances unevenness of illumination and sometimes a disturbing mottle pattern. (b) A "cleaned" image is obtained after subtraction of the background pattern. Storing and subtracting an out-of-focus frame and then applying digital contrast enhancement brings about subresolution endoplasmic reticulum and cell wall structures. (c) Sometimes, static objects such as cell wall fibers become so prominent after contrast enhancement that small details of interest in the cytoplasm in the same focal plane cannot be seen. However, if the image in (a) is stored in the frame memory while still in focus, and then subtracted as "fixed pattern noise" from the consecutive incoming video frames, a picture containing only the moving cytoplasmic elements is derived (differential image). The thin, tubular endoplasmic reticulum can be observed in this way to form a motile polygonal network that would have remained invisible otherwise.[18] Bar 5 μm.

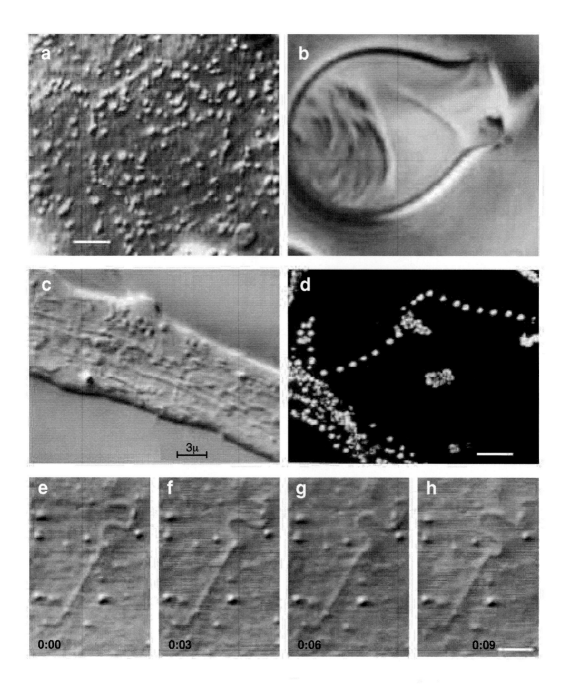

FIGURE 12.3 Examples of applications of AVEC-DIC microscopy: (a) A rotating hexagon consisting of a microfilament bundle forms in a drop of extruded cytoplasm of *Nitella*. Bar 5 μm. (b) Isolated nematocyte of *Hydra* which can be stimulated to explosively eject its tubular contents. (Courtesy of T. Holstein and W. Maile.) (c) Neurite of a N18 neuroblastoma cell in culture. Most of the organelles move either anterogradely or retrogradely with velocities of 0.5 to 2 μm/s. The smallest objects are synaptic or similarly sized membrane precursor vesicles (From Weiss, D.G., in *Recent Advances in Neurocytochemical Methods*, Calas, A., Ed., Springer-Verlag, Berlin, 1991. With permission.) (d) Organelle motion made visible by adding differential images as in Figure 12.2c to a frame memory in 1-second intervals ("trace" function of the Hamamatsu Photonic Microscope system). Organelles are spherosomes moving along actin filaments in a plant cell. Actively transported organelles form traces, while particles in Brownian motion (center) show a clearly different pattern. Bar 5 μm. (a–d) Objective lens, 100x planapochromat, NA = 1.3; additional optical magnification, 2 to 4x; microscope, Polyvar Met, Reichert (Vienna, Austria); processor, Hamamatsu Photonic Microscopy System with Chalnicon camera. (e–h) Native microtubule (diameter 25 nm) from extruded squid giant axon undergoing fishtailing motion.[22,57] Zeiss Axiomat, 100x planapochromat, NA = 1.30, oil; additional optical magnification, 4x; 50-W mercury lamp; processor, Hamamatsu Photonic Microscopy system with Chalnicon camera; sequence, 9 seconds; bar, 1.5 μm.

objects appear as resolved (i.e., as separated for the average human observer), when the depression or "trough" in the summed intensity distribution measured across the objects' Airy disks is at least 15% of the the objects' intensity. Using the Sparrow criterion, the objects may be closer together so that the intensity distribution between them may almost approach a horizontal line, because almost infinitely small intensity changes can be enhanced to visible contrast by electronic means.[4,10] However, subresolution objects are not imaged themselves but as their much larger diffraction patterns, so that very small objects appear inflated by diffraction up to the size of the resolution limit. Given a separation of more than 200 nm from neighboring structures, such objects can be clearly seen and their position and movements can be determined to nanometer accuracy.[3,22,28-30]

12.2.1.1 Equipment

When working at extremely high magnifications as is possible with VEC microscopy, the microscope requires modifications to allow for use of the technique to its full extent.[4,10] A stabilized microscope stand, a very bright illumination system and an oil immersion condenser (both NA = 1.3 or better) are recommended.[4,10] Additional magnification changers for optical magnification of 2X or 4X in addition to a 100X oil immersion objective have to be installed to reach the necessary magnification on the TV monitor of up to 10,000X (at a screen width of 25 cm). Image transfer to the camera should preferably be by direct projection onto the camera target and involve as few lens elements as possible. For polarized light techniques, a de Sénarmont compensator setting is recommended (see Step 4 below).[10,12,13]

For VEC microscopy, a geometrically distortion-free, high-resolution video camera (Chalnicon, Newvicon, or Pasecon) or a high-precision CCD (charge-coupled device) camera is required. The camera must have externally adjustable offset (black level) and gain to allow for analog contrast enhancement.[4,8,10,12,13] Automatic gain control (AGC), if present, must be disabled.

For most applications, it is very important that the image processing equipment contains a unit for *analog* contrast enhancement of the analog video signal (Figure 12.1). Applying analog contrast enhancement by manually setting gain and offset yields a high-contrast image even for very low-contrast or low-light objects. Only such enhanced images should then be digitized (at a resolution of at least 512×512 picture elements (pixels) with 256 gray levels) and subjected to arithmetic operations in the frame memory of a **digital** real-time processor (Figure 12.1). For VEC microscopy, one needs digital frame averaging to remove pixel noise, and continuous digital background subtraction to remove unwanted fixed pattern noise (mottle)[10,14] (Figure 12.2b) or, if required, non-moving image details[18] (Figure 12.2c). Digital contrast enhancement (stretching of the gray-level histogram) is another essential function usually required after image subtraction. For dual-wavelength fluorescence ratio imaging techniques (e.g., Ca^{2+} imaging with Fura-2[31-33]), the image processor must have the additional capability of fast division of full video frames. Digital microscopy as discussed here involves the processing of full gray-level images at video frequency (40 ms), so that one needs **real-time image processors**.[10,14,34] More technical details on the equipment required are published elsewhere.[4,10,34]

Video-microscopic images created by real-time image processing can be subjected to classical image analysis in order to obtain information on morphometric parameters. This is done after thresholding, that is, with binary images (usually not in real-time), and it will require conventional **image analysis systems**.[35,36]

12.2.1.2 Sample Preparation

Basically, the same samples and the same contrast techniques as for conventional light microscopy can be used in VEC microscopy. The specimen's region to be studied should be close to the cover-glass surface, where the best image is obtained. If the highest magnifications are intended, Köhler illumination may only be achieved at the surface and a few tens of micrometers below (upright microscope), since high magnification objectives are usually designed for optical imaging of objects at a distance of 170 μm from the front element. Note that spherical aberrations will be introduced because the oil-immersion objective will be focusing partially through water rather than glass. Alternatively, for imaging deep within aqueous specimens, a water-immersion objective can be employed to overcome the problem. If thicker specimens such as tissue slices, vibratome sections, or nerve bundles are to be observed, only DIC or anaxial illumination[37] techniques are recommended. The opacity of living tissue can be greatly reduced when infrared (IR) or near-infrared (near-IR) light is used.[38]

12.2.1.3 Image Generation and Improvement

Allen et al.[12,13] and Inoué[39] simultaneously described procedures of video contrast enhancement for polarized-light techniques. Allen called his techniques "Allen video-enhanced contrast" differential interference contrast and polarization (AVEC-DIC and AVEC-POL, respectively) microscopy. The AVEC techniques involve the introduction of additional bias retardation by setting the polarizer and analyzer relatively far away from extinction. Allen recommended a bias retardation of 1/9 of a wavelength (20° away from extinction) as the best compromise between high signal and minimal diffraction anomaly of the Airy pattern.[12,40] This setting has the best signal-to-noise ratio and resolution, but its use was prevented because the image was much too bright for observation by eye due to the enormous amount of stray light introduced at such settings. The strong light can, however, be removed by an appropriately large setting of the analog and/or digital offset. The steps required for image generation and improvement for the highest resolution and for visualization of subresolution objects by VEC microscopy include procedures different from those used in conventional microscopy. The best results are obtained with the procedure described below for AVEC-DIC. However, if DIC is not required, such as for bright-field, dark-field, anaxial illumination, Hoffman contrast, or fluorescence microscopy, Step 3 should simply be omitted. A more detailed description, including some theory and more technical hints, is given elsewhere.[10]

- **Step 1**   Focus the specimen. If the entire specimen consists of invisible, subresolution-size material (density gradient fractions, microtubule suspensions, unstained EM sections), it will be difficult to find the specimen plane. It may help to apply a fingerprint to one corner of the specimen side of the cover glass and use the oil droplets for focusing.
- **Step 2**   Adjust Köhler illumination. It is important to make sure that the camera receives the proper amount of light to work near saturation. Since one will apply extreme contrast enhancement later, one must start with as even an illumination setting as possible. Proper centering of the lamp and setting of the collector lens are therefore essential.
- **Step 3**   Open the condensor diaphragm fully in order to utilize the highest possible numerical aperture to obtain highest resolution. Any iris diaphragm of the objective should be fully opened. Be careful to protect the camera from high light intensity prior to this step. This setting will result in a small depth of focus, especially with DIC (optical sections of 0.3 μm or less with 100x oil objectives).
- **Step 4**   *(Only for polarized light techniques)* Set the polarizer (AVEC-POL) or the main prism or compensator (AVEC-DIC) to 1/9 of a wavelength (20° off extinction). The optical image (i.e., seen in the oculars) will disappear due to excessive stray light. The illumination might have to be reduced, to protect the camera, using neutral-density gray filters (but not by closing condenser or objective diaphragms). At this point, one must ensure that the camera still receives the proper amount of light to work near its saturation. Some manufacturers have red and green LEDs built in to indicate the illumination situation. One should see a moderately modulated image on the monitor; a **very** flat or no image indicates insufficient light.

- **Step 5**   Analog enhancement. Increase the gain on the camera to obtain good contrast. Then apply offset (pedestal). Always stop before parts of the image become too dark or too bright. Repeat this procedure several times, if necessary. Make sure that the monitor for watching the changes is not set to extreme contrast or brightness. Analog enhancement improves the contrast of the specimen but unfortunately also emphasizes dust particles, uneven illumination, and optical imperfections. These artifacts, called "mottle," are superimposed on the image of the specimen and may in some cases totally obscure it. Disturbing contributions from fixed pattern noise (mottle) or excessive amounts in unevenness of illumination can be tolerated if digital enhancement is performed later (Step 7).
- **Step 6**   Move the specimen laterally out of the field of view or (when using DIC) defocus it (preferably toward the cover glass) until it just disappears. The result is an image containing only the imperfections of one's microscope system as fixed pattern noise (background mottle pattern).
- **Step 7**   Subtract background. Store (freeze) the mottle image in a digital frame memory, preferably averaged over 8 to 64 frames, and subtract it from all incoming video frames. One should see an absolutely even and clean image, which may, however, be weak in contrast. If motile organelles need to be visualized free of immotile in-focus background objects (fixed pattern noise), an in-focus background image must be subtracted (Figure 12.2).
- **Step 8**   Perform digital enhancement. This is done by alternating between stretching a selected range of gray levels and shifting the image obtained up and down the scale of gray levels until the optimal result is found. Displaying the gray-level histogram will be helpful in selecting the upper and lower limits of the image information, which must be defined as bright white and saturated black, respectively. If the image is noisy (pixel noise), go to Step 9 or 10.
- **Step 9**   Use an averaging function in a rolling (recursive filtering) or jumping mode over two or four frames. This will allow the observation of movements in the specimen, but very fast motions and noise due to pixel fluctuations will be averaged out. Averaging over longer periods of time will filter out all undesired motion (e.g., distracting Brownian motion of small particles in suspension).
- **Step 10**   If needed, apply additional digital procedures for spatial filtering to reduce noise, enhance edges of objects, or reduce shading.[35,41,42] (see also Chapter 5).

12.2.1.4   Interpretation

Unlike in EM images, which truly resolve the submicroscopic objects depicted, the sizes of subresolution objects seen by VEC microscopy may not necessarily reflect their real size. Objects smaller than the limit of resolution (180 to 250 nm, depending on the optics and the wavelength of light used) are inflated by diffraction to the size of the resolution limit (e.g., Figures 12.3e to 12.3h). Whereas the size of the image does not enable a decision on whether one or several objects of a size smaller than the limit of resolution are present, the contrast sometimes permits such a judgment. A pair of microtubules would, for example, have the same thickness as a single one, but the contrast would be about twice as high. If large numbers of subresolution objects are separated by distances of less than 200 nm from each other (e.g., vesicles in a synapse), they will remain invisible, but they will be clearly depicted if separated by more than the resolution limit. Also remember that, if in-focus subtraction (Figures 12.2c and 12.3d) or averaging are used, the immobile or the moving parts of the specimen, respectively, may have been completely removed from the image.

12.2.2 VIDEO-INTENSIFIED MICROSCOPY (VIM)

Low-light video-microscopic techniques aimed at the visualization and quantitation of weak mono-chromatic images obtained by fluorescence microscopy are called *video-intensified microscopy (VIM)*. Fluorescence microscopy images are two-dimensional arrays of fluorescence measurements that contain information on the amounts and distribution of intracellular metabolites, dyes, antigens, or exogenously added fluorescent probes. Typical parameters for VIM measurements include Ca^{2+}-concentration, pH value, metabolites, and membrane potential as determined by the use of specifically developed fluorescence indicator dyes or the distribution of fluorescence-labeled antibodies. Specific, nontoxic fluorescent dyes are available to verify the biochemical identity of the cellular structures seen by conventional or video-microscopic techniques.[43-45]

12.2.2.1 Equipment

When video-rate observations are required for the detection of dynamic changes in real-time in illumination situations at the limit of human vision or below, a low light level camera such as a one- or two-stage silicon-intensifier target (SIT or ISIT) camera or a microchannel plate-intensified device is required.[4,10,34,46] However, due to signal noise, high sensitivity and high spatial resolution tend to be mutually exclusive features of these cameras. In contrast, cooled slow-scan CCD cameras provide high sensitivity while maintaining high spatial resolution. To ensure this, images must be recorded at a slower rate. Photon-counting cameras are required when extremely weak signals (i.e., luminescence or autofluorescence) need to be imaged at illumination conditions up to 6 orders of magnitude below the threshold of human vision or of photographic film[6,10] (see Chapter 8). Microscope setups similar to those shown in Figure 12.1 for VEC microscopy are also suitable here. The technical specifications of low light level cameras and further information on the procedures and equipment needed have been described in more detail elsewhere.[4,10,46]

12.2.2.2 Image Generation

Most high-resolution, cooled CCD cameras will record images with a dynamic range of 12 bits (4096 gray levels) instead of 8 bits (256 gray levels) used by most regular digital light microscopy systems. The high dynamic range of these cameras makes them especially suitable for recording fluorescence images. One should, however, be aware that most image processing software packages — such as PhotoShop (Adobe Systems, Inc., San Jose, CA) or Paint Shop Pro (Jasc Software, Inc., Minnetonka, MN) — can only handle 8-bit gray level and 24-bit color images. From advances in computer technology more sophisticated software can be expected in the near future that will be able to handle 12-bit gray and 36-bit color images. Keeping in mind the limitation of the current image processing software, particular care should be taken when choosing the proper exposure time. It is recommended to take several images at different exposure times.

Using cooled CCD cameras, multicolor fluorescence images can be generated from specimens stained with multiple fluorescent dyes. In this case, the correct exposure of the individual images is particularly important. First, single images for each fluorescent signal are generated. It should be noted here that the original images should always be saved uncompressed and in a file format that is universally accessible by the various types of image processing software (e.g., .tif). The individual images are then saved as 8-bit B/W images in .tif file format. Color images are then composed in, for example, PhotoShop (Adobe Systems, Inc.) or similar programs using the 24-bit RGB option of the software. Before generating a color image, the intensity range and the level of all images to be merged should be adjusted so that the whole range of intensities available for that color is used (24-bit for all three colors). The color image can then be generated by opening a new image file of appropriate size (matching that of the images to be merged) in the RGB mode and copying the single B/W images in either one of the red, green, and blue image layers.

12.2.3 Motion Analysis

Since the advent of digital light microscopy allowed the visualization of all cell organelles and some cytoskeletal elements in their vivid dynamics, the desire to analyze the cytoplasmic motion arose.[47,48] Methods to quantitatively describe intracellular motility by a large number of parameters such as velocity, straightness of path, length of excursions, reversals of direction, pauses, and others are now in use.

12.2.3.1 Equipment

Software-based systems working in conjunction with modern frame-grabber boards are capable of multiple object tracking. Usually, these devices detect objects whose brightness is above an adjustable threshold, and which are located in an adjustable region of interest. More advanced systems include object parameters such as size, shape, or color to select the objects to be analyzed. Superimposed crosshairs indicating the centers of the objects and their vertical and horizontal diameters follow automatically if the objects move. The positional coordinates are continuously collected at video frequency (1/25 or 1/30 s, depending on the video standard used). The MaxVideo system (Datacube, Inc., Peabody, MA), combined with the Area Parameter Accelerator board (APA512-MX, Vision Systems, Adelaide, Australia), provides such a system.[49,50] This combination was one of the first ones capable of describing each object by a set of up to nine basic descriptive parameters (e.g., number of pixels, perimeter, minimum and maximum x- and y-coordinates, etc.). Connected regions of either black or white pixels are recognized as objects. A set of parameters for all detected objects is compiled in a list for each video frame at video frequency. From a sequence of such particle parameter lists, the multiple trajectories can be derived by nearest neighbor analysis.[49]

More recently, PC-based systems designed as part of video-microscopic workstations have become available; for example, MetaMorph (Universal Imaging Corp., West Chester, PA) or MicroTrack (HaSoTec GmbH, Rostock, Germany). Many procedures for motion analysis are based on classical time series analysis and are therefore similar for cell organelles, single cells, microorganisms, and laboratory test animals. Additional software packages designed for applications with whole organisms and for environmental or pharmacological testing, as well as systems for the analysis of sperm motility, can therefore often be used with good results[51] (see also other chapters in this volume). Problems may, however, arise when fluorescent organelles detected by VIM are to be tracked, because of their often noisy appearance and sometimes ill-defined margins. A special system for the analysis of green-fluorescent protein-(GFP)-labeled organelles was recently described; it uses fuzzy-logic algorithms for object detection.[52]

12.2.3.2 Tracks and Trajectories

To learn more about the type of motion, the time series of the individual organelle coordinates is to be analyzed. For both recovering information on the tracks and on the specific type of motion, it is necessary to first obtain the trajectory data.

The features of the MicroTrack program serve as a typical example (Figures 12.4 and 12.5). First, one defines the conditions for recognition of the desired particles: brightness of the particles (the light intensity of the objects can be above or below the gray level of the background, in Figure 12.4 the bright particles have been selected), size limits (maximum and minimum area in pixels), and form characteristics (roundness, minimum and maximum width and height, etc). The detection of particles is performed within a user-defined region of interest (ROI). Additional experimental and control regions can also be defined. For each experimental region, a maximum number of objects to be tracked can be defined. The data for all recognized objects is then saved to the hard disk and can be analyzed, exported, and printed later. Data for each second video frame is extracted online or during playback from tape. It is useful to have a summary screen that shows the

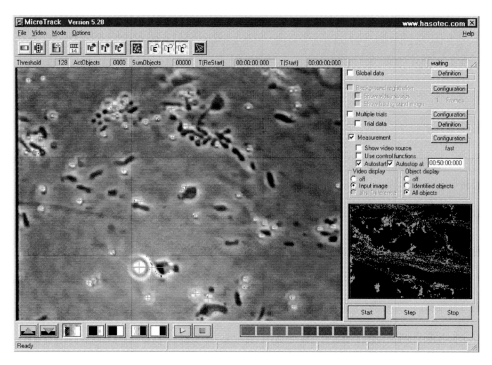

FIGURE 12.4   Detection of moving organelles. Screen shot of the MicroTrack motion analysis system during an online measurement (analysis screen) of protoplasmic streaming in a plant cell. Loading and saving data files is performed within defined regions of interest (here, entire screen). Each second video frame is being analyzed online in this example. Dimensions of all objects matching the selected criteria are marked online by crossbars. The summary screen (right-hand side) shows the accumulating data of the movements of all recognized particles. Frame grabber, FG32Path; software, MicroTrack 5.28 (both from HaSoTec GmbH, Rostock, Germany); width of the digitized image displayed, 28 μm.

accumulating positions of all recognized particles (at the right-hand side). Connecting the consecutive positional data points of the moving objects by software yields the trajectories (Figure 12.5). The example given, protoplasmic streaming of organelles in an onion epithelial cell, clearly shows regions of directed motion (center), regions of stochastic motion (top and below), and regions without moving particles.

In VEC-DIC microscopy of whole cells, the moving objects and the underlying track (microtubule or microfilament) cannot be seen simultaneously. If its position in the cell must be described, one must estimate the track by fitting a curve to the observed x, y-positions of the object. Visualization of microtubules revealed that the majority of them form straight tracks and that their diameter is small compared to the size of the organelles being transported.[53] This holds true for microtubules in axons, as well as for free microtubules. Estimating the track is hence reduced to fitting a straight line to the observed positional data of the particle. One must, however, consider that the commonly used regression model to fit a curve is not appropriate in this case because it assumes that y-values are measured with random errors, while x-values are without errors so that the estimation of the regression curve minimizes the sum of the squares of the distances of the measured y-value and the y-value of the curve. In this particular case, however, both, the x- and y-values are subject to measurement errors. Therefore, the sum of squares of the perpendicular distances of the x, y-positions to the line (track) has to be minimized.

The slope a and offset b of the fitted line $g(x) = y = ax + b$ are given by equations

$$a^2 S_{xy} + a\left(S_x^2 - S_y^2\right) - S_{xy} = 0 \tag{12.1}$$

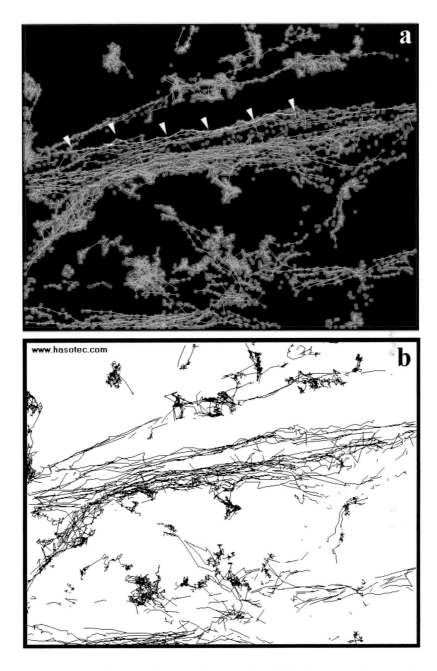

FIGURE 12.5 Generation of trajectories of moving organelles: (a) The dots show the accumulated positions of the centers of mass of the objects using a constant scale factor and the lines represent the trajectories of the respective particles. The track marked by arrowheads is further analyzed (see Figure 12.6). (b) Cumulative trajectory raw data provide a good overview of the intracellular position of tracks for the movement and of regions of Brownian motion (top and below).

and

$$\bar{y} = a\bar{x} + b \tag{12.2}$$

where S_x^2, S_y^2, and S_{xy} are the variances and the covariance, and $\bar{y}$ and $\bar{x}$ are the means of the x- and y-values. For a more detailed explanation and discussion of confidence intervals for the estimated parameters, refer to Wong.[54] Solving the above equations obtains an estimate for the track.

For any observed position of the particle, the point of intersection of the perpendicular with the line yields its estimated position on the track. If the coordinate system in transformed in such a way that the origin is translated to the estimated starting position of the particle on the track and the x-axis is rotated onto the estimated line (track), then the desired horizontal representation of the trajectory for further motion analysis is obtained. The resulting equations for the translocation and rotation were given by Weiss et al.[49] The motion of the particle is now described by two time series: one representing its position on the track (x-direction) and the other its perpendicular deviation (y-direction).

12.2.3.3 Motion Analysis

The analysis of the motions is usually based on time-series analysis similar to the procedures described by Koles et al.[55,56] and performed either with one of the above-mentioned motion analysis software packages or, alternatively, with commercial calculation programs (e.g., Microsoft Excel) or statistics and mathematics program packages that allow the programming of macros such as those developed in this laboratory.[49,50,57]

One very useful parameter to describe the motion of an organelle in more detail is the average velocity in the major direction of movement. This can be calculated in different ways, for example, as the slope of the line fitted to the positional data over time, as the average of frame-to-frame velocities, or as the ratio of distance and time for the whole tracked path. The latter method is used in the authors' work.

If the molecular mechanisms of the intracellular transport are studied, the analysis of time- and track-dependent variations of the velocity in a single trajectory are of interest because they can give insight into the mechanism of action of the underlying motor enzymes. In contrast to the average velocity, this is called "instantaneous" velocity. The methods to obtain the estimate of the instantaneous velocity vary;[55,58] therefore a more detailed description of the authors' approach is warranted here. Some difficulties will emerge when the velocity is estimated from the positional data because the coordinates are likely to be affected by measurement errors. The velocity of a particle is obtained by differentiation of the time series of x-positions. First-order differencing according to

$$v(t) = \big(x(t) - x(t-1)\big)/\Delta t \qquad (12.3)$$

(where Δt is the time between successive x-positions) is generally considered to be a rather gross approximation to differentiation. It was decided, however, to use this method for the following reasons for these studies on the existence of low-frequency velocity oscillations as claimed by Koles et al.[55,56] Due to the sampling frequencies (2 Hz to 25 Hz), the maximum resolvable frequencies of the positional variation of a particle are then 1 Hz to 12.5 Hz. In nearly all observed spectra, more than 99% of the power of the signal is well below 1 Hz. Obviously, the signal is "oversampled" and one therefore obtains, with respect to the signal, a "small" t and hence a better approximation to differentiation. Noise components at higher frequencies, induced by measurement errors, would, however, still degrade the estimate severely, because high-frequency components are "amplified" by differentiation. Therefore, one can remove critical noise components at higher frequencies before differencing the time series according to Eq. 12.3 in the following way.

First, the average velocity of the particle (trend) that is represented by the regression line is subtracted from the data set (see Figures 12.6b and c). Then transform the trend-removed time series of the x-positions into the frequency domain, using the Cooley-Tukey fast Fourier transform

(FFT) algorithm.[59] Second, a cutoff frequency for low-pass filtering, usually in the range of 0.3 Hz to 0.7 Hz, is determined from this spectrum. Third, the trend-removed time series is digitally filtered using a low-pass filter (second-order Butterworth seemed appropriate) with the respective cutoff frequency. Fourth, the trend is re-added to the filtered time series. After differencing, according to Eq. 12.3, one finally obtains the particle's instantaneous velocity, which is plotted versus time and versus the main direction of movement. This data set is now well-suited for spectral analysis by FFT analysis to determine whether the motion contains regularities in position or velocity (such as oscillations or jumps), or whether it is continuous with added stochastic fluctuations.[29,30,49] It was found that the velocity fluctuations observed in axons are stochastic and that the regular velocity oscillations described by Koles et al.[55,56] seem to be sampling artifacts.[49]

If movement episodes are short (e.g., for very fast movements in plant cells) the data sets are not suitable for filtering and FFT analyses and have to be plotted as raw data as shown in Figure 12.6 for a single organelle. Most of the parameters characterizing the movement can also be averaged for a group of particles.

12.2.4 DYNAMIC PHASE MICROSCOPY

Human eyes react to spatial intensity distributions within a limited range of wavelengths and an image is then associated with the real object by intellectual processes. However, using modern, computer-aided electro-optical techniques, it is possible to encode and represent on a monitor other physical values or functions of our optical environment that are inaccessible to direct vision, such as UV, IR, temperature fields, or the phase information of light. The nature of the chosen physical value and the information available depend on the converters and algorithms used. The thermovision technique, which is now widely used in medicine and other fields, is one example of such functional imaging.

In the 1980s, profilometry was developed as a new branch of optical instrumentation, aimed at a quantitative evaluation of subtle surface structures and roughness. This technique is based on the dependence of the phase of a reflected wave on the distance of the surface from the light source (i.e., the local surface height). The phase is shifted by 2π for each distance equal to the wavelength used, which serves here as a natural length standard. Phase measurements are performed using interferometric methods, that is, comparisons with the phase of the reference wave. Modern commercial interferometers with coherent laser sources allow one to measure extremely small optical path differences (OPD) of 10^{-6} to 10^{-8} of a wavelength so that height differences of few nanometers are detectable. The phase images of surfaces are usually obtained by scanning mode measurements and represented as 2-D pseudo-color maps of OPD h(X, Y) in image plane X, Y. They are similar to geographic maps, but with the height values on the nanometer scale and the X- and Y-distances on the scale of a few micrometers. For solid-state objects, OPD values correspond to the real height variations. For the complicated, nonuniform structure of living cells, the meaning of OPD is less well-defined because these values depend on the surface profile **and** the 3-D refractive index distribution.

Unstained and transparent biological objects are also phase retarding objects. Phase microscopy of thin biological objects can be based on local measurements of OPD values in reflection mode. Here, refractive index differences of the optically inhomogeneous internal structures of a cell, which is placed on a polished, reflecting surface, dominate the local OPD values and create the image contrast. The sensitivity (i.e., the amount of contrast and resolution) of phase microscopy methods is comparable to that of VEC microscopy. Some structures, that are barely seen in bright-field mode (light amplitude) can be clearly distinguished in phase microscopy due to the method's high sensitivity to local refractive index differences. In contrast to phase contrast and DIC microscopy, which use **relative phase differences** to create image contrast, phase microscopy as discussed here is based on **direct OPD measurements** and yields absolute values without preliminary conversion into an intensity or an enhanced differential contrast image (Figures 12.7b and c). Therefore, the measured OPD values are independent of illumination intensity and are normalized to wavelength.

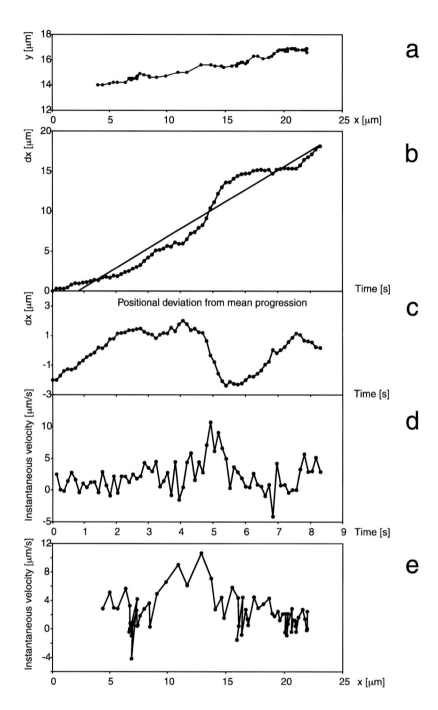

FIGURE 12.6 Steps of the motion analysis of a moving organelle: (a) Path of the organelle moving toward the right-hand side on the video screen. Coordinates measured from top left corner. The movement is a combination of linear motion and pauses with more stochastic motion. (b) Progression of the organelle from start in the main direction of movement plotted versus time. The slope along the curve indicates the instantaneous velocity. The regression line represents the average velocity, which is subtracted (trend removal) to yield the positional deviation from mean progression which is depicted in (c). (d) Instantaneous velocity calculated from the progression between two measured time points, divided by time and plotted versus time. (e) Instantaneous velocity is plotted versus position in the main direction of movement. (See also Refs. 49 and 50.)

However, the measured OPD, h(X, Y), depends on the product of geometrical height (specimen thickness) H(x, y) and local refractive index difference $\Delta n(x, y)$ in the focal plane x, y, which cannot be separated from each other easily. Therefore, interpretation of phase images requires some experience.

The fluctuation or rotation of a single isolated macromolecule between two adjacent pixels cannot be recorded as a significant change of OPD because the very small changes of refractive index accompanying this motion are averaged within the probe volume and are therefore only recorded as additional noise. However, in the case of synchronous motion of a sufficiently large number of macromolecules, their contribution to OPD may be significant and could be detected as a specific frequency component if their temporal oscillations are regular and exceed background.

12.2.4.1 Method

The new method of dynamic phase microscopy (DPM)[60,61] results from electronic periodic profile scanning along a scan-line in a phase microscopy image. The local OPD values and profiles are not constant because cells are living objects and their organelles are moving or changing their configuration in time. In order to describe and image the intracellular fluctuations, the digitized OPD data (in units of height) of a temporal series of height profiles along the scan line s (Figure 12.7b) are accumulated as a stack ("track diagram," Figure 12.7d). Sections through the track diagram yield records ("registograms") of phase height fluctuations h(X, t) for fixed scan-line points X over time (line e in Figures 12.7d and 12.7e). After Fourier transform of such time series, one obtains the corresponding spectrum of local phase height or apparent object thickness fluctuations. The whole set of such spectra representing a 2-D or 3-D position vs. frequency plot or X, f-plot (spectral pattern) shows with high spatial resolution the location of dominating frequency components along the scan-line s.[62,60]

Since cells and their organelles are optically nonuniform anisotropic objects, phase microscopy of biological objects has some peculiarities that complicate interpretation of the results. There are no simple algorithms connecting the measured phase $\varphi(X, Y)$ of the diffracted wave to the shape, that is, geometrical thickness, H(x,y), and the refractive index, $n_i(x, y, z)$, of an object region. The approximated expression for a measured local phase deviation $\Delta\varphi(X, Y)$ due to change of $n_i(x, y, z)$, is:

$$\Delta\varphi(X, Y) = 4\pi\, h(X, Y)/\lambda \approx 4\pi\, \Delta n_i(x, y)\, H(x, y)/\lambda \qquad (12.4)$$

where $\Delta n_i(x, y)$ is the local difference of the refractive indices. Therefore, the measured phase thickness, h(X, Y), will be essentially lower than its true value, H(x, y), and a small organelle cannot be identified if $\Delta n_i(x, y) \ll 1$. Typical phase images of a single mitochondrion are shown in Figures 12.7b and 12.7c.

The square of the deviation of measurements of a height standard $I(X) = \sigma^2(X)$ is used for quantitative evaluation of the intensity of fluctuations. This results in information on the intensity distribution of fluctuations along the scan-line axis X. One uncommon feature of phase images is the super-resolution. For objects with sufficiently high OPD contrast, it is possible to exceed the classical resolution limit 4 to 6 times. Evidence for super-resolution in phase images was first shown using measurements of sub-micron semiconductor structures and latex spheres[62,63] and arguments for its plausible explanation based on photon tunneling were given.[64]

12.2.4.2 Equipment

The Computer Phase Microscope "Airyscan"[65] is an interference microscope according to the Linnik layout (Figure 12.7a), with phase modulation of a reference wave. A He-Ne laser is used for coherent illumination of the object and a dissector image tube as a coordinate-sensitive photodetector. In this microscope, the measurements of thin transparent biological objects located on a polished,

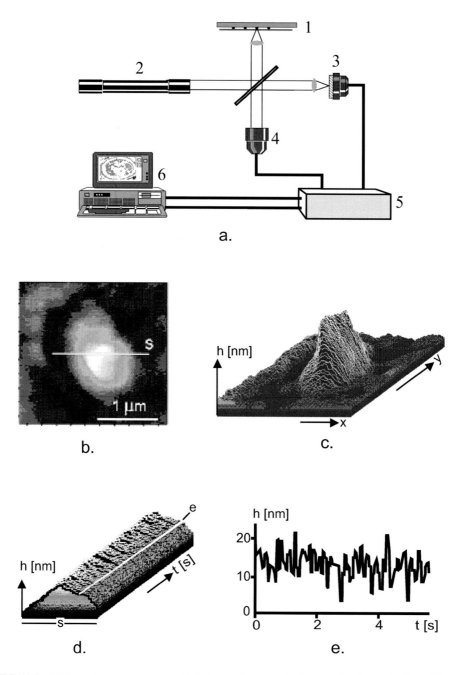

FIGURE 12.7   (a) Layout of the computer-aided phase microscope "Airyscan" for Dynamic Phase Microscopy. 1, Object on a reflecting surface with cover slip; 2, light from the zero mode He-Ne laser directed onto the beam splitter of the interferometer; 3, reference mirror on piezo-transducer for modulation of the interference pattern; 4, dissector image tube for data collection; 5, control unit; 6, computer. (b) Optical path difference (OPD) image of a single mitochondrion showing local phase height distribution. S, position of the scan-line. (c) 3-D visualization of the phase height profile in OPD units. (d) A series of phase height profiles measured along the scan-line chosen in (b). Temporal fluctuations of OPD values are obtained for a given point on the scan-line s if data are collected over time (line e) (e). If these plots are obtained for all points of the scan-line and represented in 3-D, one obtains the phase height fluctuation map that can be further subjected to spectral analysis by FFT for obtaining the X,f spectral pattern.

reflecting substrate are performed in reflected light mode. The size of the field of view can be changed from 5 to 50 µm, depending on the lens magnification. The measurements are performed with immersion optics (100x/N.A.1,3, Zeiss Jena) that yield a total optical magnification of 3500-fold. The noise-limited sensitivity is $h_{min} = 0.5$ nm, and the acquisition time is determined by the clock frequency of 1000 Hz, or 1 ms per pixel.

12.3 CELL-BIOLOGICAL APPLICATIONS

12.3.1 Visualizing Intracellular Fine Structures

It is evident from the above description that digital microscopic techniques yield most impressive image improvement in situations where the specimen has either virtually no contrast or emits almost negligible amounts of light.

With VEC-DIC microscopy, the image quality is highest in the top cell layer of tissue samples. Out-of-focus material leads to little deterioration of image quality, except in cases where strongly birefringent or depolarizing material (such as large numbers of small particles, latex beads, oil droplets, myelin, or birefringent inclusions) are in the light path. The VEC techniques allow, however, improved optical penetration of tissue slices or small organisms, especially when anaxial illumination[10,37] or light of very long wavelengths such as near-infrared is used.[38] The presence of stains or high-contrast inclusions does not allow the application of contrast enhancement to the full extent because the image often gets distorted before the small details become visible. Video enhancement, therefore, is best used with unstained material, such as living cells[18,20,66-69] or isolated cytoplasmic extracts.[22-48,70-72]

Using VIM, images can be produced at much lower illumination intensity and with lower concentrations of potentially cytotoxic dyes than previously; thus, this technique has its greatest promise in vital microscopy where the living cells need to be protected from harmful radiation or chemicals.[19,31-33,44,73-77] For VIM, the specimen needs to be thin and as close to 2-D as possible. Otherwise, out-of-focus fluorescence would make the use of confocal microscopes[6,7,78] necessary. Figure 12.8 is an example from an analysis of the dynamic behavior of microtubules labeled by transfection with Green Fluorescent Protein-(GFP)-tagged α-tubulin cDNA.

It is most advantageous when corresponding pairs of videomicrographs can be obtained[79,80] with one of the images at high resolution by VEC microscopy and one by VIM after addition of vital fluorescent dyes such as organelle-specific markers[16,44,76,81-83] or of fluorescence-labeled antibodies that report the location and amount of all kinds of antigens. The antibodies can be applied after the observation of live cells is completed and the cells fixed. Studies on live cells are, however, equally possible, if cells are microinjected with fluorescent-labeled proteins, transfected with GFP-tagged components, or permeabilized.[75,84-88] The use of 1- to 5-nm colloidal gold-coupled antibodies, which are visualized by VEC microscopy in bright-field or epi-polarization mode (Nanovid microscopy), yields non-bleaching images of higher spatial resolution than fluorescence immunocytochemistry or VEC-DIC microscopy.[25,26,89,90] These tagging techniques allow, in addition, the biochemical or molecular identification of the objects visualized by VEC techniques.

Fluorescence video micrographs are also well-suited for further quantitative image analysis of fine-morphological parameters. Although little work has been done with organelle morphometry, typical morphometric parameters can be derived if a suitable software package is applied to images stained with organelle-specific vital dyes or antibodies for endoplasmic reticulum networks, for the filamentous arrays of cytoskeletal elements (Figure 12.8) and mitochondria, or for spherical organelles such as lysosomes, secretory vesicles, or Golgi apparatus.

The resolving power of dynamic phase microscopy is also considerably better than the classical Raleigh resolution and, for suitable objects, it is in the order of tens of nanometers.[63,64]

Taken together, one can say that the various digital light microscopy techniques are best-suited for applications involving living cells or extruded cytoplasm, and that they are helpful for a better

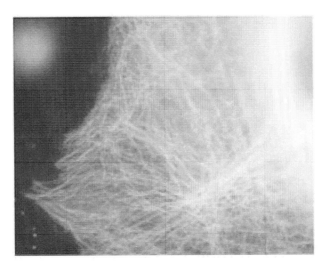

FIGURE 12.8 Green Fluorescent Protein (GFP) imaging in a living Chinese hamster ovary (CHO) cell by video-intensified microscopy (VIM). Example from an analysis of the dynamic behavior of microtubules which were for this purpose labeled by transfecting the cell with GFP-tagged α-tubulin cDNA. Microscope, Nikon Eclipse 800; CCD camera, Sensys (Photometrics, Inc., Tucson, AZ); exposure time, 100 ms; frame width, 40 μm.

understanding of the cell's physiology,[31-33,76,91,92] biochemistry,[93,94] molecular architecture,[8,74,76,95] and molecular dynamics.[29,75,96-103]

12.3.2 ANALYSIS OF INTRACELLULAR MOTILITY

12.3.2.1 Visualization of Moving Objects and Motion Analysis

Considerable progress in the field of cytoplasmic motility and fine structure had to await the advent of digital microscopy and especially of Allen video-enhanced contrast differential interference contrast (AVEC-DIC) microscopy, which allows one to see and to study **all** membrane-bounded organelles and the microtubules, together with their motions and interactions. The new technique was quickly used to study one of the most prominent systems of intracellular organelle motion, namely the transport in nerve cell processes (axoplasmic transport). From these and related studies, it became clear that organelle motion in higher animal cells proceeds along microtubules[22,23,66] and to some extent also along actin filaments.[21,71,104]

Many more breakthroughs based on these techniques followed; for example, the finding that microtubules move actively,[22,105,106] that microtubules display properties of dynamic instability *in vivo*,[57,96,97,103] that organelles move bidirectionally on a single microtubule,[22,66,105,107] that membranous organelles can switch between microtubules and actin filaments.[71] Using digital light microscopy, the force-generating enzymes or molecular motors of organelle movements could be isolated.[108-111] All biochemical studies dealing with the three rapidly growing families of motor proteins — namely kinesins, cytoplasmic dyneins, and nonconventional myosins — employ digital microscopy-based motion analysis as the only means to detect and assay these enzymes.

In plant cells, the role of the filamentous tracks is played by actin filament bundles, as could be demonstrated by video techniques.[18-20,24] Similar studies performed on various forms of cytoplasmic transport and cellular motility improved understanding of muscle contraction,[112-114] prokaryotic[115,116] and eukaryotic flagellar motion,[117] pigment granule motion,[118,119] and cytoplasmic transport in protozoans,[67,107] to mention only a few examples.

In nerve cells, it is now possible to study axonal transport in olfactory axons that are so thin that they cannot be resolved by conventional light microscopy[120] or in larger myelinated and unmyelinated axons.[47,49,50,121,122] Changes in transport to and from regenerating axons and growth cones were reported,[123,124] as well as the specific staining and morphological monitoring over months of neuromuscular endplates with nontoxic fluorescent dyes.[73,125]

With the methods described above, the desired variables to describe quantitatively and further analyze the displacement of organelles or microtubules were obtained.[29,30,47,49,52,58,92,96,98]

Quantitative characterization of the motile events in a cell is a good means to describe the cell's physiological situation, because organelle motion is a vital and very basic property of eukaryotic cells. Once this is accomplished, the motile situation can also be used as a very sensitive and multifaceted indicator to characterize cell populations in cytopathological, cytopharmacological, and cytotoxicological studies.[58,126-128] Furthermore, motion analysis of intracellular objects has been used successfully to gain more insight into the molecular mechanisms of force generation,[28-30,112,117,129,130] actin and microtubule mechanics,[131] and the assembly and disassembly dynamics of cytoskeletal filaments.[57,96-98,132,133] For the analysis of the nanometer-size steps of single molecular motor enzymes, high-resolution position measurements are required which are suitable for statistical noise analysis.[29,30,99,100]

12.3.2.2 Local Macromolecular Dynamics

Phase images can yield additional information on intracellular supramolecular morphology when used with high temporal sampling rates as in dynamic phase microscopy, with 1-ms/pixel temporal and better than 100-nm spatial resolutions. Figure 12.7 is an example of such scanning measurements of a single mitochondrion. If the scanning is performed only along a relatively short scan-line, for example, 500 nm (28 pixels), this would result in a 30-ms temporal resolution. New insight into supramolecular structures seems possible because measurements of fluctuations of optical path difference indicate a connection with intracellular activity. Measurements of mitochondria, Vero cells, red blood cells, and isolated organelles revealed characteristic dominating frequency components, the intensity of which underwent marked changes with sub-wavelength topology. The coordinates of prominent intensity fluctuations correspond to active sites in the cytoplasm or nucleus that can be further analyzed. The spatial resolution of these analyses of active sites, which have different specific frequencies ranging from 0.5 Hz to 3 Hz, is not limited by Airy disk size. The spatial correlation of the movement also determines the length or correlation radius of cooperative processes.

Among the studies already performed are the following. The distribution of the fluctuation intensity along the scan-line was measured for enzyme-loaded liposomes under ATP stimulation. The spatio-temporal patterns of liposomal fluctuations revealed a spectral pattern with dominating frequency components at approximately 2 Hz and 3 Hz. When chromatin dynamics were measured by dynamic phase microscopy, different compartments of the nuclei displayed typical dominating frequency components. This was especially seen in the nucleoli when different phases of the cell cycle were investigated.[134] The dynamic properties of DNA need to be characterized further. Intense local optical path difference fluctuations, apparently due to cooperative processes, were also measured in freshly prepared mitochondria *in vitro*.[135] ATP increased the intensity of low-frequency components, and addition of rotenone or the protonophore FCCP resulted in a decrease. These authors propose that such results can be attributed to a dependence of enzyme fluctuations on the mitochondrial membrane potential.

Since the local optical path difference is an indicator of the refractive index, its fluctuation indicates cooperative dynamic changes of the conformational state of enzyme complexes or enzyme clusters. The width (100 to 300 nm) and location of the active sites were detected with an accuracy of about 50 nm. Because the contribution of a single enzyme molecule to the optically measured activity would be negligible, the observed spatial correlation and temporal coherence of the fluctuations is attributed to cooperative effects connecting the behavior of enzymes closely adjacent to each other. Not only organelles, but also DNA or associated enzymes and mitochondrial enzymes showed such highly cooperative processes that were all ATP dependent.

12.3.3   Measuring Biochemical Parameters in the Living Cell

Microscopes are often considered to be devices that "merely" make pictures of objects. This is certainly not true for digital microscopes that generate images in which the gray or pseudo-color shades encode quantitative physical or chemical parameters of the specimen, such as motion velocity, concentrations, viscosity, birefringence, phase shift, etc. (see Table 12.1). The digital microscope can therefore assume features of a spectrophotometer, spectrofluorimeter, photon multiplier, etc. Its dynamic range may be more limited than that of some of the other devices mentioned, but this is generally compensated by the advantage that information is yielded not only in a punctual way (cuvette) but rather in two dimensions (x and y), three dimensions (x, y, z, in through-focus series, or x, y and concentration), or even in four dimensions (i.e., if video films of spatio-temporal processes are recorded). Multi-color or even true spectral information can be obtained for each pixel if time series of video frames are stored and the time series for each pixel position are subjected to spectral analysis by FFT.[136,137] This allows the simultaneous analysis of five or more fluorescent labels and, by a combination of dyes, all human chromsomes can be distinguished spectrally and displayed in different pseudo-color hues.[136]

Because the transfer function of most video cameras used for quantitative microscopy is linear instead of logarithmic, as in the case of the human eye and photographic film,[4,10] the gray-level information in video images can be used more directly to gain information on the relative amounts of fluorescent substances. Absolute values can be obtained by calibration with standard solutions or samples that are imaged at the same instrumental setting as the unknown samples, or by making cells permeable for external medium containing a known concentration of dye or ligand.[31,32]

If information on intracellular concentrations is desired, the optical pathlength (i.e., the thickness of the cell at each pixel) needs to be measured in addition to the intensity values. Such optical pathlength images are obtained when cells are loaded with a dye of homogeneous and exclusively intracellular distribution, such as fluorescein diacetate or labeled dextran and then imaged.[31,32,138,139] In this case, the image brightness codes for cell thickness in a quantitative way. "Concentration images" can be derived by dividing regular fluorescence intensity or absorption images obtained with monochromatic light by such "optical pathlength images." In such measurements, the living cell is, so to speak, converted into the biochemist's measuring cuvette.[140]

At present, a rapidly expanding application of digital microscopy in cell biology is the use of fluorescent chelators to measure concentrations of ions and their transient changes. These specially designed dyes cross the plasma membrane in the form of their more lipophilic acetoxymethyl esters and are trapped in the cell after intracellular cleavage of the ester bond. In the case of the Ca^{2+} chelator Fura-2 and some related ion indicators, the binding of the ion causes a concentration-dependent spectral shift of the chelator dye fluorescence. By dividing the images obtained at two different wavelengths and using a calibration curve (ratio values of a standard series plotted vs. concentration), one obtains an image containing the concentration information (ratio imaging).[31,32,76,138,141] This procedure is independent of the optical pathlength so that no knowledge of the cell thickness is required. Besides the concentration of Ca^{2+}, H^+ (pH-value), Mg^{2+}, Cl^-, K^+, and Na^+ concentrations can also be determined with suitable dyes.[44,76,138,142,143] Image processors can also be programmed to synchronize the filter changer and to display the ion concentration in the form of color-coded images. The temporal resolution of this technique can be less than 1 second, and if images are processed offline, data collection is possible even at video rates.

Similar to the above-mentioned extraction of coordinates from series of video frames, one can also extract intensity values of individual image points or image regions from such video series, so that concentration changes in individual cells or in subcellular regions can be followed over time (temporal analysis).[143-145] This makes studies on intracellular transfer, exchange, and metabolism of compounds possible.

Based on these techniques, improved assays for regional and temporal enzyme activity can be developed, for example, for enzymes that consume or synthesize absorbing or fluorescent compounds.[146] Direct luminescence-based enzymatic substrate measurements have been introduced for lactate, glucose, and ATP in unfixed tissue sections. The cells or cryosections are permeabilized (e.g., frozen and thawed) and exposed to the coupled enzyme system containing luciferase and suitable bioluminescent substrates and then imaged by photon counting[93,94] (Figure 12.9). An elegant technique was introduced by Langridge and collaborators to visualize gene expression in plant cells and zebrafish embryos.[147-149] In *Rhizobium*-infected soybean root nodules or in cross-sections of tobacco plants the control of gene expression by specific promoters could be directly visualized by the enzyme luciferase introduced into the plant genome as reporter gene. Light emission (i.e., activation of the promoter and expression of its associated genes) was quantitated throughout the plant, in cambium cells, or in defined subcellular regions.[147,148]

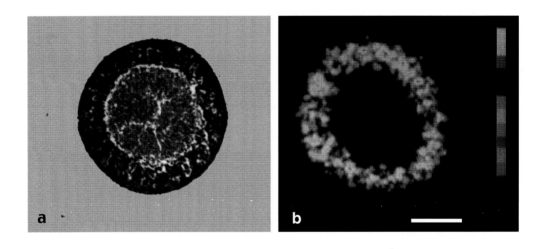

FIGURE 12.9   Application of the quantitative bioluminescence technique of regional ATP measurement by photon counting imaging to a cryostat section through a multicellular EMT6 tumor cell spheroid. (a) Unstained cryostat section. (b) Bioluminescence intensity distribution representing the local distribution of ATP concentrations. The code bar (right) spans the concentration range from 0.1 to 1.2 *mM* ATP. Bar 500 μm. (Courtesy of W. Müller-Klieser.)

The introduction of Green Fluoresent Protein (GFP) from jellyfish, which can be employed as a protein expression marker, is one of the most significant recent advances in electronic light microscopy of living cells.[85,87] A major strength of GFP is that temporal dynamics of identified molecules can be observed and measured using time-lapse imaging. GFP as an expression marker reveals not only protein synthesis in real-time *in vivo*, but can also be employed as a dynamic marker for organelles and other subcellular structures. In combination with digital light microscopy, GFP can be used to study intracellular protein trafficking and vesicle transport.[88,91,92] If different spectral variants of GFP are used, dynamic changes in the co-distribution of two components can be investigated within a living cell. GFP variants can also be utilized in the technique of fluorescence resonance energy transfer (FRET)[150] to study protein-protein interaction *in vivo* without the use of micro-injecting fluorescently labeled proteins, thereby allowing biochemical studies within a living cell.[101] Possibly, GFP can also be used for designing quantitative assays of polypeptide concentrations and dynamics.

12.4 CONCLUSION

Digital light microscopy allows one to go beyond the limits of conventional light microscopy. It enables one to see smaller objects than before, to work at lower light intensities, and to generate contrast where none could be generated by conventional techniques. Most of the new techniques and their applications yield best image improvement with unstained specimens so that live cells and their cytoplasm becomes amenable to study. As is often the case in science, new technologies bring new insight, so that already a large body of knowledge on cell structure and function has accumulated and has profoundly changed our static, electron microscopy-based understanding of the cytoskeleton and of cellular structures to a more lively, highly dynamic view. Video microscopists are presently extending the microscope's imaging power well into the molecular domain, making us eyewitness to molecular dynamics, transport, and metabolism in the living cytoplasm.

ACKNOWLEDGMENTS

The development of video-microscopic techniques in the authors' laboratories was supported by Deutsche Forschungsgemeinschaft DFG Innovations-Kolleg "Komplexe und Zelluläre Sensorsysteme" INK-27 and Schwerpunktprogramm "Neue mikroskopische Methoden" (We790/14), Hamamatsu Photonics K.K., Japan, and Ministerium für Bildung, Wissenschaft und Kultur Mecklenburg-Vorpommern. The valuable advice and help of Willi Maile and the members of the Center for Biological Visualization Techniques at the University of Rostock is gratefully acknowledged. We thank T. Holstein and W. Mueller-Klieser for their permission to reproduce images from their work.

REFERENCES

1. Shotton, D., The current renaissance in light microscopy. I. Dynamic studies of living cells by video enhanced contrast microscopy, *Proc. Roy. Microsc. Soc.*, 22, 37, 1987.
2. Webb, W.W., Light microscopy – A modern renaissance, *Ann. N.Y. Acad. Sci.*, 483, 387, 1986.
3. Weiss, D.G., Visualization of the living cytoskeleton by video-enhanced microscopy and digital image processing, *J. Cell Sci., Suppl.*, 5, 1, 1986.
4. Inoué, S. and Spring, K.R., *Video Microscopy. The Fundamentals*, 2nd ed., Plenum Press, New York, 1997.
5. Inoué, S. and Inoué, T.D., Computer-aided stereoscopic video reconstruction and serial display from high-resolution light-microscope optical sections, *Ann. N.Y. Acad. Sci.*, 483, 392, 1987.
6. Shotton, D., Ed., *Electronic Light Microscopy. The Principles and Practice of Video-Enhanced Contrast, Digital Intensified Fluorescence, and Confocal Scanning Light Microscopy*, Wiley-Liss, New York, 1993.
7. Matsumoto, B., Ed., Cell biological application of confocal microscopy, in *Methods in Cell Biology*, Vol. 38, Academic Press, San Diego, California, 1993, 380.
8. Allen, R.D., New observations on cell architecture and dynamics by video-enhanced contrast optical microscopy, *Annu. Rev. Biophys. Biophys. Chem.*, 14, 265, 1985.
9. Shotton, D.M., Review: video-enhanced light microscopy and its applications in cell biology, *J. Cell Sci.*, 89, 129, 1988.
10. Weiss, D.G., Maile, W., Wick, R.A., and Steffen, W., Video microscopy, in *Light Microscopy in Biology. A Practical Approach*, 2nd ed., Lacey, A.J., Ed., Oxford University Press, 1999, 73.
11. Weiss, D.G., Video-enhanced contrast microscopy, in *Cell Biology: A Laboratory Handbook*, 2nd ed., Vol. 3, Celis, J.E., Ed., Academic Press, San Diego, 1998, 99.
12. Allen, R.D., Travis, J.L., Allen, N.S., and Yilmaz, H., Video-enhanced contrast polarization (AVEC-POL) microscopy: a new method applied to the detection of birefringence in the motile reticulopodial network of *Allogromia laticollaris*, *Cell Motil.*, 1, 275, 1981.

13. Allen, R.D., Allen, N.S., and Travis, J.L., Video-enhanced contrast, differential interference contrast (AVEC-DIC) microscopy: a new method capable of analyzing microtubule-related motility in the reticulopodial network of *Allogromia laticollaris, Cell Motil.*, 1, 291, 1981.

14. Allen, R.D. and Allen, N.S., Video-enhanced microscopy with a computer frame memory, *J. Microscopy*, 129, 3, 1983.

15. Weiss, D.G. and Maile, W., Principles and applications of video-enhanced contrast microscopy, in *Electronic Light Microscopy: The Principles and Practice of Intensified Fluorescence, Video-Enhanced Contrast and Confocal Scanning Optical Microscopy*, Shotton, D.M., Ed., Wiley-Liss, New York, 1993, 105.

16. Lee, C. and Chen, L.B., Behavior of endoplasmic reticulum in living cells, *Cell*, 54, 37, 1988.

17. Allan, V. and Vale, R.D., Movement of membrane tubules along microtubules *in vitro*: evidence for specialised sites of motor attachment, *J. Cell Sci.*, 107, 1885, 1994.

18. Lichtscheidl, I.K. and Weiss, D.G., Visualisation of submicroscopic structures in the cytoplasm of *Allium cepa* inner epidermal cells by video-enhanced contrast light microscopy, *Eur. J. Cell Biol.*, 46, 376, 1988.

19. Quader, H., Hofmann, A., and Schnepf, E., Shape and movement of the endoplasmic reticulum in onion bulb epidermis cells: possible involvement of actin, *Eur. J. Cell Biol.*, 44, 17, 1987.

20. Allen, N.S. and Brown, D.T., Dynamics of the endoplasmic reticulum in living onion epidermal cells in relation to microtubules, microfilaments, and intracellular particle movement, *Cell Motil. Cytoskel.*, 10, 153, 1988.

21. Tabb, J.S., Molyneaux, B.J., Cohen, D.L., Kuznetsov, S.A., and Langford, G.M., Transport of ER vesicles on actin filaments in neurons by myosin V, *J. Cell Sci.*, 111, 3221, 1998.

22. Allen, R.D., Weiss, D.G., Hayden, J.H., Brown, D.T., Fujiwake, H., and Simpson, M., Gliding movement of and bidirectional transport along native microtubules from squid axoplasm: evidence for an active role of microtubules in cytoplasmic transport, *J. Cell Biol.*, 100, 1736, 1985.

23. Vale, R.D., Schnapp, B.J., Reese, T.S., and Sheetz, M.P., Organelle, bead and microtubule translocations promoted by soluble factors from the squid giant axon, *Cell*, 40, 559, 1985.

24. Kachar, B. and Reese, T.S., The mechanism of cytoplasmic streaming in Characean algal cells: sliding of endoplasmic reticulum along actin filaments, *J. Cell Biol.*, 106, 1545, 1988.

25. Mizushima, Y., Detectivity limit of very small objects by video-enhanced microscopy. *Appl. Optics*, 27, 2587, 1988.

26. Lee, G.M., Nanovid microscopy, in *Light Microscopy in Biology. A Practical Approach*, Lacey, A.J., Ed., Oxford University Press, 1999, 425.

27. Inoué, S., Imaging of unresolved objects, superresolution, and precision of distance measurement with video microscopy, in *Methods in Cell Biology*, Vol. 30, Taylor, D.L. and Wang, Y.L., Eds., Academic Press, San Diego, 1989, 85.

28. Sheetz, M.P., Turney, S., Qian, H., and Elson, E.L., Nanometre-level analysis demonstrates that lipid flow does not drive membrane glycoprotein movements, *Nature (London)*, 340, 284, 1989.

29. Mehta, A.D., Rief, M., Spudich, J.A., Smith, D.A., and Simmons, R.M., Single-molecule biomechanics with optical methods, *Science*, 283, 1685, 1995.

30. Gelles, J., Schnapp, B.J., and Sheetz, M.P., Tracking kinesin-driven movements with nanometre-scale precision, *Nature (London)*, 331, 450, 1988.

31. Parton, R.M. and Read, N.D., Calcium and pH imaging in living cells, in *Light Microscopy in Biology. A Practical Approach*, 2nd ed., Lacey, A.J., Ed., Oxford University Press, 1999, 221.

32. Silver, R.B., Ratio imaging: practical consideration for measuring intracellular calcium and pH in living tissue, *Meth. Cell Biol.*, 56, 237, 1998.

33. Tsien, R.Y., Fluorescent indicators of ion concentrations, in *Methods in Cell Biology, Fluorescence Microscopy of Living Cells in Culture, Part B: Quantitative Fluorescence Microscopy — Imaging and Spectroscopy*, Vol. 30, Taylor, D.L. and Wang, Y.-L., Eds., Academic Press, San Diego, 1989, 127.

34. Sluder, G. and Wolf, D.E., Eds., *Video Microscopy. Methods in Cell Biology*, Vol 56, Academic Press, San Diego, 1998.

35. Russ, J.C., *The Image Processing Handbook*, 2nd ed., CRC Press, Boca Raton, FL, 1994.

36. Steffen, W. and Weiss, D.G., Marktübersicht Biological Imaging. Ein Bild sagt mehr als 1000 Worte, *BioSpektrum*, 5, 118, 1999.

37. Kachar, B., Asymmetric illumination contrast: a method of image formation for video light microscopy, *Science*, 277, 766, 1985.

38. Dodt, H.-U. and Zieglgänsberger, W., Infrared videomicroscopy: a new look at neuronal structure and function, *Trends in Neurosci.*, 537, 453, 1994.

39. Inoué, S., Video image processing greatly enhances contrast, quality, and speed in polarization-based microscopy, *J. Cell Biol.*, 89, 346, 1981.

40. Hansen, E.W., Conchello, J.A., and Allen, R.D., Restoring image quality in the polarizing microscope: analysis of the Allen video-enhanced contrast method, *J. Opt. Soc. Am.*, A5, 1836, 1988.

41. Shotton, D., An introduction to digital image processing and image display in electronic light microscopy, in *Electronic Light Microscopy*, Shotton, D., Ed., Wiley-Liss, New York, 1993, 39.

42. Cardullo, R.A. and Alm, E.J., Introduction to image processing, in *Methods in Cell Biology. Video Microscopy*, Sluder, G. and Wolf, D.E., Eds., Academic Press, San Diego, 56, 1998, 91.

43. Taylor, D.L. and Wang, Y.-L., *Methods in Cell Biology. Fluorescence Microscopy of Living Cells in Culture, Part A*, Vol. 29, Academic Press, San Diego, 1989.

44. Haugland, R.P., *Handbook of Fluorescent Probes and Research Chemicals*, 6th ed., Molecular Probes, Eugene, OR, 1996.

45. Lange, B.M.H., Sherwin, T., Hagan, I.M., and Gull, K., The basics of immunofluorescence video microscopy for mammalian and microbial systems, *Trends in Cell Biol.*, 5, 328, 1996.

46. Oshiro, M., Cooled CCD versus intensified cameras for low-light video-applications and relative advantages, *Methods Cell Biol.*, 56, 45, 1998.

47. Weiss, D.G., Keller, F., Gulden, J., and Maile, W., Towards a new classification of intracellular particle movement based on quantitative analysis, *Cell Motil. Cytoskel.*, 6, 128, 1986.

48. Brady, S.T., Lasek, R.J., and Allen, R.D., Fast axonal transport in extruded axoplasm from squid giant axon, *Science*, 218, 1129, 1982.

49. Weiss, D.G., Galfe, G., Gulden, J., Seitz-Tutter, D., Langford, G.M., Struppler, A., and Weindl, A., Motion analysis of intracellular objects: trajectories with and without visible tracks, in *Biological Motion. Lecture Notes in Biomathematics*, Vol. 89, Alt, W. and Hoffmann, G., Eds., Springer Verlag, Berlin, 1990, 95.

50. Weiss, D.G., Langford, G.M., Seitz-Tutter, D., Gulden, J., and Keller, F., Motion analysis of organelle movements and microtubule dynamics, in *Structure and Functions of the Cytoskeleton, Colloque INSERM*, Vol. 171, Rousset, B.A.F., Ed., John Libbey Eurotext, Paris/London, 1988, 363.

51. Alt, W. and Hoffmann, G., Eds., *Biological Motion. Lecture Notes in Biomathematics*, Vol. 89, Springer Verlag, Berlin, 1990.

52. Tvarusko, W., Bentele, M., Misteli, T., Rudolf, R., Kaether, C., Spector, D.L., Gerdes, H.H., and Elis, R., Time-resolved analysis and visualization of dynamic processes in living cells, *Proc. Natl. Acad. Sci. U.S.A.*, 96, 7950, 1999.

53. Langford, G.M., Allen, R.D., and Weiss, D.G., Substructure of sidearms on squid axoplasmic vesicles and microtubules visualized by negative contrast electron microscopy, *Cell Motil. Cytoskel.*, 7, 20, 1987.

54. Wong, M.Y., Likelihood estimation of a simple linear regression model when both variables have error, *Biometrika*, 76, 141, 1989.

55. Koles, Z.J., McLeod, K.D., and Smith, R.S., The determination of the instantaneous velocity of axonally transported organelles from filmed records of their motion, *Can. J. Physiol. Pharmacol.*, 60, 670, 1982.

56. Koles, Z.J., McLeod, K.D., and Smith, R.S., A study of the motion of organelles which undergo retrograde and anterograde rapid axonal transport in Xenopus, *J. Physiol. (London)*, 328, 469, 1982.

57. Weiss, D.G., Langford, G.M., Seitz-Tutter, D., and Keller, F., Dynamic instability and motile events of native microtubules from squid axoplasm, *Cell Motil. Cytoskel.*, 10, 285, 1988.

58. Breuer, A.C., Lynn, M.P., Atkinson, M.B., Chou, S.M., Wilbourn, A.J., Marks, K.E., Culver, J.E., and Fleegler, E.J., Fast axonal transport in amyotrophic lateral sclerosis. An intra-axonal organelle traffic analysis, *Neurology*, 37, 738, 1987.

59. Cooley, J.W. and Tukey, J.W., An algorithm for the machine calculation of complex Fourier series, *Math. Comp.*, 19, 297, 1965.

60. Tychinsky, V., Norina, S., Odintsov, A., Popp, F., and Vyshenskaya, T., Optical phase images of living micro-objects, *Proc. SPIE Conf. Cell and Biotissue Optics*, 2100, 129, 1993.

61. Tychinsky, V., Kufal, G., Odintsov, A., and Vyshenskaja, T., Optical phase imaging of living micro-objects, *Proc. SPIE Conf.*, 2008, 160, 1993.
62. Tychinsky, V., Kufal, G., Vyshenskaya, T., Perevedentzeva, E., and Nikandrov, S., The measurements of submicrometer structures with the Airyscan laser phase microscope, *Quantum Electronics*, 27, 735, 1997.
63. Tychinsky, V., Tavrov, A., Shepelsky, D., and Shuchkin, A., The experimental evidence the phase object super-resolution, *Pisma GTPh*, 17, 80, 1991.
64. Tychinsky, V., Microscopy of sub-wavelength structures, *Uspechi Physicheskich Nauk*, 166, 1219, 1996; *Physics-Uspekhi*, 39, 1157, 1996.
65. Tychinsky, V., Masalov, I., Pankov, V., and Ublinsky, D., Computerized phase microscope for investigation of submicron structures, *Opt. Commun.*, 74, 37, 1989.
66. Hayden, J.H. and Allen, R.D., Detection of single microtubules in living cells: particle transport can occur in both directions along the same microtubule, *J. Cell Biol.*, 99, 1785, 1984.
67. Travis, J.L. and Bowser, S.S., Optical approaches to the study of foraminiferan motility, *Cell Motil. Cytoskel.*, 10, 126, 1988.
68. Taylor, D.L., Centripetal transport of cytoplasm, actin, and the cell surface in lamellipodia of fibroblasts, *Cell Motil. Cytoskel.*, 11, 235, 1988.
69. Yeh, E., Skibbens, R.V., Cheng, J.W., Salmon, E.D., and Bloom, K., Spindle dynamics and cell cycle regulation of dynein in the budding yeast, *Saccharomyces cerevisiae*, *J. Cell Biol.*, 130, 687, 1995.
70. Weiss, D.G., Meyer, M., and Langford, G.M., Studying axoplasmic transport by video microscopy and using the squid giant axon as a model system, in *Squid as Experimental Animals*, Gilbert, D.L., Adelman, W.J., Jr., and Arnold, J.M., Eds., Plenum Press, New York, 1990, 303.
71. Kuznetsov, S.A., Langford, G.M., and Weiss, D.G., Actin-dependent organelle movement in squid axoplasm, *Nature (London)*, 356, 722, 1992.
72. Steffen, W., Karki, S., Vaughan, K.T., Vallee, R.B., Holzbaur, E.L.F., Weiss, D.G., and Kuznetsov, S.A., The involvement of the intermediate chain of cytoplasmic dynein in binding the motor complex to membranous organelles of *Xenopus* oocytes, *Molec. Biol. Cell*, 8, 2077, 1997.
73. Purves, D. and Voyvodic, T., Imaging mammalian nerve cells and their connections over time in living animals, *Trends Neurosci.*, 10, 398, 1987.
74. Waterman-Storer, C.M., Sanger, J.W., and Sanger, J.M., Dynamics of organelles in the mitotic spindles of living cells: membrane and microtubule interactions, *Cell Motil. Cytoskel.*, 26, 19, 1993.
75. Wang, Y.-L., Fluorescent analog cytochemistry: tracing functional protein components in living cells, in Methods in Cell Biology, Vol. 29, chapter 1, Taylor, D.L. and Wang, Y.L., Eds., Academic Press, San Diego, 1989, 1.
76. DeBiasio, R., Bright, G.R., Ernst, L.A., Waggoner, A.S., and Taylor, D.L., Five-parameter fluorescence imaging: wound healing of living Swiss 3T3 cells, *J. Cell Biol.*, 105, 1613, 1987.
77. Whitaker, M., Fluorescence imaging in living cells, in *Cell Biology: A Laboratory Handbook,* 2nd ed., Vol. 3, Celis, J.E., Ed., Academic Press, San Diego, 1998, 121.
78. Shaw, P.J., Introduction to confocal microscopy, in *Light Microscopy in Biology. A Practical Approach*, 2nd ed., Lacey, A.J., Ed., Oxford University Press, 1999, 45.
79. Demaurex, N., Romanek, R., Rotstein, O.D., and Grinstein, S., Measurment of cytosolic pH in single cells by dual-excitiation fluorescence imaging: simultaneous visualisation using differential interference contrast optics, in *Cell Biology: A Laboratory Handbook*, 2nd ed., Vol. 3, Celis, J.E., Ed., Academic Press, San Diego, 1998, 380.
80. Foskett, J.K., Simultaneous Nomarski and fluorescence imaging during video microscopy of cells, *Am. J. Physiol.*, 255, 566, 1988.
81. Terasaki, M., Labeling of the endoplasmic reticulum with $DiOC_6(3)$, in *Cell Biology: A Laboratory Handbook,* 2nd ed., Vol. 2, Celis, J.E., Ed., Academic Press, San Diego, 1998, 501.
82. Pagano, R.E. and Martin, O.C., Use of fluorescence analogs of ceramide to study the Golgi apparatus of animal cells, in *Cell Biology: A Laboratory Handbook,* 2nd ed., Vol. 2, Celis, J.E., Ed., Academic Press, San Diego, 1998, 507.
83. Poot, M., Staining of mitochondria, in: *Cell Biology: A Laboratory Handbook,* 2nd ed., Vol. 2, Celis, J.E., Ed., Academic Press, San Diego, 1998, 513.
84. Schöpke, C. and Fauquet, C.M., Introduction of materials into living cells, in *Light Microscopy in Biology. A Practical Approach*, 2nd ed., Lacey, A.J., Ed., Oxford University Press, 1999, 373.

85. Chalfie, M., Tu, Y., Euskirchen, G., Ward, W.W., and Prasher, D.C., Green fluorescent protein as a marker for gene expression, *Science*, 263, 802, 1994.

86. Rizzuto, R., Brini, M., De Giorgi, F., Rossi, R., Heim, R., Tsien, R.Y., and Pozzan, T., Double labelling of subcellular structures with organelle-targeted GFP mutants *in vivo*, *Curr. Biol.*, 6, 183, 1996.

87. Chalfie, M. and Kain, S., *GFP, Green Fluorescent Protein: Strategies and Applications*, John Wiley & Sons, New York, 1996.

88. Wacker, I., Kaether, C., Kromer, A., Almers, W., and Gerdes, H.H., Microtubule-dependent transport of secretory vesicles visualised in real-time with a GFP-tagged secretory protein, *J. Cell Sci.*, 110, 1453, 1997.

89. Bajer, A.S., Sato, H., and Mole-Bajer, J., Video microscopy of colloidal gold particles and immuno-gold labelled microtubules in improved rectified DIC and epi-illumination, *Cell Struct. Funct.*, 11, 317, 1986.

90. De Brabander, M., Nuydens, R., Geerts, H., Nuyens, R., Leunissen, J., and Jacobson, K., Using nanovid microscopy to analyse the movement of cell membrane components in living cells, in *Optical Microscopy for Biology*, Herman, B. and Jacobson, K., Eds., Wiley-Liss, New York, 1990, 345.

91. Lang, T., Wacker, I., Steyer, J., Kaether, C., Wunderlich, I., Soldati, T., Gerdes, H.H., and Almers, W., Ca^{2+}-triggered peptide secretion in single cells imaged with Green Fluoresent Protein and eva-nescent-wave microscopy, *Neuron*, 18, 857, 1997.

92. Steyer, J.A. and Almers, W., Tracking single secretory granules in live chromaffin cells by evanescent-field fluorescence microscopy, *Biophys. J.*, 76, 2262, 1999.

93. Walenta, S., Dötsch, J., and Mueller-Klieser, W., ATP concentrations in multicellular tumor spheroids assessed by single photon imaging and quantitative bioluminescence, *Eur. J. Cell Biol.*, 52, 389, 1990.

94. Schwickert, G., Walenta, S., and Mueller-Klieser, W., Mapping and quantification of biomolecules in tumor biopsies using bioluminescence, *Experientia* 15, 460, 1996.

95. Hayden, J.H., Allen, R.D., and Goldman, R.D., Cytoplasmic transport in keratocytes: direct visual-ization of particle translocation along microtubules, *Cell Motil.*, 3, 1, 1983.

96. Seitz-Tutter, D., Langford, G.M., and Weiss, D.G., Dynamic instability of native microtubules from squid axons is rare and independent of gliding and vesicle transport, *Exptl. Cell Res.*, 178, 504, 1988.

97. Cassimeris, L., Pryer, N.K., and Salmon, E.D., Real-time observations of microtubule dynamic instability in living cells, *J. Cell Biol.*, 107, 2223, 1988.

98. deBeer, E.L., Sontrop, A.M.A.T.A., Kellermeyer, M.S.Z., Galambos, C., and Pollack, G.H., Actin-filament motion in the *in vitro* motility assay has a periodic component, *Cell Motil. Cytoskel.* 38, 341, 1997.

99. Qian, H., Sheetz, M.P., and Elson, E.L., Single particle tracking (analysis of diffusion and flow in two-dimensional systems), *Biophys. J.*, 60, 910, 1991.

100. Saxton, M.J. and Jacobson, K., Single-particle tracking: applications to membrane dynamics, *Annu. Rev. Biophys. Biomol. Struct.*, 26, 373, 1997.

101. Periasamy, A. and Day, R.N., Visualizing protein interactions in living cells using digitized GFP imaging and FRET microscopy, *Meth. Cell Biol.*, 58, 293, 1999.

102. Pierce, D.W. and Vale, R.D. Single-molecule fluorescence detection of green fluorescent protein and application to single-protein dynamics, *Meth. Cell Biol.*, 58, 49, 1999.

103. Sammak, P.J. and Borisy, G.G., Direct observation of microtubule dynamics in living cells, *Nature (London)*, 332, 724, 1988.

104. Kuznetsov, S.A. and Weiss, D.G., Use of acrosomal processes in actin motility studies, in *Cell Biology: A Laboratory Handbook*, 2nd ed., Vol. 2, Celis, J.E., Ed., Academic Press, San Diego, 1998, 344.

105. Allen, R.D. and Weiss, D.G., An experimental analysis of the mechanisms of fast axonal transport in the squid giant axon, in *Cell Motility: Mechanism and Regulation, 10. Yamada Conference, Sept., 1984, Nagoya*, Ishikawa, H., Hatano, S., and Sato, H., Eds, University of Tokyo Press, 1985, 327.

106. Keating, T.J., Peloquin, J.G., Rodionov, V.I., Momcilovic, D., and Borisy, G.G., Microtubule release from the centrosome, *Proc. Natl. Acad. Sci. U.S.A.*, 94, 5078, 1997.

107. Koonce, M.P. and Schliwa, M., Bidirectional organelle transport can occur in cell processes that contain single microtubules, *J. Cell Biol.*, 100, 322, 1985.

108. Brady, S.T., A novel brain ATPase with properties expected for the fast axonal transport motor, *Nature (London)*, 317, 73, 1985.

109. Scholey, J.M., Porter, M.E., Grissom, P.M., and McIntosh, J.R., Identification of kinesin in sea urchin eggs and evidence for its localization in the mitotic spindle, *Nature (London)*, 318, 483, 1985.

110. Vale, R.D., Reese, T.S., and Sheetz, M.P., Identification of a novel, force-generating protein, kinesin, involved in microtubule-based motility, *Cell*, 42, 39, 1985.

111. Paschal, B.M., Shpetner, H.S., and Vallee, R.B., MAP 1C is a microtubule-activated ATPase which translocates microtubules *in vitro* and has dynein-like properties, *J. Cell Biol.*, 105, 1273, 1987.

112. Yanagida, T., Nakase, M., Nishiyama, K., and Oosawa, F., Direct observation of motion of single F-actin filaments in the presence of myosin, *Nature (London)*, 307, 58, 1984.

113. Sellers, J.R. and Kachar, B., Polarity and velocity of sliding filaments: control of direction by actin and of speed by myosin, *Science*, 249, 406, 1990.

114. Kron, S.J. and Spudich, J.A., Fluorescent actin filaments move on myosin fixed to a glass surface, *Proc. Natl. Acad. Sci. U.S.A.*, 83, 6272, 1986.

115. Block, M., Fahrner, K.A., and Berg, H.C., Visualization of bacterial flagella by video-enhanced light microscopy, *J. Bacteriol.*, 173, 943, 1991.

116. Steinberger, B., Petersen, N., Petermann, H., and Weiss, D.G., Movement of magnetic bacteria in time-varying magnetic fields, *J. Fluid Mechanics*, 273, 189, 1994.

117. Vale, R.D. and Toyoshima, Y.Y., Rotation and translocation of microtubules *in vitro* induced by dyneins from *Tetrahymena* cilia, *Cell*, 52, 459, 1988.

118. McNiven, M.A. and Porter, K.R., Microtubule polarity confers direction to pigment transport in chromatophores, *J. Cell Biol.*, 103, 1547, 1986.

119. Rogers, S.L., Tint, I.S., Fanapour, P.C., and Gelfand, V.L., Regulated bidirectional motility of melanophore pigment granules along microtubules *in vitro*, *Proc. Natl. Acad. Sci. U.S.A.*, 94, 3720, 1997.

120. Weiss, D.G. and Buchner, K., Axoplasmic transport in olfactory receptor neurons, in *Molecular Neurobiology of the Olfactory System*, Margolis F.L. and Getchell T.V., Eds., Plenum Press, New York, 1988, 217.

121. Allen, R.D., Metuzals, J., Tasaki, I., Brady, S.T., and Gilbert, S.P., Fast axonal transport in squid giant axon, *Science*, 218, 1127, 1982.

122. Llinás, R., Sugimori, M., Lin, J.-W., Leopold, P.L., and Brady, S.T., ATP-dependent directional movement of rat synaptic vesicles injected into the presynaptic terminal of squid giant synapse, *Proc. Natl. Acad. Sci. U.S.A.*, 86, 5656, 1989.

123. Forscher, P. and Smith, S.J., Actions of cytochalasins on the organization of actin filaments and microtubules in a neuronal growth cone, *J. Cell Biol.*, 10, 1505, 1988.

124. Goldberg, D.J. and Burmeister, D.W., Looking into growth cones, *Trends Neurosci.*, 12, 503, 1989.

125. Herrera, A.A. and Banner, L.R., The use and effects of vital fluorescent dyes: observation of motor nerve terminals and satellite cells in living frog muscles, *J. Neurocytol.*, 19, 67, 1990.

126. Weiss, D.G., Videomicroscopic measurements in living cells: dynamic determination of multiple end points for *in vitro* toxicology, *Molec. Toxicol.*, 1, 465, 1987.

127. Maile, W., Lindl, T., and Weiss, D.G., New methods for cytotoxicity testing: quantitative video microscopy of intracellular motion and mitochondria-specific fluorescence, *Molec. Toxicol.*, 1, 427, 1987.

128. Geisler, B., Weiss, D.G., and Lindl, T., Video-microscopic analysis of the cytotoxic effects of 2-hydroxyethyl methacrylate in diploid human fibroblasts, *In-vitro Toxicology*, 8, 367, 1996.

129. Kishino, A. and Yanagida, T., Force measurements by micromanipulation of a single actin filament by glass needles, *Nature (London)*, 334, 74, 1988.

130. Kamimura, S. and Takahashi, K., Direct measurement of the force of microtubule sliding in flagella, *Nature (London)*, 293, 566, 1985.

131. Felgner, H., Frank, R., and Schliwa, M., Flexural rigidity of microtubules measured with the use of optical tweezers, *J. Cell Sci.*, 109, 509, 1996.

132. Mikhailov, A. and Gundersen, G.G., Relationship between microtubule dynamics and lamellipodium formation revealed by direct imaging of microtubules in cells treated with nocodazole or taxol, *Cell Motil. Cytoskel.*, 41, 325, 1998.

133. Kurachi, M., Kikumoto, M., Tashiro, H., Komiya, Y., and Tashiro, T., Real-time observation of the disassembly of stable neuritic microtubules induced by laser transection: possible mechanisms of microtubule stabilization in neurites, *Cell Motil. Cytoskel.*, 42, 87, 1999.

134. Vyshenskaja, T.V., Onishchenko, G.E., Petrashchuk, O.M., Tychinsky, V.P., and Nikandrov, S.L., Chromatin dynamics measured by phase microscopy in different cell cycle stages, *Eur. J. Cell Biol.*, 78(Suppl. 49), 29, 1999.

135. Tychinsky, V.P. Yagudzinsky, L.S., Leterrier, J.-F., Odensjö-Leterrier, M., and Weiss, D.G., Real-time measurements of mitochondrial activity using the dynamic phase microscopy method, *Eur. J. Cell Biol.*, 78(Suppl. 49), 79, 1999.

136. Schröck, E., du Manoir, S., Veldman, T., Schoell, B., Wienberg, J., Ferguson-Smith, M.A., Ning, Y., Ledbetter, D.H., Bar-Am, I., Soenksen, D., and Ried, Y.G.T., Multicolor spectral karyotyping of human chromosomes, *Science*, 273, 494, 1996.

137. Budde, A., Mosenheuer, M., Weiss, D.G., and Gemperlein, R., Spectral image analysis in oocytes and cultured cells by a Fourier-interferometric modulated light technique, *Eur. J. Cell Biol.* 78(Suppl. 49), 81, 1999.

138. Giuliano, K.A., Nederlof, M.A., DeBiasio, R., Lanni, F., Waggoner, A.S., and Taylor, D.L., Multi-mode light microscopy, in *Optical Microscopy for Biology*, Herman, B. and Jacobson, K., Eds., Wiley-Liss, New York, 1990, 543.

139. Luby-Phelps, K., Taylor, D.L., and Lanni, F., Probing the structures of cytoplasm, *J. Cell Biol.*, 102, 2015, 1986.

140. Weiss, D.G. and Steffen W., Lichtmikroskopie heute: Die lebende Zelle als Meßküvette, *BioSpektrum*, 4, 67, 1998.

141. Siegumfeldt, H., Rechinger, K.B., and Jakobsen, M., Use of fluorescence ratio imaging for intracellular pH determination of individual bacterial cells in mixed cultures, *Microbiology*, 145, 1703, 1999.

142. Baju, B., Murphy, E., Levy, L.A., Hall, R.D., and London, R.E., A fluorescent indicator for measuring cytosolic free magnesium, *Am. J. Physiol.*, 256, C540, 1989.

143. Prpic, V., Cowlen, M.S., and Adams, D.O., Application of digital imaging microscopy to studies of ion fluxes in murine peritoneal macrophages, in *Optical Microscopy for Biology*, Herman, B. and Jacobson, K., Eds., Wiley-Liss, New York, 1990, 337.

144. Wedekind, P., Kubitscheck, U., Heinrich, O., and Peters, R., Line-scanning microphotolysis for diffraction limited measurements of lateral diffusion, *Biophys. J.*, 71, 1621, 1996.

145. Wade, M.H., Trosko, J.E., and Schindler, M., A fluorescence photobleaching assay of gap junction-mediated communication between human cells, *Science*, 232, 525, 1986.

146. Van Noorden, C.J.F., *In situ* measurements of enzyme reactions, *Eur. Micr. Anal.*, 7, 11, 1990.

147. Langridge, W.H.R., Fitzgerald, K.J., Koncz, C., Schell, J., and Szalay, A.A., Dual promoter of *Agrobacterium tumefaciens* mannopine synthase genes is regulated by plant growth hormones, *Proc. Natl. Acad. Sci. U.S.A.*, 86, 3219, 1989.

148. Langridge, W.H.R. and Szalay, A.A., Bacterial and coelenterate luciferases as reporter genes in plant cells, *Meth. Molec. Biol.*, 82, 385, 1998.

149. Mayerhofer, R., Araki, K., and Szalay, A.A., Monitoring of spatial expression of firefly luciferase in transformed zebrafish, *J. Biolumin. Chemilumin.*, 10, 271, 1995.

150. Bastiaens, P.I.H. and Jovin, T.M., Fluorescence resonance energy transfer microscopy, in *Cell Biology: A Laboratory Handbook,* 2nd ed., Vol. 3, Celis, J.E., Ed., Academic Press, San Diego, 1998, 136.

151. Brown, A.F. and Dunn, G.A., Microinterferometry of the movement of dry matter in fibroblasts, *J. Cell Sci.*, 93, 56, 1989.

152. Montag, M., Wild, A., Spring, H., and Trendelenburg, M.F., Visualization of chromatin arrangement in giant nuclei of mouse trophoblast by videomicroscopy and laser scanning microscopy, in *Nuclear Structure and Function,* Harris, J.R. and Zbarsky, J.B., Eds., Plenum Press, New York, 1990, 272.

153. Forscher, P., Kaczmarek, L.K., Buchanan, J., and Smith, S.J., Cyclic AMP induces changes in distribution and transport of organelles within growth cones of *Aplysia* bag cell neurons, *J. Neurosci.*, 7, 3600, 1987.

154. Petermann, H., Weiss, D.G., Bachmann, L., and Petersen, N., Motile behaviour and measurement of the magnetic moment of magnetotactic bacteria in rotating magnetic fields, in *Biological Motion, Lecture Notes in Biomathematics*, Vol. 89, Alt, W. and Hoffmann, G., Eds., Springer-Verlag, Berlin, 1990, 387.

155. Gross, D. and Loew, L.M., Fluorescent indicators of membrane potential: microspectrofluorometry and imaging, in *Methods in Cell Biology*, Vol. 30, chap. 7, Taylor, D.L. and Wang, Y.-L., Eds., Academic Press, San Diego, 1989, 193.

156. Obaid, A.L., Koyano, T., Lindstrom, J., Sakai, T., and Salzberg, B.M., Spatiotemporal pattern of activity in an intact mammalian network with single-cell resolution: optical studies of nicotinic activity in an enteric plexus, *J. Neurosci.*, 19, 3073, 1999.
157. Mitchison, T. J., Sawin, K.E., and Theriot, J.A., Caged fluorescent probes for monitoring cytoskeleton dynamics, in *Cell Biology: A Laboratory Handbook*, 2nd ed., Vol. 2, Celis, J.E., Ed., Academic Press, San Diego, 1998, 127.

13 Imaging Cilia and Flagella

Michael E. J. Holwill, Helen C. Taylor, and Hervé Delacroix

CONTENTS

13.1 INTRODUCTION

The key structure in a eukaryotic flagellum or cilium is the axoneme; the axoneme consists of a cylindrical array of nine doublet microtubules surrounding a pair of singlet microtubules, together with a variety of microtubule linkages and projections.[1] The mechanism responsible for the movement of these organelles is associated with actively sliding microtubules,[2] in which the dynein arms on one peripheral doublet of the axoneme exert a tipward force on the neighboring doublet.[3,4] (For convenience, the two types of organelle are collectively referred to as cilia.) The shearing forces produced by the dynein–tubulin interaction are converted into bending moments by basal structures and interdoublet linkages that oppose free sliding, and the coordinated activity of the dynein–microtubule interaction results in the bending patterns seen on the intact organelles. While there is considerable evidence to support the sliding microtubule model, the mechanism by which bends are initiated and propagated has yet to be described in detail, although many of the physical constraints required can be specified.

The relationships between structure and function in cilia are investigated through images obtained by light and electron microscopy, and a range of techniques of image processing and analysis has been used to interpret the data. High-resolution images are recorded photographically, but recent technical advances have allowed images to be generated electronically from both optical and electron microscopes. Computational analysis requires information in digital form, which can be produced from photographic images by a variety of techniques. Many of the images will require the removal of noise or artifacts prior to analysis. To interpret the two-dimensional raw data obtained by both light and electron microscopy, three-dimensional computer models have been developed and tested by comparing equivalent views of the real system and the model. This chapter reviews all these procedures and comments on their general effectiveness in relation to the study of cilia.

13.2 LIGHT MICROSCOPE IMAGES

The beat patterns of cilia observed through the light microscope and recorded photographically or electronically reflect the mechanochemical events that generate the internal forces needed to deform the organelle. A knowledge of the precise shape adopted by a cilium, together with its beat frequency, can therefore provide information about the macromolecular mechanisms responsible for motility. The beat frequency is a parameter that can be measured directly using a stroboscope,[5] photodetector methods,[6] or laser doppler techniques.[7] Monitoring the movement of gold beads attached to doublets[8] in a reactivated axoneme has provided data consistent with microtubule sliding. The determination of bend shape requires sophisticated interpretive techniques, usually involving the use of a computer. To apply the techniques, it is necessary to supply the spatial coordinates of the flagellar image to the computer. This can be achieved by tracking the image manually (e.g., using a bitpad); this is a time-consuming process, which can be prone to operator error and bias, but one that has been used to good effect.[9] To acquire coordinates without the introduction of this bias, various techniques for the partially automated collection of data have been devised. Because flagellar images are often recorded by dark-field microscopy, so that an image appears as a bright line on a dark background, simple thresholding procedures can be employed to detect the position of the line. However, the images are often imperfect, with variations in width and intensity along their lengths, and foreign material in the neighborhood of the flagellum may produce light scattering that degrades the image. Further, the image of the region where the flagellum is attached to the cell is usually obscured by the high intensity of light scattered by the cell body. These factors render it difficult to employ a totally automatic system for digitization, and the techniques described for this procedure generally require the intervention of an operator at some stage.

An electro-optical curve follower[10,11] designed specifically to record coordinates of flagellar waveforms utilized a four-quadrant photodiode to monitor the intensity of light from an image on a back-projection screen. When the detector was positioned manually at one end of the flagellar image, the intensities of light falling on the quadrants provided feedback to stepping motors in such a way that the detector followed the image. Cartesian coordinates of the flagellum were recorded as voltages from linear potentiometers at the edges of the screen. While the technique was found to be as accurate as manual tracking methods, and has been used successfully,[12] the development of image digitization techniques has led to alternative procedures that allow coordinate recording to be achieved more rapidly than, and as accurately as, the mechanical tracking system.

Digitized images of cilia are derived either from photographs,[13,14] using an appropriate 35-mm or 16-mm camera, or from high-speed video recordings using the appropriate computer interface and software.[15] In extracting information from a digitized image, advantage can be taken of the fact that the representation of the cilium should be a fine line on a contrasting background. In the technique developed by Baba and Mogami,[13] an arc-shaped strip approximately normal to the cilium was generated using a computer program, the pixels within the strip were examined, and those with intensities greater than a set threshold were identified. The distribution of average intensities across the strip when it straddles a line image has a maximum in the central region and drops to zero on either side; such an intensity profile serves to identify the cilium image and was used as a comparator in the computer program. The starting position of the strip on the cilium image was chosen by an operator, and the coordinates of the center of the cilium were calculated from the average position of the significant pixels. By moving the strip along the cilium under computer control, the coordinates of the ciliary centerline could be established. If the intensity distribution differed from that expected, the operator was alerted so that appropriate action could be taken.

In an alternative approach that provides data in a form that requires no further processing for certain analyses, Brokaw[14] has fitted the digitized image of a sperm with a model consisting of a stylized head joined to a series of connected straight-line segments 0.5 μm or 1 μm long representing the sperm tail. In fitting the model to the image, the optimum position of the head is first established by monitoring the distribution of intensities in the image. Subsequently, using a similar intensity criterion, the angle between adjacent segments is altered until an optimum fit is found. Data are therefore recorded as intrinsic coordinates, that is, the distance (s) of a segment along the image together with the angle (ϕ) of that segment relative to a reference line, in this case in the direction of the basal segment.

A digitized image is not a perfect reproduction of the original because of noise introduced by the recording techniques used. It is therefore usual to perform some averaging by repeatedly tracking an individual cilium, sometimes with associated rotation of the digitizing camera to compensate for nonuniformities in its detector.

Having obtained the coordinate data, several analyses have been directed toward establishing the precise form of the bend shape on a cilium, because this has implications for the internal mechanochemistry. If it has not already been done as part of the digitization process, the data is converted into the intrinsic coordinates (ϕ, s) described above. The nature of the (ϕ, s) curve has been investigated graphically in terms of the variation of curvature (= $d\phi/ds$) with s, and interpreted in terms of the internal mechanism of motility. In early studies, visual curve-fitting procedures led to ciliary wave-shapes being described as arc-line,[16] meander-like,[17] and sine-generated;[18] these shapes appear similar when plotted in Cartesian coordinates (Figure 13.1a), but their intrinsic forms (Figure 13.1b) are significantly different from one another. Each shape has a different implication for the internal mechanism responsible for motility. An arc-line shape, with its transition from the circular to the linear configuration, or vice versa, may reflect abrupt conformational changes at the macromolecular level within the axoneme. The meander is the shape adopted by a bent elastic rod freely hinged at both ends to minimize its potential energy; while such constraints might not be truly representative of ciliary dynamics, if the cilium is found to adopt the shape of a meander, it may imply that the elasticity of the axoneme plays a major role in wave formation or propagation. The simple relation that characterizes the sine-generated wave can, if this is the shape of a cilium, have implications for the molecular oscillator involved in bend generation.

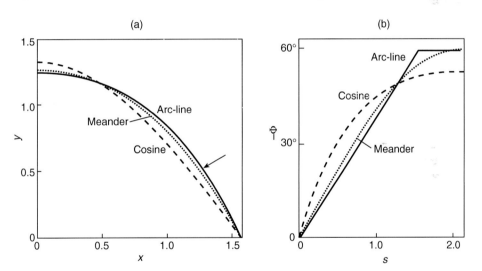

FIGURE 13.1 Plots of arc-line, meander, and cosine curves in (a) Cartesian and (b) intrinsic coordinates. The arrow in (a) indicates the junction between the arc and line. The cosine curve is included for comparison purposes, as flagellar waves generally do not assume this shape. The sine-generated curve is indistinguishable from the meander at this resolution.

Because the distinctions between forms are relatively subtle, it is desirable to have an objective analytical procedure to investigate wave-shapes, rather than a visual curve-fitting approach, which can be somewhat subjective. One such objective approach is to consider the coefficients of the Fourier series that represents the (ϕ, s) curve.[19] Since the flagellar bends tend to be asymmetric (Figure 13.2A), the Fourier approach is not directly applicable. To overcome the analytical problems, and also to allow investigation of changes in wave-shape as a wave propagates, the digitized flagellum is divided into quarter-waves (Figure 13.2B); the symmetric waves generated by appropriate reflection and translation of the quarter-waves are then analyzed using the Fourier technique.

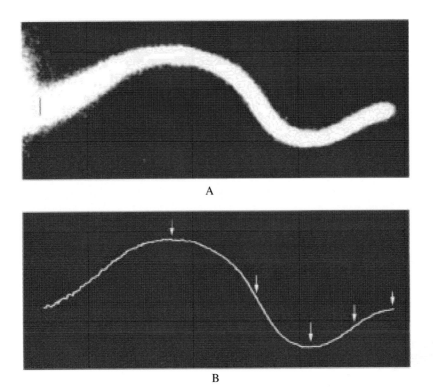

FIGURE 13.2 (A) dark-field image of the flagellum of *Crithidia oncopelti*; the short vertical line at the base of the flagellum indicates the position from which digitization is initiated. (B) digitized representation of the image in the panel above; the vertical arrows indicate the division of the flagellum into quarter-waves in preparation for Fourier analysis.

A given curve has a unique set of values for the coefficients, so that ciliary bend-shapes can be described unambiguously and compared with known theoretical wave-shapes (Table 13.1). In *Crithidia*, for example, this analysis has shown that bends on *in vivo* flagella fit an arc-line,[12] while those on demembranated flagella fit the meander better.[20] This implies that the internal elastic and external viscous constraints are different for the two types of preparation. The differences are presumably associated with structures, such as the membrane, removed from the living flagellum to produce the *in vitro* preparation. The bends on the hispid flagellum of *Ochromonas* are matched closely by the arc-line.[21] The mechanical loading on this flagellum, with its rigid, hair-like projections, is different from that on the smooth flagellum of *Crithidia*, which has the same bend-shape, suggesting that the arc-line form is imposed by the intact axoneme, which is therefore not affected significantly by external forces. The Fourier method tends to smooth out irregularities in the (ϕ, s) curves, which can be useful where the irregularities originate from noise (see below).

Additional information is often available from the analysis of bend forms. For example, Baba and co-workers[13,22] report abrupt changes in the curvature of bends, which they interpret in terms of a quantization of the bending process.

The analytical procedures described above have been applied to images of flagella with two-dimensional waveforms, so that a sharp image is recorded either photographically or electronically. In many recordings of cilia, parts of the image are blurred, indicating that the beat pattern is not truly planar, but three-dimensional. Woolley[23,24] has provided a quantitative description of the three-dimensional nature of hamster sperm tail undulations by analyzing the blurred regions of dark-field images of the flagellum. Woolley used the fact that the width of the blurred region of a flagellar image varies linearly with the perpendicular distance of the flagellum from the focal plane. The direction of the displacement was determined by covering one half of the microscope condenser

TABLE 13.1
Relative Values of Fourier Coefficients
for Different Wave Shapes

	Fourier Coefficient			
Curve	1st	2nd	3rd	4th
Sinusoid	100	13.5	3.9	1.0
Arc-line	100	−10.41	3.26	−1.31
Meander	100	0.3	0	0
Sine-generated	100	0	0	0

with a colored filter; this resulted in a coloration of one edge of an underfocused image but the opposite edge of an overfocused image. In this way, the bends on the hamster sperm were shown to be essentially planar, but the planes of successive bends were not the same.

Sugino and Machemer[25] have reconstructed the three-dimensional beat pattern of the proximal portion of *Stylonychia* cilia by analyzing the two-dimensional geometry of the motion recorded by high-speed cinephotography, with the viewing direction, on average, along the axis of the cilium. The sequence of images obtained suggested that the proximal region of the cilium moved in the surface of a cone with an elliptical cross-section and with an axis that was not perpendicular to the cell surface. To describe the cone quantitatively, it was assumed that the cilium image of maximum length represented the true length of the organelle, and determined the angular extent of the beat in the image plane from the extreme angles made by the proximal portion of the cilium with the anterio-posterior axis of the cell. From these parameters, as well as the minimum length of the two-dimensional image of the cilium, Sugino and Machemer obtained expressions for the angle of inclination of the cone to the cell surface and the eccentricity of the cross-section. In one example, the cone axis was inclined at $63°$ to the cell surface, while the ratio of the lengths of the major and minor axes of the ellipse was 6.5. They also showed that the proximal region moved continuously in one direction to sweep out the conical surface. If the axoneme does not twist along the cilium, the orientation of the microtubules relative to the cell axis remains unaltered during the ciliary beat. By assuming that the dynein arms on five of the microtubules are active at any one time, the moments needed to generate the three-dimensional movement observed can be produced by the unidirectional progression of dynein activation and deactivation around the axoneme. Both the microtubule sliding rate and the rate of progression of dynein activation/deactivation were found to vary with the stage of the ciliary cycle. The analysis is limited to the behavior of the proximal region of the axoneme, and more extensive studies are required to determine the three-dimensional beat pattern of the entire cilium.

13.3   ELECTRON MICROSCOPE IMAGES

Electron microscopy is capable of producing images with a higher resolution than those generated by the optical microscopy discussed in Section 13.2, and has been used in the analysis of axonemal morphology. In preparing cilia samples for the high-vacuum environment of the electron microscope, two general procedures are used: one based on dehydration and fixation, the other on rapid freezing. Fixed samples may be viewed whole or after sectioning, but in both cases heavy-metal stains are used to generate contrast in the image. Rapidly frozen samples are split open and a metal replica made of the exposed surface; the replica is examined in the electron microscope. The general axonemal structure revealed by electron microscopy is well-understood, but details of structures such as the dynein motor domains are controversial. Electron microscopy of disintegrating axonemes has demonstrated that the motor molecules, arranged in two parallel rows (inner and outer arms) on each peripheral microtubule doublet, produce the forces that cause relative sliding of the microtubule doublets within the axoneme.[26] The motor molecules are about 10 nm in diameter and require careful manipulation to minimize damage during specimen preparation. Many researchers believe that the structural damage that might occur during these preparations leads to the introduction of artifacts. Dynein is an ATPase that undergoes a cycle of activity involving ATP hydrolysis and changes in molecular conformation. During the cycle, it interacts with the neighboring microtubule doublet and applies the force needed to cause inter-doublet sliding. Differences in the conformational states of the outer dynein arms have been observed on micrographs of axonemes exposed to different ATP concentrations. However, micrographs of samples exposed to the same ATP concentration, but prepared for electron microscopy using the different techniques of freeze-etch, negative-stain, and thin-section, show the outer arms in different relationships to other axonemal structures.[27] Critical interpretation of the observations has led to the conclusion that the dynein arm is arrested in a particular phase of its activity and, therefore, with a specific structural configuration depending on the preparative technique used.[28]

As with the light microscope images, objective techniques have been applied to the interpretation of electron micrographs. Many of these techniques aim to increase the signal-to-noise ratio of the images, that is, to produce a reconstructed image that contains less spurious information than the original. This noise can originate from a number of different sources, such as specimen distortion or uneven deposition of metal during the preparative process. These techniques provide enhanced two-dimensional images that have been used in several structural-evaluation studies. Other processes have been developed to obtain three-dimensional structural information from a series of two-dimensional micrographs.

13.3.1 ENHANCEMENT OF TWO-DIMENSIONAL IMAGES

Fourier filtering of images of periodic structures has been used to increase the signal-to-noise ratio. In early studies,[29] optical diffraction patterns obtained from the electron micrographs were used to determine periodicities in ciliary microtubules. The diffraction pattern, recorded on film, is masked to retain only those intensities due to the repeating structure and obscure intensities originating from "noise" in the micrograph; reconstruction of the repeating structure is then possible by placing the masked pattern in a suitable optical system.[30,31] The modern and widely used equivalent of this procedure is to filter the Fourier transform of those areas of a digitized micrograph that contain repeating structures, and to compute the inverse Fourier transform, thereby creating a "reconstructed" image of the average structure.[32] Burgess et al.[27] have applied this approach to the study of outer dynein arm structure in the domestic fowl sperm flagellum, and have produced drawings showing different views of the arms in relaxed and rigor states.

The use of image averaging is not restricted to images with linear periodicity but can also be performed on images of structures such as cross-sections of a ciliary axoneme. In this case, image averaging requires a number of digitized micrographs of equivalent specimen sections that must first be aligned with respect to one another by appropriate rotation and translation. These axonemal sections cannot usually be aligned directly because of distortion introduced by sectioning the specimen. To overcome this problem, each of the images has to be stretched, centered, and oriented. Afzelius et al.[33] used a manual procedure to determine the axes of the ellipse into which the normally circular axoneme had been deformed prior to stretching the section to restore it to its circular shape. Nitschké et al.[34] have developed a semi-automated method in which the coordinates that identify the position of each peripheral microtubule are retrieved from the cross-correlation map (Figure 13.3B) between the original image and an ideal tubule (Figure 13.3A); these coordinates are then used to compute the parameters of the geometric transformation that will stretch, rotate, and translate the components of the axonemal section to give an image in which the doublets lie on a circle with a specified radius (Figure 13.3C). By repeating this procedure for a number of cross-sections, a set

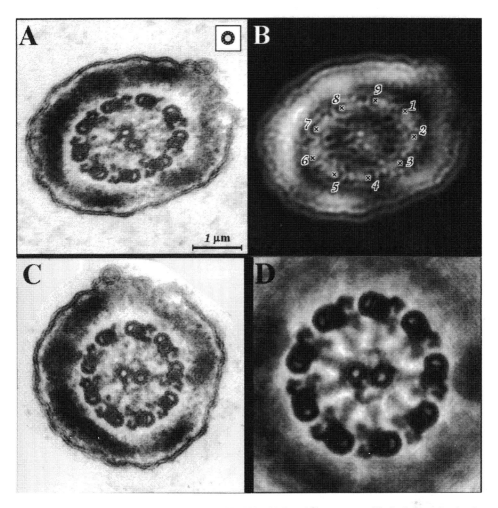

FIGURE 13.3 Steps in the procedure developed by Nitschké et al.[34] to correct elliptical distortion in electron micrographs of ciliary cross-sections. (A) Original distorted image of a human sperm flagellum with idealized microtubule inset; (B) cross-correlation of image in (A) with idealized microtubule. Each numbered cross indicates the location of the center of one of the microtubules in a doublet; (C) corrected image, with the microtubules arranged in a circle; and (D) average image derived from 11 corrected cross-sections.

of matched images, suitable for image averaging, is generated; Figure 13.3D shows an average obtained from 11 cross-sections of human sperm flagella.

Alignment prior to image averaging was also used by O'Toole et al.[35] in their studies of peripheral microtubule doublets and the attached dynein arms. In these studies,[36,37] images of clearly-defined microtubule doublets were extracted and positioned in a 70 × 90 pixel rectangle. Transformations were applied to overlay between 50 and 100 doublets, and both low-pass and high-pass filtering were performed to enhance various components of the image before image averaging proceeded. Stain densities of thinly sectioned samples from the same species were found to vary from day to day, and reliable statistical comparisons required three independent preparations. The three averaged images were aligned using linear transformations and the intensities of the images normalized (using the outer dynein arm intensity for comparison) before a "grand average" was calculated. Studies comparing the grand averages obtained from doublets of wild-type and mutant *Chlamydomonas* are discussed below.

Reinforcement techniques can be used to investigate the angular periodicity of doublets in the axoneme cross-section. Reinforcement can be achieved by the physical rotation of a negative at the appropriate level in an optical system, such as a photographic enlarger.[31] When studying axonemes, such techniques are only effective if the microtubule doublets lie on a circle. As noted in the previous paragraph, distortion can result from the preparative techniques and each cross-section may need to be corrected using a digital method such as that discussed previously[34] before image reinforcement can be undertaken. Since the axoneme has nine-fold symmetry, the image is rotated through 40° about the axoneme center and superimposed on the original. This superposition is repeated nine times, a procedure that increases the prominence of the periodic structure but yields a blurred region at the axoneme center since the central pair does not have nine-fold symmetry.

Electron microscopy, often combined with image averaging techniques, has been used by a number of researchers in the structural studies of the axonemes of different cells. Afzelius et al.[33] superimposed correspondingly oriented images of a number of doublets to reveal a nonuniform arrangement of protofilaments that can be associated with attachment sites of the spokes and other doublet structures. The motility of several *Chlamydomonas* mutants has been observed and electron microscopy used to show whether outer or inner dynein arms were present.[38] This important study suggested that outer dynein arms controlled beat frequency, while the inner arms influenced the bend-shape of the beating flagella. Other mutants with various degrees of motility have been examined using electron microscopy and image analysis.[36,37,39] In many of the mutants, various inner-arm structures were found to be absent. To assess these cases, average images of cross-sections and longitudinal sections, each 60 nm in thickness, of doublet microtubules were produced for the different mutants. Gardner et al.[37] determined the locations of the missing inner-arm structures by considering the variance of these images with that of equivalent average images of wild-type *Chlamydomonas*. The procedure is illustrated in Figure 13.4, which shows average images of the inner arms from (A) wild-type and (B) mutant *Chlamydomonas* flagella; Figure 13.4C shows the difference between the images in Figure 13.4A and 13.4B, thereby indicating the structures missing in the mutant axoneme. Genetic analysis of these mutants has enabled many of the inner-arm components to be identified chemically. The result is a two-dimensional map showing the location of seven isoforms of inner-arm dynein and a dynein regulatory complex (Figure 13.9 of Gardner et al.[37]).

The selection of a suitable portion of an axonemal or doublet image for these studies is to some extent subjective, and it is therefore possible to eliminate data relating to actual structures rather than to artifacts. For example, averaging all nine peripheral doublets together assumes there is no radial asymmetry within the axoneme. By comparing the variance of averaged images of individual doublets, King et al.[39] suggest that there may be differences in the inner-arm structure between doublets in a *Chlamydomonas* mutant.

The Fourier and image-averaging techniques are valuable tools for the study of periodic structures that have identical configurations over long distances. However, they have limited power in studies of active axonemes where dynein arms and other structures may adopt different conformations. Application of a Fourier transform forces a periodicity onto the arms irrespective of the nature of the original image, and each repeating element will be represented in precisely the same way by an average of the corresponding elements in the original image. Similarly, in averaging 50 to 100 cross-sections of 60-nm thickness, average structures for the outer arms and the three different inner-arm complexes will be obtained. As a consequence, neither technique will provide information about the conformation changes that occur during the cycles of individual inner and outer arms.

An objective method of image classification based on multivariate statistical analysis has been applied to images of outer dynein arms to investigate conformational changes resulting from differing axonemal preparations.[40] In this study, "rigor," "relaxed," and "active" domestic fowl sperm axonemes were prepared using differing concentrations of ATP, and freeze-etch replicas of longitudinal slices showing outer dynein arms arranged between doublets were obtained. The

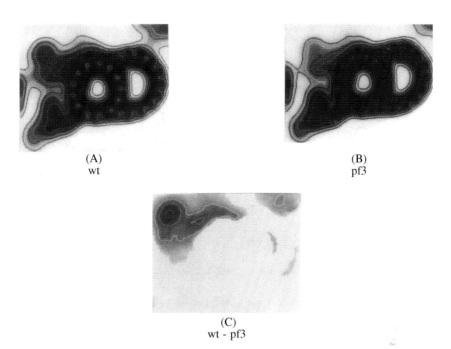

(A)
wt

(B)
pf3

(C)
wt - pf3

FIGURE 13.4 Illustration of the procedure used by Gardner et al.[37] to identify structures present in wild-type axonemes but missing from mutant axonemes. Average images of the inner arms from (A) wild-type and (B) mutant *Chlamydomonas* flagella; (C) difference between the images in (A) and (B), showing the structures missing from the mutant axoneme. (From Gardner et al., *J. Cell Biol.*, 127, 1318, 1994. With permission from the Rockefeller University Press.)

negatives were digitized using a densitometer and the images coded to conceal their identities. Individual outer arms were windowed, assigned to bins on the basis of their orientation on the axonemal surface, and aligned. Using discriminant analysis, it was shown that the arms could be classified objectively as rigor-like, relaxed-like, or active-like. The two-dimensional morphologies of arms in the relaxed and active states were found to be very similar but differed from the rigor state. On the basis of this study, Burgess[40] has generated a diagram to show the proposed two-dimensional conformational change that occurs between the relaxed and rigor states.

13.3.2 THREE-DIMENSIONAL IMAGE RECONSTRUCTION

Three-dimensional tomographic reconstructions can be made directly, either from a stack of serial sections or from a tilt-angle series of images.[41] Organelles like the cilium, basal body, or centriole have a complex structure that can be retrieved more accurately from sets of serial sections than from a tilt-angle series. Procedures are required to accurately align the serial sections and hence to preserve the shape of the organelle. For this purpose, latex beads, with diameters that are very large in comparison to the mean depth of the sections, are included in axoneme preparations. After sectioning, it is possible to accurately arrange successive sections by aligning the corresponding beads. The overall structure of the human centriole, an organelle similar to the basal body of the cilium, has been determined using this technique (Figure 13.5).[42] The centriole is organized around a complex, doubly-twisted network of microtubule triplets. Each 2-D section (Figure 13.5A) in a regularly spaced series is processed to enhance the contrast (Figure 13.5B). Specially developed software utilizes the information from the enhanced series of images to generate a 3-D computer model showing the twisted microtubular core of the centriole (Figure 13.5C) and a solid reconstruction of the structure (Figure 13.5D).

Another approach to the interpretation of electron micrographs involves the construction of three-dimensional models of the structure. Models made from materials such as plasticine are time-

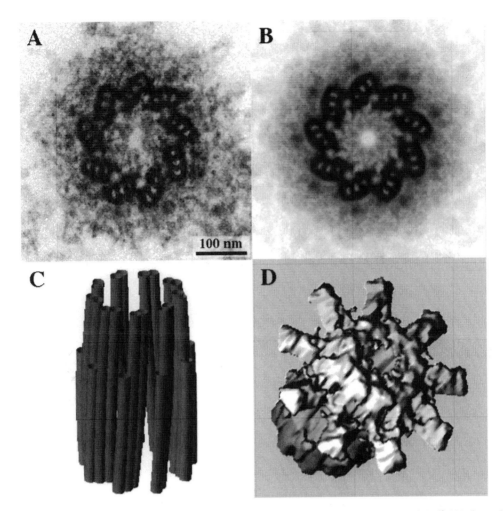

FIGURE 13.5 Stages in the reconstruction of a three-dimensional model of the centriole.[42] (A) One of a set of serial cross-sections of the centriole; (B) contrast-enhanced version of (A); (C) three-dimensional representation of the microtubular network of an isolated centriole; (D) solid view of the complete centriole.

consuming to construct, tend to be inaccurate, and cannot readily represent the dynamic properties of the system. Nevertheless, such methods have been employed to obtain a 3-D reconstruction of inner arms from a tilt-angle series of micrographs.[43] With the development of computer graphics has come the capability to construct, using appropriate software, complex three-dimensional models to be viewed on a visual display monitor; the images thus produced are two-dimensional, but the three-dimensional character is revealed by presenting different orientations of the structure on the monitor screen. This type of modeling allows for easy comparison of the proposed configuration with the electron micrograph, as both are two-dimensional projections of a three-dimensional structure.

A computer-modeling software package called SURREAL (SURface REndering ALgorithms) has been specifically designed to investigate the structure of ciliary axonemes.[44] SURREAL has been used to generate images of a model axoneme at 4-nm resolution with the aim of reconciling different interpretations of electron micrographs. The original model,[44] published in 1991, was constructed using evidence obtained from electron microscopy and optical diffraction of electron micrographs. This model is reassessed against experimental evidence as it becomes available. An example of how the model has developed can be seen in the cut-away model of the axoneme shown

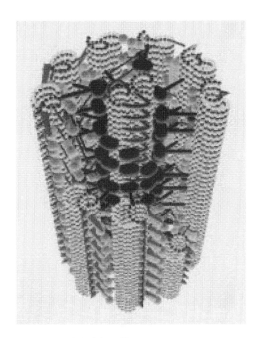

FIGURE 13.6 Computer-generated model of the ciliary axoneme, cut away to expose the internal structures. The nine peripheral doublet microtubules are built from α- and β-tubulin, and are shaded differently in the model. Radial spokes project from each peripheral doublet toward the center, and both the inner and outer dynein arms, with their spherical motor subunits, are clearly visible.

in Figure 13.6, which includes three-dimensional inner-arm structures that are consistent with all the electron micrographs and image-processed data published before 1999. In the inner-arm study, Taylor et al.[28] used several data sources to build a model of the three inner-arm complexes. The model was then viewed from angles appropriate for comparison with additional data and the model refined until it satisfied all the available information. The resulting three-dimensional model of inner dynein arms mounted on a doublet was found to agree with micrographs that were previously thought to give conflicting information. Computer graphics shareware is now freely available over the Internet. Although this software has generally been designed to give images which are visually pleasing rather than structurally accurate, it has been used successfully to produce a model of the axoneme. More modern software has several advantages over SURREAL in terms of flexibility and sophistication and is replacing SURREAL as the rendering tool.

Computer modeling can also be used to investigate dynamic aspects of axonemal behavior; a possible outer dynein arm cycle has been designed on the basis of distinct arm conformations determined from electron micrographs,[44] while tentative cycles for the inner arms have also been proposed[28] (Figure 13.7). The dynamics of arm activity can be assessed by viewing animations of the cycle.[45] These animations have enabled the interactions between the dynamic structures and adjacent axonemal structures to be studied.[46]

Computer modeling provides a convenient structural representation of the axoneme and complements experimental studies using electron microscopy and other techniques designed to reveal the axonemal structure. To generate the computer model, a precise knowledge of the relative positions, sizes, and shapes of all the individual structures is required, so that this approach reveals the areas in which the data are deficient and suggests appropriate experiments to rectify the deficiencies. In addition to providing structural information about the axoneme, computer modeling has the potential to provide interpretations relating to functional aspects of the system.[28] It is, for example, currently being used to study possible relations between the axonemal motors.[46]

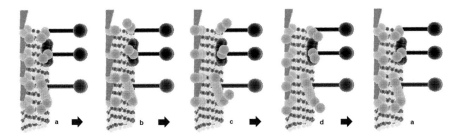

FIGURE 13.7 (a)–(d) Proposed conformations of inner and outer dynein arms at different stages of their mechanical cycles, during which they interact with the neighboring microtubule and exert a force on it directed toward the flagellar tip. The neighboring microtubule would lie between the reader and the arms. The cycles are presented in diagrammatic form, and are not synchronized as might be suggested by the figure.

13.4 FUTURE PERSPECTIVES

The investigation of ciliary movement involves a wide range of imaging procedures — some well-established, others, such as atomic force microscopy, just beginning to be applied to the problem. The established techniques of image analysis, together with some procedures designed specifically for the study of the axoneme, are yielding important information about the mechanism of motility. It is anticipated that these techniques, as well as further developments in image processing, in combination with appropriate experimental approaches, will be utilized in the solution of the many unresolved problems associated with the mechanochemical events that are responsible for the cyclic motion of the cilium.

The application of modern image analysis procedures relies on the availability of a digitized image, that is, one that can be stored numerically in a computer. Such images are recorded most conveniently by video and other electronic recording techniques. In many studies of cilia, the full potential of these recording procedures has yet to be realized. As previously discussed, digitization for the study of beat forms has usually been achieved by processing a photographic image. Although the cilium image is known to be a simple curve, and as such can be recorded accurately despite some discontinuities in the image, a procedure involving the processing of an intermediate image has the inherent disadvantage that additional noise and distortion can be introduced.

In light microscopy, photographic rather than video recording is used because standard video framing rates of 25 or 30 s^{-1} do not allow adequate time resolution of the ciliary beat (which usually has a frequency of about 20 Hz, but can be considerably higher). High-speed video techniques are available and have been used to record images of cilia at framing rates of 200 per second;[20,47] higher speeds can be achieved by modifying the line scanning control of the camera, but this degrades the image resolution. In addition, limited use has been made of image processing capabilities, such as contrast enhancement and background subtraction, since the recordings are made in a non-standard format. Technological developments in SIT and CCD video cameras, with appropriate modifications to allow higher than standard recording rates to be achieved, and the appropriate adaptation of image-processing software will allow high-quality, time-resolved video recordings of cilia to be made routinely.

Electronic image recording is becoming more common in electron microscopy, and image-enhancement techniques and analyses such as that of Afzelius et al.[33] have great potential for the provision of new structural details, and need to be explored further. With its higher resolution, the electron microscope also has the potential to provide improved wave-shape information compared with that available from optical microscopy. Surprisingly, this potential has been little exploited. It depends on accurate preservation of waveforms during preparation for electron microscopy, but there are good indications that this can be achieved. Studies of whole-mount specimens — either in scanning or transmission electron microscopy — should, in principle, give the detail required and have the potential to provide, simultaneously, indications of internal structural changes.

Prospects for time-resolved imaging at high magnification using either X-ray microscopy or time-resolved X-ray diffraction lie in the future, but have great potential for providing information about structural changes during bend generation and propagation that is unobtainable at present. Additionally, the scanning probe microscope used in various modes is capable of yielding structural data on the cilium at resolutions comparable to that achieved by electron microscopy, but with the sample in a liquid environment similar to the *in vivo* condition. Additionally, this instrument can be used to investigate the mechanical properties of the system, including its elasticity and force-generating capacity, with the added exciting prospect of a direct link between structure and function.

REFERENCES

1. Holwill, M. E. J. and Taylor, H. C., Mechanisms of flagellar propulsion, in *The Flagellates*, Leadbeater, B. S. C. and Green J. C., Eds., Taylor & Francis, London, 49, 2000.
2. Satir, P., Switching mechanisms in the control of ciliary motility, *Mod. Cell Biol.*, 4, 1, 1985.
3. Sale, W. S. and Satir, P., Direction of active sliding of microtubules in *Tetrahymena* cilia, *Proc. Natl. Acad. Sci. U.S.A.*, 74, 2045, 1977.
4. Woolley, D. M. and Brammall, A., Direction of sliding and relative sliding velocities within tripsinized sperm axonemes of *Gallus domesticus, J. Cell Sci.*, 88, 361, 1987.
5. Brokaw, C. J., Movement of the flagella of *Polytoma uvella, J. Exp. Biol.*, 40, 149, 1963.
6. Sanderson, M. J. and Dirksen, E. R., A versatile and quantitative computer-assisted photoelectronic technique used for the analysis of ciliary beat cycles, *Cell Motility*, 5, 267, 1985.
7. Lee, W. I. and Verdugo, P., Laser light-scattering spectroscopy. A new application in the study of ciliary activity, *Biophys. J.*, 16, 1115, 1976.
8. Brokaw, C. J., Direct measurements of sliding between outer doublet microtubules in swimming sperm flagella, *Science*, 243, 1593, 1989.
9. Brokaw, C. J., Automated methods for estimation of sperm flagellar bending parameters, *Cell Motility*, 4, 417, 1984.
10. Silvester, N. R. and Johnston, D., An electo-optical curve follower with analogue control, *J. Phys. E: Sci. Inst.*, 9, 990, 1976.
11. Johnston, D. N. and Silvester, N. R., A digitally controlled curve follower, *J. Phys. E: Sci. Inst.*, 12, 235, 1979.
12. Johnston, D. N., Silvester, N. R., and Holwill, M. E. J., An analysis of the shape and propagation of waves on the flagellum of *Crithidia oncopelti, J. Exp. Biol.*, 80, 299, 1979.
13. Baba, S. A. and Mogami, Y., An approach to digital image analysis of bending shapes of eukaryotic flagella and cilia, *Cell Motility*, 5, 475, 1985.
14. Brokaw, C. J., Computerized analysis of flagellar motility by digitization and fitting of film images with straight segments of equal length, *Cell Motil. Cytoskel.*, 17, 20, 1990.
15. Glazzard, A. N., Hirons, M. R., Mellor, J. S., and Holwill, M. E. J., The computer assisted analysis of television images as applied to the study of cell motility, *J. Submicrosc. Cytol.*, 15, 305, 1983.
16. Brokaw, C. J., Non-sinusoidal bending waves of sperm flagella, *J. Exp. Biol.*, 43, 155, 1965.
17. Rikmenspoel, R., Contractile mechanisms in flagella, *Biophys. J.*, 11, 446, 1971.
18. Hiramoto, Y. and Baba, S., A quantitative analysis of flagella movement in echinoderm spermatozoa, *J. Exp. Biol.*, 76, 85, 1978.
19. Silvester, N. R. and Holwill, M. E. J., An analysis of hypothetical flagellar waveforms, *J. Theor. Biol.*, 35, 505, 1972.
20. Marchese-Ragona, S. P., Glazzard, A. N., and Holwill, M. E. J., Motile characteristics of 9+2 and 9+1 flagellar axonemes of *Crithidia oncopelti, J. Exp. Biol.*, 145, 199, 1989.
21. Holwill, M. E. J., Bend shapes and molecular mechanisms of flagella, in *Cilia, Mucus and Mucociliary Interactions*, Baum, G. L., Priel, Z., Roth, Y., Liron, N., and Ostfeld, E. J., Eds., Marcel Dekker, New York, 1998, 563.

22. Baba, S. A., Mogami, Y., and Nonaka, K., Discrete nature of flagellar bending detected by digital image analysis, in *Biological Motion (Lecture Notes in Biomathematics)*, Alt, M. and Hoffmann, G., Eds., Springer-Verlag, Berlin, 1991.

23. Woolley, D. M., A method for determining the three-dimensional form of active flagella, using two-colour darkground illumination, *J. Microscopy*, 121, 241, 1981.

24. Woolley, D. M. and Osborn, I. W., Three-dimensional geometry of motile hamster spermatozoa, *J. Cell Sci.*, 67, 159, 1984.

25. Sugino, K. and Machemer, H., Axial-view recording: an approach to assess the third dimension of the ciliary cycle, *J. Theor. Biol.*, 125, 67, 1987.

26. Satir, P., Landmarks in cilia research from Leeuwenhoek to us, *Cell Motil. Cytoskel.*, 32, 90, 1995.

27. Burgess, S. A., Dover, S. D., and Woolley, D. M., Architecture of the outer arm dynein ATPase in an avian sperm flagellum, with further evidence for the B-link, *J. Cell Sci.*, 98, 17, 1991.

28. Taylor, H. C., Satir, P., and Holwill, M. E. J., Assessment of inner dynein arm structure and possible function in ciliary and flagellar axonemes, *Cell Motil. Cytoskel.*, 43, 167, 1999.

29. Amos, L. A., Linck, R. W., and Klug, A., Molecular structure of flagellar microtubules, in *Cell Motility*, Vol. 3, Goldman, R., Pollard, T., and Rosenbaum, J., Eds., Cold Spring Harbor Laboratory, New York, 1976, 847.

30. Klug, A. and DeRosier, D. J., Optical filtering of electron micrographs: reconstruction of one-sided images, *Nature*, 212, 29, 1966.

31. Beeston, B. E. P., Horne, R. W., and Markham, R., Electron diffraction and optical diffraction techniques, in *Practical Methods in Electron Microscopy*, Vol. 1, Pt. 2, Glauert, A. M., Ed., Amsterdam:North Holland, 1972.

32. Misell, D. L., Image analysis, enhancement and interpretation, in *Practical Methods in Electron Microscopy*, Vol. 7, Glauert, A. M., Ed., Amsterdam:North Holland, 1978.

33. Afzelius, B. A., Bellon, P. L., and Lanzavecchia, S., Microtubules and their protofilaments in the flagellum of an insect spermatozoon, *J. Cell Sci.*, 95, 207, 1990.

34. Nitschké, P., Pignot-Paintrand, I., Iftode, F., and Delacroix, H., DEFPARAM: a program package for aligning elliptical sections of biological objects containing an n-fold symmetry, *CABIOS*, 11, 553, 1995.

35. O'Toole, E., Mastronarde, D., McIntosh, J. R., and Porter, M. E., Computer-assisted analysis of flagellar structure, in *Cilia and Flagella,* Dentler, W. and Witman, G., Eds., Academic Press, New York, 1995, 183.

36. Mastronarde, D. N., O'Toole, E., McDonald, K. L., McIntosh, J. R., and Porter, M. E., Arrangement of inner dynein arms in wild-type and mutant flagella of *Chlamydomonas*, *J. Cell. Biol.*, 118, 1145, 1992.

37. Gardner, L. C., O'Toole, E., Perrone, C. A., Giddings, T., and Porter, M. E., Components of a dynein regulatory complex are located at the junction between the radial spokes and the dynein arms in *Chlamydomonas* flagella, *J. Cell Biol.*, 127, 1311, 1994.

38. Brokaw, C. J. and Kamiya, R., Bending patterns of *Chlamydomonas* flagella. IV. Mutants with defects in inner and outer dynein arms indicate differences in dynein arm function, *Cell Motil. Cytoskel.*, 8, 68, 1987.

39. King, S., Inwood, W., O'Toole, E., Power, J., and Dutcher, S., The *bob2-1* mutation reveals radial asymmetry in the inner dynein arms region of *Chlamydomonas reinhardtii*, *J. Cell Biol.*, 126, 1255, 1994.

40. Burgess, S. A., Rigor and relaxed outer dynein arms in replicas of cryofixed motile flagella, *J. Mol. Biol.*, 250, 52, 1995.

41. Radermacher, M., Three-dimensional reconstruction of single particles in electron microscopy, in *Image Analysis in Biology,* Häder, D.-P., Ed., CRC Press, Boca Raton, FL, 1991, 219.

42. Nitschké, P., Ravisé, S., Paintrand, M., Bornens, M., and Delacroix, H., The centriole: twist and pitch. Towards a 3-D reconstruction, *J. Trace and Microprobe Tech.*, 13, 383, 1995.

43. Muto, E., Kamiya, R., and Tsukita, S., Double-rowed organization of inner dynein arms in *Chlamydomonas* flagella revealed by tilt-series thin-section electron microscopy, *J. Cell Sci.*, 99, 57, 1991.

44. Sugrue, P., Avolio, J., Satir, P., and Holwill, M. E. J., Computer modelling of *Tetrahymena* axonemes at macromolecular resolution: interpretation of electron micrographs, *J. Cell Sci.*, 98, 5, 1991.

45. Holwill, M. E. J., Foster, G., Guevara, E., Hamasaki, T., and Satir, P., Computer modelling of the ciliary axoneme, *Cell Motil. Cytoskel.*, Video Suppl. 5, 1998.

46. Holwill, M. E. J., Taylor, H. C., Guevara, E., and Satir, P., Computer modelling: a versatile tool for the study of structure and function in cilia, *Eur. J. Protistol.*, 34, 239, 1998.

47. Cosson, J., Cachon, M., Cachon, J., and Cosson, M.-P., Swimming behaviour of the unicellular biflagellate *Oxyrrhis marina*: *in vivo* and *in vitro* movement of the two flagella, *Biol. of the Cell*, 63, 117, 1988.

14 3-D Color Video Microscopy of Intact Plants

Kenji Omasa

CONTENTS

14.1 INTRODUCTION

Recent advances in computerized light microscopy systems have enabled the three-dimensional (3-D) analysis of the structure of objects.[1-6] To obtain information on the 3-D architecture of cells and tissues at high magnification, the confocal laser-scanning microscope (CLSM) has been effectively used.[4-8] In this system, the 3-D image is typically constructed by stacking numerous two-dimensional (2-D) images, which are obtained at consecutive confocal planes. The CLSM has fluorescence imaging capability, and it can provide monochromatic or pseudo-color images. However, this system cannot be used to obtain suitable 3-D full real color RGB images. In addition, using the CLSM for *in situ* observation of cells and tissues over a wide magnification range under natural growing conditions is difficult. This problem results because, in this situation, the laser is operated at a narrow working distance and must be adjusted, thereby affecting the physiological reactions of the target cells.

Traditionally, conventional stereomicroscopes have been used for 3-D color observation in biological applications. However, only low-magnification observation has been possible due to the needs of a large focal depth and a wide working distance. Stereo-paired images, necessary for three-dimensionality, are obtained by dual video cameras (instead of dual eye lenses) or by a single camera and a shifting microscope stage.[5,8,9] Many algorithms have been developed for determining a range image (i.e., depth image) from stereo-paired images. However, these algorithms have a drawback in automatically matching corresponding points between stereo-paired images, while also lacking practicality due to the high computational expense required for calculating a precise range image.[10-12]

Range images can even be reconstructed from monocular images through the "shape-from-x" algorithms, in which x is the focus, shading, texture, or contour.[13-16] These methods can be applied to monocular light microscopy. The shape-from-focus (SF) method evolved through studies on automatic focusing techniques and for cases in which limited measurements of an object's depth are available.[17-19] The SF method is one of the most practical means for automatically reconstructing range images from fine-texture images. Nayar and Nakagawa[14] first proposed an SF method, applying it to calculate the range image of a fine-texture steel ball from a series of 2-D monochromatic images that were obtained at consecutive focused planes. However, this algorithm cannot precisely determine the range image of objects with coarse or glossy texture (e.g., plant cells, tissues, and seedlings).

In an attempt to identify an algorithm that would be effective for reconstructing range images from monocular light microscope images of plant cells, tissues, or seedlings, Omasa et al. recently compared the previously described SF algorithm, which is based on a sum-modified Laplacian (SML) operator[14] to modified shape-from-focus (MSF) algorithms that were based on either a linear regression (LR) or max-min (MM) operator.[16] It was found that the LR operator-based method was the most effective for obtaining good-quality range images from coarse-texture images (e.g., seedlings). However, it was unsuitable for imaging glossy textures (e.g., plant cells). Combining active illumination (e.g., in a checked pattern) with the SF algorithm yielded more exact depth maps of a solder joint on a circuit board,[20] but this type of lighting had not been applied to measuring plant cells with color and glossy textures. This chapter extends Omasa's work by incorporating the enhanced MSF algorithm into a new computerized 3-D light microscopy system. The combined system was designed for imaging the color, texture, and shading of intact plants and seedlings under natural growing conditions and over a wide range of magnification.[21]

14.2   COMPUTERIZED LIGHT MICROSCOPY SYSTEM

The computerized light microscope system used (Figure 14.1) was comprised of a modified light microscope (FS-60F, Mitsutoyo) (Figure 14.2), a stepping motor control system for auto-focusing and controlling each axis of the microscope stage and lens tube, a charge-coupled device (CCD) color video camera (XC-999, SONY or C5310, chilled type, Hamamatsu), and a personal computer system (desktop or notebook type) for analyzing the resulting 3-D images. Because of the inherent wide working distances (13.0 mm with a 100x objective, 20.5 mm with a 50x objective, and 34.0 mm with a 2x objective), intact plants can be viewed under growing conditions similar to those in their natural environment. The microscope's focal depths (0.6 mm with 100x objective, 0.9 mm with 50x objective, and 91 mm with 2x objective) were adjusted to obtain a series of 2-D images at consecutive focused planes; these images were used to reconstruct the 3-D images. Signals from the stepping motor control system automatically translated the microscope stage along the x-, y-, and z-axes at 0.02 μm per pulse. Movement along the z-axis of the lens tube was also controlled for auto-focusing and measurement of large microscopic objects, being a rapid but rough adjustment. The image was measured by the CCD camera and digitized by a built-in video A/D converter. Each series of RGB digitized images (640 × 480 pixels; 8-bit color) was stored until subjected to 3-D analysis.

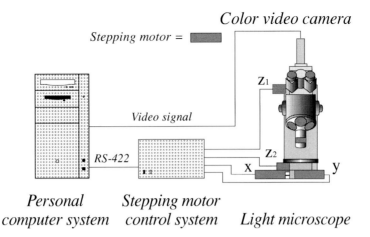

FIGURE 14.1 Computerized light microscopy system for 3-D measurement of shape with color texture and growth of intact plants under natural growth conditions over a wide range of magnification. Movement along the x-, y-, and z-axes of the microscope stage is controlled by the stepping motors x, y, and z_2. Movement along the z-axis of the lens tube is also controlled by the stepping motor z_1, being a rapid albeit inexact adjustment. (From Omasa, K. and Kouda, M., *Environment Control in Biology*, 36, 4, 1998. With permission.)

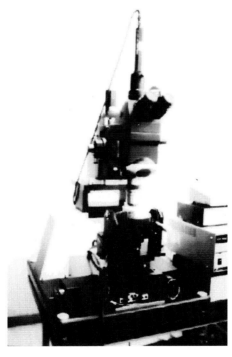

FIGURE 14.2 Photograph of the light microscopy system developed and used in the present series of experiments. (From Omasa, K. and Kouda, M., *Environment Control in Biology*, 36, 4, 1998. With permission.)

14.3   MSF ALGORITHM

Using the LR operator-based MSF algorithm, focused 3-D RGB images were reconstructed from a series of limited-focus planes, which were obtained by z-axis traversal of the microscope stage and lens tube; organ growth was also measured (Figure 14.3). Basically, the algorithm interpolates the depth data such that 3-D shapes with color texture which can be visualized in any direction on the upper side using a wire frame-outline filled in with a focused color image. For 3-D analysis of organ growth (e.g., leaf area), the organ image is extracted from the texture mapping images.

14.3.1   Focus Measure Operator

The LR operator, effective for focus measure, was used to obtain focus measure images from a series of original color images (Figure 14.3a), which were collected at limited-focus planes. Data was calculated on each axis of four square-line masks ($M_k(i, j)$, $k = 1, 2, 3$, and 4; see Figure 14.3a); i, j indicates the center of the $Ms \times Ms$ line masks, a methodology that reduces computation time. If the intensity at each point (x, y) of the original RGB image is denoted as $I_R(x, y)$, $I_G(x, y)$, and $I_B(x, y)$, then $I_{max}(x, y)$ is

$$I_{max}(x, y) = \max\{I_R(x, y), I_G(x, y), I_B(x, y)\} \tag{14.1}$$

The focus measure $f(i, j)$ can subsequently be determined using the LR operator, in which the sum of the absolute error between $I_{max}(x, y)$ and the regression line of $I_{max}(x, y)$ is calculated using the least-squares method on the axes (p_k, $k = 1, 2, 3$, and 4) of each line mask. That is,

$$f(i, j) = \sum_{k=1,2,3,4} \sum_{x,y \in Mk(i,j)} \left| I_{max}(x, y) - \{a_k(i, j) + b_k(i, j)p_k(x, y)\} \right| \tag{14.2}$$

where $a_k(i, j)$ and $b_k(i, j)$ represent the regression coefficients on line mask $M_k(i, j)$ in which $x, y \in M_k(i, j)$ indicates that (x, y) exists in $M_k(i, j)$. The LR operator is better suited for processing images with coarse texture than are SML or MM operators.[16] (see Appendix.)

14.3.2   Gaussian Interpolation and Median Filter

The number of focused planes, N, was limited to ten or less in order to reduce computation requirements. Although a discrete value of range of depth could be estimated from the focus measure, adding Gaussian interpolation provided smoother, more accurate range estimates (Figure 14.3a). The focus measure calculated by assuming the distribution of image data at each image point (i, j) of continuous focused planes can be approximated as a Gaussian distribution in which the optimum focus point, the focused range $z_p(i, j)$, is the peak of the Gaussian distribution. This assumption allows $z_p(i, j)$ to be determined from only three focus measures and their distances from a single reference point on the z-axis, z_n.[14] Thus,

$$z_p = \frac{\left(\ln f_n - \ln f_{n+1}\right)\left(z_n^2 - z_{n-1}^2\right) - \left(\ln f_n - \ln f_{n-1}\right)\left(z_n^2 - z_{n+1}^2\right)}{2\Delta z\left\{\left(\ln f_n - \ln f_{n-1}\right) + \left(\ln f_n - \ln f_{n+1}\right)\right\}} \tag{14.3}$$

where coordinates (i, j) are omitted for simplicity, and f_n is the largest value in a series of focus measures f_h($h = 1, \ldots, n - 1, n, n + 1, \ldots, N$) estimated by Eq. 14.2 at limited-focused planes, f_{n-1}, f_{n+1}, z_{n-1}, and z_{n+1}; the neighboring values of f_n and z_n, and Δz; the distance between neighboring focused planes. The range image estimated by Eq. 14.3 included strong, spike-like noise, which was removed using a median filter (i.e., the z_p of each pixel was replaced by the median of all z_p

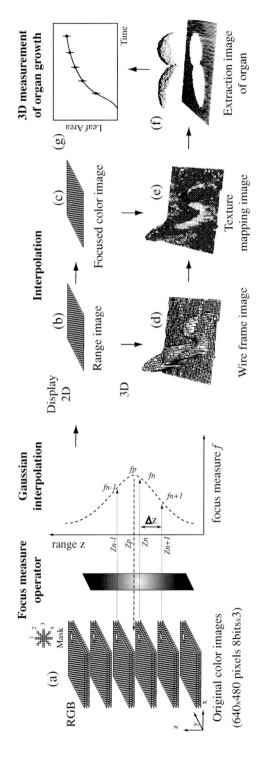

FIGURE 14.3 Algorithm sequence for 3-D measurement and texture mapping. (From Omasa, K. and Kouda, M., *Environment Control in Biology*, 36, 4, 1998. With permission.)

in the line mask). This filter is particularly effective when the noise pattern consists of strong, spike-like components and the characteristic to be preserved is edge-sharpness.[22]

14.3.3 COMPOSITION OF FOCUSED COLOR IMAGE AND TEXTURE MAPPING

A focused color image was generated after interpolating the original RGB intensities using the information on the focused range at each image point. That is, the intensity $I_{R,p}(i, j)$ of the R image focused at an image point (i, j) was estimated by

$$I_{R,p} = I_{R,n-1} \frac{z_n - z_p}{\Delta z} + I_{R,n} \left\{ \frac{z_p - z_{n-1}}{\Delta z} \right\} \qquad (14.4)$$

where coordinates (i, j) are omitted for simplicity, and $I_{R,n}$ and $I_{R,n-1}$ are the intensities of the R image measured at z_n and z_{n-1}. Equation 14.4 is correspondingly applied to obtain $I_{G,p}(i, j)$ and $I_{B,p}(i, j)$, thereby enabling composition of the focused color image. Using newly developed software, a 3-D wire frame-display incorporating color information was generated, as well as a texture mapping display in any direction on the upper side; 3-D shape and color tones can be easily recognized from the computer display. This software incorporates a "rapid preview" function so that the view position can be easily verified.

14.3.4 MEASUREMENT OF ORGAN GROWTH

Changes in the area of cotyledons, an indicator of seedling growth, were automatically measured from 3-D images. Because the textured-mapping images of cotyledons were predominantly green in color, they could easily be extracted from those of other organs and the background. That is, the cotyledon image was extracted by a set of image points satisfying

$$I_{G,p}(i, j)/\{I_{R,p}(i, j) + I_{G,p}(i, j) + I_{B,p}(i, j)\} \geq I_{GTH} \qquad (14.5)$$

and

$$I_{B,p}(i, j)/\{I_{R,p}(i, j) + I_{G,p}(i, j) + I_{B,p}(i, j)\} \leq I_{BTH} \qquad (14.6)$$

where I_{GTH} and I_{BTH} are predetermined, fixed, threshold values. Small holes or uneven edges in the extracted image were filled and smoothed by expanding or shrinking the image as necessary, achieved by dividing the 3-D image into a large number of triangles and then applying Heron's formula to compute leaf area after smoothing the uneven surface. The measurements of a metallic plate with an uneven surface confirmed the suitability of this method. Automatically extracting an image of a target organ is difficult when its tone and brightness are similar to those of other organs and the background. However, as long as there is even a slight difference in tone and brightness, applying the unsharp masking processing before the thresholding is effective.[23] The difference in texture is also used for segmentation and thresholding.[5]

14.4 THREE-DIMENSIONAL MICROSCOPY OF GROWING PLANTS/CELLS

14.4.1 PLANT MATERIALS

Petunia (*Petunia hybrida* Vilm cv. Mitchell) plant seedlings (which have a coarse texture) were selected for low-magnification observations and measurements. After germination, seedlings were grown at 25°C in a Petri dish covered with moistened filter paper; the dish was then placed on the microscope stage. Illumination during growth was automatically controlled at 60 μmol photons $m^{-2} s^{-1}$ (PPFD) in a 12 h:12 h light:dark cycle. A 2x objective (numerical aperture [NA] = 0.055; depth-of-focus [DF] = 91 μm) and camera relay lens of 0.5x were used. These conditions allowed visualization of organ growth and of shape (with color texture) under light from the surroundings.

High-magnification microscopy was carried out on cells of an intact pothos (*Epipremnum aureum*) plant using a 50x objective (NA = 0.55, DF = 0.9 μm) and 1x relay lens. To facilitate reconstruction of the 3-D shape of these glossy-texture cells, a fine texture was illuminated on the glossy-texture cells through a checked glass filter and polarizing filter connected to the microscope via an optical cable adapter in addition to light from the surroundings. Pothos plants were grown under typical indoor conditions.

14.4.2 THREE-DIMENSIONAL RECONSTRUCTION OF PETUNIA SEEDLINGS WITH COARSE TEXTURE

A series of original color images of a petunia seedling (Figure 14.4) reflects nine focused planes, with the interval $\Delta z = 0.3$ mm. The plane of focus in the first image (1) is the wet filter paper, that for the last (9) is just above the seed leaf. From these color images, the focus measure was estimated using Eq. 14.2 and the focused range using Eq. 14.3; the mask size (including the median filter) for these operations was $Ms = 13$. Despite the small number of discrete original images, the resulting wire-frame 3-D range image (Figure 14.5) is smooth and exact.

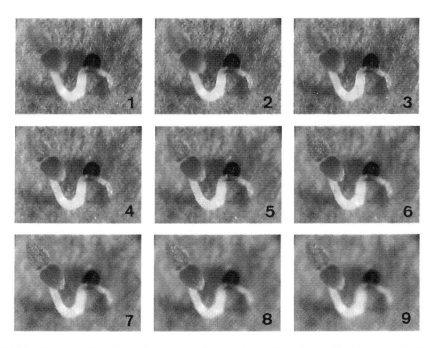

FIGURE 14.4 A series of original color images of a petunia seedling obtained by changing the focus planes from wet filter paper (1) to a plane just above the cotyledon (9) at consecutive intervals of 0.3 mm. (From Omasa, K. and Kouda, M., *Environment Control in Biology*, 36, 4, 1998. With permission.)

The ultimate accuracy of the reconstructed images depends on the surface texture (which is related to the focused color image) of petunia seedlings; however, a more precise range image can be obtained by optimizing the mask size. For example, if spike noise in the range image leads to choosing incorrect focused planes, the focused color image becomes unclear at the image points.

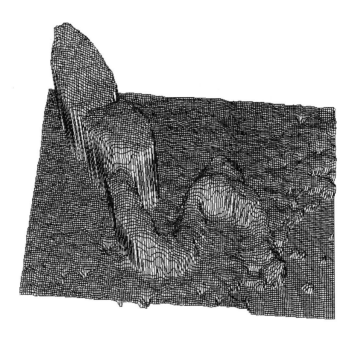

FIGURE 14.5   Wire-frame range image reconstructed from the original color (RGB) images in Figure 14.4 by using the LR operator-based algorithm. (From Omasa, K. and Kouda, M., *Environment Control in Biology*, 36, 4, 1998. With permission.)

Too large a mask size also leads to unclear image points,[5,22] especially at seedling edges. Applying the LR operator-based algorithm to the focused color image obtained using Eq. 14.4 (Figure 14.6)* reduced image degrading effects produced by spike noise and edges. In fact, with a same mask size (N = 13), this operator yielded clearer images than did the SML or MM operators (see Figures 14.6 and 14.7a and 14.7b).[16] This operator also was more effective when using RGB images in place of a monochromatic image (see Figures 14.6 and 14.7c).[16]

The degrading effect produced by edges was evaluated using three test images filled with specks (Figure 14.8). The test chart A is a focused image with a black stripe, and B and C are images defocused in order; the changes in gray level on a horizontal axis of the images is shown in the lower side. Figure 14.9 shows a comparison between focus measures computed by the MM operator and the LR operator. The alphabetical order of test image with the largest value in three focus measures at each position was selected as a focused plane. Although the MM operator mistook the image selection on the surroundings of edges, the LR operator reached a correct result. The degrading effect of specks was removed by selecting a median filter of suitable size.

A texture mapping image (Figure 14.10), generated by combining the 3-D range image and the focused color image, clearly shows the 3-D tone, texture, and shape of the seedling. By manipulating the computer mouse, the observation of the reconstructed texture mapping image is allowed from any direction on the upper side. The underside of the seedling cannot be similarly evaluated because the 3-D image is reconstructed from color images (collected through the lenses of the light microscope) of the upper side, but slanting the microscope stage facilitates imaging of this region to some extent.

The mean area of cotyledons (correlated with growth), as determined from 2-D and 3-D images acquired from the upper side of leaves, was used to document the growth history of an *in situ* petunia seedling over 9 d after providing water supply (Figure 14.11). Following germination after

---

* Color Figure 14.6 follows page 288.

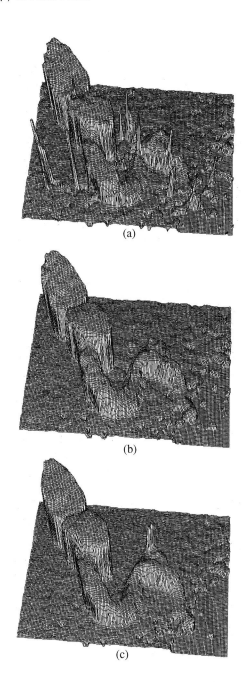

FIGURE 14.7 Comparison among wire-frame range images reconstructed using the SML operator-, the MM operator-, and the LR operator-based algorithms. (a) SML operator, G image (N = 13) (b) MM operator, RGB image (Ms = 13) (c) LR operator, G image (Ms = 13).

4 d, favorable growth is apparent. Note that values from 2-D images are markedly (26% to 38%) smaller than those from 3-D images; the accuracy of leaf area as determined from the 3-D shape was dependent on the angle between the objective and the leaf surface and the texture (unevenness) of the leaf surface. In experiments incorporating a metal disk that had an uneven texture (similar to that for the leaf surface), decreasing the angle to the objective's face increased the error; in Figure 14.12, the error is about 18% from 10° to 30°. However, this error was reduced to ≈5%

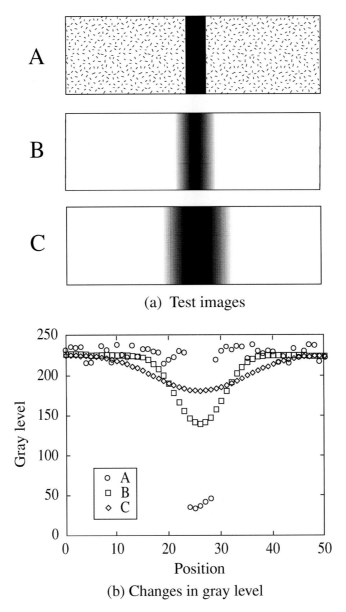

FIGURE 14.8   A series of test images and changes in gray level on the horizontal axis of the images. (a) Test images. (b) Changes in gray level. (From Omasa, K. et al., *Trans. Soc. Instrument Control Eng.*, 33, 752, 1997. With permission.)

using a smoothing filter, which determined the average value for an area of 5×5 pixels. In tests that imaged a metallic hemisphere with 2-mm radius, the error in surface area was within 3% (data not shown).

14.4.3   THREE-DIMENSIONAL RECONSTRUCTION OF POTHOS CELLS WITH GLOSSY TEXTURE

Determining the range focus in the glossy parts of cells of pothos plants was difficult with the previously described method, and illuminating the checked pattern (Figure 14.13A) on the cells had some detrimental effects on imaging. However, at a suitable mask size ($Ms = 15$), which reflects

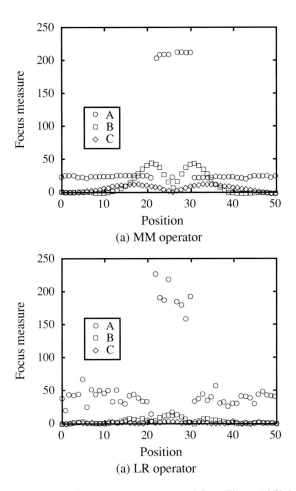

FIGURE 14.9 Comparison between focus measures computed from Figure 14.8b by the MM operator and the LR operator. (a) MM operator. (b) LR operator. (From Omasa, K. et al., *Trans. Soc. Instrument Control Eng.*, 33, 752, 1997. With permission.)

the width of the checking, using the median filter tended to erase the checked pattern (Figure 14.13B). In particular, intensity and saturation along red lines were averaged after filtering, whereas the hue was only slightly affected (Figure 14.14). The reconstructed texture mapping image of pothos cells obtained from a series of eight original color images ($\Delta z = 3$ µm) incorporates checked illumination and median filtering (Figure 14.15). The well-rounded cells are smooth; the color tone changed slightly.

14.5 CONCLUSIONS

Results show that the presented computerized light microscopy system can perform 3-D imaging of shape and color texture of plant cells. Because of its wide working distances even at high magnification (>13 mm for a 100x objective), this newly developed equipment can be used to monitor *in situ* growth of intact plants under natural conditions. This new system is effective over wide ranges of magnification — from a low magnification, which is obtained using a 2x objective and a 0.5x relay lens, to the high level, which incorporates a 100x objective and a 1x relay lens. The performance of the system presented in this chapter is better than that of the authors' previously described remote-controlled light microscopy system, which was developed for observing stomatal

FIGURE 14.10 Texture mapping image generated by combining the 3-D range image (Figure 14.5) and the focused color image (Figure 14.6). (From Omasa, K. and Kouda, M., *Environment Control in Biology*, 36, 4, 1998. With permission.)

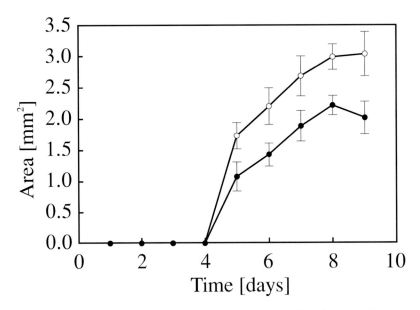

FIGURE 14.11 Changes in area of cotyledons in an intact petunia seedling after providing water supply to the seed. ●: Mean value of leaf area estimated by using a 2-D image measured from the upper side. ○: Mean value of leaf area estimated by using the 3-D shape. Vertical bars indicate ±1 standard deviation. (From Omasa, K. and Kouda, M., *Environment Control in Biology*, 36, 4, 1998. With permission.)

movements.[24-26] Another advantage of the present system is its rapid processing time. Measuring a series of original color images and reconstructing the corresponding 3-D image with color texture can be accomplished automatically within several minutes, which is likely less time than required

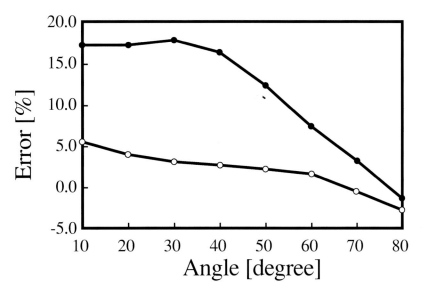

FIGURE 14.12 Errors in surface area estimated using the 3-D shape depend on the angle to the objective's face and the unevenness of the surface. ●: Before use of a 5 × 5 smoothing filter. ○: After use of the filter. (From Omasa, K. and Kouda, M., *Environment Control in Biology*, 36, 4, 1998. With permission.)

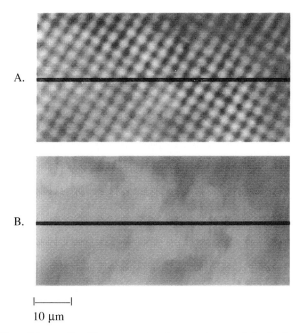

10 μm

FIGURE 14.13 Effect of the median filter on erasing the superimposed checked pattern. (A) Original image, obtained after illuminating cells in a checked pattern. (B) Image after filtering. (From Omasa, K. and Kouda, M., *Environment Control in Biology*, 36, 4, 1998. With permission.)

for other methods (e.g., analysis of stereo paired images[10,12]). Moreover, by incorporating newly developed software, the resulting 3-D image can be observed from any direction (but from above) by the click of a mouse.

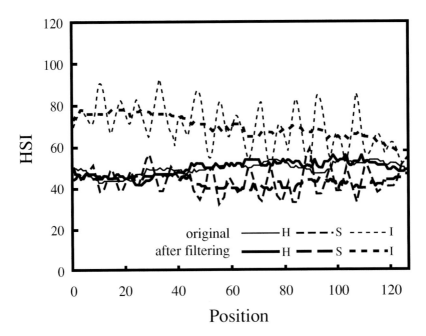

FIGURE 14.14 Hue (H), intensity (I), and saturation (S) values of image points along lines in Figure 14.13 before and after filtering. (From Omasa, K. and Kouda, M., *Environment Control in Biology*, 36, 4, 1998. With permission.)

50 μm

FIGURE 14.15 Texture mapping image of pothos cells obtained using a series of eight original color images that were illuminated in a checked pattern. (From Omasa, K. and Kouda, M., *Environment Control in Biology*, 36, 4, 1998. With permission.)

The modified LR operator-based shape-from-focus (MSF) algorithm used was well-suited for processing images from plants with coarse texture.[16] However, this method was ill-suited for processing images of objects with a glossy texture (e.g., plant cells). This limitation was mitigated by illuminating cells with a checked pattern; the only drawback of this technique was a slight change in cell color tone. Using the MSF algorithm also prevented incorrect range estimates due to the unfocused areas around the edges of the object in the original color image.

The surface area of each organ of intact plants has traditionally been estimated through the use of 2-D images;[23,27,28] this practice may lead to unacceptable errors due to lack of information about the 3-D shape. Using the new imaging system presented in this chapter, the surface area can be determined within an error of 5% at objective angles from $10°$ to $80°$ through direct 3-D measurement. In light of its overall performance, this system is likely to be an effective tool for assessing 3-D changes in the shape and color tone of plants growing under natural conditions over a wide range of magnification.

ACKNOWLEDGMENTS

Sincere gratitude is extended to Dr. M. Onoe, Emeritus Professor of the University of Tokyo and executive advisor of Ricoh Co., Ltd., for his valuable guidance. This research was supported in part by The New Technology Development Foundation of Japan.

APPENDIX A

A.14.1 SML OPERATOR

If the intensity at each point (x, y) of the original image is denoted as $I(x, y)$, the discrete approximation to the modified Laplacian[14] $L_M(x, y)$ is given by

$$L_M(x, y) = \left|2I(x, y) - I(x - step, y) - I(x + step, y)\right| + \left|2I(x, y) - I(x, y - step) - I(x, y + step)\right| \tag{14.7}$$

where *step* is a variable spacing between the pixels used to compute the derivatives, and $I(x, y)$ uses $I_G(x, y)$. The focus measure $f(i, j)$ is computed as the sum of $L_M(x, y)$ in $N \times N$ mask $M_N(i, j)$

$$f(i, j) = \sum_{x,y \in MN (i,j)} L_M(x, y), \text{ if } L_M(x, y) \geq T_1 \tag{14.8}$$

where $x, y \in M_N(i, j)$ means that x and y exists in $M_N(i, j)$ and $x = i - (N - 1)/2, \ldots, i + (N - 1)/2$, $y = j - (N - 1)/2, \ldots, j + (N - 1)/2$. The typical size of N is 3 or 5, the *step* = 1 or 2, and $T_1 = 7$.[14]

A.14.2 MM OPERATOR

If the intensity at each point (x, y) of the original RGB image is denoted as $I_R(x, y)$, $I_G(x, y)$, and $I_B(x, y)$, then $I_{min}(x, y)$ is

$$I_{min}(x, y) = \min\{I_R(x, y), I_G(x, y), I_B(x, y)\} \tag{14.9}$$

where data are calculated on each axis of four square-line masks ($M_k(i, j)$, $k = 1, 2, 3$, and 4; see Figure 14.3A).[16] The focus measure $f(i, j)$ is calculated from Eq. 14.1 and Eq. 14.9 in the $M_k(i, j)$. That is,

$$f(i, j) = \max_{x,y \in Mk(i,j)} \left\{I_{max}(x, y)\right\} - \min_{x,y \in Mk(i,j)} \left\{I_{min}(x, y)\right\} \tag{14.10}$$

REFERENCES

1. Erhardt, A., Zinser, G., Komitowski, D., and Bille, J., Reconstructing 3-D light-microscopic images by digital image processing, *Appl. Optics*, 24, 194, 1985.
2. Hiraoka, Y., Sedat, J. W., and Agard, D. A., The use of a charge-coupled device for quantitative optical microscopy of biological structures, *Science*, 238, 36, 1987.
3. Hiraoka, Y., Minden, J. S., Swedlow, J. R., Sedat, J. W., and Agard, D. A., Focal points for chromosome condensation and decondensation revealed by three dimensional *in vivo* time-lapse microscopy, *Nature*, 342, 293, 1989.
4. Pawley, J. B., Ed., *Handbook of Biological Confocal Microscopy*, Plenum Press, New York, 1990, 232.
5. Russ, J. C., *The Image Processing Handbook*, 2nd ed., CRC Press, Boca Raton, FL, 1994, 696.
6. Gu, M., *Principles of Three Dimensional Imaging in Confocal Microscopes*, World Scientific, Singapore, 1996, 337.
7. Knebel, W. and Schnepf, E., Confocal laser scanning microscopy of fluorescently stained wood cells: a new method for three-dimensional imaging of xylem elements, *Trees*, 5, 1, 1991.
8. Rigaut, J. P., Carvajal-Gonzales, S., and Vassy, J., Confocal image cytometry-quantitative analysis of three-dimensional images obtained by confocal scanning microscopy, in *Image Analysis in Biology*, Häder, D.-P., Ed., CRC Press, Boca Raton, FL, 1992, 109.
9. Hiraoka, Y., Agard, D. A., and Sedat, J. W., Temporal and spatial coordination of chromosome movement, spindle formation, and nuclear breakdown during prometaphase in *Drosophila melanogaster* embryos, *J. Biol.*, 111, 2815, 1990.
10. Inokuchi, S. and Sato, K., *3D Imaging Techniques for Measurement*, Syokoudo, Tokyo, 1990, 189 (in Japanese).
11. Faugeras, O., *Three-Dimensional Computer Vision: A Geometric Viewpoint*, MIT Press, Cambridge, MA, 1993, 165.
12. Chellappa, R. and Rosenfeld, A., Vision engineering: designing computer vision systems, in *Handbook of Pattern Recognition & Computer Vision*, Chen, C. H., Pau, L. F., and Wang, P. S. P., Eds., World Scientific, Singapore, 1993, 805.
13. Kanatani, K., *Group-Theoretical Methods in Image Understanding*, Springer-Verlag, Berlin, 1990, 239.
14. Nayar, S. K. and Nakagawa, Y., Shape from focus: an effective approach for rough surfaces, in *Proc. '90 IEEE Int. Conf. Robotics & Automation*, 1990, 218.
15. Jähne, B., *Spatio-Temporal Image Processing: Theory and Applications*, Springer-Verlag, Berlin, 1993, 39.
16. Omasa, K., Kouda, M., and Ohtani, Y., 3-D microscopic measurement of seedlings using a shape-from-focus method, *Trans. Soc. Instrument Control Engineers*, 33, 752, 1997 (in Japanese with English abstract).
17. Krotkov, E., Focusing, *Int. J. Computer Vision*, 1, 223, 1987.
18. Grossmann, P., Depth from focus, *P attern Recog. Lett.*, 5, 63, 1987.
19. Darrell, T. and Wohn, K., Pyramid based depth from focus, in *Proc. CVPR*, 1988, 504.
20. Noguchi, M. and Nayar, S. K., Microscopic shape from focus using active illumination, in *Proc. '94 IEEE Int. Conf. Pattern Recognition*, 1994, 147.
21. Omasa, K. and Kouda, M., 3-D color video microscopy of intact plants: a new method for measuring shape and growth, *Environment Control in Biology*, 36, 217, 1998.
22. Gonzalez, R. C. and Woods, R. E., *Digital Image Processing*, Addison-Wesley, Reading, MA, 1992, 191.
23. Omasa, K. and Onoe, M., Measurement of stomatal aperture by digital image processing, *Plant Cell Physiol.*, 25, 1379, 1984.
24. Omasa, K., Hashimoto, Y., and Aiga, I., Observation of stomatal movements of intact plants using an image instrumentation system with a light microscope, *Plant Cell Physiol.*, 24, 281, 1983.
25. Omasa, K., Hashimoto, Y., Kramer, P. J., Strain, B. R., Aiga, I., and Kondo, J., Direct observation of reversible and irreversible stomatal responses of attached sunflower leaves to SO_2, *Plant Physiol.*, 79, 153, 1985.
26. Omasa, K. and Croxdale, J. G., Image analysis of stomatal movements and gas exchange, in *Image Analysis in Biology*, Häder, D.-P., Ed., CRC Press, Boca Raton, FL, 1992, 171.

27. Matsui, T. and Eguchi, H., Image processing of plants for evaluation of growth in relation to environment control, *Acta Horticulturae*, 87, 283, 1978.
28. Omasa, K., Image instrumentation methods of plant analysis, in *Modern Methods of Plant Analysis*, Vol. 11, Linskens, H. F. and Jackson, J. F., Eds., New Series, Springer-Verlag, Berlin, 1990, 203.

15 Three-Dimensional Electron Microscopy of Biological Macromolecules: Quaternion-Assisted Angular Reconstitution and Single Particles

Gregory J. Czarnota, Daniel R. Beniac, Neil A. Farrow, George Harauz, and F. Peter Ottensmeyer

CONTENTS

15.1   THREE-DIMENSIONAL STRUCTURE DETERMINATION OF BIOLOGICAL MACROMOLECULES AND THEIR COMPLEXES BY ELECTRON MICROSCOPY

The recent advent of three-dimensional (3-D) image reconstruction methods in electron microscopy (EM) that permit the determination of the structure of virtually any biological macromolecule or macromolecular complex, isolated or in regular arrays, has brought about a renaissance in which electron microscopy plays a more prominent role than ever before in the study of biological structure and function. Historically, a biologist wishing to pursue electron microscopy and image analysis had been limited to interpreting two-dimensional (2-D) projection images of objects, whereas it was their 3-D structures that were intimately related to their function. Only in special cases, when biological specimens had high internal symmetry, such as icosahedral viruses or helical particles, and when the specimen could be experimentally coaxed to crystallize in two- or three-dimensional arrays or fortuitously exhibited preferred orientations, could 3-D structures be determined. It is now possible to add a final and general approach to these techniques to study the 3-D structure of any pure biological macromolecule or macromolecular complex with a method known as quaternion-assisted angular reconstitution.[1-4] This method requires several hundred to a few thousand images of the biological macromolecule of interest, depending on the desired level of detail, and consequently demands only small amounts of purified biological substance. It uses images of sets of single particles of the biological macromolecule at random orientations; crystals or preferred orientations are not required. The approach can be and has been applied to both bright-field images of stained macromolecules and to images of biological particles cryo-prepared and imaged unstained in vitreous ice. It can be and has been applied to dark-field scanning transmission electron microscope images of stained particles or unstained freeze-dried proteins,[4,5] and to microanalytical energy-filtered images of macromolecules and macromolecular complexes.[6,7] Additionally, in comparison to other approaches in structural biology such as x-ray crystallography, NMR spectroscopy, and even electron microscopy-based crystallography, the required time for analysis by the quaternion-assisted angular reconstitution approach is significantly shorter. The following sections first introduce in more detail the different types of electron microscopy that have been used with the quaternion-assisted three-dimensional image reconstruction procedure. Advances in the quaternion-assisted angular reconstitution method are subsequently presented and discussed, with a brief comparison to other electron microscopic determination methods. Examples of the quaternion-assisted reconstitution method are also given in the context of its application to biological questions related to: the storage of genetic information, the expression of genetic information through the process of transcription, the synthesis of proteins, and protein trafficking to appropriate cellular compartments. These examples also demonstrate how the use of quaternion-assisted angular reconstitution has shed light on the functions and related conformational changes of macromolecular complexes that have eluded structural analysis using other methods.

15.2   TYPES OF ELECTRON MICROSCOPY USED IN QUATERNION-ASSISTED THREE-DIMENSIONAL IMAGE RECONSTRUCTION

15.2.1   DARK-FIELD AND BRIGHT-FIELD ELECTRON MICROSCOPY

Three-dimensional (3-D) electron microscopic image reconstruction from single particles has been carried out using almost all forms of electron microscopy currently available. Dark-field electron microscopy has been particularly suited to probing the structures of biological macromolecules on thin support films because of the high contrast or visibility available with this technique. It has been used previously to determine the structures of large biological macromolecules such as ribosomes, proteasomes, and ribonucleoprotein particles.[8] Small macromolecules such as myokinase, protamine, polypeptides such as vasopressin, and even 3-D reconstructions of the poly-lysine alpha-helix have also been investigated with this method.[9-11] In addition to high contrast, advantages of this type of microscopy include its intrinsically higher resolution compared to bright-field electron microscopy, which arises in part from the absence of phase contrast and its dependence on the primary beam, and because the heavy-atom stains typically employed in bright-field microscopy to enhance image contrast are not required. Such stains typically limit resolution in bright-field studies to approximately 20 Å due to the size of individual stain molecules, as well as to the tendency of stain to occlude underlying protein structure.

Bright-field studies of unstained macromolecules, combined with cryo-preservation and electron diffraction from 2-D crystals, have resulted in biological structures being solved with resolutions as high as 3.4 Å.[12-14] Very large numbers of molecular images, on the order of tens of thousands, have been used to achieve such high-resolution structures in bright-field. In dark-field microscopy, the number of images required is much less. Theory suggests the number of images using the two techniques should be only about four-fold different. However, in practice more particles than the theoretical minimum have been used in bright-field, probably due to a less than optimal use of the image information in classification and averaging. For example, in the determination of the structure of the skeletal muscle calcium release channel, Serysheva et al.[15] used 3000 images in conjunction with image classification and averaging, obtaining a resolution of 30 Å. Theoretically, such a resolution should be achievable using only a few tens of images of individual particles.[16,17] In contrast, the structure of the 54-kDa signal sequence-binding protein of SRP54, presented further below, was determined at a resolution of 15 Å using 200 dark-field images of the unstained protein.[4] Nevertheless, both types of microscopy have been used in conjunction with quaternion-assisted angular reconstitution utilizing both stained and unstained bright-field images of biological macromolecules[5,18,19] and several forms of dark-field microscopy.[5-7,19]

In dark-field electron microscopy, images are formed using only scattered electrons; thus, amplitude contrast contributes solely to the formation of the image. Hence, in dark-field images, image intensity is proportionally related to electron density. In comparison, since bright-field EM images are formed with both amplitude contrast and phase contrast, image intensity is not as readily interpreted. It is defined by a contrast transfer function that is instrument specific and related to the amount of defocus in the image. It describes how spatial features in images are related to positive and negative image intensities. This characteristic of bright-field EM — that structural features of different sizes appear with different intensities in images, including contrast reversal — must in general be complemented by image processing where the effect of the instrumental contrast transfer function is corrected in part, and where the combination of information from series of images taken at different defocus values results in an overall positive contrast function akin to the situation in dark-field EM. However, bright-field EM in the fixed-beam transmission electron microscope has the advantage, in comparison to dark-field microscopy in such an instrument, that it requires a four- to tenfold lesser radiation dose to form an image of comparable signal-to-noise ratio.[10] In a scanning transmission electron microscope, however, the advantage in terms of dose and signal-to-noise ratio can be on the side of dark-field microscopy (see below).

15.2.2 ENERGY-FILTERED ELECTRON MICROSCOPY

Electron energy loss imaging, a special type of dark-field electron microscopy, permits moderately high-resolution structural studies to be carried out in conjunction with elemental microanalysis and is made possible in the transmission electron microscope using an electron spectrometer such as the prism-mirror-prism (PMP) system or an omega filter.[21] Due to the early availability of a commercial PMP system, most biological results published have been obtained with such an energy filter. The approach has been used to investigate numerous nucleoprotein complexes including ribosomes, transcription factors such as UBF and TFIIIA bound to promoter DNA,[22-24] and nucleosomes in different ionic environments and physiological conditions.[7,20,24-28] Although the technique requires a higher exposure of the specimen to the electron beam than non-microanalytical microscopic methods, the stability of DNA-protein complexes permits two-dimensional (2-D), image-based analyses at moderately high resolutions.[23-30] The technique has very high sensitivity, and is easily capable of detecting the 300 P atoms in a single nucleosome, with a lower limit of 30 atoms at doses of 1 to 2 C/cm^2 and at a signal-to-noise ratio of 5, the latter considered to be the threshold of certain detectability.[31] It has been used in conjunction with quaternion-assisted three-dimensional (3-D) image reconstruction methods.[5-7,20,32,33] Recent advances in spectrometer design, including a 20-fold improvement in energy resolution for this type of spectrometer, now make the use of the energy-filtered transmission electron microscope extremely attractive.[34,35] With such advances, it is now possible not only to obtain high-resolution elemental maps, but also to detect and use valence excitations of molecules equivalent to optical absorptions or colors in a full image format. Such excitations permit the determination of chemical distributions within a specimen. This approach, in conjunction with 3-D reconstruction, could delineate the location of chemical substrates within molecular complexes.

15.2.3 SCANNING TRANSMISSION ELECTRON MICROSCOPY

The relatively high electron dose required in the fixed-beam transmission electron microscope for dark-field electron microscopy of biological macromolecules is almost entirely due to the limiting requirements of the imaging objective lens, which results in a relatively inefficient collection of scattered electrons. For example, as few as 7% of the electrons scattered elastically for carbonaceous biological macromolecules fall into the objective aperture of such a microscope to form the dark-field image.[10] In comparison, due to differences in instrument design, the scanning transmission electron microscope (STEM) is capable of dark-field image formation using about 70% of the elastically scattered electrons and provides additional advantages over a conventional transmission instrument.

In the STEM, images are acquired digitally. An electron probe as small as 3 Å in diameter, produced by a field emission gun, a condenser, and objective lens, is scanned in raster fashion across the specimen. The consequent electron scattering from the specimen is measured simultaneously by a number of fixed electronic detectors. An annular dark-field detector and an on-axis bright-field detector can be used. Alternatively, the predominantly elastic electrons can be captured by the annular detector, while the inelastic scatter signal can be separated via an electron spectrometer and captured simultaneously. This latter combination, with the addition of the two signals, permits dark-field images to be obtained using a lower radiation dose than in a fixed-beam transmission electron microscope. This is feasible using STEM images because the absence of postspecimen lenses obviates the formation of chromatic aberrations and loss of resolution typically associated with inelastically scattered electrons in conventional transmission electron microscopes. The addition of elastic and inelastic dark-field images is an appreciable gain in carbonaceous specimens, such as biological material, since the ratio of inelastic scattering to elastic scattering in carbon (Z = 6) is approximately 3:1.[36] This image addition improves the signal-to-noise ratio twofold at a constant dose or permits the acquisition of images at a constant signal-to-noise ratio with a fourfold lower dose.

15.3 ELECTRON MICROSCOPIC THREE-DIMENSIONAL RECONSTRUCTION TECHNIQUES

15.3.1 CRYSTALLOGRAPHIC AND RANDOM CONICAL TILT METHODS

As stated above, many different computational techniques and experimental approaches permit the determination of the 3-D structures of biological macromolecules and their complexes using electron microscopy. These structures range in resolution from relatively low (30 Å) as typically seen in reconstructions of large virions or macromolecular complexes,[37,38] to sufficiently high resolution (3.4 Å) permitting the atomic coordinates to be fitted in the chlorophyll *a/b*-protein complex of the plant light-harvesting complex II.[17] Techniques to determine the 3-D structures of biological macromolecules using electron microscopy include 2-D electron crystallography, which can be used when one has planar crystals of a biological macromolecule in conjunction with tilt-series (similar to computerized axial tomography), and the random conical tilt method, which is used for specimens that exhibit preferred orientations (reviewed in Ref. 39). More recently, angular reconstruction techniques were extended[1,2] and can be used in reconstructions of randomly oriented, single asymmetric particles.[4,40] A detailed discussion follows.

Two-dimensional electron **crystallographic techniques** that rely on electron diffraction and the ability of a biological specimen to form a two-dimensional crystal have typically yielded structures with the greatest resolution. This method combines crystallography with an image-based approach to derive the phase information necessary to solve crystallographic structures. These methods are used most often with stained specimens.[39,41] However, they can be applied in analyses of micrographs of frozen hydrated unstained specimens imaged at high resolutions. An example is the first structure determined by electron crystallography: bacteriorhodopsin, at a resolution of 3.5 Å, considered to be the modern triumph of this technique.[17]

A number of reconstruction approaches make use of helical symmetry in biological structures or their arrangements, as exemplified by the structure of *Limulus* tropomyosin,[42] *Dictyostelium* myosin S1 decorated F-actin,[43] and recently the acetylcholine receptor channel imaged in the closed and open states[44,45] at 9 Å resolution using an electron dose of 10 e/Å[2].

Another method, applicable to samples that exhibit preferred orientations, is known as the **random conical tilt method** and has been used to resolve the structures of a number of macromolecules and their complexes.[46,51] Akin to structures determined using tilt series and 2-D electron crystallography, the resolution in the structures determined by this approach is best in the azimuthal or in-plane dimension, but is approximately 1.5 times worse in the dimension perpendicular to the plane in which the molecules lie. This technique requires two images to be taken, the second only being used to determine the azimuthal rotations of each molecule. This application is nevertheless limited by the requirement that the specimen adopt recognizable preferred orientations with respect to the support film. In this method, a single micrograph is required and the dose to the sample can be as low as a sparing 0.4 e/Å[2], whereas 2-D crystallography requires a higher dose often due to multiple tilt exposures.[52] Nevertheless, in the crystallographic approach, damage effects from the electron dose can be averaged out more precisely due to the exact alignment of molecules within the crystal.

Recently, a number of reconstructions of particles have been carried out using the random conical tilt method or the angular reconstruction method on images of sets of single particles, frozen-hydrated in order to preserve their macromolecular conformation. In the initial stages of analysis, the corresponding resolutions of these reconstructions are of the order of 20 to 30 Å, possibly because of unfilled gaps in the contrast transfer functions, image aberrations caused by specimen charging, and/or radiation damage caused in part by radiolysis of ice.[15,48,53]

15.3.2 QUATERNION-ASSISTED ANGULAR RECONSTITUTION

Quaternion-assisted angular reconstitution methods are being used with increasing frequency in the determination of structures of biological macromolecules that do not crystallize, do not exhibit any symmetry, and do not adopt a preferred orientation. These methods are based on the principle of angular reconstitution,[54-56] which has been augmented by us using quaternion mathematics to permit orientation determination of individual noisy images.[57] This method has been used in solving the structures of several biological macromolecules — the procedure in total being referred to as IQAD, for Iterative Quaternion-assisted Angle Determination.[1-7,20] The method permits structural determinations of macromolecules under many different conditions, including different ionic environments, different post-translational modifications, and different physiological states. In comparison, crystallographic studies are often limited to conditions that promote crystallization. Like all electron microscopic investigations, the quaternion-assisted angular reconstitution method is not limited by a maximum specimen size, giving it an advantage over NMR spectroscopy, which presently has an upper size limit of approximately 25 kDa for structural studies of monomeric protein macromolecules or their domains. A protocol similar to the quaternion-assisted method[1-4,58] was developed later and used bright-field microscopy combined with 2-D image averaging instead of quaternion mathematics to reduce the effects of image noise.[15] When the quaternion approach is not used, 2-D images must be averaged to reduce noise before angle determination is sufficiently accurate via angular reconstitution. This requires a larger number of initial images, cluster analysis to determine closely related sets of images, and subsequent averaging within each set. The average image of each set has a higher signal-to-noise ratio, but generally lower spatial resolution from the inclusion of non-identical orientations in the set.

The angular reconstitution approach is based on the central section theorem, originally applied in EM reconstructions[59] to determine the relative angular orientations of randomly oriented, highly symmetric icosahedral viruses. It is stated in the central section theorem that the 2-D Fourier transforms (central sections) of any 2-D projections of a 3-D density distribution will have an identical 1-D Fourier transform (common line) where the central sections intersect. Once the positions of the common line are determined, they define axes about which the pairs of projections are constrained to rotate. An extension of the intercomparison to three images fixes the relative orientations of the images and permits a series of equations to be defined to determine the projection orientations with respect to one another. Equivalently, this concept can be expressed in real space by the statement that any two projection images of a 3-D density distribution will have identical 1-D projections (common axes). Schematics of the central section theorem are presented in Figure 15.1. Our implementation, carried out in real space, makes use of sinograms of images[56] that are collections of consecutive line projections around the image, and correlation functions of pairs of sinograms to determine the orientations of common axes (Figure 15.2). For example, in order to find the common line projection between two 2-D electron microscopic projection images, two such sinograms — one for each projection — must be calculated. Each sinogram represents the set of all computed line projections for a specific image. Each line projection in one sinogram is then compared to every line projection in the other sinogram. The comparison is carried out by calculating a cross-correlation coefficient (a measure of similarity) between them. The distribution of the coefficient is a two-dimensional function referred to as a sinogram correlation function. The position of the maximum in this correlation function gives the position of the common axis within each of the 2-D projection images. In processing a set of electron microscopic images, after computerized automatic image selection,[8] one computes a sinogram for each image and for each sinogram typically seven sinogram correlation functions, ultimately determining the positions of common axes between pairs of images. These common axis positions are then used in combination with quaternion mathematics to optimize the fit of one image with respect to the others.[1,2]

In simplified terms, quaternion mathematics is a four-dimensional, vector-based algebra and is applicable to the angular reconstitution procedure since the unique features of these mathematics reformulate the complex problem of calculating an optimal fit among many common axes, the orientational interrelationship of many images, into a problem readily solvable by standard Eigenvector-Eigenvalue procedures.[57] Moreover, the use of quaternion mathematics facilitates the use of angular reconstitution directly with noisy images of single particles. It thus obviates the otherwise necessary cluster analysis and 2-D averaging of much larger numbers of particles used to produce class averages with higher signal-to-noise ratios but with an inevitable worsening of resolution.

15.4 LIMITS AND LIMITATIONS

15.4.1 IMAGE QUANTITY

The number of images to use in any image reconstruction procedure must always be considered because the resolution of the reconstruction is inherently related. Low resolutions in the range of

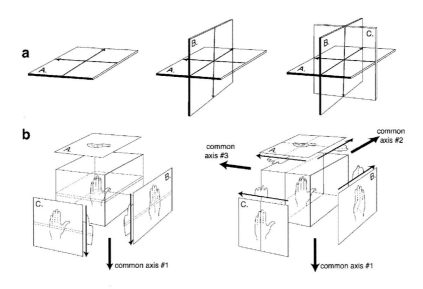

FIGURE 15.1 Demonstration of the Fourier real-space interpretations of the common-axis method. A, B, and C demark three individual projection images of the same particle. Each image shares a common axis with each of the other images. (a) In Fourier space, the common axes given in projection image A, which are represented by arrows, are apparent when images B and C are introduced (from left to right). (b) The principle of determining common axes. If one has an object (in this case a hand) from which three projection images (A, B, and C) are generated, then the common axis between B and C can be computed because any slice of the three-dimensional hand has the same integrated line intensity regardless from which projection image of the hand it is computed. Hence, the slice of the hand in image B (b, left) has the same integrated intensity as the slice of the hand in image C (b, left) since they originate from the same slice of the three-dimensional hand. Hence, if the object in question is sliced from top to bottom in images B and C (as illustrated) integrated line intensities will be generated that are equivalent (common axis #1). Similar processes may be carried out to calculate common axis #2 (between images B and A) and common axis #3 (between images A and C) (b, right). These common axes are the same as illustrated in (a).

25 to 40 Å have been obtained for bright-field images of particles imaged in vitreous ice. Significantly higher resolutions of 15 Å have been obtained for particles imaged using dark-field electron microscopy and the STEM with pixel sizes of 5 Å. In all cases, the number of images has been relatively low, typically ranging from 100 to 300 images,[4,6] although in one case as many as 4000 images classified and averaged into 50 class representations have been used.[19] Arguments have been presented for phase-contrast electron microscopy that conclude that 3-D structures can only be obtained with certainty for unstained proteins greater than 400 kDa in mass, and then only by using more than 12,800 images.[17] Those same arguments, applied in the case of angular reconstitution using dark-field imaging using both elastic and inelastic electron scattering signals, indicate that for 3-Å resolution at a dose of 5 e/Å,[2] only 2500 images are required at a molecular mass of only 100 kDa.[16]

15.4.2 RESOLUTION MEASURES

The assessment of the resolution of a 3-D reconstruction of a biological macromolecule is often difficult because there is no clear consensus among electron microscopists as to what the precise mathematical definition of resolution should be. In light of this, measures of resolution for

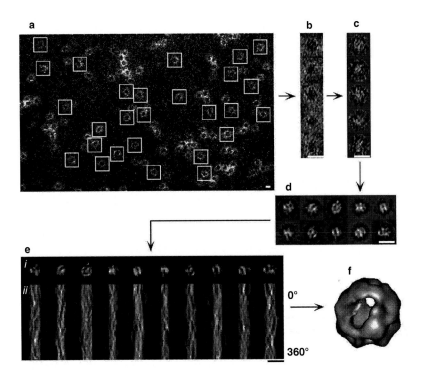

FIGURE 15.2   Overview of the computer image analysis process, using bovine myelin basic protein (MBP) on lipid monolayers as an example. First, digitized single particles of MBP: lipid complexes are interactively selected from a large, single image (a), and subsequently stored in individual images with a single particle centrally located in each image (b). The individual images can then undergo a pretreatment process, which includes contrast reversal, variance normalization, and finally bandpass filtering (c). Once all the particles are pretreated, they can undergo several rounds of reference alignment, multivariate statistical analysis, and hierarchical ascendant classification, to produce final image averages (d), although this is not necessary in the case of dark-field images which can be pretreated (c) and then advance directly to (e). The next step (e), involves the removal of all background signal, leaving only the single particle (e(*i*)), and then for each particle, a sinogram is calculated (e(*ii*)). The sinograms are then utilized in the angular reconstitution process to determine the relative angular orientations of the 2-D images to produce the final 3-D reconstruction (f). Scale bars represent 10 nm.

reconstructions determined using quaternion-assisted angular reconstitution have been calculated using a well-established phase residual-based approach. Each 3-D reconstruction is taken and divided into two sets in terms of its constituent images. Each set of images, the total number in each set being half of the original, is used independently to redetermine angular orientations of the images and to produce two additional 3-D reconstructions. These two reconstructions are then used to determine resolution by computing 3-D Fourier transforms for each of the reconstructions and calculating the average differences in phase between the Fourier transforms. The phase residual computation is carried out in radial shells from the center of each transform outward. Both unweighted phase differences, and phase differences weighted by the average of the pairs of normalized spatial frequency specific amplitudes from each of the Fourier transforms, are typically calculated. Similar measures can also be calculated in two dimensions using 2-D central sections of 3-D Fourier transforms, which correspond in real space to 2-D projections of the 3-D recon- structions. A phase difference, or phase residual, of 0° for identical spatial frequencies from two aligned molecules means exact overlap of structure. In contrast, a phase difference of 90° represents complete randomness (i.e., the noise limit), between the structures at this level of detail. The spatial frequency at which, on average, a phase difference of 90° first occurs has been taken as a measure of the resolution limit in 3-D reconstructions.[59] Although more conservative arbitrary cut-offs, such as 45° or 60°, have also been used for the resolution limit, a separate measure, the Fourier ring correlation at the noise limit between two structures, has produced the same resolution limit as the noise-limited 90° phase residual.[48,49] Results can be displayed in a more intuitive form as the cosine of the phase difference with respect to spatial frequency or equivalent resolution. This cosine is a measure of the overlap between the reconstructions at each spatial frequency, as determined by Eq. 15.1:

$$\frac{\int_{0°}^{180°} \sin(\theta - \theta_R)d\theta}{\int_{0°}^{180°} \sin(\theta)d\theta} = \cos(\theta_R) \tag{15.1}$$

where the denominator is the integral for the reference structure, and is the contribution of the Fourier structure amplitude with phase θ integrated over a sampling interval (defined as half a wavelength at the given spatial frequency). The numerator refers to the structure being compared, and is the contribution of the structure amplitude shifted in phase by θ_R (the phase difference or phase residual) integrated over the same sampling interval. The ratio represents the fractional overlap or the degree of identity between the structures. The calculated phase at any spatial frequency in the 3-D Fourier spectrum of the molecule therefore represents the positioning of the density variations in the structure at that level of detail.

For example, in a resolution analysis of the 54-kDa signal-sequence protein reconstruction determined using quaternion-assisted angular reconstitution,[4] a comparison between two independent 100-image reconstructions indicated a similarity in 3-D structure up to a limit of 14.9 Å (90° phase residual or no overlap of structure). This limit between the 200-image reconstruction and each of the 100-image reconstructions had an average value of 12.4 Å (12.8 Å and 11.4 Å). These values were close to the theoretical resolution limit of 10 Å based on the Shannon sampling theorem,[60] since the original images were acquired with a 5-Å pixel size. Weighting of the phase differences by the corresponding amplitudes in the Fourier transform had little effect on the outcome: in the first comparison above leaving the 12.4-Å resolution limit unchanged, reducing the resolution of 14.9 Å to 15.7 Å in the second, and improving the phase difference in the central sections from 79° to 76°. The reconstruction illustrates the potential of STEM for high-resolution electron microscopic structural analyses. Other single-particle microscopy-based analyses of asymmetric biological macromolecules have yet to obtain this level of detail, at present.

15.5 BIOLOGICAL MACROMOLECULES STUDIED USING QUATERNION-ASSISTED ANGULAR RECONSTITUTION AND THREE-DIMENSIONAL IMAGE RECONSTRUCTION

15.5.1 Macromolecules in DNA Synthesis: Klenow Fragment of DNA Polymerase

DNA polymerase is responsible for replicating DNA in the cell. The Klenow or large 68-kDa fragment of DNA polymerase contains both the polymerase activity of this enzyme and its $3' \rightarrow 5'$ proofreading exonuclease activity, which corrects replication DNA-basepair-mismatch errors. The Klenow fragment was selected for use during the development of the quaternion-assisted approach because its structure is known from X-ray crystallographic analysis.[61] In tests, the quaternion-assisted algorithm was robust even with the addition of significant noise to images derived from the crystallographic structure of the DNA polymerase.[1-3] It was also used successfully in the 3-D reconstruction of the structure of this polymerase using micrographs obtained by conventional room-temperature dark-field TEM,[3] although limited in resolution to about 20 Å by the high dose required by that EM technique (Figure 15.3). Nevertheless, the similarity between the two structures is remarkable, especially given the high doses involved in that electron microscopic study.

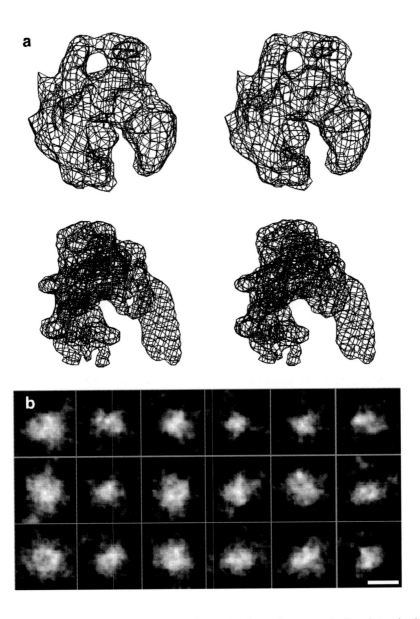

FIGURE 15.3 X-ray crystallographic and quaternion-assisted angular reconstitution determined electron microscopic structures for the Klenow fragment of DNA polymerase. (a) Two stereo images: the upper image shows the 3-D reconstruction of the Klenow fragment produced by angular reconstitution, and the lower image shows the structure as determined by x-ray crystallography; (b) 18 of the 318 dark-field TEM images of the unstained Klenow fragment of DNA polymerase used for angular reconstitution; scale bar, 3.0 nm

15.5.2 MACROMOLECULES IN PROTEIN TRAFFICKING: SRP54

Following up on the potential realized on the known structure of the DNA polymerase above, SRP54 was the first unknown structure determined by quaternion-assisted angular reconstitution.[4] A major improvement on the method, in comparison to the previous study, was the use of low-dose STEM micrographs of the freeze-dried protein imaged at liquid nitrogen temperatures. Quick-freezing to liquid nitrogen temperatures ensures immobilization of the molecules in a solid glass-like environment similar to the aqueous state, while freeze-drying results in a contrast-rich specimen that has not been distorted structurally by phase boundary forces during dehydration. Dark-field electron micrographs of single-particle spreads of molecules were acquired digitally with 5-Å pixel steps at a dose between 20 and 30 e/Å^2, corresponding, at this low temperature, to an optimal dose for imaging at a resolution of 10 Å.[62] Separate signals from electrons scattered at low angles and at high angles were recorded simultaneously. These were combined during image processing to increase the signal-to-noise ratio. The result is a reproducible two-domain structure consistent with biochemical and genetic analyses. Additionally, the structures of the individual domains exhibit strong correlations with proteins of similar function that exhibit amino acid sequence homology to SRP54 (Figure 15.4).* The resolution of the reconstruction (15 Å, discussed above) was high enough to permit the computationally assisted fitting of the structures of biologically related proteins. The GTP-binding domain of SRP fit well with GTP-binding domains of related proteins, while its putative peptide binding domain exhibited similar structure to other related peptide binding proteins. Additionally, the structure of SRP54 proved to be readily incorporated as a model in the cycle of SRP action.[4]

15.5.3 MACROMOLECULES IN HUMAN DISEASE: MYELIN BASIC PROTEIN

Myelin basic protein, MBP, is a constituent of the myelin sheath that surrounds the axons of the central and peripheral nervous systems.[63,64] It is strongly associated with the disease multiple sclerosis and is the responsible prime agent in an animal model of multiple sclerosis, experimental allergic encephalomyelitis.[65,67] To date, MBP has been refractory to X-ray crystallographic analyses, making it an ideal specimen for electron microscopic angular reconstitution structure determination. Its structure was determined from single particles adsorbed to lipid monolayers believed to resemble its environment *in vivo* (Figure 15.5).[68] Features of this structure served as constraints for a predicted atomic resolution structure[18] and support the hypothesis that in multiple sclerosis, the modifications of MBP that lead to a reduced charge distribution result in a weakened attachment to the myelin sheath membrane.

15.5.4 MACROMOLECULES WITH ENZYMATIC ACTIVITY: PHOSPHOENOLPYRUVATE SYNTHASE

Phosphoenolpyruvate synthase (EC 2.7.9.2) from the hyperthermophilic archaeon *Staphylothermus marinus*, which inhabits hot deep aquatic sulfur springs, forms what is the largest known multimeric enzyme. The enzyme is a complex of 24 identical 93-kDa subunits with a total mass of 2.25 MDa.[69] In this unique reconstruction, 4000 bright-field cryo-electron microscope images of the particle embedded in vitreous ice were classified into 40 class averages. These class averages then served as input images into the quaternion-assisted image alignment procedure. The obtained reconstruction (Figure 15.6) exhibited a resolution of 40 Å, limited in this case by instrument defocus, ideally selected for such a massive specimen, and the first zero of the contrast-transfer function. The reconstruction reveals a unique octahedral architecture with elemental 4-fold, 3-fold, and 2-fold symmetry axes, supporting the hypothesis that its elaborate quaternary arrangement represents an adaptation to an extreme environment.[19,69]

---

* Color Figure 15.4 follows page 288.

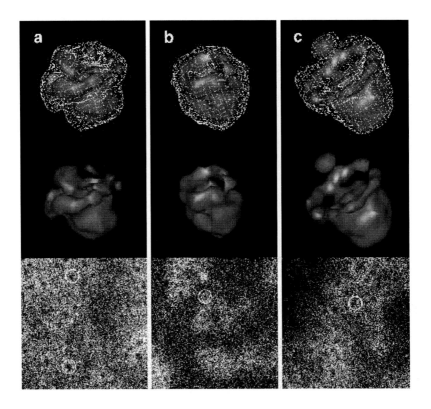

FIGURE 15.5   Angular reconstitution of human myelin basic protein (MBP) complexed with lipid. MBP was incubated with lipid monolayers, negatively stained with methylamine vanadate, and imaged by STEM. Three different charge isomers of the 18.5-kDa human MBP were incubated with the lipid. These were C1 with a net charge of +20 (a), C8 with a net charge of +14 (b), and MC8 with a net charge of +2 (c). The bottom row of this figure shows three sample fields of view in which several single particles have been circled. In these images, the negative stain provides the majority of the contrast and appears as a white halo around the dark single particles. For each sample (a, b, c), 5000 individual single particles were analyzed by 2-D image analysis, and the image averages were used to produce the 3-D reconstructions shown in the upper two rows. In the top row, the reconstruction is shown with a white envelope surrounding it, that defines the outer boundary of the reconstruction, equivalent to the boundary in Figure 15.2(f). In the middle row, only the interior is shown, depicting structure of slightly higher density. All three reconstructions have a ring-like structure and a compact base region, but they differ in the degree of compactness. These three different charge isomers of MBP are implicated in multiple sclerosis (MS): C1 is associated with healthy myelin, C8 is associated with chronic MS, and MC8 is associated with acute Marburg type MS. At present, the exact lipid-protein stoichiometry is unknown for these complexes; however, they are too large to consist of a single MBP molecule. MBP is believed to play a role in the compaction of the myelin sheath; and since MS is a disease in which the lipid myelin sheath is broken down, the present results provide insight into the ability of different charge isomers of MBP to associate and compact lipid. Scale bar represents 10 nm.

15.5.5   NUCLEOPROTEIN COMPLEXES: RIBOSOMES

The ribosome is the complex protein-synthesizing machinery in the cell. In the prokaryote *Escherichia coli*, the ribosome is composed of two subunits: 21 proteins and a 16S ribosomal RNA (rRNA) in the smaller subunit, and 32 proteins and 5S and 23S rRNAs in the larger subunit. The structure

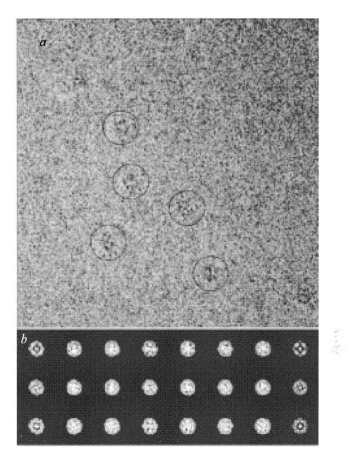

FIGURE 15.6   Angular reconstitution of the 2.25-MDa *Staphylothermus marinus* phosphoenolpyruvate (PEP) synthase. (a) Cryo-electron microscopy using a Philips CM20FEG with a field emission gun and CompuStage. The purified protein was embedded in vitreous ice and micrographed under low-dose conditions, of the order of 500 to 600 nm² per scan. At an accelerating voltage of 120 kV and an instrumental magnification of 20,000x, digital images comprising 1024^2 pixels of 0.477 nm ¥ 0.477 nm size at the specimen level were recorded digitally at defocus values of approximately 3000 nm. The particles could not be easily distinguished in the ice at lower defocus values. The size of each square is 1024 pixels, which is equivalent to 488 nm at the object level. (b) Euler angles defining 99 individual Pps complexes relative 3-D orientations were determined using angular reconstitution, and 3-D reconstructions were computed by back-projection using exact filtering. Reproducible spatial resolutions of these 3-D reconstructions were of the order of 3.5 nm, as ascertained by Fourier shell correlation. Octahedral symmetry was imposed on the reconstructions, which are displayed here as shaded surface views. Columns 1 and 8 showed the reconstruction being viewed down 4-fold rotational axes; column 5 is roughly along a 3-fold (quasi-6-fold), and column 4 is roughly along a 2-fold rotational symmetry axis. The three rows represent three different 3-D reconstructions, in which input images have undergone different degrees of "eigen-filtering" as a noise reduction scheme. High-density thresholds were chosen to emphasize symmetry. Here each image side corresponds to 34 nm. Scale bar represents 25 nm.

of the ribosome has been recently determined using electron spectroscopic imaging and quaternion-assisted angular reconstitution to a resolution of 38 Å.[5]

Three-dimensional reconstructions have been determined for both the large and small ribosomal subunits of *Escherichia coli*. Using image subtraction methods in conjunction with angular reconstitution, the rRNA distributions within these nucleoprotein complexes have been determined (Figure 15.7).* This represents the first time that the rRNA structure was directly visualized in its

---

* Color Figure 15.7 follows page 288.

three-dimensional configuration using microanalytical electron microscopy. Additionally, these structures of the ribosomal subunits provide a new and corroborative interpretation of the rRNA distribution within the subunits of the ribosome. The structures correspond favorably with independent non-microanalytical 3-D reconstructions of the *E. coli* ribosome from bright-field cryo-electron micrographs and with models of translation.[53,70]

15.5.6 NUCLEOPROTEIN COMPLEXES: NUCLEOSOME STRUCTURE AND IONIC ENVIRONMENT

The nucleosome is a nucleoprotein complex composed of DNA wrapped about a core of histone proteins.[71,72] Its canonical role in the cell is the packaging of DNA, although a number of recent biochemical and genetic studies have shown that the nucleosome is also an active macromolecular complex involved in the facilitation, modulation, and repression of gene expression.[72-79] Prior to electron microscopic analysis by angular reconstitution, nucleosome structure had been investigated by a number of diverse biophysical techniques, including crystallographic studies of nucleosome particles using X-ray and neutron diffraction.[80-87] These studies consistently resulted in only one structure: an oblate ellipsoid. In contrast, past studies using electron microscopy and 3-D reconstruction techniques indicated a different form consistent with a prolate ellipsoid.[20,25,88,89] Three-dimensional reconstruction using particles prepared in different ionic environments revealed that the structure of the nucleosome changes dramatically with ionic environment, existing in several ionic strength-dependent conformations (Figure 15.8).[20] One of these corresponds to the 3-D electron microscopic structure previously determined,[25] while another corresponds to the crystallographic configuration of this particle. In total, the particle exhibited over 11 conformations with changes in ionic environment, ranging from 0.1 to greater than 1000 mM NaCl, consistent with the other biophysical techniques that have indicated changes in structure for this macromolecular complex.[90-95] In the analysis of nucleosome structure using electron microscopic image reconstruction methods, the structures obtained not only resolved a long-standing structural debate concerning nucleosome conformation, but also illustrated the potency of microscopic analyses, especially in situations where different conformations of a particle are related to its dynamic role in the cell.

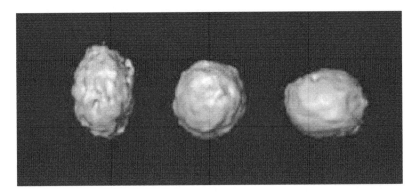

FIGURE 15.8   Three-dimensional structures for nucleosomes in three different ionic environments, as determined by electron microscopy and image reconstruction techniques. The left, middle, and right reconstructions are structures for the nucleosome prepared in the presence of 10, 30, and 150 mM NaCl, respectively. These reconstructions have heights and diameters of (108 Å, 72 Å), (99 Å, 90 Å) and (80 Å, 110 Å), respectively. All conformations are essentially circular when viewed from the top. Structures shown correspond to the theoretical volume for the combined nucleic acid and protein components of the nucleosome. (From Czarnota, G.J. and Ottensmeyer, F.P., *J. Biol. Chem.*, 271, 3677, 1996. With permission.)

15.5.7 NUCLEOPROTEIN COMPLEXES: NUCLEOSOME STRUCTURE AND GENE EXPRESSION

A number of studies have addressed changes in nucleosome structure associated with gene expression.[96-103] As detailed above, crystallographic investigations of nucleosome particles have shown that the structure of the nucleosome from transcriptionally quiescent genes is consistent with a compact oblate ellipsoid. In contrast, the results of a full three-dimensional structural characterization using spectroscopic electron microscopy and quaternion-assisted three-dimensional image reconstruction methods to characterize nucleosome particles chromatographically purified from actively transcribing chromatin have revealed a significantly different conformation.[6,7] The 32-Å resolution structure reveals an open C-shaped and elongated conformation that exhibits an asymmetric mass distribution and an accessible core (Figure 15.9),* consistent with physicochemical and biophysical studies. A corresponding three-dimensional phosphorus map determined using image subtraction techniques[5,22,25] indicates that the DNA in this type of nucleosome has a conformation consistent with particle unfolding and can be represented by a bent spring model. The open configuration of this nucleosome is believed to permit more readily interactions with transcription factors and RNA-polymerase-II-holoenzyme and is putatively related to the high degree of histone acetylation in nucleosomes associated with transcriptionally active genes. Both the DNA and the histones in the particle have an altered conformation compared to the canonical nucleosome, and the electron microscopic analysis illustrates how this macromolecular complex and its changes in structure not only participate in, but facilitate gene expression.[6,7]

15.6 SUMMARY AND FUTURE DIRECTIONS

The future of electron microscopy hinges on its ability to solve structures not amenable to study by X-ray crystallography or NMR spectroscopy and on its potential ability to provide high-resolution structural information at atomic detail. Electron microscopic approaches utilizing two-dimensional crystallography have already shown the latter to be possible,[12,44,45] whereas the angular reconstitution approach to determining three-dimensional macromolecular reconstructions has already made the former a reality in a number of cases.[4-7] The analysis of the 54-kDa subunit of SRP showed that, using images acquired with 5-Å pixels in two dimensions, obtaining 15-Å resolution in three dimensions was possible. The scanning transmission electron microscope can readily produce low-dose images of cryo-prepared specimens with a beam diameter of 3 Å, at pixel sizes of 2 or 1 Å. Thus, the potential exists to obtain significantly higher resolutions in three dimensions. Such structures could be utilized to phase X-ray data and help solve the crystallographic phase problem. The difference between the theoretically possible 10-Å resolution[60] in the SRP54 study and the 15-Å resolution obtained in the SRP study is believed to be a result of image misalignment in terms of three-dimensional angular orientations. What is clearly needed is an improved orientation determination to circumvent this particular limitation. Investigations into the calculation of sinograms and their correlation functions have indicated that due to the noise in the images, the highest peak in the correlation function is not always the correct peak. A number of alternative modified correlation functions are presently being explored and should lead to minimizing errors in the calculation of image orientations.

Another avenue, as yet relatively unexplored by electron microscopists, is the structural determination of large multiparticle macromolecular complexes such as the transcription initiation complex or chromatin remodeling factors. Electron microscopists to date generally have been satisfied with obtaining structures of single particles, and the few complexes that have been characterized have been done so mainly by electron crystallography. Biology is on the precipice of moving from simply being satisfied with identifying and cataloging structures to studying how

---

* Color Figure 15.9 follows page 288.

they interact in concert. The quaternion-assisted angular reconstitution approach provides an excellent vehicle for investigating the structure-function relationships and changes in structure of macromolecular complexes such as RNA-polymerase-II-holoenzyme, or the chromatin activator SWI/SNF. Such complexes can be assembled in step-wise fashion, imaged in active conformations in the presence of substrate, and studied synchronously with the 3-D electron microscopic reconstruction procedure, thus revealing how they are assembled and how they function.

In conclusion, since its earliest inception[54-56] to its first uses in solving known and unknown structures,[3,4,15] angular reconstitution has matured rapidly. The procedure offers an approach to determining molecular structure that complements higher resolution approaches such as X-ray and electron crystallography and NMR spectroscopy. It is powerful because it can be applied in cases where the biological macromolecule is not tractable for analysis by those methods, but relatively low quantities of specimen are present (as only microgram quantities are required). Angular reconstitution is also versatile because it can be used with a number of different types of electron microscopy. The three-dimensional electron microscopic approach to studying the structures and functions of biological macromolecules will inevitably play an important role in the future of modern biology.

ACKNOWLEDGMENTS

We thank Karl Varga for assistance with the manuscript. We thank Allan Fernandes and Brenda Rutherford for technical support. The writing of this chapter was supported by awards from the University of Toronto Faculty of Medicine to GJC. The many studies detailed in this chapter were made possible through grants and awards made to the authors from the MRC Canada, NCIC Canada, the Multiple Sclerosis Society of Canada, NSERC, PENCE, and the University of Toronto.

REFERENCES

1. Farrow, N. and Ottensmeyer, F. P., *A posteriori* determination of relative projection directions of arbitrarily oriented macromolecules, *J. Opt. Soc. Am. A.*, 9, 1749, 1992.
2. Farrow, N. and Ottensmeyer, F. P., Automatic 3-D alignment of projection images of randomly oriented objects, *Ultramicroscopy*, 52, 141, 1993.
3. Ottensmeyer, F. P. and Farrow, N., Three-dimensional reconstruction from dark-field electron micrographs of macromolecules at random unknown angles, *Proc. 50th Annu. Meeting Electron Microscopy Soc. Am.*, 1992, 1058.
4. Czarnota, G. J., Andrews, D. W., Farrow, N. A., and Ottensmeyer, F. P., A structure for the signal sequence binding protein SRP54: 3-D reconstruction from STEM images of single molecules, *J. Struct. Biol.*, 133, 35, 1994.
5. Beniac, D. R., Czarnota, G. J., Rutherford, B. L., Ottensmeyer, F. P., Ottensmeyer, F. P., and Harauz, G., The *in situ* architecture of *Escherichia coli* ribosomal RNA derived by electron spectroscopic imaging and three-dimensional reconstruction, *J. Microscopy*, 88, 29, 1997.
6. Bazett-Jones, D. P., Mendez, E., Czarnota, G. J., Ottensmeyer, F. P, and Allfrey, V. G., Visualization and analysis of unfolded nucleosomes associated with transcribing chromatin, *Nucl. Acids Res.*, 24, 321, 1996.
7. Czarnota, G. J., Bazett-Jones, D. P., Mendez, E., Allfrey, V. G., and Ottensmeyer, F. P., High resolution microanalysis and three-dimensional nucleosome structure associated with transcribing chromatin, *Micron*, 28, 419, 1997.
8. Andrews, D. W., Yu, A. H. C., and Ottensmeyer, F. P., Automatic selection of molecular images from dark-field electron micrographs, *Ultramicroscopy*, 19, 1, 1986.
9. Ottensmeyer, F. P., Andrew, J. W., Bazett-Jones, D. P., Chan, A. S. K., and Hewitt, J., Signal-to-noise enhancement in dark-field electron micrographs of vasopressin: filtering of arrays of images in reciprocal space, *J. Microscopy*, 109, 259, 1977.

10. Misell, D. L., *Image Analysis, Enhancement, and Interpretation*, North-Holland Publishing, Oxford, 1978.

11. Ottensmeyer, F. P., Electron spectroscopic imaging: parallel energy filtering and microanalysis in the fixed-beam electron microscope, *J. Ultrastruct. Res.*, 88, 121, 1984.

12. Henderson, R., Baldwin, J. M., Ceska, T. A., Zemlin, F., Beckmann, E., and Browning, K. H., Model for the structure of bacteriorhodopsin based on high-resolution electron cryo-microscopy, *J. Mol. Biol.*, 213, 899, 1990.

13. Baumeister, W. and Typke, D., Electron crystallography of proteins: state of the art and strategy for the future, *MSA Bull.*, 23, 11, 1993.

14. Kühlbrandt, W., Wang, D., and Fujiyoshi, Y., Atomic model of plant light-harvesting complex by electron crystallography, *Nature*, 367, 614, 1994.

15. Serysheva, I. I., Orlova, E. V., Chiu, W., Sherman, M. B., Hamilton, S. L., and van Heel, M., Electron cryomicroscopy and angular reconstitution used to visualize the skeletal muscle calcium release channel, *Nature Struct. Biol.*, 2, 18, 1995.

16. Ottensmeyer, F. P., Fernandes, A. B., Timmer, M., Kroft, J., Varga, K., and Moremen, K. W., 3-D reconstruction of mannosidase II from single particle distributions: noise reduction approaches for higher resolution, *Proc. 14th Int. Con. Electron Microscopy*, 1998.

17. Henderson, R., The potential and limitations of neutrons, electrons and X-rays for atomic resolution microscopy of unstained biological macromolecules, *Quart. Rev. Biophys.*, 28, 171, 1995.

18. Ridsdale, R. A., Beniac, D. R., Tompkins, T. A., Moscarello, M. A., and Harauz, G., Three-dimensional structure of myelin basic protein. II. Molecular modelling and considerations of predicted structures in multiple sclerosis, *J. Biol. Chem.*, 272, 4269, 1997.

19. Harauz, G. and Li, W., Three-dimensional cryoelectron microscopic reconstruction of the 2.25-MDa homomultimeric phosphoenol pyruvate synthase from *Staphylothermus marinus*, *Biochem. Biophys. Res. Commun.*, 241, 599, 1997.

20. Czarnota, G. J. and Ottensmeyer, F. P., Structural states of the nucleosome, *J. Biol. Chem.*, 271, 3677, 1996.

21. Henkelman, R. M. and Ottensmeyer, F. P., An energy filter for biological electron microscopy, *J. Microsc. (Oxford)*, 102, 72, 1974.

22. Bazett-Jones, D. P., LeBlanc, B., Herfort, M., and Moss, T., Short-range DNA looping by the *Xenopus* HMG-box transcription factor, xUBF, *Science*, 264, 1134, 1994.

23. Neil, K. J., Ridsdale, R., Rutherford, B., Taylor, L., Larson, D. E., Gilbeitil, M., Rothblum, L. I., and Harauz, G., Structure of recombinant rat UBF by electron image analysis and homology modelling, *Nucl. Acids. Res.*, 24, 1472, 1996.

24. Bazett-Jones, D. P. and Brown, M. L., Electron microscopy reveals that transcription factor TFIIIA bends 5S DNA, *Mol. Cell. Biol.*, 9, 336, 1989.

25. Harauz, G. and Ottensmeyer, F. P., Nucleosome reconstruction via phosphorus mapping, *Science*, 226, 936, 1984.

26. Locklear, L., Jr., Ridsdale, J. A., Bazett-Jones, D. P., and Davie, J. R., Ultrastructure of transcriptionally competent chromatin, *Nucl. Acids Res.*, 18, 7015, 1990.

27. Oliva, R., Bazett-Jones, D. P., Mezquita, C., and Dixon, G. H., Factors affecting nucleosome disassembly by protamines *in vitro*. Histone hyperacetylation and chromatin structure, time dependence, and the size of the sperm nuclear proteins, *J. Biol. Chem.*, 262, 17016, 1987.

28. Oliva, R., Bazett-Jones, D. P., Locklear, L., and Dixon, G. H., Histone hyperacetylation can induce unfolding of the nucleosome core particle, *Nucl. Acids Res.*, 18, 2739, 1990.

29. Bazett-Jones, D. P. and Ottensmeyer, F. P., Phosphorus distribution in the nucleosome, *Science*, 211, 169, 1981.

30. Adamson-Sharpe, K. M. and Ottensmeyer, F. P., Spatial resolution and detection sensitivity in microanalysis by electron loss selected imaging, *J. Microscopy*, 122, 309, 1982.

31. Rose, A., *Vision: Human and Electronic*, Plenum Press, New York, 1973.

32. Beniac, D. R., Czarnota, G. J., Rutherford, B. L. Ottensmeyer, F. P., and Harauz, G., Three-dimensional architecture of *Thermomyces lanuginosus* small subunit ribosomal RNA, *Micron*, 28, 13, 1997.

33. Beniac, D. R. and Harauz, G., Structures of small subunit ribosomal RNAs *in situ* from *Escherichia coli* and *Thermomyces lanuginosus*, *Mol. Cell. Biochem.*, 148, 165, 1995.

34. Jiang, X. G. and Ottensmeyer, F. P., Optimization of a prism-mirror-prism imaging energy filter for high resolution electron microanalysis, *Optik*, 94, 88, 1993.

35. Barfels, M. M. G, Jiang, X., Heng, Y. M., Arsenault, A. L., and Ottensmeyer, F. P., Low energy loss electron microscopy of chromophores, *Micron*, 29, 97, 1998.

36. Burge, R. E., Mechanisms of contrast and image formation of biological specimens in the transmission electron microscope, *J. Microscopy (Oxford)*, 98, 251, 1973.

37. Dryden, K. A., Wang, G., Yeager, M., Nibert, M. L., Coombs, K. M., Furlong, D. B., Fields, B. N., and Baker, T. S., Early steps in reovirus infection are associated with dramatic changes in supramolecular structure and protein conformation: analysis of virions and subviral particles by cryoelectron microscopy and image reconstruction, *J. Cell Biol.*, 122, 1023, 1993.

38. Phipps, B. M., Typke, D., Hegerl, R., Volker, S., Hoffman, A., Stetter, K. D., and Baumeister, W., Structure of a molecular chaperone from a thermophilic archaebacterium, *Nature*, 361, 475, 1993.

39. Stewart, M., Introduction to the computer image processing of electron micrographs of two-dimensionally ordered biological structures, *J. Electron Microscopy Technique*, 9, 325, 1988.

40. Czarnota, G. J. and Ottensmeyer, F. P., Chromatin substructure: structural states and conformational transitions of the nucleosome, *Proc. Int. Congr. Electron Microscopy*, 1994, 441.

41. Darst, S., Edwards, A. M., Kubalek, E., and Kornberg, R. D., Three-dimensional structure of yeast RNA polymerase II at 16 A resolution, *Cell*, 66, 121, 1991.

42. Lehman, W., Craig, R., and Vibert, R., Ca(2+)-induced tropomyosin movement in *Limulus* thin filaments revealed by three-dimensional reconstruction, *Nature*, 368, 65, 1994.

43. Schröder, R. R., Manstein, D. J., Jahn, W., Holden, H., Rayment, I., Holmes, K. C., and Spovich, J. A., Three-dimensional atomic model of F-actin decorated with *Dictyostelium* myosin S1, *Nature*, 364, 171, 1993.

44. Unwin, N., Nicotinic acetylcholine receptor at 9 Å resolution, *J. Mol. Biol.*, 229, 1101, 1993.

45. Unwin, N., Acetylcholine receptor channel imaged in the open state, *Nature*, 373, 37, 1995.

46. Radermacher, M., Wagenknecht, T., Verschoor, A., and Frank, J., Three-dimensional structure of the large ribosomal subunit from *Escherichia coli*, *EMBO J.*, 6, 1107, 1987.

47. Radermacher, M., Electron microscopy three-dimensional reconstruction of single particles from random and nonrandom tilt series, *J. Electron Microscopy Technique*, 9, 359, 1988.

48. Frank, J., Penczek, P., Grassucci, R., and Srivastava, S., Three-dimensional reconstruction of the 70S *Escherichia coli* ribosome in ice: the distribution of ribosomal RNA, *J. Cell. Biol.*, 115, 597, 1991.

49. Frank, J., Radermacher, M., Wagenknecht, T., and Verschoor, A., A new method for three-dimensional reconstruction of single macromolecules using low-dose electron micrographs, *Ann. N.Y. Acad Sci.*, 483, 77, 1986.

50. Frank, J., Carazo, J. M., and Radermacher, M., Refinement of the random-conical reconstruction technique using multivariate statistical analysis (MSA) and classification, *Eur. J. Cell Biol.*, 48, 143, 1988.

51. Hinshaw, J. E., Carragher, B. D., and Milligan, R. A., Architecture and design of the nuclear pore complex, *Cell*, 69, 1133, 1992.

52. Wagenknecht, T., Carazo, J. M., Radermacher, M., and Frank, J., Three-dimensional reconstruction of the ribosome from *Escherichia coli*, *Biophys. J.*, 55, 455, 1988.

53. Frank, J., Zhu, J., Penczek, P., Li, Y., Srivasta, S., Verschoor, A., Radermacher, M., Grassucci, R., Lata, R. K., and Agarwal, R. K., A model of protein synthesis based on cryo-electron microscopy of the *E. coli* ribosome, *Nature*, 376, 441, 1995.

54. Vainsthein, B. K. and Goncharov, A. B., Determination of the spatial orientation of arbitrarily arranged identical particles of an unknown structure from their projections, *Proc. XI*[th] *Int. Cong. Electron Microscopy*, Kyoto, Japan, 1986, 459.

55. Goncharov, A. B., Vainshtein, B. K., Ryskin, A. I., and Vagin, A. A., Three-dimensional reconstruction of arbitrarily oriented identical particles from their electron photomicrographs, *Kristallografiya*, 32, 858, 1987.

56. van Heel, M., Angular reconstitution: *a posteriori* assignment of projection directions for 3-D reconstruction, *Ultramicroscopy*, 21, 111, 1987.

57. Harauz, G., Representation of rotations by unit quaternions, *Ultramicroscopy*, 33, 209, 1990.

58. Ottensmeyer, F. P., Czarnota, G. J., Andrews, D. W., and Farrow, N. A., Three-dimensional reconstruction of the 54 kDa signal recognition protein SRP54 from STEM dark-field images of the molecule at random orientations, *Proc. 13th Int. Cong. Electron Microscopy*, 1994, 509.

59. Crowther, R. A., DeRosier, D. J., and Klug, A., Procedures for three-dimensional reconstruction of spherical viruses by Fourier synthesis from electron micrographs, *Proc. Roy. Soc. London Ser. A.*, 317, 319, 1970.

60. Shannon, C. E., Communication in the presence of noise, *Proc. Inst. Radio Engrs.*, 37, 10, 1949.

61. Olis, L., Brick, P., Hamlin, R., Xuong, N. G., and Steitz, T. A., Structure of large fragment of *Escherichia coli* DNA polymerase I complexed with dTMP, *Nature*, 313, 762, 1985.

62. Hayward, S. B. and Glaeser, R. M., Radiation damage of purple membrane at low temperature, *Ultramicroscopy*, 4, 201, 1979.

63. Kirschner, D. A., Inouye, Y., Ganser, A. L., and Mann, V., Myelin membrane structure and composition correlated: a phylogenetic study, *J. Neurochem.*, 53, 1599, 1989.

64. Deber, C. M. and Reynolds, S. J., Central nervous system myelin: structure, function, and pathology, *Clin. Biochem.*, 24, 113, 1991.

65. Kies, M. W., Chemical studies on an encephalitogenic protein from guinea pig brain, *Ann. N.Y. Acad. Sci.*, 122, 161, 1965.

66. Carnegie, P. R., Amino acid sequence of the encephalitogenic basic protein from human myelin, *Biochem. J.*, 123, 57, 1971.

67. Eylar, E. H., Brostoff, S., Hashim, G., Callam, J., and Burnett, P., Basic A1 protein of the myelin membrane, *J. Biol. Chem.*, 246, 5770, 1971.

68. Beniac, D. R., Luckevich, M. D., Czarnota, G. J., Tompkins, T. A., Ridsdale, R. A., Ottensmeyer, F. P., Moscarello, M. A., and Harauz, G., Three-dimensional structure of myelin basic protein. I. Reconstruction via angular reconstitution of randomly oriented single particles, *J. Biol. Chem.*, 272, 4261, 1997.

69. Harauz, G., Cicicopol, C., Hegerl, R., Cejka, Z., Goldie, K., Santorius, U., Engel, A., and Baumeister, W., Structural studies on the 2.25 MDa homomultimeric phosphoenolpyruvate synthase from *Staphylothermus marinus*, *J. Struct. Biol.*, 116, 17, 1996.

70. Stark, K., Mueller, F., Orlova, E. V., Schatz, M., Duve, P., Erdemir, T., Zemlin, F., Brimacombe, R., and van Heel, M., The 70S *Escherichia coli* ribosome at 23 Å resolution: fitting the ribosomal RNA, *Structure*, 3, 815, 1995.

71. Kornberg, R. D. and Klug, A., The nucleosome, *Sci. Am.*, 244, 52, 1981.

72. van Holde, K. E., What happens to nucleosomes during transcription?, in *Chromatin*, Springer-Verlag, New York, 1988.

73. Grunstein, M., Histone function in transcription, *Annu. Rev. Cell Biol.*, 6, 643, 1990.

74. Grunstein, M., Histones as regulators of genes, *Sci. Am.*, 68, October 1992.

75. Kornberg, R. D. and Lorch, Y., Irresistible force meets immovable object: transcription and the nucleosome, *Cell*, 67, 833, 1991.

76. Felsenfeld, G., Chromatin as an essential part of the transcriptional mechanism, *Nature*, 355, 219, 1992.

77. Grunstein, M., Durrin, L. K., Mann, R. K., Fisher-Adams, G., and Johnson, L. M., in *Transcriptional Regulation*, McKnight, S. L. and Yamamoto, K. R., Eds., Cold Spring Harbor Laboratory Press, Cold Spring Harbor, New York, 1992, 1295.

78. Lewin, B., Chromatin and gene expression: constant questions, but changing answers, *Cell*, 79, 397, 1994.

79. Wolffe, A., Transcription: in tune with the histones, *Cell*, 77, 13, 1994.

80. Finch, J. T., Lutter, L. C., Rhodes, D., Brown, B., Levitt, M. D., and Klug, A., Structure of nucleosome core particles of chromatin, *Nature*, 269, 29, 1977.

81. Bentley, G. A., Lewit-Bentley, A., Finch, J. T., Podjarny, A. D., and Roth, M., Crystal structure of the nucleosome core particle at 16 Å resolution, *J. Mol. Biol.*, 176, 55, 1984.

82. Burlingame, R.W., Love, W. E., and Moudrianakis, E. N., Crystals of the octameric histone core of the nucleosome, *Science*, 223, 413, 1984.

83. Richmond, T. J., Finch, J. T., Rushton, B., Rhodes, D., and Klug, A., Structure of the nucleosome core particle at 7 Å resolution, *Nature*, 311, 532, 1984.

84. Uberbacher, E. C. and Bunick, G. J., Structure of the nucleosome core particle at 8 Å resolution, *J. Biomol. Struct. Dynam.*, 7, 1033, 1989.

85. Arents, G., Burlingame, R. W., Wang, B. C., Love, W. E., and Moudrianakis, E. N., The nucleosomal core histone octamer at 3.1 Å resolution: a tripartite protein assembly and a left-handed superhelix, *Proc. Natl. Acad. Sci. U.S.A.*, 88, 10148, 1991.

86. Struck, M. M., Klug, A., and Richmond, T. J., Comparison of X-ray structures of the nucleosome core particle in two different hydration states, *J. Mol. Biol.*, 224, 253, 1992.

87. Arents, G. and Moudrianakis, E. N., Topography of the histone octamer surface: repeating structural motifs utilized in the docking of nucleosomal DNA, *Proc. Natl. Acad. Sci. U.S.A.*, 90, 10489, 1993.

88. Czarnota, G. J., Conformational characterization of nucleosome structure by EM, *Proc. 51st Annu. Meeting Microscopy Society of America*, 1993, 194.

89. Zabal, M. M. Z., Czarnota, G. J., Bazett-Jones, D. P., and Ottensmeyer, F. P., Conformational characterization of nucleosomes by principal component analysis of their electron micrographs, *J. Microscopy*, 172, 205, 1993 .

90. Burch, J. S. and Martinson, G., The roles of H1, the histone core and DNA length in the unfolding of nucleosomes at low ionic strength, *Nucl. Acids Res.*, 8, 4969, 1980.

91. Dieterich, A. E. and Cantor, C. R., Kinetics of nucleosome unfolding at low ionic strength, *Biopolymers*, 20, 111, 1981.

92. Libertini, L. J. and Small, E. W., Effects of pH on low-salt transition of chromatin core particles, *Biochemistry*, 21, 3327, 1982.

93. Chung, D. and Lewis, P. N., Conformations of the core nucleosome: effects of ionic strength and high mobility group protein 14 and 17 binding on the fluorescence emission and polarization of dansylated methionine-84 of histone H4, *Biochemistry*, 24, 8028, 1985.

94. Oohara, I. and Wada, A., Spectroscopic studies on histone-DNA interactions. I. The interaction of histone (H2A, H2B) dimer with DNA: DNA sequence dependence, *J. Mol. Biol.*, 196, 399, 1987.

95. Oohara, I. and Wada, A., Spectroscopic studies on histone-DNA interactions. II. Three transitions in nucleosomes resolved by salt-titration, *J. Mol. Biol.*, 196, 389, 1987.

96. Allfrey, V. G., Faulkner, R. M., and Mirsky, A. E., Acetylation and methylation of histones and their possible role in the regulation of RNA synthesis, *Proc. Natl. Acad. Sci. U.S.A.*, 51, 786, 1964.

97. Chen, T. A. and Allfrey, V. G., Rapid and reversible changes in nucleosome structure accompany the activation, repression, and superinduction of murine fibroblast protooncogenes *c*-fos and *c*-myc, *Proc. Natl. Acad. Sci. U.S.A.*, 84, 5252, 1987.

98. Allfrey, V. G. and Chen, T. A., Nucleosomes of transcriptionally active chromatin: isolation of template-active nucleosomes by affinity chromatography, *Meth. Cell Biol.*, 35, 315, 1991.

99. Johnson, E. M., Sterner, R., and Allfrey, V. G., Altered nucleosomes of active nucleolar chromatin contain accessible histone H3 in its hyperacetylated forms, *J. Biol. Chem.*, 262, 6943, 1987.

100. Sterner, R., Boffa, L. C., Chen, T. A., and Allfrey, V. G., Cell cycle-dependent changes in conformation and composition of nucleosomes containing human histone gene sequences, *Nucl. Acids Res.*, 15, 4375, 1987.

101. Chen, T. A., Sterner, R., Cozzolino, A., and Allfrey, V. G., Reversible and irreversible changes in nucleosome structure along the *c*-fos and *c*-myc oncogenes following inhibition of transcription, *J. Mol. Biol.*, 212, 481, 1990.

102. Chen, T. A, Smith., M. M., Sy, L., Sternglanz, R., and Allfrey, V. G., Nucleosome fractionation by mercury affinity chromatography. Contrasting distribution of transcriptionally active DNA sequences and acetylated histones in nucleosome fractions of wild-type yeast cells and cells expressing a histone H3 gene altered to encode a cysteine 110 residue, *J. Biol. Chem.*, 266, 6489, 1991.

103. Walker, J., Chen, T. A., Sterner, R., Berger, M., Winston, F., and Allfrey, V. G., Affinity chromatography of mammalian and yeast nucleosomes. Two modes of binding of transcriptionally active mammalian nucleosomes to organomercurial-agarose columns, and contrasting behavior of the active nucleosomes of yeast, *J. Biol. Chem.*, 265, 5736, 1990.

16 Three-Dimensional Reconstruction of Single Particles in Electron Microscopy

Michael Radermacher

CONTENTS

16.1 INTRODUCTION

For an understanding of the function of many biological particles, knowledge of their three-dimensional (3-D) architecture is a prerequisite. For many smaller molecules, the answer can often be found using the techniques of X-ray crystallography, which are capable of resolving structure to the atomic level.[1,2] Electron microscopy of noncrystalline biological specimens, however, yields structural information at a lower resolution, typically in the range of 2 to 3 nm in three dimensions; and although images of two-dimensional crystals have been recorded in the electron microscope with a resolution of 0.3 nm,[3,4] such fine details cannot yet be resolved in images of single particles. The advantage of electron microscopic structure determination of single particles lies in the fact that no crystallization is necessary and virtually all larger macromolecules are amenable to such an approach. In a preparation of single molecules, no binding sites are blocked (e.g., by a crystalline arrangement), which make this approach uniquely suitable for studying biological function. One of the best examples in the reconstruction of the *Escherichia coli* ribosome, where first the two subunits have been reconstructed followed by the complete ribosome. Constructs of the ribosome with tRNAs bound have been analyzed and these studies continue so that all ligands will be localized in the ribosomes at different stages of translation.[5,6]

Throughout the development of 3-D reconstruction in electron microscopy, however, one goal has been to determine the architecture of biological particles with higher and higher resolution. There are two major effects that limit resolution. One limit is set by the artifacts that occur during preparation of macromolecules for electron microscopy. The effects of negative staining and subsequent drying have been widely discussed in the literature.[7] Only with the use of frozen hydrated preparation techniques[8-11] can these artifacts be avoided. A second serious limitation of the resolution is the radiation sensitivity of biological objects.[12] Single atoms within some inorganic specimens can be imaged with the electron microscope, but the electron dose required for these images would reduce the biological material to ashes. For biological specimens, extremely low electron doses must be used. In fact, the higher the intended resolution, the lower the electron dose must be in order to avoid destruction of fine structures. It is widely accepted that at resolutions on the order of 2 nm, electron doses of 10 e/Å^2 are acceptable. At this level, however, the signal-to-noise ratio of a single image is less than 1 meaning that the noise in an image is stronger than the signal.

The first 3-D reconstruction in electron microscopy was done of a helical particle (the tail of bacteriophage T4).[13] Such particles provide in a single image all views necessary to determine the 3-D structure. Although the primary reason for choosing such a specimen was the simplicity of the data collection, this approach also minimized the effects of radiation damage.

For the first reconstructions of single asymmetrical particles, the different views were obtained from a tilt series, where the specimen is tilted by small increments and an image is recorded for each specimen orientation.[14-17] Even the use of extremely low doses for each single image cannot avoid a high cumulative electron exposure of the particle.

In the processing of two-dimensional (2-D) images of asymmetrical single particles, averaging techniques have been used for a long time to increase the signal-to-noise ratio while keeping the electron dose as low as possible.[18-20] Averaging techniques have only later expanded successfully into three dimensions. In these techniques, the different views needed for a 3-D reconstruction are obtained from images of identical particles in different orientations. The first attempts were made at the same time by Verschoor et al.[21,22] and van Heel[23] with the reconstruction of the 30S ribosomal subunit from *Escherichia coli*; yet in these studies, the inaccuracy of the measurement of the viewing angles was great, which led to a low resolution in the final 3-D structure. Major progress, however, was achieved with the development of the reconstruction technique from a single-exposure, random conical tilt series,[24,25] which for the first time allowed the combination of images of many particles to form one common conical tilt series in which all projecting directions could be measured with great accuracy. Until now, this technique is the most reliable approach for the reconstruction of single asymmetrical particles, and a large number of structures have been determined with this method in a comparably short period of time.[26]

The theory underlying all 3-D reconstruction techniques was originally developed in 1917 by Radon,[27] who analyzed what now is called the Radon transform, an integral transform and its inversion. The same theory was developed independently much later by Cormack,[28,29] who at the time was not aware of the existence of Radon's theory. Other, equivalent approaches originate from reconstruction techniques used in X-ray crystallography, taking the point of view that a single particle is a crystal degenerated into a single unit cell. This approach led to Fourier inversion algorithms based on the projection theorem for Fourier transforms.[13,14] Mathematically, reconstruction algorithms based directly on the inversion of the Radon transform and those based on Fourier interpolation and inversion are two equivalent formulations of the same theory.

Based on the different theoretical approaches, there exist different reconstruction algorithms. In electron microscopy, the most popular reconstruction schemes are based on Fourier transforms. These include the Fourier-Bessel reconstruction techniques,[70,31] Fourier interpolation techniques,[32] and weighted or deconvoluted back-projection algorithms.[33-38]

A second class of methods uses an algebraic approach and treats the reconstruction problem as a matrix inversion. Given a number of measurements (the points in the projection) and a number of unknowns (the points in the volume), a system of equations can be created and solved iteratively, provided that sufficient measurements are available. The first iterative technique developed and applied to electron micrographs was the algebraic reconstruction technique (ART).[39,40] Claims about the performance of this method at that time created controversy between the group that had developed the first iterative algorithm and groups using Fourier inversion-type algorithms.[41,42] To date, both algorithms are being used. Whereas Fourier methods maintain a linear relationship between the reconstruction and the projections, iterative techniques allow a simple implementation of *a priori* knowledge. The first formulation of the algorithm contained a constraint postulating positive densities in the reconstruction; many other constraints have been added since then.[43] For noniterative reconstruction techniques, the adding of *a priori* knowledge can be done in a second step after the reconstruction, using a method known as Projection Onto Convex Sets (POCS).[44-46]

16.2   THE RADON TRANSFORM AND THE PROJECTION THEOREM

An object can be reconstructed in three dimensions if all its projections are known. This statement is based on the theory of Radon transforms or on the projection theorem, both of which are equivalent formulations of the same mathematical problem.

The Radon transform is an integral transform defined as follows.[27,47] Let f(**r**) be a function of vector **r** in an n-dimensional Euclidean space, and let f obey certain regularity conditions; then the integral transform:

$$\hat{f}(p, \xi) = \int f(\mathbf{r})\, \delta(p - \xi \cdot \mathbf{r})\, d\mathbf{r} \tag{16.1}$$

exists and has an inverse transform; ξ is a unit vector defining a direction in n dimensions, p is a positive real number, and δ is the delta function.

In two dimensions, ξ is a function of a single angle ϕ

$$\xi(\phi) = (\cos\phi,\ \sin\phi) \tag{16.2}$$

and Eq. 16.1 describes the set of line integrals in any direction in the plane. The integral extends over the whole plane and the δ-function selects a line in the direction of $\hat{\xi}$. The function $\hat{f}(p, \xi)$ is the set of all one-dimensional projections along the directions $\xi(\phi)$.

In three dimensions, Eq. 16.1 is an integral over planes, and in n dimensions an integral over (n − 1) dimensional hyperspaces. For three-dimensional reconstructions, only the situations n = 2 and n = 3 are of importance.

The inverse Radon transform for even dimensions (n ≥ 2) is:

$$f(\mathbf{r}) = \frac{C_n}{i\pi} \int\limits_{|\xi|=1} d\xi \int\limits_{\infty}^{-\infty} dp \frac{\left(\frac{\partial}{\partial p}\right)^{n-1} \hat{f}(p, \xi)}{p - \xi \cdot \mathbf{x}} \tag{16.3}$$

The inverse of the Radon transform for odd dimensions (n > 3) is:

$$f(\mathbf{r}) = C_n \int\limits_{|\xi|=1} \left(\frac{\partial}{\partial p}\right)^{n-1} \hat{f}(p, \xi) d\xi \tag{16.4}$$

$$\text{with } C_n = \frac{(-1)^{(n-1)/2}}{2(2\pi)^{(n-1)}} \tag{16.5}$$

Equation 16.3 states for n = 2 that if all one-dimensional projections of an object f(**r**) are known, then the object can be recovered from its projections. Equation 16.4 states that if the sums over all planes in all directions of a three-dimensional object are known, then the object can be recovered. Although the projections obtained in the microscope are not sums over planes but sums along lines (the projecting electron beam), the sum over all planes can be determined in theory by a subsequent two-dimensional Radon transform of the two-dimensional projections. The object can then be recovered using Eq. 16.4.

An equivalent formulation of the mathematical problem is the projection theorem for Fourier transforms. It states that the Fourier transform of a projection is a central section through the Fourier transform of the object. If all projections are known, then all central sections through the object's Fourier transform are known and, by an appropriate inverse Fourier transform of the set of projection data, the object can be recovered. This statement is valid for any dimensions. The following proves the projection theorem in three dimensions. Let f(**r**) be the three-dimensional object and let

$$p(\mathbf{q}) = \int f(\mathbf{r}) \, dz \tag{16.6}$$

be its projection, for simplicity assumed to be along the z-direction. Let

$$F(\mathbf{r}^*) = \int f(\mathbf{r}) e^{-2\pi i \left(xx^* + yy^* + zz^*\right)} \, dx \, dy \, dz \tag{16.7}$$

be the Fourier transform of f(**r**). The central section through the Fourier transform at z* = 0 can be written as:

$$F(\mathbf{r}^*) = \int f(\mathbf{r}) e^{-2\pi i \left(xx^* + yy^* + z0\right)} \, dx \, dy \, dz \tag{16.8}$$

$$= \int f(\mathbf{r}) dz \, e^{-2\pi i \left(xx^* + yy^*\right)} \, dx \, dy \tag{16.9}$$

The first integral over z in Equation 16.9 is the projection operation, which then is Fourier transformed in two dimensions. Thus, the central section through the Fourier transform (Eq. 16.8) is the Fourier transform of a projection (Eq. 16.9).

The equivalence between the Radon transform and the central section theorem can most easily be seen for n = 2. Let f(**r**), **r** ∈ **R**^{2} be a function in two dimensions. Its Radon transform is

$$\hat{f}(p, \xi) = \int f(\mathbf{r}) \, \delta(p - \xi \cdot \mathbf{r}) \, d\mathbf{r} \tag{16.10}$$

The Fourier transform of f(**r**) is

$$F(\mathbf{r}^*) = \int f(\mathbf{r}) \, e^{-2\pi i \mathbf{r} \cdot \mathbf{r}^*} \, d\mathbf{r} \tag{16.11}$$

which can be rewritten as

$$\int f(\mathbf{r}) \, e^{-2\pi i t} \, \delta(t - \mathbf{r}^* \cdot \mathbf{r}) \, dt \, d\mathbf{r} \tag{16.12}$$

Substitution of **r*** by sξ and t by s·p results in

$$F(s\xi) = |s| \int_{-\infty}^{\infty} \int f(\mathbf{r}) \, \delta(sp - s\xi \cdot \mathbf{r}) \, e^{-2\pi i sp} \, d\mathbf{r} \, dp$$

$$= \int_{-\infty}^{\infty} \int f(\mathbf{r}) \, \delta(p - \xi \cdot \mathbf{r}) \, d\mathbf{r} \, e^{-2\pi i sp} \, dp \tag{16.13}$$

The inner integral over **r** is the Radon transform of f(**r**), which describes the one-dimensional projection of f in the direction of the unit vector ξ. The integral over p is the Fourier transform of the projections. The left-hand side F(sξ) is a section through the Fourier transform of f in direction ξ. Equation 16.13 states that the one-dimensional Fourier transform along p of the two-dimensional Radon transform yields the two-dimensional Fourier transform of the original function f(**r**) sampled along radii in directions $\xi(\phi)$.

16.3 TILT GEOMETRIES

For the determination of the 3-D structure, projections of the object from many different directions spanning a large angular range must be available. Several different data collection schemes are used in electron microscopy, and the choice between them depends on the properties and size of the particle that is being analyzed, as well as on the reconstruction algorithm used.

16.3.1 SINGLE-AXIS TILTING

The single-axis tilt geometry was the first one used in electron microscopy. Here, the specimen grid is mounted on a stage that can be rotated around one axis. For side entry stages, this tilt axis is along the specimen rod. Normally, the tilt angle ranges from –60° to +60°, and the images are taken at small angular intervals (Figure 16.1), typically between 1° and 10°, depending on the resolution desired in the reconstruction (see Section 16.5). The Fourier transforms of the projections provide measurements of the 3-D Fourier transform of the object. For the complete sampling of the 3-D Radon transform or Fourier transform, a tilt range of 180° would be needed; however, for most microscopes, only a range of 120° (±60°) can be achieved. Because of this limitation in a single-axis tilt series, a double wedge is missing in the Fourier transform of the object which is not covered by measurements. If the tilt range is ±α and the 3-D Fourier transform of the object is enclosed in a sphere of radius R*, then the volume of this "missing wedge" is:[48]

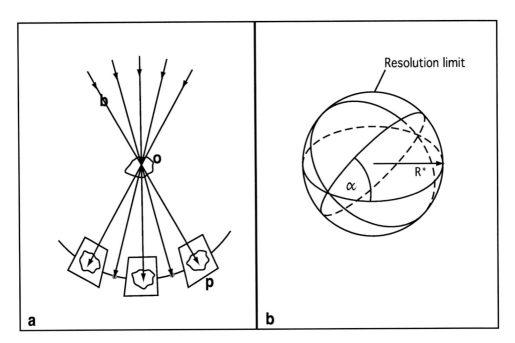

FIGURE 16.1 Single-axis tilt geometry. (a) Real space: o, object; p, projection; b, electron beam. (b) Coverage of the Fourier space: R*, radius of the resolution sphere, α, maximum tilt angle. (From Radermacher, M., *Dreidimensionale Rekonstruktion bei kegelförmiger Kippung im Elektronenmikroskop*, Ph.D. thesis, Technische Universität München, FRG, 1980. Copyright CRC Press LLC, Boca Raton, FL.)

$$V_w = \frac{8}{3} R^{*3} \left(\frac{\pi}{2} - \alpha \right), \alpha \text{ in radians} \tag{16.14}$$

With the volume of the complete sphere being

$$V_{sp} = \frac{4}{3} \pi R^{*3} \tag{16.15}$$

the ratio between the volume of the missing wedge and the volume of the complete Fourier transform is

$$\frac{V_{sp}}{V_w} = \frac{\pi - 2\alpha}{\pi} \tag{16.16}$$

If the Fourier transform is assumed to be enclosed in a cylinder along the tilt axis with radius R*, instead of a sphere, the same volume ratio applies. With a tilt range of ±45°, 50% of the Fourier transform is undetermined; and for ±60°, 33% of the Fourier coefficients are still unknown. This missing wedge leads to artifacts in the reconstruction.

16.3.2 CONICAL TILTING

It was noted previously that in a single-axis tilt experiment with limited tilt range, a double wedge in the 3-D Fourier transform of the object cannot be measured. The volume of the undetermined part of the Fourier transform can be decreased if a conical tilt geometry is employed (Figure 16.2). Here, the specimen grid is tilted by one fixed angle, which should be as large as possible. The

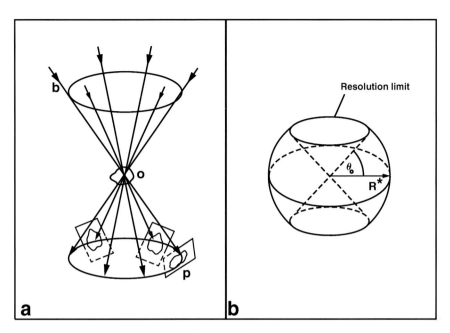

FIGURE 16.2   Conical tilt geometry. (a) Real space: o, object; p, projection; b, electron beam. (b) Coverage of the Fourier space: R^*, radius of the resolution sphere; θ_0, fixed tilt angle. (From Radermacher, M., *Dreidimensionale Rekonstruktion bei kegelförmiger Kippung im Elektronenmikroskop*, Ph.D. thesis, Technische Universität München, FRG, 1980. Copyright CRC Press LLC, Boca Raton, FL.)

tilted specimen is then rotated within its plane by small angular increments, and for each position an image is recorded. This procedure is equivalent to changing the illumination direction along the surface of a cone. The undetermined region in the Fourier transform of the object is now a double cone, the volume of which depends on the size of the fixed tilt angle θ_0.

Assuming again a spherical limitation of the 3-D Fourier transform of the object, the volume of the missing double cone is:[48]

$$V_{dk} = \frac{4}{3}\pi R^{*3}\left(1 - \sin\theta_0\right)$$
(16.17)

and the volume ratio is now

$$\frac{V_{dk}}{V_{sp}} = \left(1 - \sin\theta_0\right)$$
(16.18)

As an example, the undetermined volume of the Fourier transform is 29.3% for $\theta_0 = 45°$ and only 13.4% for $\theta_0 = 60°$. Because this volume is much smaller than in a data set obtained by single-axis tilting, the artifacts in the reconstruction are substantially reduced. The disadvantage of this method is that for a given object size and resolution, about twice as many images need to be available as compared to a single-axis tilt geometry, and the problems of radiation damage are magnified.

16.3.3   RANDOM CONICAL TILTING

The random conical tilt method[18,24,25] has the advantage of the conical tilt geometry in that a larger part of the Fourier transform can be measured than with single-axis tilting, and at the same time it overcomes the problem of specimen damage by large cumulative electron doses that are encountered in the tilt experiments described above. The data collection scheme starts from the premise that many particles with identical structure are available. Most macromolecular assemblies have the tendency to attach to the carbon support film in a specific preferred orientation such that only one rotational degree of freedom is left, namely a rotation within the specimen plane. If such a specimen is tilted, typically by 50° to 60°, then the set of all particle images forms a conical tilt series (Figure 16.3). Here, in contrast to the conical tilting scheme explained above, the rotation within the tilted specimen plane is not achieved by a mechanical rotation of the specimen, but by the random in-plane orientation of the different particles. As a result, the azimuthal angles in such a conical tilt series are distributed randomly. In addition to the image of the tilted specimen, a second image of the same specimen area is recorded without tilt. This second image is used for alignment of the particle images. The reconstruction method implicitly is an averaging procedure because the tilt series is comprised of images of many different particles. As in averaging techniques used in two dimensions also in this technique all particles in a series have to be analyzed for identical shape and orientation with respect to the specimen plane. This analysis is done with images extracted from a second — zero-degree — micrograph. The methods used are the same methods (multivariate statistical analysis and classification[21,49-53]) used in standard 2-D averaging and alignment methods for single particles. The 0° micrograph is recorded after the image of the tilted specimen and thus does not add to the cumulative dose of the projections that are used for the actual reconstruction. If particles show more than one preferred orientation, then separate reconstructions (one for each orientation) can be performed. Their relative orientation in three dimensions can then be determined by comparison of the reconstructed volumes. Subsequently, all images can be used for a single, merged reconstruction. As the merged reconstruction is a combination of several conical tilt data sets, the geometry is referred to as a multi-cone geometry.[54] However, before combining several projection series, one needs again to ascertain the structural identity of the particles as distortions can often be observed that depend on the particle's orientation relative to the plane of the support film.[55] This effect is especially strong in stained and dried preparations, but cannot be excluded in frozen hydrated specimens, where these distortions could be caused by the air-water interface.

The random conical tilt geometry is a technique that allows a reconstruction of asymmetrical particles under very low dose conditions. Exposures of less than $10e/Å^2$ can easily be achieved. It has been used extensively and can be set up in a fashion that also allows the inexperienced user to successfully apply it within a relatively short time. Even with the advent of techniques for reconstruction from randomly oriented particles, the random conical reconstruction scheme is being used for a first starting model, as it allows fewer degrees of freedom and questions like the handedness of the structure can be answered uniquely. In practice, resolutions of about 2.7 nm can easily be achieved. This is not an intrinsic limit of the method. This resolution limit is mostly caused by a limit in image quality.

16.3.4 RANDOM TILTING

A number of experimental setups are conceivable that yield projections of particles in entirely random directions. If surface effects are kept minimal, images of particles in random orientations can most easily be obtained from a specimen of particles suspended in vitreous ice.[11] The technique is being used extensively in the reconstruction of symmetric particles, especially viruses with icosahedral symmetry, where, because of the symmetry, the orientation of each particle in space can be found using the method of common lines.[56] Because of the symmetry, only a small number of images may be needed to provide sufficient data for a 3-D reconstruction; for example, an image of a single icosahedral particle may correspond to 60 different views of this object. For asymmetrical particles, on the other hand, a reconstruction from a set of entirely randomly oriented projections poses a number of problems. First, the number of images needed is very large as compared to the

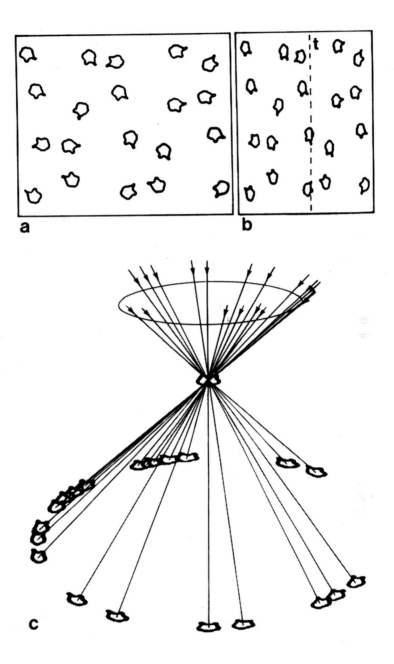

FIGURE 16.3 Geometry of the single-exposure random conical tilt collection scheme. (a) Specimen with identical particles in the same in-plane orientation without tilt; (b) same specimen area as in (a) tilted by 50°; (c) images extracted from (b) in their position in an equivalent conical tilt series. (From Radermacher, M., et al., *J. Microscopy*, 146, 113, 1987. With permission.)

reconstruction of symmetrical particles. Second, because there is no preferred orientation, there exists as yet no technique for monitoring structural variations.

Early schemes to calculate a reconstruction from randomly oriented particles were proposed by Kam.[57] A more recent technique is the reconstruction method developed by Provencher and

Vogel,[58,59] where, starting with approximate angular estimations, the correct tilt angle is fitted as part of the reconstruction algorithm. However, application of this method to particles without symmetry requires several micrographs of the same specimen area to be recorded with different tilt angles.

Other techniques for 3-D reconstruction from randomly oriented projections are being developed that use a common line approach to determine the orientation of particles in three dimensions. This can most easily be achieved by one-dimensional (1-D) cross-correlation of the 2-D Radon transforms of the projections.[60,61] A 2-D Radon transform is often referred to as a sinogram. Each line in a 2-D Radon transform is a 1-D projection of the 2-D image. Any two 2-D projections of a 3-D object have at least one 1-D projection in common. Finding these common 1-D projections in a set of at least three non-coaxial projections uniquely determines their projecting directions; or, because the projections originate from different particles, these common lines determine uniquely the orientation of the three associated particles. Although the method is quite straightforward, it is extremely noise sensitive and does not produce the desired results when applied to single low-dose images. Improvements can only be achieved by pre-averaging of a number of projections that show the same particle orientation, thus increasing the signal-to-noise ratio, or — as subaverages might be difficult to obtain from totally random oriented particles — a technique needs to be found that is able to find all intersections between all projections simultaneously and use redundancies between them to increase the accuracy of the technique.

16.3.5   DEFINITION OF THE ROTATION MATRICES

Let r^j be a vector in the coordinate system of the projection p_j that is obtained by projecting the object tilted by the three angles ψ_j, θ_j, and ϕ_j along the direction z^j. The transformation between r^j and a vector r in the coordinate system fixed to the object then is:

$$r^j = \omega_j r$$

$$\text{with } \omega_j = D_{\psi_j} \cdot D_{\theta_j} \cdot D_{\phi_j} \tag{16.19}$$

$$D_{\psi_j} = \begin{pmatrix} \cos\psi_j & \sin\psi_j & 0 \\ -\sin\psi_j & \cos\psi_j & 0 \\ 0 & 0 & 1 \end{pmatrix} \tag{16.20}$$

$$D_{\theta_j} = \begin{pmatrix} \cos\theta_j & 0 & -\sin\theta_j \\ 0 & 1 & 0 \\ \sin\theta_j & 0 & \cos\theta_j \end{pmatrix} \tag{16.21}$$

$$D_{\phi_j} = \begin{pmatrix} \cos\phi_j & \sin\phi_j & 0 \\ -\sin\phi_j & \cos\phi_j & 0 \\ 0 & 0 & 1 \end{pmatrix} \tag{16.22}$$

For a conical and a random conical tilt series, $\psi \equiv 0$. The tilting is done by first rotating the object around its z-axis by the angle ϕ_j and then tilting it by θ_j around the new y-axis. Throughout a conical tilt series, θ_j = const. for all projections.

For single-axis tilting, $\psi \equiv 0$ and $\phi \equiv 0$; thus, the data collection is done by tilting the object around its y-axis by $-\theta_j$ and then projecting along z^j.

16.4 RECONSTRUCTION ALGORITHMS

The reconstruction algorithms used in electron microscopy can be divided into three groups: Fourier reconstruction methods, weighted or deconvoluted back-projection methods, and iterative direct-space methods. The largest choice of algorithms is available for single-axis tilt geometry, in part because it was the first tilt method used and in part because the reconstruction problem can be separated into a set of reconstructions of 2-D slices perpendicular to the tilt axis from 1-D projections. This reduction to a lower dimensionality simplifies the reconstruction problem and reduces the amount of data that in some types of algorithms needs to be handled simultaneously, thus allowing computer implementations of algorithms that could not be implemented for a true 3-D problem because of the large amount of data involved.

16.4.1 FOURIER AND FOURIER-BESSEL RECONSTRUCTION ALGORITHMS

Fourier reconstruction algorithms are based on the projection theorem. The Fourier transform of a projection is a central section through the Fourier transform of the object. Thus, the three-dimensional Fourier transform can be measured from a set of projections.

The points in the Fourier transform that are known from the projections most often do not coincide with the sampling points needed to carry out the inverse Fourier transform, and an interpolation from the measured points onto the appropriate coordinate system is necessary.[30,32] For a Cartesian grid, the correct interpolation method is a Whittaker-Shannon interpolation.[62] The values in the Fourier transform of an image that is sampled optimally can change very rapidly, and thus, for example, a bilinear interpolation can lead to substantial errors.

The Whittaker-Shannon interpolation is based on Shannon's sampling theorem.[62] Assume that the object is enclosed into a rectangular box with edge lengths a, b, c. The Fourier transform of the object is then completely determined if values on grid-points spaced 1/a, 1/b, 1/c apart are known. All points (x^*, y^*, z^*) not lying on the sampling grid (here, the points known from the projections) are determined through the equation

$$F\left(x^*, y^*, z^*\right) = \sum_h \sum_k \sum_l F_{hkl} C_{hkl,h'k'l'} \tag{16.23}$$

with $x_j^* = h' \cdot 1/a$, $y_j^* = k' \cdot 1/b$, $z_j^* = l' \cdot 1/c$, where h', k', and l' are real numbers and

$$C_{hklj} = \frac{\sin \pi \left(ax_j^* - h\right) \cdot \sin \pi \left(by_j^* - k\right) \cdot \sin \pi \left(cz_j^* - l\right)}{\pi \left(ax_j^* - h\right) \pi \left(by_j^* - k\right) \pi \left(cz_j^* - l\right)} \tag{16.24}$$

$F_{hkl} = F(h \cdot 1/a, k \cdot 1/b, l \cdot 1/c)$ are the values of the Fourier transform on the Cartesian grid. To solve the interpolation problem, the matrix C_{hkl} needs to be inverted, and the Fourier values F_{hkl} can then be calculated.

$$F_{hkl} = \sum_{h'} \sum_{k'} \sum_{l'} C^{-1}_{h'k'l',hkl} F\left(x^*, y^*, z^*\right) \tag{16.25}$$

For an object with edge dimension 64, this matrix would have 4096^3 elements and a solution can only be attempted if the problem can be reduced in dimensionality. As pointed out previously, a reduction is possible if the reconstruction is done from a single-axis tilt series because in this case, the interpolation can be carried out across slices perpendicular to the tilt axis. Still, the inversion of a matrix of size 4096^2 is a formidable task.

For the 2-D reconstruction problem posed by the single-axis tilt geometry, a more appropriate transform to use is the Fourier-Bessel transform.[30] Such a transform requires that the measurement points lie on a polar grid, which is in fact the grid obtained when the set of images is recorded with equal angular increments. In this case, no interpolation is necessary. According to the projection theorem, the Fourier transform of each projection is a 1-D central section through the Fourier transform of a slice. The Fourier coefficients of this slice then are known on a star-shaped polar grid. The radial sampling depends on the size and resolution of the projections, the angular sampling on the tilt increment.

Let (r, θ, y) be cylindrical polar coordinates in real space, y being the coordinate along the tilt axis, and (r^*, θ, y^*) the corresponding coordinates in Fourier space. Let $f(r, \theta, y)$ be the object density and let $F(r^*, \theta, y^*)$ be its Fourier transform. The expansion of f and F into cylinder functions yields:

$$f(r, \theta, y) = \sum_n \int g_n(r, y^*) e^{in\theta} e^{2\pi i y y^*} dy^* \tag{16.26}$$

$$F(r^*, \theta, y^*) = \sum_n G_n(r^*, y^*) e^{in\theta + \pi/2} \tag{16.27}$$

where $G_n(r^*, y^*)$ is the Fourier-Bessel transform of $g_n(r, y^*)$.

$$G_n(r^*, y^*) = \int_{-\infty}^{+\infty} g_n(r, y^*) J_n(2\pi r^* r) 2\pi r dr \tag{16.28}$$

$$g_n(r, y^*) = \int_{-\infty}^{+\infty} G_n(r^*, y^*) J_n(2\pi r^* r) 2\pi r^* dr^* \tag{16.29}$$

In the application to a single-axis tilt series with equal angular increments, Eq. 16.27 can be solved analytically for G_n. Using Eq. 16.29 followed by Eq. 16.26, the three-dimensional object can be recovered. If the tilt series is not recorded with equal angular increments, Eq. 16.27 can be regarded as observational equations that may be solved by the method of least-squares.[30]

16.4.2 SUMMATION OR SIMPLE BACK-PROJECTION

An algorithm that forms the central part of modified back-projection methods, and of iterative methods to be explained later, is the summation or simple back-projection algorithm. It is best explained as a heuristic approach that produces an approximation to the three-dimensional object. From each of the projections in a tilt series, which may have been obtained through the use of any tilt geometry, back-projection bodies[15] are created by smearing each projection perpendicular to the projection plane into a three-dimensional cube. Mathematically, this operation can be described as a convolution of the two-dimensional projection with the three-dimensional function:

$$l(x, y, z) = \delta(x, y) \cdot c(z) \tag{16.30}$$

with

$$c(z) = \begin{cases} 1 & \text{for} \quad -a \le z \le +a \\ 0 & \text{otherwise} \end{cases} \tag{16.31}$$

2a is chosen larger than the diameter of the object in any direction. Usually, 2a is chosen equal to one of the dimensions of the projection. A convolution of the projection with the function $l(x, y, z)$ (Eq. 16.30) replaces each point in the projection by a line perpendicular to the projection plane, where every point along this line has the density value of the projection. The 3-D object created in this way is the back-projection body. If $p(x, y)$ is the projection, then the back-projection body $\hat{p}(x, y, z)$ is:

$$\hat{p}(x, y, z) = \iint p(x, y) \cdot l(x - x', y - y', z) dx\, dy \qquad (16.32)$$

In a simple back-projection procedure, these 3-D back-projection bodies are rotated in space using the tilt angles of the projections and then added.[34] The result is a 3-D density distribution that is an approximation of the object's density. A reconstruction by a simple back-projection, however, contains systematic errors. Part of the problem is illustrated in Figure 16.4, which shows a plane object containing three disks reconstructed from just three projections. Obviously, there are density variations visible where no density had been in the original. The intersections of the rays created in smearing out the projections create a star-shaped structure around each disk, and the intersections of these rays can form spurious high-density areas that can be mistaken for additional objects. However, if many projections are used, the situation improves considerably.

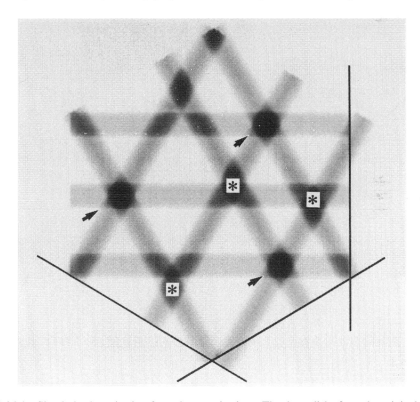

FIGURE 16.4 Simple back-projection from three projections. The three disks from the original object are marked by arrows. Note the rays around each reconstructed point and spurious densities that could be mistaken for additional disks, marked with *.

Although simple back-projection results in only crude approximation of the object, the technique is important to understand. Because it is a linear algorithm, the reconstruction that is created by its use can be analyzed using the concept of point-spread and transfer functions. The density distributions calculated can therefore be corrected by a deconvolution process to yield the accurate reconstruction. Simple back-projection is also used as the first approximation in many iterative techniques.

16.4.3 Weighted and Deconvoluted Back-Projection Methods

The simple back-projection algorithm represents a linear process, and the shape of each reconstructed point is independent of its location in the volume. Thus, according to the theory of linear systems,[63,64] the performance of this algorithm can be described by a single point-spread function, from which a convolution kernel can be determined to deconvolute the simple back-projection and obtain the correct reconstruction.

Two different types of algorithms can be found in the literature: (1) weighted back-projection methods, in which this deconvolution is achieved by dividing the Fourier transform of the reconstructed object by the appropriate transfer function (the Fourier transform of the point-spread function); and (2) deconvoluted back-projection methods, which carry out this deconvolution in real space.[25,33,36,37,65]

In the following, the transfer function of the simple back-projection is derived by making use of the projection theorem and by comparing the Fourier transforms of the original object with that of the object obtained by simple back-projection. The Fourier transform $\hat{P}$ of a back-projection body $\hat{p}$ is:

$$\hat{P}\left(x^*, y^*, z^*\right) = \int_{-\infty}^{+\infty} \hat{p}(x, y, z)e^{-2\pi i\left(xx^* + yy^* + zz^*\right)} dx\, dy\, dz \qquad (16.33)$$

Substitution of Eq. 16.32 into Eq. 16.33 yields

$$\hat{P}\left(x^*, y^*, z^*\right) = \iint \int_{-\infty}^{+\infty} \iint p(x', y')\delta(x' - x, y' - y)$$

$$\cdot c(z)dx'\, dy' \cdot e^{-2\pi i\left(xx^* + yy^* + zz^*\right)} dx\, dy\, dz \qquad (16.34)$$

$$= \int_{-\infty}^{+\infty}\int_{-\infty}^{+\infty}\int_{-a}^{+a} p(x, y)e^{-2\pi i\left(xx^* + yy^*\right)} dx\, dy\, e^{-2\pi i\left(zz^*\right)} dz$$

$$= P\left(x^*, y^*\right) \cdot 2a \cdot \text{sinc}\left(2a\pi z^*\right)$$

with $\sin c(x) = \sin(x)/x$. The back-projection algorithm consists of a summation of all back-projection bodies placed in their proper orientation.

$$b(x, y, z) = \sum_j \hat{p}\left(x^j, y^j, z^j\right) \qquad (16.35)$$

Using Eq. 16.34, the Fourier transform of the back-projected object becomes

$$B\left(x^*, y^*, z^*\right) = \sum_j P\left(x^{*j}, y^{*j}\right) \cdot 2a \cdot \text{sinc}\left(2a\pi z^{*j}\right) \qquad (16.36)$$

where $r^j = (x^j, y^j, z^j) = \omega^j r$, with ω as defined in Eq. 16.19. For determination of the point-spread function, the algorithm is applied to a set of projections of a single point. The Fourier transform of an image or a 3-D volume containing a single point in its origin is constant and for simplicity assumed to be 1. Setting $P(x^{*j}, y^{*j}) \equiv 1$ yields

$$H\left(x^*, y^*, z^*\right) = \sum_j 2a \cdot \text{sinc}\left(2a\pi z^{*j}\right) \tag{16.37}$$

where H is the Fourier transform of a single point reconstructed by simple back-projection. H in Eq. 16.37 is the transfer function of a simple back-projection. As no assumptions were made about the specific tilt geometry, H is the transfer function for a back-projection from projections with arbitrary tilt geometry. To obtain the correct reconstruction, which for a single point in three dimensions is 1 everywhere, the back-projection must be divided by H. H is real-valued and thus equivalent to a weight applied to the Fourier transform. A reconstruction obtained by back-projection with subsequent division by H is called a weighted back-projection. According to the convolution theorem for Fourier transforms, a division of the 3-D Fourier transform of the back-projection by H is equivalent to a deconvolution of the object with the point-spread function of the back-projection. The division by H is possible as long as $H \neq 0$. However, wherever $H = 0$, no information is present, as is the case, for example, within the missing wedge for single-axis tilt series with limited angular range; and without the use of *a priori* knowledge, no information can be recovered within this range. The artifacts caused by the missing information are not specific for a weighted back-projection but occur in any reconstruction algorithm that does not make use of *a priori* knowledge.

The division by H can either be applied to the Fourier transform of the 3-D volume after back-projection or to the Fourier transforms of the projections before back-projection. Because of the projection theorem, the two different modes of application are mathematically equivalent. The amount of calculations necessary and the accuracy of the reconstruction, however, may be quite different, depending on the data set and tilt geometry used. If only a few projections are available, then an application of the weighting function to the projections is more efficient. However, if the number of projections becomes very large, then the amount of calculations will be smaller when the weighting function is applied to the 3-D Fourier transform of the back-projected object. In both cases, however, better accuracy is achieved when the weighting function is applied to the Fourier transform of the projections.

From Eq. 16.37, special analytical forms of the weighting function can easily be derived. If the diameter "a" of the reconstruction volume in Eq. 16.37 is assumed to be infinite, then the sinc-function becomes a δ-function. For a conical tilt series where $\psi = 0$ and $\theta_j = \theta_0 =$ constant, the right-hand side of Eq. 16.37 is solely a function of ϕ_j. If, furthermore, the angle ϕ is considered to be continuous, the sum in Eq. 16.37 can be replaced by an integral, resulting in:

$$H\left(x^*, y^*, z^*\right) = \int_{-\pi}^{+\pi} \delta\left(z'^*(\phi)\right) d\phi \tag{16.38}$$

Expressing $z'^*(\phi)$ by the coordinates in the object coordinate system using the transformation in Eq. 16.19, Eq. 16.38 becomes:

$$H\left(x^*, y^*, z^*\right) = \int_{-\pi}^{+\pi} \delta\left[x^* \sin(\theta_0)\cos(\phi) + y^* \sin(\theta_0)\sin(\phi) + z^* \cos(\theta_0)\right] d\phi \tag{16.39}$$

After x^*, y^*, z^* are expressed in cylindrical coordinates r^*, Γ, z^*, Eq. 16.39 can be solved with the aid of the relation:

$$\delta\big(f(x)\big) = \sum_n \frac{1}{\big|f'(x_n)\big|}\delta(x - x_n), \quad \text{with } f'(x_n) = \frac{df(x)}{dx}\bigg|_{x=x_n} \tag{16.40}$$

Equation 16.40 is valid for $f'(x_n) \neq 0$. x_n are the zeros of $f(x)$ within the integration range. The resulting transfer function of a back-projection for conical geometry is:[37,66,67]

$$H\big(r^*, \Gamma, z^*\big) = \frac{2}{\sqrt{r^{*2}\sin^2(\theta_0) - z^{*2}\cos^2(\theta_0)}}\, r^* > z^* \cot\theta_0 \tag{16.41}$$

and the three-dimensional weighting function W_c for conical tilting is:

$$W_c = 1\big/H\big(r^*, \Gamma, z^*\big) = 1/2 \cdot \sqrt{r^{*2}\sin^2(\theta_0) - z^{*2}\cos^2(\theta_0)} \tag{16.42}$$

In an analogous fashion, the transfer function for single-axis tilting with equal angular increments can be determined. In a single-axis tilt series, ψ and ϕ in the matrix ω (Eq. 16.19) are 0, and θ is assumed to be continuous. Again, the "a" in Eq. 16.37 is assumed to be infinity.

The transfer function of a simple back-projection from a single-axis tilt series therefore becomes:

$$H(r^*, \Gamma, z^*) = 1/r^* \tag{16.43}$$

and the weighting function W_s for single-axis tilting is simply

$$W_s = r^* \tag{16.44}$$

which leads to the so-called r^*-weighted back-projection.[33-36] Because the single-axis tilt geometry was the first used in electron microscopy, the r^*-weighted back-projection was also the first to be applied. The weighting function for conical tilting was derived much later,[37,66] and the weighting function in Eq. 16.37 for arbitrary tilt geometry was first applied in 1986 in connection with the reconstruction from a single-exposure random conical tilt series.[25]

For single-axis tilt geometries, the deconvolution has not only been done by a multiplication of the Fourier transform with r^*, but also by the equivalent convolution applied to the real-space data, the convolution kernel being the inverse Fourier transform of r^*. However, as r^* is not square integrable, as is required for the Fourier transform to exist, assumptions about a band limit of the function must be made. The convolution kernels found in the literature differ because of the different band-limiting functions used.

One of the earliest deconvolution kernels was derived by Bracewell and Riddle[33] in connection with 3-D reconstruction in radio astronomy. Let $F(x^*, y^*)$ be the Fourier transform of $f(x, y)$.

$$f(x, y) = \int_{-\infty}^{+\infty}\int_{-\infty}^{+\infty} F\big(x^*, y^*\big)\, e^{2\pi i\big(x^* x + y^* y\big)} dx^* dy^* \tag{16.45}$$

By converting the spatial frequency variables to polar coordinates and assuming a maximum Fourier radius M as a band limit, Eq. 16.45 becomes:

$$f(x, y) = \int_0^\pi \int_{-M}^M F\big(r^*, \theta\big)\, e^{2\pi i r^*(x\cos\theta + y\sin\theta)} r^*\, dr^*\, d\theta \tag{16.46}$$

The integral over θ in Eq. 16.46 can be interpreted as the continuous form of the back-projection $F(r, \theta)$ as the set of Fourier transforms of the 1-D projections at angles θ, and r^* as the weighting function for single-axis tilting. In Bracewell and Riddle,[33] the r^* is replaced by a band-limited function

$$K\left(r^*\right) = \Pi\left(\frac{r^*}{2M}\right) - \Lambda\left(\frac{r^*}{M}\right)$$

with

$$\Pi(x) = \begin{cases} 1 & \text{for} \quad \left|r^*\right| < 1/2 \\ 0 & \text{for} \quad \left|r^*\right| > 1/2 \end{cases}$$

and

$$\Lambda\left(r^*\right) = \begin{cases} 1 - \left|r^*\right| & \text{for} \quad \left|r^*\right| < 1 \\ 0 & \text{for} \quad \left|r^*\right| > 1 \end{cases} \tag{16.47}$$

The inverse Fourier transform of $K(r^*)$ is the real-space convolution kernel of a weighted back-projection from a single-axis tilt series.

$$\int_{-\infty}^{+\infty} K(r)\, e^{2\pi i r^* r}\, dr = 2M \sin c\left(2\pi M r^*\right) - M \sin c^2\left(M\pi r^*\right) = k\left(r^*\right) \tag{16.48}$$

While for the derivation of Eq. 16.48 a cutoff filter has been used to limit the Fourier transform, a number of other convolution kernels can be found that are derived using other band-limiting functions.[68,69]

16.4.4 ITERATIVE RECONSTRUCTION METHODS

Another class of methods includes the iterative reconstruction algorithms. These were introduced by Gordon, Bender, and Herman[40] in application to a single-axis tilt geometry and modified later by Goitein[70] and by Gilbert.[71,72] They have been extended to conical tilt geometry by Colsher,[73,74] wherein an extensive review of these techniques can also be found. These techniques are undergoing constant development and some of the more recent ones are COMET[75] and "ART with blobs."[76]

The three major types of iterative algorithms are ART (Arithmetic Reconstruction Technique).[40] SIRT (Simultaneous Iterative Reconstruction Technique,[71] and ILST (Iterative Least-Square Technique).[70]

All three techniques start with an estimate of the object, which can either be a uniform density distribution or a simple back-projection of the object. The calculated 3-D density is projected in the direction of the original[40] projections, and the differences between the experimental projections and the calculated images are determined. From these differences, corrections to the points in three dimensions are calculated. In ART, these calculations are carried out sequentially for one projection at a time, whereas in SIRT, the corrections from all projecting directions are applied simultaneously. In ART and SIRT, the corrections are based on the difference between the calculated and measured projections. In ILST, the error is the squared difference divided by an estimate of the variance of the noise along each ray.

In a conical geometry, the direction of a projection is determined by the angles θ_0 and ϕ_1. The value of a point at location (m, n) in the projection at angles (θ_0, ϕ_1) is calculated using the ray-sum

$$R^q(\theta_0, \phi_1, m, n) = \sum_{\substack{\text{all-points-contained} \\ \text{in-ray-}(\theta_0, \phi_1, m, n)}} D^q(i, j, k) \tag{16.49}$$

$D^q(i, j, k)$ is the three-dimensional density. The index q indicates the q-th iteration. Let $P(\theta_0, \phi_1, m, n)$ be the value of the experimental projection at point m, n; then, in ART and SIRT, the error at this point is calculated as

$$E^q(\theta_0, \phi_1, m, n) = P(\theta_0, \phi_1, m, n) - R^q(\theta_0, \phi_1, m, n) \tag{16.50}$$

In additive ART, the error is distributed over the projecting ray, and the density at each voxel is changed according to

$$D^{q+1}(i, j, k) = D^q(i, j, k) + E^q(\theta_0, \phi_1, m, n)/N_{lmn} \tag{16.51}$$

for (i, j, k) in ray (l, m, n). N_{lmn} is the number of points within the ray. In multiplicative ART, the voxel value is changed according to:

$$D^{q+1}(i, j, k) = D^q(i, j, k) \left(\frac{P(\theta_0, \phi_1, m, n)}{R^q(\theta_0, \phi_1, m, n)} \right) \tag{16.52}$$

again for voxel (i, j, k) within the ray (l, m, n).

Equations 16.51 and 16.52 are applied for each projection one at a time. As a result, differences between the last ray-sums and the values of the points in the last-used projection will be minimal, whereas larger differences can be observed with respect to earlier processed images.

In SIRT, the errors for all projections are calculated simultaneously and the corrections that are applied to each point in the three-dimensional density are calculated from the combined error of all projections.

In the third iterative method, ILST,[70] the residual between all experimental and all calculated projections, defined as

$$\Delta^q = \sum_l \sum_m \sum_n \left[P(\theta_0, \phi_1, m, n) - R^q(\theta_0, \phi_1, m, n) \right]^2 / \sigma^2(l, m, n) \tag{16.53}$$

is minimized. σ is the variance of the noise in the ray at location (m, n) on the l-th projection. The solution to Eq. 16.53 is found by an iterative relaxation technique.

16.5 RESOLUTION OF A THREE-DIMENSIONAL RECONSTRUCTION

Before a three-dimensional (3-D) structure determination is attempted, the resolution should be estimated that is required to provide an answer to the functional problem being studied. For example, if the existence of a channel with a diameter of 2 nm is expected, then the resolution of the reconstruction should be in that range. The required resolution determines aspects of the reconstruction process from the very beginning, the image recording in the microscope, and sometimes even the biochemical preparation. As shown below, the final resolution in three dimensions depends on the ratio between the size of the complete object being reconstructed and the number of projections available. Given a fixed number of projections, the resolution deteriorates when the size of the object gets larger. If the detail in question is smaller than the predicted resolution, two options are available: either the number of images can be increased, or a biochemical preparation should be tried that contains only a fraction of the particle. For example, if a particle consists of a number of identical subunits, then their relative location can be found in a reconstruction with rather low resolution. If details at higher resolution are required, the subunits should be isolated and analyzed separately.

Although the number of projections and the object size are the primary limitations of the resolution, other factors contribute. The resolution of each image in a tilt series should be better than the resolution expected in the reconstruction. Further limitations of resolution can result from inaccuracies in the numerical procedures used for the alignment of the projections toward a common origin, inaccuracies in the measured tilt angles, and, in reconstructions from random or random conical tilt series, from structural variabilities among particles that are assumed to be identical. The accuracy of the alignment is also dependent on the signal-to-noise ratio of the images, which in turn depends on the electron dose and on the specimen contrast.

16.5.1 RESOLUTION AS A FUNCTION OF THE NUMBER OF PROJECTIONS

The relation between the number of projections and the size of the object is determined by Shannon's sampling theorem,[62] which states that at least two measurements per resolution element need to be available. Assuming equally spaced measurements and a resolution of d, the spacing between the sampled values should not exceed d/2. For the Fourier transform, Shannon's sampling theorem states that for an object of diameter D, the measurement points in the Fourier transform should not be further apart than 1/D.

The number of projections needed for a specific resolution in a reconstruction from a single-axis tilt series is[30,33]

$$N = \pi \frac{D}{d} \tag{16.54}$$

Or, if expressed in terms of the angular increment $\Delta\theta = \pi/N$ between tilts, the resolution can be determined as:

$$d = \Delta\theta \cdot D \tag{16.55}$$

For a conical tilt series with equal azimuthal angular increments, the relation between the number of projections, the resolution, and the diameter of the reconstructed object is[48,67]

$$d = 2\pi \cdot D/N \cdot \sin\theta_0 \text{ for N even} \tag{16.56}$$

and

$$d = \frac{\pi D}{2N} \tan\theta_0 \cdot \sqrt{16\cos^2\theta_0 - (d/D)^2} \quad \text{for N odd, } \theta_0 < \pi/2 - \Delta\phi/4 \tag{16.57}$$

$\Delta\phi$ being the azimuthal angular increment. For large N, both formulas result in approximately the same value for d. If θ_0 in Eq. 16.56 is set to $\pi/2$, which would make the conical tilting scheme degenerate into a single-axis tilt geometry, then, given a resolution d, the resulting number of projections is twice as large as found with Eq. 16.54. The reason is that the azimuthal angle ϕ in a conical tilt series has a range of 360° instead of just 180° and thus half of the projections contain redundant information.

Because the resolution formula above do not take into account the missing wedge in a limited-angle, single-axis tilt series or the missing cone in a conical tilt series, the resolution d is the resolution in the direction of the x-axis or within the x-y plane, respectively.

16.5.2   EFFECTS OF THE MISSING DATA

Because of the missing wedge in the Fourier transform of a reconstruction from single-axis tilt series and the missing cone in a reconstruction from a conical tilt series, the point-spread function does not have a circular or spherical symmetry. Figure 16.5 shows such a point-spread function for a reconstruction from a conical tilt series. It can be seen that the image point is elongated in the z-direction, the direction of the electron beam in the microscope. A closer analysis shows that side minima of this point can also be seen, as well as traces of the projecting rays. A complete analytical description of this point-spread function has only been obtained for a single-axis tilt geometry.[77] For conical geometry, such a description was obtained by numerical integration.[78] The effect of the missing wedge/cone on the resolution can be found using a quadratic approximation of the image point, essentially describing it as an ellipsoid.[48,66,67] From the ratio between the principal axes of this ellipsoid, an estimate for the resolution in the z-direction can be obtained. The resolution in the z-direction can then be expressed as an elongation factor — the ratio between the long principal axis divided by the short principal axis — multiplied with the resolution in the x-y plane (for conical tilting) or in the x-direction for single-axis tilting around y. The elongation factor for single-axis tilting is:

$$e_{xz} = \sqrt{\frac{\alpha + \sin \alpha \cos \alpha}{\alpha - \sin \alpha \cos \alpha}} \qquad (16.58)$$

with α being the maximum tilt angle. For conical tilting,

$$e_{xz} = \sqrt{\frac{3 - \sin^2 \theta_0}{2 \sin^2 \theta_0}} \qquad (16.59)$$

with θ_0 being the fixed cone angle. The above formula only describe the effect on the resolution. Some other effects can be more serious and are difficult to judge. Strands that appear in a reconstruction without connection may in reality be connected, and features may appear that are not present in the real structure. This effect decreases with higher tilt angle. Structural artifacts are quite strong if only a maximum tilt angle of 30° is possible and become rather small above 45° in a conical tilt series and at 60° in a single-axis tilt series.

16.6   RESTORATION

The distortions caused by the missing data in the 3-D Fourier transform of a 3-D reconstruction from a single-axis tilt series with limited tilt range or from a conical tilt series cannot be corrected by deconvoluting the reconstruction with a convolution kernel inverse to the point-spread function because the point spread-function is $\equiv 0$ within the missing data range.

The distortions can be reduced, however, if additional information is available that can be used to provide values for the Fourier transform in the missing region. In iterative procedures *a priori*

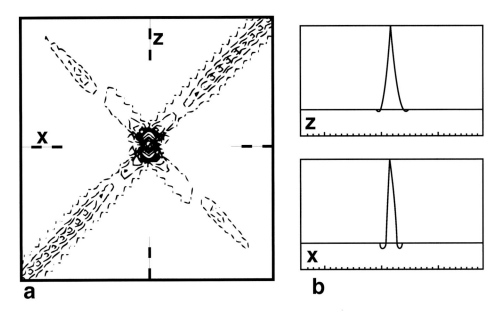

FIGURE 16.5 Section through a three-dimensional image point reconstructed by weighted back-projection from a conical tilt series. Tilt angle 45°, point reconstructed from 25 projections. Because of the missing cone, the point is elongated and shows side-minima. Contour levels relative to image maximum (from outside to inside) 1% < 0, 1% > 0, at levels 10%, 20%, 40%, …, 100%. (b) Cross-sections through (a) as indicated. (From Radermacher, M., *Dreidimensionale Rekonstruktion bei kegelförmiger Kippung im Elektronenmikroskop*, Ph.D. thesis, Technische Universität München, FRG, 1980. Copyright CRC Press LLC, Boca Raton, FL.)

information can be built into the error criterion that is being minimized. For other reconstruction methods, *a priori* information can be applied to the finished reconstruction using the method of projections onto convex sets (POCS).[44-46] Because a detailed description of this method is beyond the scope of this chapter, only the basic ideas are outlined here.

Normally, in a reconstruction from projections that do not cover the complete angular range, all points in regions of the Fourier transform where no data are available are set to 0. This choice selects one specific solution out of many possible ones. Any combination of values chosen for the Fourier transform in the non-measured range would result in a 3-D volume whose calculated projections would be equally consistent with the measured projection. Thus, for a set of given projections, an infinite set of valid solutions can be found. This solution set can be reduced by the introduction of *a priori* knowledge, such as the limited volume that the structure occupies — in practice, defining a boundary around a reconstructed particle outside of which the values in the volume are assumed to be 0 — or the condition that the range of densities in the reconstruction is limited. Both conditions are applied to the real space data and will affect every point in the Fourier transform of the object, thus creating inconsistencies with the measured projections. To obtain an object that again is consistent with the projection data, the 3-D density distribution is now Fourier transformed and every point in the Fourier transform for which measured data are available is replaced by its original value. The result is a real-space density distribution that again is in agreement with the projections, but now satisfies the imposed conditions only approximately. The procedure is iterated until the changes to the object become minimal. The final result is a structure that is consistent with the projections as well as consistent with the imposed conditions. The theory of the method of POCS shows that a unique solution can be found if the solution sets defined by *a priori* knowledge are convex and form a common intersection.

16.7   PRACTICAL ASPECTS OF THREE-DIMENSIONAL RECONSTRUCTION

In practice, three-dimensional reconstruction of single particles requires more than just a choice of reconstruction algorithm. In the following, an outline is given of the single steps required to carry out a reconstruction. Assume, as an example, a single particle, approximately globular, with a diameter of 30 nm, that is to be reconstructed with a resolution better than 3 nm.

16.7.1   RECONSTRUCTION FROM SINGLE-AXIS AND CONICAL TILT SERIES

As most of the steps are the same for single-axis and a conical tilt series, both with equal angular increments, the two geometries are treated in the same section.

According to Eq. 16.54, N = 31 projections are needed in a single-axis tilt series to obtain 3-nm resolution, which is equivalent to an angular increment of 5.8°. The next smaller integer step is a 5° increment. Assuming a goniometer tilt range of 60°, a total of 25 images need to be recorded. This will result in a reconstruction that has a resolution of 2.6 nm visible in the direction parallel to the specimen plane and approximately 4 nm in the perpendicular direction (calculated from Eq. 16.58).

If the object is to be reconstructed from a conical tilt series, using the maximum tilt angle the microscope allows, assumed to be 60°, then, according to Eq. 16.57, 55 images are needed, which corresponds to an azimuthal angular increment of 6.5°. Choosing 6° instead leads to 60 images and in turn to a resolution of 2.7 nm within planes parallel to x-y. Using Eq. 16.59, the resolution in the z-direction can be estimated to be 3.3 nm.

Low-dose techniques should be used during microscopy.[79,80] A dose of about 10 e/Å^2 can easily be achieved in a single exposure so that the cumulative exposure of the specimen is 250 e/Å^2 for a single-axis tilt series and 600 e/Å^2 for a conical tilt series.

As a next step, the images are analyzed in an optical diffractometer. The micrograph is illuminated with a coherent parallel laser beam and the diffraction pattern is observed. All images in the tilt series should be in the underfocus range,[81] and the first zero transition of the microscope's transfer function should be at a radius that corresponds to a resolution better than 3 nm — the resolution aimed for in the reconstruction. If a higher resolution is anticipated, then this first zero transition must be at a larger radius, or the transfer function must be corrected in the computer.[82]

Subsequently, the images are digitized. For 3-nm resolution in the reconstruction, each pixel in the image should correspond to less than 1.5 nm on the object scale. The digitization should be done with a pixel size that is again smaller by a factor of about 2 because, in the course of image processing, the image quality may deteriorate due to interpolations used in rotating and shifting the images. Oversampling by a factor of 2 ensures that the final resolution will not be limited by these operations.

The direction of the tilt axis, which in most microscopes is not parallel to either of the micrograph axes, must be separately determined for each image, and the images in the tilt series must be aligned to a common origin in three dimensions. This is done using markers, either gold particles that are applied to the specimen during specimen preparation or other small features that can be followed through the whole tilt series. The locations of these gold markers are measured in every image of the tilt series and the geometrical transformation between the images is fitted to these marker locations.[83] As an alternative, cross-correlation techniques can be used. As analyzed in detail in Guckenberger,[84] the images are aligned in a series relative to each other starting with the 0° image, centered in such a way that its center of gravity coincides with the image center. If the angular increment between the projections in a single-axis tilt series is $\Delta\phi = \pi/N$, then the projection p_i with angle $i \cdot \Delta\phi$ is stretched perpendicular to the tilt axis by a factor of $1/\cos(i \cdot \Delta\phi)$, and the images are successively cross-correlated with each other and shifted such that the center of gravity of the cross-correlation peak coincides with the center of the image. This procedure can be modified by changing the order in which images are aligned to minimize the accumulation of alignment errors.[85] Analogously, for a conical tilt series, each image is stretched by $l/\cos(\theta_0)$, rotated by the azimuthal angular difference to its predecessor, and successively cross-correlated, again such that the center of gravity of the cross-correlation peak coincides with the image center after alignment. The aligned tilt series then are ready to be used in one of the reconstruction algorithms described in this chapter.

16.7.2 RECONSTRUCTION FROM A RANDOM CONICAL TILT SERIES

Reconstruction from a single-exposure random conical tilt series is a technique that allows the determination of the structure of asymmetrical single particles with minimal radiation damage. Because only one image of each particle is used in the reconstruction, the cumulative dose can easily be kept to less than 10 e/Å^2, as compared to 250 to several thousand e/Å^2, which even under low-dose imaging conditions are accumulated in the collection of a regular tilt series.

During microscopy, one image is recorded under low-dose conditions with the specimen tilted by a large angle (>45°). A second 0° image of the same specimen area is taken afterward and is only used for the determination of particle orientations and particle variability.

The parts that apply to the diffractometry and digitization of regular tilt series described above also apply for a reconstruction from a single-exposure random conical tilt series. The tilt image is first analyzed in the diffractometer. Usually, it contains a whole range of defocus values, depending on the location of each particle in the micrograph and ranging from underfocus to overfocus. Again, only that area in underfocus is used, which also shows the desired resolution. The direction of the tilt axis and the tilt angle is again determined from marker coordinates or small features seen in the image that can serve as markers in a comparison of the tilt image and the 0° image.

Along with the fixed tilt angle, the azimuthal angle of each particle in the image field must also be determined. This angle is the rotation angle of each particle within the specimen plane and is determined from images extracted from the 0° micrograph using rotational and translational cross-correlation procedures.[24] Because a reconstruction from a random conical tilt series inherently is an averaging procedure, all particles used in the reconstruction should be of identical shape. Furthermore, they should have the same orientation relative to the specimen support film because the same tilt angle θ_0 is assumed for all particles. As a pattern recognition problem, this translates into the condition that the images of all particles in the 0° projection must be identical after rotational alignment. In general, this will not be the case, and the subset of particles that fulfills this condition must be selected. Multivariate statistical analysis, correspondence analysis, nonlinear mapping, and classification techniques[49-53,86] can be used to achieve this selection.

All angles are known after the 0° images are aligned, and the selected particles now form a conical tilt series with known, random azimuthal angles. The alignment of the tilt images to a common origin is best achieved by cross-correlation of each tilt image with its 0° counterpart, again after stretching the tilt image by $1/\cos(\theta_0)$. To increase the significance of the cross-correlation peak, a large area surrounding the particle can be used in this cross-correlation, since not only the particles in the 0° image and the tilted image have the same shape, but also the surrounding areas are projections of the same part of the support film. For images from a frozen hydrated sample, the translational alignment is refined further by cross-correlating the tilt images with reprojections of the 3-D reconstruction.[87]

The aligned tilt images can then be used in a reconstruction using the weighted back-projection algorithm for arbitrary geometry, or any of the iterative methods.

16.8 EXAMPLES OF APPLICATIONS

Reconstruction from a single-axis tilt series was originally developed for the reconstruction of single macromolecules[15] but, because of the problem of radiation damage, macromolecules are now mostly reconstructed using the random conical reconstruction technique.[25] However, reconstructions from single-axis tilt series are still important for particles that do not occur in multiple identical copies, precluding the application of averaging techniques. Its main application is therefore found in the study of subcellular structures in the high-voltage microscope.

16.8.1 RECONSTRUCTION OF THICK SECTIONS IN HIGH-VOLTAGE ELECTRON MICROSCOPY

The main application of the single-axis tilt geometry can be found in electron microscopy of sectioned material.[88,89] Here, the structures analyzed differ from one preparation to the next and only the principal building plane can be conserved. Second, because the structures are larger and the required resolutions are lower, radiation damage is less of a problem.

One of the first objects studied by high-voltage electron microscopy and 3-D reconstruction techniques was the cilium of newt lung tissue embedded in a 0.25-μm plastic section.[89] Here, 53 projections were recorded within an angular range of −54° to +50° in 2° increments (Figures 16.6 and 16.7). With a diameter of the reconstruction volume of 320 nm, the final resolution was 12 nm. Clearly visible were the central microtubules, the outer double microtubules, and the dynein arms.

16.8.2 RECONSTRUCTION OF MACROMOLECULES

The first reconstruction of a macromolecular assembly was the reconstruction of the bacteriophage T4,[13] followed later by the reconstruction of the fatty acid synthetase,[15-17] which was done using a single-axis tilt series. Subsequent reconstructions from single-axis tilt series included the reconstruction of the 30S and 50S ribosomal subunit from *Escherichia coli*,[90,91] the 30-nm chromatin fiber,[92] and the Balbiani ring ribonucleoprotein particles.[93]

The only macromolecular structure reconstructed from a regular conical tilt series was the reconstruction of fatty acid synthetase.[66] Because of the limited availability of truly eucentric tilt-rotation stages for electron microscopy, and more so because of the high cumulative dose in collecting such a tilt series, the technique has not been used further for the reconstruction of macromolecules.

Most reconstructions of macromolecular assemblies have been carried out using the single-exposure, random conical tilt technique because of the very low radiation damage of the specimen and also because of the ease of data collection (only very few tilt pairs are needed). The first such reconstruction was of the 50S ribosomal subunit from *Escherichia coli* (Figure 16.8).[24,94] This reconstruction was followed by the structure determination of the 50S subunit depleted of protein L7/L12,[95] the 70S ribosome from *E. coli*,[96] the 40S ribosomal subunit from rabbit reticulocytes,[97] the 80S ribosome from rabbit reticulocytes,[98] the calcium release channel from sacroplasmic reticulum,[99] and the reconstruction of the *Androctonus australis* hemocyanin.[100] Reconstructions of the ribosome of *E. coli* in different states of translation have recently been done and are a prime example of the power of these techniques.[5] Similar progress, yet at an earlier stage, can be seen in the reconstruction of the calcium-release channel where the native structure has been determined and through 3-D difference imaging ligands (calmodulin and FK binding protein) have been localized (Figure 16.9).[101-103]

The first successful reconstruction of a single asymmetrical particle from random tilts was the reconstruction of the 50S ribosomal subunit from *Escherichia coli* in a frozen hydrated preparation.[59]

16.9 CONCLUSION

Within the last 20 years, three-dimensional reconstruction of macromolecular assemblies from electron micrographs has increasingly gained importance. A large number of reconstruction algorithms are available, some specialized for specific geometries or symmetries, others with general

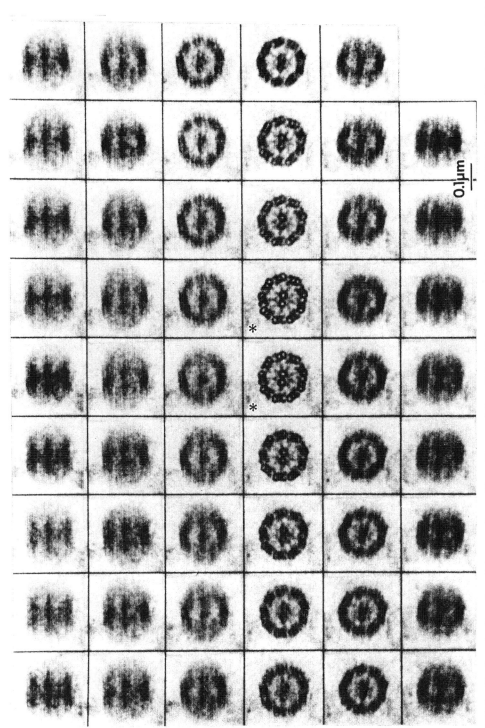

FIGURE 16.6 Single-axis tilt series of the newt lung cilium. Range −54° to 50° in 2° increments. Axial view at 8° and 10° tilt indicated by *. (From McEwen, B., Radermacher, M., Rieder, C. L., and Frank J., *Proc. Natl. Acad. Sci., U.S.A.*, 83, 1986. With permission.)

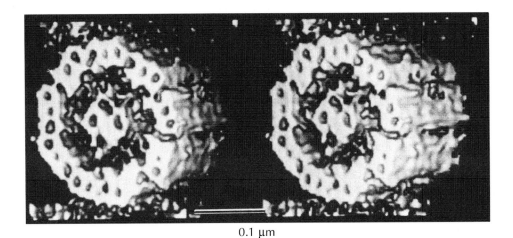

0.1 μm

FIGURE 16.7 Three-dimensional reconstruction of the newt lung cilium. (From Frank, J., McEwen, B., Radermacher, M., and Rieder, C. L., in *Proc. 44th Annu. Meet. EMSA*, Bailey, G. W., Ed., San Francisco Press, San Francisco, CA, 1986. With permission.)

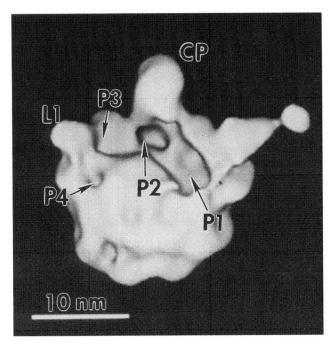

FIGURE 16.8 Reconstruction of the 50S ribosomal subunit from *Escherichia coli*, clearly showing the central protuberance CP; a large interface canyon and several deep pockets P1, P2, P3 and P4; P1 and P3 in the location where antibodies had been mapped that affect the peptidyl transferase activity[107-110] P2 is connected to a channel that leads to the back of the subunit[111,112] and, consistent with the immunoelectron microscopical findings by Ryabova et al.,[113] was concluded to be the location of the peptidyl transferase center. (From Frank, J., Verschoor, A., Radermacher, M., and Wagenknecht, T., in *The Ribosome: Structure, Function and Evaluation*, Hill, W. E., Moore, P. B., Dahlberg, A., Schlessinger, D., Garret, R. A., and Warner, J. R., Eds., American Society for Microbiology, Washington, D.C., 1990, 107. With permission.)

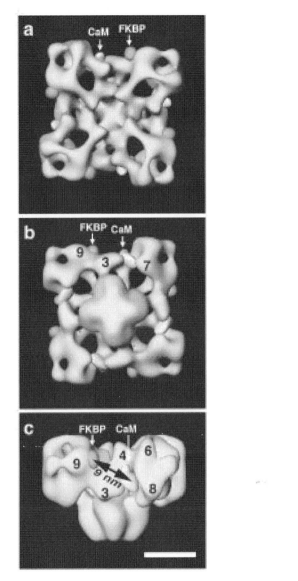

FIGURE 16.9 Localization of Calmodulin (CaM) and the FK506-binding protein (FKBP) on the calcium release channel. The image is a superposition of surface representations of the three-dimensional differences between a control reconstruction, a reconstruction of a channel with bound Calmodulin, and a reconstruction of the channel with bound FK506-binding protein. Numbers indicate globular protein domains (Adapted from Wagenknecht, T., Radermacher, M., Grassucci, R., Berkowitz, J., Xin, H.-B., and Fleischer, S., *J. Biol. Chem.*, 272, 32463, 1997. With permission.)

applicability. Both Fourier methods and weighted back-projection methods have the property that their quality can be described by a point-spread function, whereas for iterative reconstruction methods, the quality of the result is less predictable and can depend on the object. Originally, only iterative methods allowed an implementation of *a priori* information. Later, this has also been possible for the non-iterative reconstruction methods by the use of the method of POCS. *A priori* knowledge about macromolecules in electron microscope preparations is very limited, in contrast to reconstructions in medical X-ray tomography. Therefore, an ideal reconstruction technique would rely mostly on measured data and, whenever possible, techniques should be used that are able to measure the full angular range. One of these techniques is the multi-random conical tilt scheme, where several random conical tilt series of particles with different orientations relative to the specimen support are combined. Much progress has been made over the last 5 years in the development of reliable methods for reconstruction from projections of entirely randomly oriented particles,[60,61,104] and several successful reconstructions have been published.[105,106]

Future developments of reconstruction techniques for single-particle specimens are geared toward three major areas. The first goal is an improvement of the resolution in three dimensions. The second goal is the design of methods that allow fast three-dimensional difference imaging for the study of different functional states of macromolecules. The third goal is the merging of structural information from other techniques with the structures obtained from electron microscopy. Resolution limitations caused by staining, drying, and the associated particle deformations have been overcome by the development of frozen hydrated preparation techniques. Macromolecules can now be preserved in their native physiological state. Image quality is improving via the more general introduction of electron microscopes with field-emission guns and the higher stability of cryostages on the market today. Image processing techniques are being improved to make use of all information found in the electron micrograph. Whereas currently the best results are obtained by processing vast numbers of images, more sophisticated processing techniques will hopefully allow one to obtain the same quality results from less data. It is clear, however, that the resolution also depends on the specimen itself. Some specimens are variable and any effort to obtain a high-resolution structure might fail. As with all experimental and new techniques, the skill of the people doing the biochemistry for preparing the specimens and of those who use and improve the electron microscope and the programs to process the images in the end determine the highest resolutions achievable. The achievement of high resolution, however, is not the only goal of these techniques. The value of these methods in finding answers to biological questions cannot be overestimated. At this time (1999), the techniques have matured to a point where, down to resolutions of about 25 Å, they can be used as standard tools and should be part of any structural study of unknown larger macromolecules or macromolecular complexes. Through comparison of macromolecules in different functional states, they can aid in the understanding of biological function like few other techniques.

ACKNOWLEDGMENTS

The author acknowledges support by grant NSF DBI 9515518, and thanks the Max Planck Society for its financial support. The author also would like to thank T. Ruiz and F. Langheinrich for critical reading and practical help with this chapter.

REFERENCES

1. Blundell, T. L. and Johnson, L. N., *Protein Crystallography*, Academic Press, New York, 1976, 381.
2. Smith, J. L., Corfield, P. W. R., Hendrickson, W. A., and Low, B. W., Refinement at 1.4 Å resolution of a model of erabutoxin b. Treatment of ordered solvent and discrete disorder, *Acta Crystallogr., Sect. A*, 44, 357, 1988.
3. Kühlbrandt, W., Wang, D. N., and Fujiyoshi, Y., Atomic model of plant light-harvesting complex by electron crystallography, *Nature*, 367, 614, 1994.
4. Henderson, R., Baldwin, J. M., Ceska, T. A., Zemlin, F., Beckmann, E., and Downing, K. H., Model for the structure of bacteriorhodopsin based on high-resolution (electron) cryomicroscopy, *J. Mol. Biol.*, 213, 899, 1990.
5. Malhotra, A., Penczek, P., Agrawal, R. K., Gabashvili, I. S., Grassucci, R. A., Jünemann, R., Burkhardt, N., Nierhaus, K. H., and Frank, J., *Escherichia coli* 70S ribosome at 15 Å resolution by cryo-electron microscopy: localization of fMet-tRNA and fitting of L1 protein, *J. Mol. Biol.*, 280, 103, 1998.

6. Dube, P., Bacher, G., Stark, H., Mueller, F., Zemlin, F., van Heel, M., and Brimacombe, R., Correlation of the expansion segments in mammalian rRNA with the fine structure of the 80S ribosome; a cryoelectron microscopic reconstruction of the rabbit reticulocyte ribosome at 21 Å resolution, *J. Mol. Biol.*, 279, 403, 1998.

7. Moody, M. F., Structure of the sheath of bacteriophage T4, *J. Mol. Biol.*, 25, 167, 1967.

8. Taylor, K. A. and Glaeser, R. M., Electron microscopy of frozen hydrated biological specimens, *J. Ultrastruct. Res.*, 55, 448, 1976.

9. Dubochet, J., Lepault, J., Freeman, R., Berriman, J. A., and Homo, J.-C., Electron microscopy of frozen water and aqueous solutions, *J. Microscopy*, 128, 219, 1982.

10. Lepault, J., Booy, F. P., and Dubochet, J., Electron microscopy of frozen biological suspensions, *J. Microscopy*, 129, 89, 1983.

11. Dubochet, J., Adrian, M., Chang, J.-J., Lepault, J., McDowall, A. W., and Schultz, P., Cryo-electron microscopy of vitrified specimens, *Q. Rev. Biophys.*, 21, 129, 1988.

12. Glaeser, R. M., Limitations to significant information in biological electron microscopy as a result of radiation damage, *J. Ultrastruct. Res.*, 36, 466, 1971.

13. DeRosier, D. J. and Klug, A., Reconstruction of three dimensional structures from electron micrographs, *Nature*, 217, 130, 1968.

14. Hoppe, W., Langer, R., Knech, G., and Poppe, C., Proteinkristallstrukturanalyse mit Elektronenstrahlen, *Naturwissenschaften*, 55, 333, 1968.

15. Hoppe, W., Schramm, H. J., Sturm, M., Hunsmann, N., and Gaßmann, J., Three-dimensional electron microscopy of individual biological objects. I. Methods, *Z. Naturforsch.*, 31a, 645, 1976.

16. Hoppe, W., Schramm, H. J., Sturm, M., Hunsmann, N., and Gaßmann, J., Three-dimensional electron microscopy of individual biological objects. II. Test calculations, *Z. Naturforsch.*, 31a, 1370, 1976.

17. Hoppe, W., Schramm, H. J., Sturm, M., Hunsmann, N., and Gaßmann, J., Three-dimensional electron microscopy of individual biological objects. III. Experimental results on yeast fatty acid synthetase, *Z. Naturforsch.*, 31a, 1370, 1976.

18. Frank, J., Goldfarb, W., Eisenberg, D., and Baker, T. S., Reconstruction of glutamine synthetase using computer averaging, *Ultramicroscopy*, 3, 283, 1978.

19. Frank, J. and Goldfarb, W., Methods for averaging of single molecules and lattice fragments, in *Electron Microscopy at Molecular Dimensions*, Baumeister, W. and Vogell, W., Eds., Springer-Verlag, Berlin, 1980, 261.

20. Frank, J., Verschoor, A., and Boublik, M., Computer averaging of electron micrographs of 40S ribosomal subunits, *Science*, 214, 1353, 1981.

21. Verschoor, A., Frank, J., Radermacher, M., Wagenknecht, T., and Boublik, M., Three-dimensional reconstruction of the 30S ribosomal subunit from randomly oriented particles, in *Proc. 41st Annu. Meet. EMSA*, Bailey, G. W., Ed., San Francisco Press, San Francisco, 1983, 758.

22. Verschoor, A., Frank, J., Radermacher, M., Wagenknecht, T., and Boublik, M., Three-dimensional reconstruction of the 30S ribosomal subunit from randomly oriented particles, *J. Mol. Biol.*, 178, 677, 1984.

23. van Heel, M., Three-dimensional reconstruction of the 30S ribosomal subunit, in *Proc. 41st Annu. Meet. EMSA*, Bailey, G. W., Ed., San Francisco Press, San Francisco, 1983, 460.

24. Radermacher, M., Wagenknecht, T., Verschoor, A., and Frank, J., Three-dimensional reconstruction from a single-exposure random conical tilt series applied to the 50S ribosomal subunit of *Escherichia coli*, *J. Microscopy*, 146, 113, 1987.

25. Radermacher, M., Wagenknecht, T., Verschoor, A., and Frank, J., A new 3-D reconstruction scheme applied to the 50S ribosomal subunit of *E. coli*, *J. Microscopy*, 141, RP1, 1986.

26. Frank, J., Verschoor, A., Wagenknecht, T., Radermacher, M., and Carazo, J.-M., A new non-crystallographic image-processing technique reveals the architecture of the ribosome, *Trends Biochem. Sci.*, 13, 123, 1988.

27. Radon, J., Über die Bestimmung von Funktionen durch ihre Integralwerte längs gewisser Mannigfaltigkeiten, *Berichte über die Verhandlungen der Königlich Sächsichen Gesellschaft der Wissenschaften zu Leipzig, Math, Phys. Klasse*, 69, 262, 1917.

28. Cormack, A. M., Representation of a function by is line integrals, with some radiological applications, *J. Appl. Phys.*, 34, 2722, 1963.

29. Cormack, A. M., Representation of a function by is line integrals, with some radiological applications, *J. Appl. Phys.*, 35, 2908, 1964.

30. Crowther, R. A., DeRosier, D. J., and Klug, A., The reconstruction of a three-dimensional structure from projections and its application to electron microscopy, *Proc. Roy. Soc. London Ser. A*, 317, 319, 1970.

31. Smith, P. R., Peters, T. M., and Bates, R. H. T., Image reconstruction from a finite number of projections, *J. Phys. A. Math. and General*, 6, 361, 1973.

32. Hoppe, W., Das Endlichkeitspostulat und das Interpolationstheorem der dreidimensionalen elektronen-mikroskopischen Analyse aperiodischer Strukturen, *Optik*, 29, 617, 1969.

33. Bracewell, R.N. and Riddle, A. C., Inversion of fan-beam scans in radio astronomy, *Astrophys. J.*, 150, 427, 1969.

34. Vainshtein, B. K., Finding the structure of objects from projections, *Sov. Phys. Crystallogr.*, 15, 781, 1971.

35. Vainshtein, B. K. and Orlov, S. S., Theory of the recovery of functions from their projections, *Sov. Phys. Crystallogr.*, 17, 253, 1972.

36. Gilbert, P. F. C., The reconstruction of a three-dimensional structure from projections and its application to electron microscopy. II. Direct methods, *Proc. Roy. Soc. London, Ser. B*, 182, 89, 1972.

37. Radermacher, M. and Hoppe, W., 3-D reconstruction from conically tilted projections, *Proc. 9th Int. Congr. Electron Microsc.*, 1, 218, 1978.

38. Radermacher, M., Weighted backprojection methods, in *Electron Tomography*, Frank J. Ed., Plenum Press, New York, 1992, 91.

39. Bender, R., Bellman, S. H., and Gordon, R., ART and the ribosome, *J. Theor. Biol.*, 29, 483, 1970.

40. Gordon, R., Bender, R., and Herman, G. T., Algebraic reconstruction techniques (ART) for three dimensional electron microscopy and X-ray photography, *J. Theor. Biol.*, 29, 471, 1970.

41. Crowther, R. A. and Klug, A., ART and science, or conditions for 3-D reconstruction from electron microscope images, *J. Theor. Biol.*, 32, 199, 1971.

42. Bellman, S. H., Bender, R., Gordon, R., and Rowe, J. E., ART is science being a defense of algebraic reconstruction techniques for three-dimensional electron microscopy, *J. Theor. Biol.*, 32, 205, 1971.

43. Trussel, H. J., *A priori* knowledge in algebraic reconstruction methods, in *Advances in Computer Vision and Image Processing*, Vol. 1, JAI Press, Greenwich, CT, 265, 1984.

44. Tam, K. C. and Perez-Mendez, V., Tomographical imaging with limited-angle input, *J. Opt. Soc. Am.*, 71, 582, 1981.

45. Sesan, M. I., Image restoration by the method of convex projections. II. Applications and numerical results, *IEEE Trans. Med. Imag.*, 1, 95, 1982.

46. Carazo, J. M. and Carracosa, J. L., Information recovery in missing angular data cases: an approach by the convex projections method in three dimensions, *J. Microscopy*, 145, 23, 1987.

47. Deans, S. R., *The Radon Transform and Some of Its Applications*, John Wiley & Sons, New York, 1983.

48. Radermacher, M. and Hoppe, W., Properties of 3-D reconstructions from projections by conical tilting compared to single axis tilting, *Proc. 7th Eur. Congr. Electron Microscopy*, 1, 132, 1980.

49. van Heel, M. and Frank, J., Use of multivariate statistics in analyzing the images of biological macromolecules, *Ultramicroscopy*, 6, 187, 1981.

50. Frank, J. and van Heel, M., Correspondence analysis of aligned images of biological particles, *J. Mol. Biol.*, 161, 134, 1982.

51. Radermacher, M. and Frank, J., Use of nonlinear mapping in multivariate image analysis of molecule projections, *Ultramicroscopy*, 17, 117, 1985.

52. Frank, J., Classification of macromolecular assemblies studied as single particles, *Q. Rev. Biophys.*, 23, 281, 1990.

53. Carazo, J.-M., Rivera, F. F., Zapata, E. L., Radermacher, M., and Frank, J., Fuzzy sets-based classification of electron microscopy images of biological macromolecules with an application to ribosomal particles, *J. Microscopy*, 157, 187, 1990.

54. Frank, J., Carazo, J.-M., and Radermacher, M., Refinement of the random conical reconstruction technique using multivariate statistical analysis and classification, *Eur. J. Cell Biol.*, 48(Suppl. 25), 143, 1988.

55. Carazo, J.-M., Wagenknecht, T., and Frank, J., Variations of the three-dimensional structure of the *Escherichia coli* ribosome in the range of overlap views, *Biophys. J.*, 55, 465, 1989.

56. Crowther, R. A., Procedures for three-dimensional reconstruction of spherical viruses by Fourier synthesis from electron micrographs, *Philos. Trans. Roy. Soc. London Ser. B*, 261, 21, 1971.

57. Kam, Z., The reconstruction of structure from electron micrographs of randomly oriented particles, *J. Theor. Biol.*, 82, 15, 1980.

58. Provencher, S. W. and Vogel, R., Three-dimensional reconstruction from electron micrographs of disordered specimens. I. Method, *Ultramicroscopy*, 25, 209, 1988.

59. Vogel, R. and Provencher, S. W., Three-dimensional reconstruction from electron micrographs of disordered specimens. II. Implementation and results, *Ultramicroscopy*, 25, 223, 1988.

60. van Heel, M., Angular reconstitution: *a posteriori* assignment of projection directions for 3D reconstruction, *Ultramicroscopy*, 21, 111, 1987.

61. Goncharov, A. B. and Gelfand, M. S., Determination of mutual orientation of identical particles from their projections by the moments method, *Ultramicroscopy*, 25, 317, 1988.

62. Shannon, D. E., Communication in the presence of noise, *Proc. IRE*, 37, 10, 1949.

63. Goodman, J. W., *Introduction to Fourier Optics*, McGraw-Hill, New York, 1968.

64. Papoulis, A., System and transforms with applications, in *Optics*, Robert, E. and Krieger, F., Eds., McGraw-Hill, New York, 1986.

65. Ramachandran, C. N. and Lakshminarayanan, A. V., Three-dimensional reconstruction from radiographs and electron micrographs: application of convolution instead of Fourier transforms, *Proc. Natl. Acad. Sci. U.S.A.*, 68, 2236, 1971.

66. Radermacher, M., Dreidimensionale Rekonstruktion bei kegelförmiger Kippung im Elektronenmikroskop, Ph.D. thesis, Technische Universität München, Germany, 1980.

67. Radermacher, M., Three-dimensional reconstruction of single particles from random and nonrandom tilt series, *J. Electron Microsc. Technol.*, 9, 359, 1988.

68. Kwok, Y. S., Reed, I. S., and Truong, T. K., A generalized $|\omega|$ - filter for 3D-reconstruction, *IEEE Trans. Nucl. Sci.*, 24, 1990, 1977.

69. Budinger, T. F., Gullberg, G. T., and Huesman, R. H., Emission computed tomography, in *Image Reconstruction from Projections, Topics in Applied Physics*, Vol. 32, Herman, G. T., Ed., Springer-Verlag, Berlin, 1979, 147.

70. Goitein, M., Three-dimensional density reconstruction from a series of two-dimensional projections, *Nucl. Instr. Methods*, 101, 509, 1972.

71. Gilbert, P. F. C., Iterative methods for the three-dimensional reconstruction of an object from projections, *J. Theor. Biol.*, 36, 105, 1972.

72. Budinger, T. F. and Gullberg, G. T., Three-dimensional reconstruction in nuclear medicine emission imaging, *IEEE Trans. Nucl. Sci.*, 21, 2, 1974.

73. Colsher, J. G., Iterative Three-Dimensional Reconstruction from Projections: Application in Electron Microscopy, Ph.D. thesis, Lawrence Livermore Laboratory, University of California, Livermore, 1976.

74. Colsher, J. G., Iterative three-dimensional image reconstruction from tomographic projections, *Comput. Graphics Image Process.*, 6, 513, 1977.

75. Skoglund, U., Ofverstedt, L. G., Burnett, R. M., and Bricogne, G., Maximum-entropy three-dimensional reconstruction with deconvolution of the contrast transfer function: a test application with adenovirus, *J. Struct. Biol.*, 117, 173, 1996.

76. Marabini, R., Herman, G. T., and Carazo, J.-M., 3D reconstruction in electron microscopy using art with smooth spherically symmetric volume elements (Blobs), *Ultramicroscopy*, 72, 53, 1998.

77. Tam, K. C. and Perez-Mendez, V., Limited-angle 3-D reconstruction using Fourier transform iterations and Radon transform iterations, *Opt. Eng.*, 20, 586, 1981.

78. Chiu, M. Y., Barrett, H. H., Simpson, R. G., Chou, C., Arendt, J. W., and Gindi, G. R., Three-dimensional radiographic imaging with a restricted view angle, *J. Opt. Soc. Am.*, 69, 1323, 1979.

79. Williams, R. C. and Fischer, H. W., Electron microscopy of tobacco mosic virus under conditions of minimal beam exposure, *J. Mol. Biol.*, 52, 121, 1970.

80. Knauer, V., Schramm, H. J., and Hoppe, W., A minimal dose technique in 3-D electron microscopy, in *Electron Microscopy*, Vol. 2, Sturgess, J. M., Ed., Microscopical Society of Canada, Toronto, Canada, 1978, 4.

81. Typke, D. and Radermacher, M., Determination of the phase of complex atomic scattering amplitudes from light optical diffractograms of electron microscope images, *Ultramicroscopy*, 9, 131, 1982.

82. Hoppe. W., Use of zone correction plates and other techniques for structure determination of aperiodic objects at atomic resolution using a conventional electron microscope, *Philos. Trans. Roy. Soc., Ser. B*, 261, 71, 1971.

83. Luther, P. K., Lawrence, M. C., and Crowther, R. A., A method for monitoring the collapse of plastic sections as a function of electron dose, *Ultramicroscopy*, 24, 7, 1988.

84. Guckenberger, R., Determination of a common origin in the micrographs of tilt series in three-dimensional electron reconstruction, *Ultramicroscopy*, 9, 167, 1982.

85. Frank, J., McEwen, B. F., Radermacher, M., Turner, J. N., and Rieder, C. L., Three-dimensional tomographic reconstruction in high-voltage electron microscopy, *J. Electron Microsc. Tech.*, 6, 193, 1987.

86. van Heel, M., Multivariate statistical classification of noisy images (randomly oriented biological macromolecules), *Ultramicroscopy*, 13, 165, 1984.

87. Radermacher, M., Srivastava, S., and Frank, J., The structure of the 50S ribosomal subunit from *E. coli* in frozen hydrated preparation reconstructed with SECReT, in *Proc. 10th Eur. Congr. Electron Microscopy*, Megias-Megias, L., Rodriguez-Garcia, M. I., Rios, A., and Arias, J. M., Eds., Granada, 1992, Vol. 3, 19.

88. Olins, A. L., Olins, D. E., Levy, H. A., Durfee, R. C., Margle, S. M., Tinnel, E. P., Hinerty, B. E., Dover, S. D., and Fuchs, H., Modeling Balbiani ring gene transcription with electron microscope tomography, *Eur. J. Cell Biol.*, 35, 129, 1984.

89. McEwen, B. F., Radermacher, M., Rieder, C. L., and Frank, J., Tomographic three-dimensional reconstruction of cilia ultrastructure from thick sections, *Proc. Natl. Acad. Sci., U.S.A.*, 83, 9040, 1986.

90. Knauer, V., Hegerl, R., and Hoppe, W., Three-dimensional reconstruction and averaging of 30S ribosomal subunits of *Escherichia coli* from electron micrographs, *J. Mol. Biol.*, 163, 409, 1983.

91. Oettl, H., Hegerl, R., and Hoppe, W., Three-dimensional reconstruction and averaging of 50S ribosomal subunits of *Escherichia coli* from electron micrographs, *J. Mol. Biol.*, 163, 431, 1983.

92. Subirana, J. A., Munoz-Guerra, S., Radermacher, M., and Frank, J., Three-dimensional reconstruction of chromatin fibers, *J. Biomol. Stereodyn.*, 1, 705, 1983.

93. Skoglund, U., Anderson, K., Standberg, B., and Daneholt, B., Three-dimensional structure of a specific pre-messenger RNP particle established by electron microscope tomography, *Nature*, 319, 560, 1986.

94. Radermacher, M., Wagenknecht, T., Verschoor, A., and Frank, J., Three-dimensional structure of the large ribosomal subunit from *Escherichia coli, EMBO J.*, 6, 1107, 1987.

95. Carazo, J.-M., Wagenknecht, T., Radermacher, M., Mandiyan, V., Boublik, M., and Frank, J., Three-dimensional structure of 50S *Escherichia coli* subunits depleted of protein L7/L12, *J. Mol. Biol.*, 201, 393, 1988.

96. Wagenknecht, T., Carazo, J.-M., Radermacher, M., and Frank, J., Three-dimensional reconstruction of the ribosome from *Escherichia coli, Biophys. J.*, 55, 455, 1989.

97. Verschoor, A., Zhang, N. Y., Wagenknecht, T., Obrig, T., Radermacher, M., and Frank, J., Three-dimensional reconstruction of mammalian 40S ribosomal subunit, *J. Mol. Biol.*, 209, 115, 1989.

98. Verschoor, A. and Frank, J., Three-dimensional structure of the mammalian cytoplasmic ribosome, *J. Mol. Biol.*, 214, 737, 1990.

99. Wagenknecht, T., Grassucci, R., Frank, J., Saito, A., Inui, M., and Fleischer, S., Three-dimensional architecture of the calcium channel/foot structure of sarcoplasmic reticulum, *Nature*, 338, 167, 1989.

100. Boisset, N., Taveau, J.-C., Lamy, J., Wagenknecht, T., Radermacher, M., and Frank, J., Three-dimensional reconstruction of native *Androctonus australis, J. Mol. Biol.*, 216, 743, 1990.

101. Wagenknecht, T., Radermacher, M., Grassucci, R., Berkowitz, J., Xin, H.-B., and Fleischer, S., Locations of calmodulin and FK506-binding protein on the three-dimensional architecture of the skeletal muscle ryanodine receptor, *J. Biol. Chem.*, 272, 32463, 1997.

102. Wagenknecht, T. and Radermacher, M., Ryanodine receptors: structure and macromolecular interactions, *Curr. Opin. Struct. Biol.*, 7, 258, 1997.

103. Samso, M. and Wagenknecht, T., Contributions of electron microscopy and single-particle techniques to the determination of the ryanodine receptor three-dimensional structure, *J. Struct. Biol.*, 121, 172, 1998.

104. Radermacher, M., Three-dimensional reconstruction from random projections: orientational alignment via Radon transforms, *Ultramicroscopy*, 53, 121, 1994.

105. Serysheva, I. I., Orlova, E. V., Chiu, W., Sherman, M. B., Hamilton, S. L., and van Heel, M., Electron cryomicroscopy and angular reconstitution used to visualize the skeletal muscle calcium release channel, *Nature Struct. Biol.*, 2, 18, 1995.

106. Martin, C. S., Radermacher, M., Wolpensinger, B., Engel, A., Miles, C. S., Dixon, N. E., and Carazo, J. M., Three-dimensional reconstructions from cryoelectron microscopy images reveal an intimate complex between helicase DnaB and is loading partner DnaC, *Structure*, 6, 501, 1998.

107. Stöffler, G., Bald, R., Kastner, B., Lührman, R., Stöffler-Meilicke, M., and Tischendorf, G., Structural organization of the *Escherichia coli* ribosome and localization of functional domains, in *The Ribosome: Structure, Function and Genetics*, Chambliss, G., Craven, G. R., Davies, J., Davies, K., Kahan, L., and Nomura, M., Eds., University Park Press, Baltimore, MD, 1980, 171.

108. Lührman, R., Bald, R., Stöffler-Meilicke, M., and Stöffler, G., Localization of the puromycin binding site on the large ribosomal subunit of *Escherichia coli* by immunoelectron microscopy, *Proc. Natl. Acad. Sci., U.S.A.*, 78, 7276, 1981.

109. Olson, H. M., Grant, P. D., Cooperman, B. S., and Glitz, D. G., Immunoelectron microscopic localization of puromycin binding on the large subunit of the *Escherichia coli* ribosome, *J. Biol. Chem.*, 257, 2649, 1982.

110. Olson, H. M., Nichoolson, A. W., Cooperman, B. S., and Glitz, D. G., Localization of sites of photoaffinity labeling of the large subunit of *Escherichia coli* by arylazide derivative of puromycin, *J. Biol. Chem.*, 260, 10236, 1985.

111. Radermacher, M., Frank, J., and Wagenknecht, T., The probable exit site of the polypeptide in the ribosome: analysis of densities in three-dimensional reconstruction, in *9th Eur. Congr. Electron Microscopy*, Dickinson, H. G. and Goodhew, P. J., Eds., Institute of Physics, Bristol, 1988, Vol. 3, 323.

112. Frank, J., Vershoor, A., Radermacher, M., and Wagenknecht, T., Morphologies of eubacterial and eucariotic ribosomes as determined by three-dimensional electron microscopy, in *The Ribosome: Structure, Function and Evolution*, Hill, W. E., Moore, P. B., Dahlberg, A., Schlessinger, D., Garrett, R. A., and Warner, J. R., Eds., American Society for Microbiology, Washington, D.C., 1990, 107.

113. Ryabova, L. A., Selivanova, O. M., Baranov, V. I., Valiliev, V. D., and Spirin, A. S., Does the channel for the nascent polypeptide exist in the ribosome? Immune electron microscopy study, *FEBS Lett.*, 226, 255, 1988.

17 Computer-Aided Three-Dimensional Reconstruction from Serial Section Images

Norio Baba

CONTENTS

17.1 INTRODUCTION

The serial section reconstruction technique has been utilized for a long time as a basic method in biological fields in which, in the old-fashioned way, a model is built by stacking materials like balsa wood sheets or plastic plates cut along contour lines of objects. Thus, the object is reconstructed from section images. Since the 1970s, computer processing and graphics were introduced to assist in this work.[1-30] Subsequently, commercial image processors, including systems based on personal computers and software packages, provided for reconstruction, although investigators sometimes regarded these as insufficient, depending on the level of their research work.[3] Furthermore, with improvements in the quality of online image detectors, even in electron microscopy, a total reconstruction system including an electron microscope for online image acquisition was developed.[31] Computer-aided techniques have become more convenient for investigators,[32-41] although the preparation of serial sections and their images remains cumbersome.

A partial solution to the problem of the difficulty and time involved in preparing and working with long series of serial sections is the use of thicker sections, limited, of course, by the specific research purpose. Domon and Wakita[42] adopted a technique of alternating semithin and ultrathin sections by which the three-dimensional (3-D) structure of an osteoclast, especially the clear zone, was reconstructed. They pointed out that to use semithin sections made the tedious work easier. Marko et al.[34] developed a useful reconstruction system for thick serial section images taken with a high-voltage electron microscope using stereo-pair images. The principle is based on the stereo-photogrammetry for making terrain maps. The stereo-pair image of a section is displayed on the graphical terminal, and contour lines are drawn using a cursor viewed in stereo. They furthermore improved the system named "Stereocon," which is quite functional. In parallel to these improvements on sectioning, the advances in the image processing method are important.

This chapter describes the method of computer-aided serial section reconstruction from a technological point of view. During the last few years, the environment of personal computers has been revolutionized. Multiple tasks can be performed on an integrative computer system: writing texts, drawing figures, processing images, manipulating 3-D data, rendering 3-D images, making movies, sending the data to other people in the world and communicating with them, and, possibly, producing a real 3-D model with laser lithography. It seems advantageous for the serial section reconstruction work to utilize this environment of multiple processing. Researchers can get various representations of a 3-D model with various application software programs that are convenient for analytical observation, measurement, fusion with artificial models, and 3-D modification. Therefore, it becomes more important to develop methods to extract data of contour lines or object images from serial section images, including image alignment and correction of image distortion due to physical deformation of thin sections. If these 3-D data are not adequately acquired, it is difficult to observe the reconstructed model in three dimensions. This chapter describes the important methods for image alignment and correction of image distortion. Some methods to obtain the 3-D data set of contour lines and object images through the serial section images are also described in the above-mentioned background of the personal computer environment.

There are two basic representation methods for the serial section reconstruction: (1) the polygon-based surface model representation, and (2) the voxel representation. The former uses a data set of contour lines and the latter needs a serial section image data set. By analogy to the picture element (pixel), the volume raster element is termed a *voxel*. For clear analysis of the 3-D morphological surface shape, the polygon-based method may be superior; but for 3-D density distribution itself and an object consisting of very fine complicated parts, the voxel method must be used because contour lines are no longer drawn in section images. The voxel method seems to serve all purposes, but actually it does not. If a data set of necessary image areas is not adequately extracted from serial section images, the 3-D representation by the voxel method often becomes meaningless. Showing practical applications — throughout this review — it is stressed that it is important to choose between the voxel representation and the polygon-based method when considering their merits and weaknesses.

17.2   WORKING TOOLS IN COMPUTER PROCESSING

Only several years ago a dedicated image processor or a high-priced computer system was necessary for 3-D reconstruction. Due to the recent rapid evolution of computer technology, this equipment is being now replaced by a personal computer. Peripherals for image acquisition and image storage, namely online CCD camera, image reader using a CCD line sensor, pen-tablet, hard disk, magneto-optical disk, and DVD RAM also rapidly evolved and became available for personal use. Because many (tens of images or more) serial section images must be processed and stored, such devices are very useful. There are many application software solutions for handling images and drawing lines and shapes. Using this software, one can manually align the serial images. This is facilitated using the mouse to rotate the image so that it is superimposed on a neighboring section image while viewing both images in real-time using the layer function. By properly changing the transparency in the layer function, the operating efficiency is improved. If there are mark points for alignment in the images, the operation can be easily performed. Application software with an image-stacking function generates an animation of the serial images, which is useful for checking the precision of alignment and imagining a rough 3-D model. For example, the program "NIH image" (public domain software) has a function that changes the speed of the animation by simple key control and a menu that generates a 3-D voxel image of the stack data. If, for a simple case, the gray-level stack images are converted to a set of binary images by the thresholding method, a solid model can be produced quickly and observed in three-dimensional rotation. (Of course, software of this type, including that for 3-D Computer Tomography (CT) and confocal microscope imaging, can produce a 3-D image using the live-gray stack images without the thresholding processing. However, in this author's experience, it is not so simple to generate an interpretation model by the live data. Generally, a careful modification of the image data is needed.)

In the conventional manner of the computer-aided serial section reconstruction, contour lines are constructed stacking along the z-axis perpendicular to the section plane using surface polygons linking between neighboring contour lines above and below. Therefore, and first of all, a researcher must trace the contour lines in every section image. At present, and in most cases, this still relies on manual work. Many programs on the market are convenient for such trace works. A Bezier function for curved line drawing is useful because the line can be smoothly drawn from relatively few discrete points and partial redrawing, and correction is simple. A pen-tablet is also useful; with a pencil-like device, one can draw contour lines directly on a section image displayed by a liquid-crystal monitor. Recent 3-D computer graphics (CG) software on the market is well-developed, and it is possible to produce a surface polygon model that constructs the input contour lines through every set of surface polygons between a pair of neighboring contour lines above and below; this must be done by manual operations. For example, in the software STRATA STUDIO Pro (Strata, Inc.), a function called "skin" produces such a set of polygons by a manual procedure. However, once the entire surface polygon data is produced, a great variety of 3-D representation methods can be used (e.g., rendering the 3-D images by various ray tracing techniques, smoothing the model, cutting the model, Boolean calculation between multiple 3-D models).

Because manual polygon production by such general-purpose CG software is very cumbersome and generally not appropriate for a precise reconstruction, one should use an exclusive application software for the serial section, contour line reconstruction. Then, it is recommended that the surface polygon data made by this software be fed into such general-purpose CG software. Since, in general, recent application software shows a high degree of compatibility, the data transfer between software packages is without problems. Many useful applications for the 3-D representation and the data manipulation mentioned above can be performed.

Furthermore, it is possible to produce a real model from the 3-D voxel data or the 3-D surface polygon data by laser lithography. The laser lithography apparatus stacks up each slice made of photo-hardening resin and builds up the entire model. Although several hours or more may be required for this process, a very complicated and fine real model can be constructed. The slice thickness is about 0.2 to 0.5 mm, and the spot diameter at making the slice sheet by the laser beam is about 0.1 mm (from SLP-4000/5000, Denken Engineering Co., Ltd.). Thus, a realistic reconstruction of the 3-D structure can be obtained.

17.3   IMAGE ALIGNMENT AND DISTORTION CORRECTION

Although serial section reconstruction is similar to the reconstructions in 3-D confocal microscopy and that from serial sectional images obtained by computer tomography (CT), there is a major difference between them. In the former, an object is physically sectioned and every section is photographed, in contrast to the optical sectioning used in the other reconstructions. Therefore, the section images must be aligned and corrected for distortion before use in the 3-D reconstruction. If possible, it is preferable to "nick" a specimen with a laser or other suitable device before cutting to facilitate the alignment.[31] However, in high magnifications, such marking may be difficult.

In this section, two methods are described that made progress toward automatic alignment: (1) the use of the image cross-correlation function (XCF) between neighboring section images; and (2) the use of local image features. The former method is generally used for measuring the displacement of any pair of images and alignment of them. However, in addition, the rotational mismatch must be measured. The latter is closely related to the method of matching points in stereo-pair images (in digital stereophotogrammetry). There are many algorithms for matching points using local image features. In this review, however, the cross-correlation coefficient[43,44] is employed as the basic tool for this purpose because of its simplicity and performance. In this application, the latter method was used to align images more accurately after the manual alignment or the alignment by the image XCFs. Some results were obtained and are the basis for the automatic alignment.

Second, the problem of the correction for local deformations in section images is discussed. This problem can be solved using the elastic model proposed by Durr et al.[45] This model is based on the results already established in the fields of material strength or elasticity.[46] In this chaper, the method is described and adapted for practical use.

17.3.1   ALIGNMENT BY IMAGE CROSS-CORRELATION

Consider a pair of neighboring section images in which a rotational difference and a relative displacement must be corrected. In this process, successive XCFs between their images are calculated in which one of two images is successively rotated by a small angle. At each image rotation, a maximum value of the XCF is calculated and stored together with its position. Further, among these successive maxima, the final maximum is identified, which means that the rotational difference and the displacement are measured. This process is performed by the FFT (fast Fourier transform) algorithm and associate transformations. The XCF is calculated by the convolution theorem using FFTs. If a modern personal computer is used, the computing time is around tens of seconds for an image size of 512×512 pixels, an angle of about $10°$, and an interval of $1°$. Note that in the XCF calculation, the image function $f(x, y)$ is not directly used: only $f(x, y) - f_{ave}$ should be used, where f_{ave} is the average value of the image. An example of this processing is shown in Figure 17.1.

17.3.2   ALIGNMENT BY A SET OF CORRESPONDING POINTS

The second method is to search for a set of corresponding points in a neighboring image pair. The normalized cross-correlation coefficient[43,44] is defined as:

$$\rho = \frac{\sum_{i}^{n} g_{1i}g_{2i} - \frac{1}{n}\left(\sum_{i}^{n} g_{1i}\right)\left(\sum_{i}^{n} g_{2i}\right)}{\sqrt{\left\{\sum_{i}^{n} g_{1i}^{2} - \frac{1}{n}\left(\sum_{i}^{n} g_{1i}\right)^{2}\right\}\left\{\sum_{i}^{n} g_{2i}^{2} - \frac{1}{n}\left(\sum_{i}^{n} g_{2i}\right)^{2}\right\}}} \qquad (17.1)$$

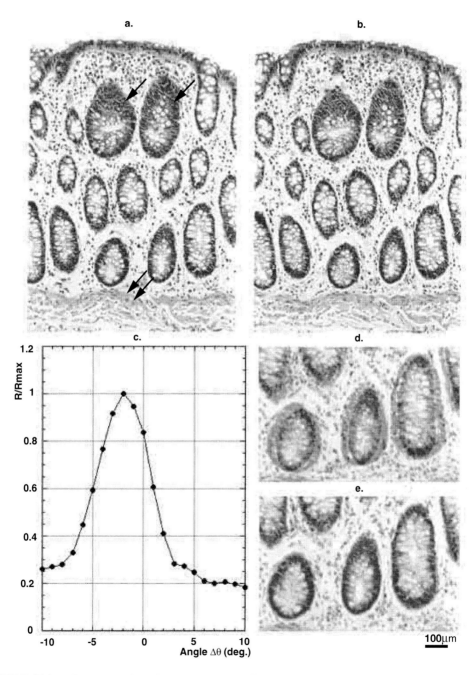

FIGURE 17.1   Alignment with the image cross-correlation; (a) and (b) a pair of neighboring section images to be aligned, typical sections of a single-gland adenoma (arrow) surrounded by crypts in human tissues from familial adenomatosis coli (double arrows: muscularis mucosae); (d) and (e), superimposed images (a) and (b) before and after the alignment, respectively; (c) successive maxima of the XCF peak values for the angle $\Delta\theta$, the image of (b) was displaced and rotated to make the correlation highest as shown in (e). (From Baba, N., Satoh, H., and Nakamura, S., *Bioimages*, 1, 105, 1993. With permission.)

where g_{1i} is a gray value in a window called the template window (W) (in one of two images); g_{2i} that in the larger window is called the search area (S) (in the other image); and n the number of gray values. The center of the template window represents the image point to be correlated. The value of ρ lies in the interval between -1 and $+1$. $\rho = +1$ means that both window data sets are identical; $\rho = 0$ means that they are independent of each other; and $\rho = -1$ describes a reciprocal relation. This ρ is calculated for every position of W within S. The best fit of W within S is given at the position where ρ has its maximum value.

Following three technical points may be useful in the practical use of this tool:

1. Simply, a variance of the image data W is available to check whether it has sufficient image features to be correlated.
2. The window size of W should be made variable in order to be able to automatically enlarge its size until its image variance satisfies a threshold level.
3. In the determination of the best-fit position, this algorithm should be improved to be sure to obtain it, as follows. If there is a best-fit position for W within S, the distribution of ρ at the vicinity of the position may be like a quadratic parabola. Therefore, first, the algorithm obtains the distribution of ρ within S and captures several such quadratic parabola distributions within the area of S. Then, the best-fit position is selected where ρ is highest among several maxima. Depending on the situation, the best position is determined at an accuracy better than 1 pixel (by a quadratic parabola interpolation).[44] This method is applicable to cases where the rotational difference and the relative displacement between images are small. In an examination reported here, a pre-alignment was necessary, either manually or with the XCF method described above.

Once a set of corresponding positions (x, y) and (x', y') is obtained by either the above way or manually, the angle θ and displacement (a, b) giving a least-squares fit to these can be determined as follows:[47] Let the transformation from one picture to the other be described by

$$\begin{bmatrix} x' \\ y' \end{bmatrix} = \begin{bmatrix} \cos\theta & \sin\theta \\ -\sin\theta & \cos\theta \end{bmatrix} \begin{bmatrix} x \\ y \end{bmatrix} + \begin{bmatrix} a \\ b \end{bmatrix} \tag{17.2}$$

and define a sum square deviation between observed and predicted positions (x', y')

$$S = \sum_{i}^{n} \left[\left(x_i' - x_i \cos\theta - y_i \sin\theta - a \right)^2 + \left(y_i' - y_i \cos\theta + x_i \sin\theta - b \right)^2 \right] \tag{17.3}$$

where n is the number of corresponding positions. Then, from the minimization conditions,

$$\partial S/\partial a = 0, \; \partial S/\partial b = 0, \; \text{and} \; \partial S/\partial \theta = 0 \tag{17.4}$$

and a, b, and θ are mathematically deduced as follows:

$$a = \frac{1}{n} \left[\sum_{i}^{n} x_i' - \cos\theta \sum_{i}^{n} x_i - \sin\theta \sum_{i}^{n} y_i \right]$$

$$b = \frac{1}{n} \left[\sum_{i}^{n} y_i' - \cos\theta \sum_{i}^{n} y_i - \sin\theta \sum_{i}^{n} x_i \right] \tag{17.5}$$

$$\theta = \tan^{-1}(X/Y), \tag{17.6}$$

with

$$X = \frac{1}{n^2}\left[n\sum_i^n y_i' \cdot x_i - n\sum_i^n x_i' \cdot y_i + \sum_i^n y_i \sum_i^n x_i' - \sum_i^n y_i' \sum_i^n x_i \right]$$

$$Y = \frac{1}{n^2}\left[n\sum_i^n x_i' \cdot x_i - n\sum_i^n y_i' \cdot y_i + \sum_i^n x_i \sum_i^n x_i' - \sum_i^n y_i' \sum_i^n y_i \right] \tag{17.7}$$

For θ, the twofold ambiguity is resolved by taking whichever direction gives the lower value of S. Of course, these expressions are advantageous for a large displacement and rotation.

17.3.3 CORRECTION OF IMAGE DISTORTION

The fundamental problem for the imaging of serial sections is the expansion or contraction of section images or local deformations of sections arising during specimen preparation or observation with a microscope. With the elastic model,[45] the local displacement of an image pair is approximated by the deformation of the 2-D plate. For the model, the displacements of corresponding points serve as boundary conditions of a deformed plate. The basic principle for the calculation of the deformation is the minimization of the deformation energy E:

$$E = \frac{1}{2}\iint \left(\sigma_x \varepsilon_x + \sigma_y \varepsilon_y + \tau_{xy}\gamma_{xy}\right) dxdy \rightarrow \min \tag{17.8}$$

where $\sigma(\sigma_x, \sigma_y, \tau_{xy})$ is the stress vector and $\varepsilon(\varepsilon_x, \varepsilon_y, \gamma_{xy})$ is the strain vector.[46] This equation is written using the displacements (u, v) as follows:

$$E \propto \iint \left\{ \frac{1}{1-\nu}\left[\left(\frac{\partial u}{\partial x}\right)^2 + \left(\frac{\partial v}{\partial y}\right)^2 + 2\nu\left(\frac{\partial u}{\partial x}\right)\left(\frac{\partial v}{\partial y}\right)\right] + \frac{1}{2}\left(\frac{\partial u}{\partial x} + \frac{\partial v}{\partial y}\right)^2 \right\} dxdy \rightarrow \min \tag{17.9}$$

where ν is the Poisson ratio. The solution of this function is a system of coupled partial differential equations for the displacements (u, v) of the deformed image plane. The numerical calculation of this equation system leads to a displacement map of corresponding pixels that allows the correction of the distorted sampling geometry of digitized images. Before discussion of its practical use, an outline of the numerical calculation is described, and the Poisson ratio is also mentioned. In this case, the iterative calculation method is used. Consider the regular net plane shown in Figure 17.2f. First, displacement vectors of corresponding pass points are given as boundary conditions; these pass points are determined by the corresponding positions in the pair of images (see Figures 17.2a and 17.2b). At every lattice point, the partial differential equations

$$\partial E/\partial u_{i,j} = 0 \tag{17.10}$$

and

$$\partial E/\partial v_{i,j} = 0 \tag{17.11}$$

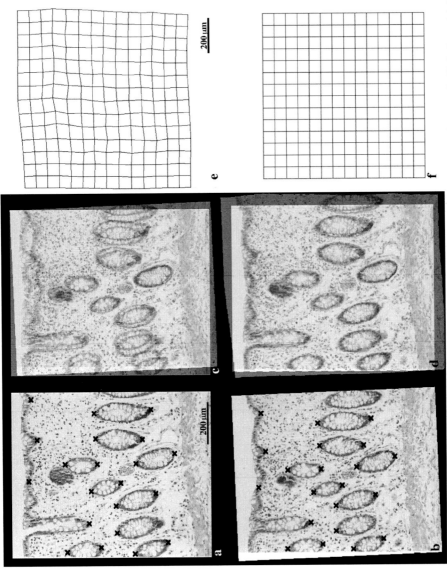

FIGURE 17.2   Correction for image distortion; (a and b) a pair of neighboring section images. The distortion of image (b) was corrected compared with the image (a). Corresponding points are marked by symbol x (see text for details); (c and d), superimposed images of (a and b) before and after the correction, respectively; (e and f), net planes used for the resampling the pixel data. A regular net of f was set on the image (b). The deformed net plane e was calculated based on the elastic model under the boundary conditions given by the corresponding points. (From Baba, N., Satoh, H., and Nakamura, S., *Bioimages*, 1, 105, 1993. With permission.)

are approximately solved using numerical differential equations, where $(u_{i,j}, v_{i,j})$ means the displacement at the lattice point (i, j), and the relative deformation energy E is approximated using the neighboring displacements (u, v) (a simple case includes only five displacements at (i, j), (i − 1, j), (i, j − 1), (i + 1, j), (i, j + 1)). Iteration of this calculation a sufficient number of times results in a deformed net plane that indicates a displacement map satisfying the boundary conditions as shown in Figure 17.2b. Here, two typical cases of Poisson's ratio ν are mentioned: (1) ν = 0.0 and (2) ν is close to 1.0. The former case corresponds to a very hard material, and the latter a soft one. For ν = 0.0, the displacement set as boundary conditions has influence all over the plane; but for latter case only, local areas were influenced. It was found that to set ν = 0.0 is of an advantage because even if the number of pass points is small, areas between pass points can be deformed uniformly. Of course, the accuracy of the method can be improved by enlarging the set of pass points.

The correction for distortion is divided into the following steps:

1. the search for and determination of corresponding points between neighboring section images
2. the calculation of the displacement field with the elastic plate model
3. the correction or resampling of the image geometry with the calculated displacement field

Because the calculated deformed net plane means the deformed resampling geometry, the distorted image is corrected by resampling the pixel data according to the deformed net plane. The interpolation technique is used for resampling within a mesh.

17.4   CONTOUR LINE DRAWING AND OBJECTIVE IMAGE AREA EXTRACTION

In conventional serial section reconstruction, contour lines of an object are stacked, and must therefore be extracted from section images. In the voxel-processing-based reconstruction, this kind of contour line extraction may sometimes be necessary because simple stacking of serial section images will not create an interpretable result. As described in Section 17.2, there are convenient tools like a pen-tablet and application programs that are useful to manually draw contour lines directly on a section image displayed on a monitor using the layer function. Although these are convenient, it is far more advantageous if contour lines could be automatically drawn by image processing. At present, this is generally impossible; however, there are methods to assist the work or facilitate automatic drawing. Some of these are described below.

17.4.1   CONVENTIONAL METHODS

The easiest method among the various image processing techniques is to produce a binary image by thresholding. In this author's experience, there are some cases where this simple thresholding is applicable to low-magnification light or electron microscope images in material science. The reason is that, in these cases, each material appears in a distinguishable uniform gray level. (Shading compensation image processing and some similar basic techniques assisting the thresholding may be needed.) If an image object is converted into a binary image, the contour line is easily created by existing commercial application software. However, in the case of biological section images, this technique is of little use because such uniformity does not exist in biological images; rather, fine structure appears due to the staining.

As an alternative, there is a method of thresholding using filtering. In this method, fine structures that are meaningless for contour line acquisition are blurred by image filtering. Although there are various filter types, in this case a nonlinear type such as **median filtering** may be useful rather than a **convolution** type because the latter may produce some artificial structures or the edge of an image object may be blurred. Another method for contour line acquisition is a sequential method called "edge tracking." If the contrast of the contour image is sufficient and noise is negligible, the image processor can semi-automatically trace the contour line. The operator indicates a position on the contour line with a mouse and the algorithm searches in a direction along which the slope of the contrast is lowest.[35] In the contour line drawing shown in Figure 17.3, two contours (indicated by arrows) were traced by the thresholding jointly using median filtering and the edge-tracking method.

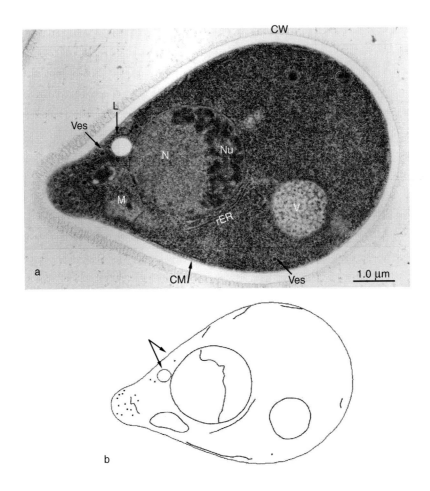

FIGURE 17.3 (a) Section of a yeast target cell induced by the mating pheromone (α-factor) taken with a 200-kV electron microscope. This section is a typical one among the serial section micrographs. (b) Contour lines of organelles and positions of lipid droplets and small vesicles. Abbreviations: CM, cell membrane; CW, cell wall; L, lipid droplet; M, mitochondrion; N, nucleus; Nu, nucleolus; V, vacuole; Ves, vesicle; rER, rough endoplasmic reticulum.

17.4.2 TEXTURE SEGMENTATION

In many cases, it is impractical to apply the methods discussed above to biological thin-section images because of noisy granularity due to staining. However, one can extract arbitrary biological objects from the whole image area. This may be because the human eye discriminates differences of image texture based on the spatial granular distribution. An investigation into the discrimination based on texture analysis in biological thin-section images is meaningful.[48] There are various image processing methods for texture analysis described in the literature.[49,50] A standard approach in texture analysis is reviewed here.

The method of texture analysis is principally divided into two approaches: statistical and structural. For biological section images, the statistical approach is appropriate because the image is normally not periodical like a crystal. In the statistical approach, there are various ways to measure the features of the texture. Weszka et al.[51] tested the discriminating power of various tools: spatial gray-level dependence method (SGLDM), gray-level difference method (GLDM), gray-level run length method (GLRLM), and power spectrum method (PSM). The SGLDM is employed here because it was found to be superior to the other methods. Although Weszka's test images were not microscope images, those results may be a good guideline. This method basically depends on an evaluation of a gray-level co-occurrence matrix P(i, j) which is calculated with a joint probability density function f(i, j; d,θ). f(i, j; d,θ) is a probability that a gray level of a certain pixel is "j," where this pixel is at a distance of "d" and a direction of θ from another pixel possessing "i" gray level. The probability f() is obtained by measuring this event within a certain image area (see Ref. 48 for details). This square image area is called a "window" in this section; and, for example, one side is a few tens of pixels, which is called "window size." The texture analysis is performed by repeating the measurement of the P(i, j) moving the "window" pixel by pixel. For biological thin section images taken with a microscope, the P(i, j) may represent differences of size and spatial distribution of staining granules. For relatively coarse granularity and short distances of "d" relative to the size of granules, high values may appear near the diagonal elements of the matrix P(); and for fine granularity and comparable distances of "d" to the granular size, values may be almost equally distributed over the matrix. To evaluate this probability distribution in the matrix P(), four standard estimators — called **inverse difference moment, angular second moment, entropy, and contrast** — were practically employed in this test and are widely used in general image analysis.[52] These estimators are mathematically defined with the co-occurrence matrix P() (e.g., for "angular second moment," the estimator is $\Sigma\Sigma_{ij}P^2(i, j)$; see Ref. 48 for details), and they discriminate the image texture of each biological object in the section image.

Figure 17.4a shows an electron microscope image that is suitable as a sample to investigate the discriminating ability of the texture-based method. This ultra-thin section of the yeast cell, *Saccharomyces cerevisiae* was obtained by the freeze-substitution fixation method.[53] The image shows the morphology of a vacuole in a multiple protease-deficient cell under nitrogen starvation (see Refs. 54 and 55 for details on specimen preparation and biology). Many spherical autophagic bodies (AB) appear in the vacuole (V). The texture of the bodies and the vacuole resemble each other, which is appropriate for this test. To obtain the spatial distribution of staining granules at high resolution, photographic negatives taken with a microscope (direct magnification is × 10,000) were read at 840 dpi (about 30 μm/pixel) by a commercial flatbed CCD scanner. The whole cell image was recorded at 2014 × 1948 pixels. The examined parameters of the "window size" and "d" needed for the feature measurement were 10 × 10 to 50 × 50 pixels and 1 to 5 pixels, respectively. Original images were scanned at 256 gray levels, but the number of gray levels in the measurement of P(i, j) that corresponds to the size of the matrix was 16.

The standard four estimators — inverse difference moment, angular second moment, entropy, and contrast — and a compound estimator by multiplying angular second moment and contrast were applied to practical thin section images. Typical results are shown here. Figures 17.4b, c, d show various feature maps of the inverse difference moment (Figure 17.4b), the angular second

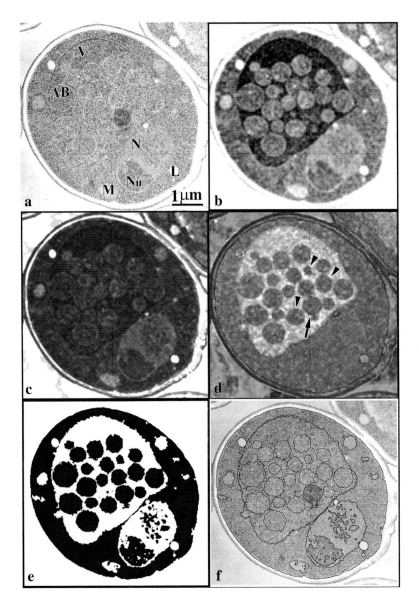

FIGURE 17.4 Application of the texture analysis-based method to a biological thin section image of (a) (a yeast cell shows morphology of a vacuole in a multiple protease-deficient cell under nitrogen starvation). Many spherical autophagic bodies (AB) appear in the vacuole (V). Note: these textures of the bodies and the vacuole resemble each other, which is appropriate to this study. (b), (c), and (d) are computed feature maps by standard estimators of "inverse difference moment," "angular second moment," and the compound estimator, respectively. (e and f): an example of the object extraction image (binary image) and the contour line image superimposed on the original image of (a), which are obtained by the simple thresholding of feature maps by the "angular second moment"-estimator (c) and the compound estimator (d). These results show that the texture-based method has an ability to discriminate differences of spatial staining granular distributions (see text for details, including explanations on arrows and arrowheads). Parameters: the "window size" is 30×30 pixels, the distance $d = 1$ pixel, and the number of gray levels in the feature measurement is 16, where the whole image is processed at 2014×1948 pixels. Abbreviations: AB, autophagic body; L, lipid; M, mitochondrion; N, nucleus; Nu, nucleolus; V, vacuole. (From Baba, N., Ichise, N., and Tanaka, T., *J. Electron Microsc.*, 45. 298, 1996. With permission).

moment (Figure 17.4c), and the compound estimator (Figure 17.4d). The value of each estimator is represented in gray levels in the figures. For example, the angular second moment discriminates organelles such as a nucleus, nucleolus, lipid droplets, mitochondria, and the vacuole, but they hardly discriminate the autophagic bodies. Although the inverse difference moment extracts the autophagic bodies better than the angular second moment, it can not extract all but misses one body (see the arrow in Figure 17.4d). However, this disadvantage has been dissolved by the compound estimator. In Figure 17.4d, all autophagic bodies are clearly extracted. The marked one (indicated by the arrow), which is not discriminated by the angular second moment and the inverse difference moment, is obviously detected. Some autophagic bodies (indicated by arrowheads) that are difficult to discriminate even by human observation are more clearly detected by some estimators. Moreover, the compound estimator has a useful characteristic, that is, in the computed feature image shown in Figure 17.4d, edge profiles of objects occur as sharp contours. Because of this characteristic, autophagic bodies are clearly extracted. The compound estimator also succeeded in the extraction of autophagic bodies from the vacuole image area, but it was not successful in that of other organelles. If one wishes to extract all objects, it is necessary to use a combination of estimators. Figure 17.4e shows an example of image area extraction by simple thresholding of feature maps of the compound estimator and the angular second moment, and Figure 17.4f is a superimposed image of the contour lines on the original image (Figure 17.4a). The present results suggest that the texture-based method is a useful assistant tool for image area extraction processing. Of course, the extracted contour lines are not exactly equal to those traced by biologists; however, the image processing tools presented here have superior discrimination power when compared with other methods such as thresholding, thresholding combined with image filtering, and edge tracking.

17.5 POLYGON-BASED SURFACE MODEL REPRESENTATION

17.5.1 ANALYSIS OF RELATIONS BETWEEN SERIAL CONTOUR LINES

Each contour line is linked to one or more neighboring contour lines of the sections above and below. The computer needs the information as to which pair of contour lines (or multiple pairs of contour lines) should be linked between neighboring sections. In reconstructions with many contour lines and sections, the operator is probably reluctant to define the relations between all contour lines. A simple method is feasible to obtain the information from the overlapping rate between each neighboring pair of contour lines.[56] Figures 17.5b and 17.5c illustrate the method using a pair of neighboring contour line images as a typical example. Figure 17.5a shows a section image taken with a scanning electron microscope (SEM) corresponding to the upper contour lines. The method is described below. In Figures 17.5b and 17.5c, the upper contour lines are drawn as heavy lines and the lower ones are drawn as thin lines.

Every pair of contour lines between the upper and lower sections is listed in a table by the computer. For example, if there are n contour lines in the upper section and m contour lines in the lower one, a total of $n \times m$ pairs are listed in the table. The table consists of n rows and m columns. Each pair is examined for possible linkage according to the criterion described below, and the result is put into the table. This is exemplified by one pair which consists of two contour lines in the upper and lower sections, respectively (see A and B in Figure 17.5c). Subsequently, binary images are created, and each area is determined for the upper (Su) and lower (Sl) contour line, respectively: for a closed contour line, the inside is defined as 1 and the outside as 0. In the case of an open contour line, it is dilated homogeneously (this dilation process is similar to that seen in R1 to R5 in Figure 17.5b, as described below), and the algorithm sets the dilated region to 1 inside and 0 outside. For example, in Figure 17.5c, Su is an area surrounded by line A and Sl is an area surrounded by line B. An overlapping part (common part) between two binary images is generated by a logical operation, and the area of the common part, Sc, is also calculated. For example, Sc is the area indicated by C in Figure 17.5c. Finally, the ratios Sc:Su, and Sc:Sl are compared, and the greater

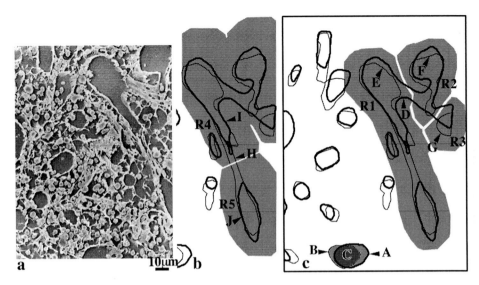

FIGURE 17.5 Explanatory illustration of automatic analysis of relations between contour lines and automatic branching processing using a practical case. (b and c) Neighboring contour lines in which the upper section contours are drawn in heavy lines and the lower section contours in thin lines; (a) section of the red pulp in the splenomegaly in rat taken with SEM corresponding to the upper contours. For example, a pair of contour lines A and B in (c) is examined to determine whether these should be linked, based on the degree of overlapping between the pair. Contour line D in (c) is judged to be branched to three contour lines, E, F, and G, and automatically divided into three parts. Similar binary processing is performed for contour line H in (b). H is divided according to two regions, R4 and R5, generated by dilating areas I and J, respectively (see text for details).

ratio is used for further comparison using a predefined threshold. If the measured ratio is greater than the threshold, the pair of contour lines is judged to be linked.

If a single contour line has been found to be related to several contour lines in the upper or lower section, using the method described above, the structure is branched, and the single contour line is appropriately divided into several lines. For example, contour line D shown in Figure 17.5c (upper section) is judged to be branched to three contour lines, E, F, and G (lower section). Contour line D is therefore divided into three parts according to the boundary and three isolated regions (R1, R2, and R3) are generated by "dilation processing," as shown in the figure. The dilation extends the three regions surrounded by the E, F, and G lines homogeneously, but every boundary line between every isolated region is preserved. A similar binary processing is performed for the lower section contour line H in Figure 17.5b. Line H is divided according to the two regions, R4 and R5, generated by dilating areas I and J, respectively.

This processing is executed for all contour lines through all serial sections to produce a large final table that shows all relations between the serial contour lines. Therefore, the computer can output the numbers of subsets of contour lines that define isolated solids and branch points, etc., together with the measured surface areas and volumes of every subset or solid, as described below. An application of this automatic subdivision processing of contour lines is shown below. Further development of such analytical methods is desirable so that they can be extended to other types of 3-D analyses and allow easier 3-D data manipulation.

17.5.2 SURFACE AND SOLID MODELINGS

At this point, a 3-D model has been constructed by establishing links between contour lines of the same group. To define such connections, fitting triangular patches between neighboring contour lines is generally used. This fitting is also referred to as "smooth tiling."[37] The algorithm for the automatic connection used here is based on the following basic idea. A contour line consists of contour points; for each contour point, the algorithm tries to locate corresponding points on the adjacent contour lines that would be the most appropriate ones for connection. Furthermore, the algorithm examines whether the tiling is continuous or smooth.[40] As an example, Figure 17.6a shows part of a "wire frame" model of a yeast cell membrane constructed by this algorithm.

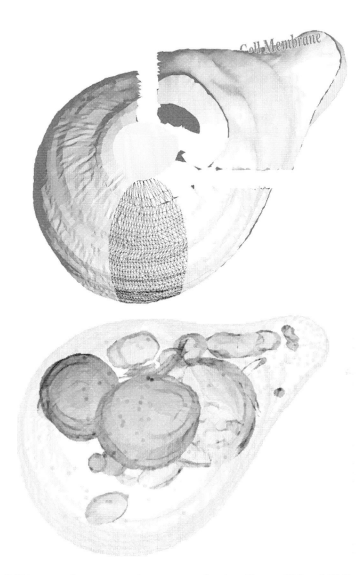

FIGURE 17.6   3-D images of reconstructed yeast organelles (see Figures 17.3 and 17.7) were rendered by commercial 3-D graphics software STRATA STUDIO Pro (Strata, Inc.) as an example, where the original reconstructed surface polygon data was once converted to the DXF file format, well-known to be a compatible 3-D data format. In the view above, the cell membrane polygon data was rendered in the "wire frame" mode, a simple ray tracing mode, and a smooth remodeling mode mixed together, in which an arbitrary cell membrane part was deliberately pulled away. There are many 3-D data manipulation menus. In the view below, the reconstructed organelles were rendered so as to be seen through the membrane.

There is a point that must be treated carefully: if the size of an object is comparable to the thickness of the section or a tube-like object, which is curved three dimensionally, neighboring contour lines to be linked frequently deviate considerably, and reasonable links cannot be generated using the current method. A simple solution is presented in the following.[56] First, two binary images are created, filling the inside of each contour line with 1 and the outside with 0. Second, each weight center is calculated, and one contour line is moved so that it is coincident with the other at the position of the weight center. Finally, the two aligned contour lines are linked by the current method, and preserving the generated links, the moved contour line returns to the original position. This improved technique can perform "smooth tiling."

17.5.3 VARIOUS 3-D COMPUTER GRAPHICS

The performance of recent commercial software has been improved over previous versions. Therefore, it is advisable to upgrade to these programs. Once the whole surface polygon data are produced, various 3-D computer graphics can be used.

It is necessary to generate a smooth surface image of the reconstructed model because surface roughness is an obstacle in the visualization of the 3-D model. Surface roughness is caused by insufficient numbers of sections with respect to the complexity of the structure to be investigated or by artifacts arising during specimen preparation or photography with a microscope. Smooth shading is generally used to calculate the smooth brightness distribution on the displayed 3-D surface.[57] To further ameliorate the effect of smooth shading, it is possible to smooth the surface of the reconstructed model itself.[40,41] There are various methods for this purpose established in the field of computer graphics.[57,58] Figure 17.6a shows a demonstration of the smoothing of the surface polygon model by the commercial software, STRATA STUDIO Pro (Strata, Inc.). A reconstructed yeast cell membrane model is used. Two divided surface parts are shown; one surface is drawn according to the original surface polygon data and the other according to the remodeled smooth surface data.

Some techniques are available to cut through a reconstructed model. Two methods are basically thinkable: one is to use a Boolean calculation and the other a regrouping of surface polygons. In the former method, an arbitrary primitive model like a cube is produced, and it is overlapped with the original model. Subsequently, a logic product or subtraction is performed, and then a cut model is created. If appropriately chosen, the primitive shape is refined and the cut model is adjusted in position, direction, and size. In the latter method, the original surface polygons are first released to individual ones, and the regrouping is performed by designing a 3-D region to be cut. A rough cut is shown in Figure 17.6, in which the latter method was used.

In some cases, certain structural objects can be displayed in the final 3-D image in a simplified form such as a sphere or a rod, if this is adequate for the research purpose or from the viewpoint of resolution. Computer graphics is suitable for this purpose. In the application of intracellular structures in yeast target cells reported here (see Figures 17.6 and 17.7), vesicles and lipid droplets were modeled as spheres, and their sizes were roughly approximated. There are several 3-D graphics in addition to the above. Commercial software has many ways of ray tracing. One example among those is shown in Figure 17.6, in which the intracellular organelles are seen through the cell membrane.

17.5.4 APPLICATIONS

17.5.4.1 Reconstruction of Yeast Target Cells

Figure 17.3 shows an electron micrograph of a sectioned yeast target cell (*Saccharomyces cerevisiae* X2180-1A (MATa))[59] induced by a mating pheromone, called α-factor. Specimens were prepared for transmission electron microscopy by freeze substitution. This fixation method preserves well the intracellular organelles and the membrane systems.[53] The thickness of the section was estimated to be 70 to 100 nm, as judged from the interference colors and by determining the number of sections that covered a spherical body of a given diameter. As shown in the micrograph, after a-factor treatment, cells developed a pointed projection, which is a unique pattern of oriented cell surface growth. To understand the process of morphological changes in the cells, serial thin sections of a total of 30 cells were prepared. The cells were classified into four groups by the shape of the projection. A typical cell from each stage was chosen and reconstructed (Figure 17.7).* The following organelles are contained: filasome, Golgi body, lipid granule shown as spherical model, mitochondrium, nucleus, nucleolus, rough endoplasmic reticulum, and membrane vesicle in a particle model (for more results and details, see Refs. 41 and 60). The reconstructed views in Figure 17.7 represent cells in different stages (each column). At stage I, cells were spherical without a projection. In stage II, cells had just started projection formation and contained more membrane vesicles than at the earlier stage. Only a small number of vesicles were located near the projection. The nucleolus was located at the side opposite the projection. In stage III, cells had a short projection and more membrane vesicles, which were notably concentrated at the tip of the projection. In stage IV, cells having an elongated projection showed apparently concentrated membrane vesicles in the projection. The nucleus moved close to the neck portion of the projection. Some interesting results were obtained for the topological relationship between various organelles in the process of projection formation. The results are described in detail by Baba et al.[60]

Once reconstructed data is completed, it is possible to use that data for various purposes. Using the surface polygon data, the volume and surface area of the reconstructed object can be measured.[61] The volume of the object is calculated by summing up every volume of planar plates formed between neighboring contour lines, in which the binary processing that calculates areas surrounded by arbitrary curved lines is utilized. The surface area of the reconstructed object is obtained by summing up all areas of the polygon patches constructing the 3-D contour of the object. The accuracy of these calculations depends on the accuracy of the estimation of section thickness and on errors in the lowest and the highest sections. The latter errors may arise from having a boundary of the object in the lowest or highest section not cut perpendicular, but more or less parallel to its long axis. Organelle volumes and their ratios to total cell volumes were analyzed quantitatively during projection formation (see Ref. 60). There is a special purpose for the reconstructed data: since these data can be converted to those in a general-purpose data format, they can be processed for a special purpose. For example, it is possible to produce an original computer animation that illustrates processes in cell biology. Furthermore, it is possible to generate an interactive program in which, by selecting any menu, a biological story is presented using 3-D images. Dallman[62] now produces such an innovative CD-ROM product, which will be published in the near future.

17.5.4.2 Reconstruction of Splenic Sinuses of Splenomegaly in Portal Hypertensive Rats

Figure 17.5a shows a section of the red pulp of splenomegaly in portal hypertensive rat taken with a SEM. Endothelial cells and ring fibers compose splenic sinuses. Misawa et al.[63] tried to reconstruct a structure of the splenic sinuses. This morphological study was started to clarify the relationship between the development of splenomegaly and portal hypertension.[63] The study was performed mainly by scanning electron microscopy in liver-cirrhotic rats. Liver cirrhosis was produced by intraperitoneal injections of 4% thioacetamide (TAA) in saline. The details for material and specimen preparation are described by Misawa et al.[63] Aliquots of fixed splenic tissue were alcohol-freeze-fractured and routinely processed for the SEM. Other aliquots were embedded in paraffin and serial sections 6 mm in thickness were prepared for the SEM. Figure 17.8 shows the reconstructed result of the splenic sinuses in rat from animals with highly elevated portal pressure. Figures 17.8c and 17.8d represent stereo-pair images. As described above, the system first analyzed

---

* Color Figure 17.7 follows page 288.

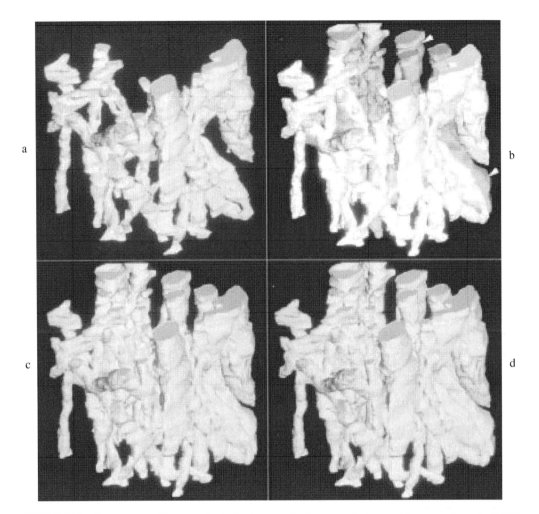

FIGURE 17.8   Reconstructed images of splenic sinuses of splenomegaly in portal hypertensive rats. (a) The result before smoothing by increasing pseudo intermediate sections; (b, c, and d) the results of smoothing. In (b), the same part as in (a) is displayed distinguishing the bright part, which corresponds to the model in (a), and the dark parts indicated by arrows, which correspond to the other parts isolated from the bright part. A stereo pair image can be seen in (c) and (d).

the relations between input serial contour lines and then divided them into several sets of contour lines. One set corresponded to a single isolated solid. This analysis was performed by investigating every relationship between neighboring contour lines. Figure 17.8a shows the largest isolate found by computer, and this had no connection with any other part in the reconstructed whole model (cf. other photos in the same figure). Next, the software performed the smoothing of the model and reduced the surface roughness. The surface roughness due to the large variation in the ratio between neighboring contours was a problem to be solved. This was remarkable since the structure is very complex. However, the surface roughness was smoothed by the following improved manner. First, interpolated contour lines were generated and inserted into the original contour line data to increase the outward resolution.[64,65] In this case, the number of 36 original sections was increased to 71 sections. Second, simple binary voxel processing was introduced to smooth the surface. All planar plates formed by contour lines were located in a 3-D binary memory space (i.e., the original model was constructed in space) and the 3-D averaging processing was performed. Finally, new contour line data converted from the smoothed model by the averaging was reread, and the final image visualized. By smoothing, the quality of the model image was increased remarkably. Figures 17.8b to 17.8d show the result. In Figure 17.8b, the same part is displayed as in Figure 17.8a with a remarkable difference: the bright part in Figure 17.8b corresponds to the model in Figure 17.8a, and the dark parts in Figure 17.8b indicated by arrows correspond to other parts isolated from the bright part. Because the smooth shading processing was not finished, the surface roughness was not completely removed, but the stereo-pair images (Figures 17.8c and 17.d) revealed the complex structure well-resolved in three dimensions.

17.6 VOXEL REPRESENTATION

Most of the recent work is based on contour line acquisition and polygon-based surface or solid model reconstruction. With advances in computer hardware and software technology, an alternative method has become practicable and is based on a voxel representation of the object using no contour lines.[66] This method generates a result more faithful to the originally recorded images than that produced with the contour line-based method. A realistic 3-D image can be generated because of a voxel representation of the object based on live image data. Since all section image data is kept available, an interactive parameter setting is possible in many cases. In the computer, the voxels are stored as a stack of section images. The method frees researchers from the cumbersome task of tracing contour lines. This reconstruction method was tested and applied to a sample, a light micrograph series of minute colonic adenomas surrounded by crypts of human familial adenomatosis coli. Details of specimen preparation and serial sectioning are described by Nakamura and Kino.[67]

17.6.1 PREPROCESSING

First, the section images were input through an image reader. Some input images (about 512×512 pixels) are shown in Figures 17.1 and 17.2, which show the single-gland adenoma (Figure 17.1, arrows) surrounded by colonic crypts, the surface epithelium, and muscularis mucosae (Figure 17.1, double arrows). Some preprocessing is necessary for the final rendering of 3-D images.[68] Gray levels through the serial images must be equalized. To do this, the histogram modification technique was used.[49] In each neighboring pair of section images, a histogram of gray levels of one image was modified to equalize it with that of the other image. In this case, there were a few micrographs whose gray-level histograms were slightly different from those of the neighboring section micrographs. Those images were processed with the histogram modification method to equalize them.

The alignment of the serial images was performed with the methods described previously. To avoid migration of the sections out of the reconstruction area during the serial alignment, it was performed from the middle section image toward the last one and from the same middle section toward the first one, respectively. On the whole, the images were aligned by comparing the image XCFs between each pair of neighboring sections. Figure 17.1 shows a typical example of this process. A pair of neighboring sections (Figure 17.1a and b) were aligned according to the result of the image XCFs shown in Figure 17.1c for the successive rotation of the image Figure 17.1b. The peak values for every XCF are compared as shown in Figure 17.1c. In this case, the rotational difference was $-2°$ and the relative displacement was small. Superimposition of the two images is shown in Figure 17.1d and e before and after the alignment, respectively. Since unexpectedly small amounts of rotational difference resulted from the manual rough alignment, time should be allowed for processing. However, for cases in which the rotational difference is large, the section image is distorted or changed, and the speed of the processor is slow, an alternative manual manner may be convenient, in which the user marks the corresponding points visually in the image pair. The relative displacement and the angle are obtained by substituting coordinates of the points into Eqs. 17.5–17.7.

The method of matching points by a computer was examined. One of the results is shown in Figures 17.2a and b. In this case, one of the paired images was distorted (Figure 17.2b), but similar image features clearly existed at local areas between the images. After the prealignment by the XCF method, the user marked characteristic positions in one image (Figure 17.2a). The computer searched for the corresponding positions with the method based on the correlation coefficient ρ (Eq. 17.1) and marked those positions in the other image (see symbols x in Figure 17.2b). The template window size was around 20 to 30 pixels and the search window size around 50 to 60 pixels. A few positions in Figure 17.2b were the second higher maxima of the distribution ρ, which seemed to be more correct than the positions where ρ had the highest maximum, but the differences between them were small. This method was also examined for the case in which the change of image between neighboring sections was small, such as that shown in Figures 17.1a and b; and, as predicted, the correct matching positions were found. Of course, the method is applicable provided that the displacement angle between the images is not large. Because this condition can be achieved with the image XCF method, and the image positions of the template can be chosen using some estimators based on image gradient or image variance, etc., an algorithm for the automatic alignment can be designed using both methods provided that the speed of the processor is sufficient. For example, in this examination and in our experience with digital stereophotogrammetry, the image points (the template windows) can be appropriately chosen using the local variance of the image provided that the noise is negligible. The threshold value for the selection depends on the image contrast; for example, the value for these section images was about 40 to 50 (in gray level) in standard deviation. Further, to avoid correlation failure, a threshold value for the correlation coefficient ρ was set; for example, in these sections the value was around 0.65.

The method based on the elastic model was applied to correct the image distortion due to deformations of sections. Figure 17.2 shows an example of the result. Since the section image of Figure 17.2b seemed to be distorted, first, a set of corresponding positions was determined, as described above. Subsequently, as shown in Figures 17.2e and f, a distorted resampling geometry for the correction was obtained by numerical calculations under boundary conditions given by the set of corresponding positions. If the corresponding points did not coincide with the lattice points, a linear interpolation method was used to determine the boundary conditions for the initial net (Figure 17.2f). The image distortion could effectively be corrected by resampling the pixel data according to the calculated resampling geometry. The accuracy of the correction can be improved by enlarging the set of points and increasing the number of meshes.

In the stacking of section images, image data between neighboring sections must be interpolated according to an estimation of the section thickness. In this application, the linear interpolation was used. In general, a simple stacking of the serial section images may generate an unsatisfactory result because even one image has considerable image data. Depending on the research purpose, unnecessary image parts in 3-D representation should be omitted to generate a reconstructed object that is simple to interpret. Further, noise must be eliminated. Figure 17.9a shows the simple stacking of the serial section images. Figure 17.9b shows the result of the noise removal. The noise elimination clearly exposed the objects to be shown. These 3-D images were generated with the solid modeling and surface shading methods. For solid modeling, the image data was simply selected with a density threshold value. In this application, the value was determined so as to leave the image parts of muscularis mucosae. This level could also be checked by observing 3-D solid representation with surface shading. Stained small dots and small areas in the images were eliminated in binary processing with an area threshold. In 3-D voxel processing, this was performed similarly with a volume threshold. Further, in the 3-D representation, the reconstructed result was smoothed with a Gaussian filter.

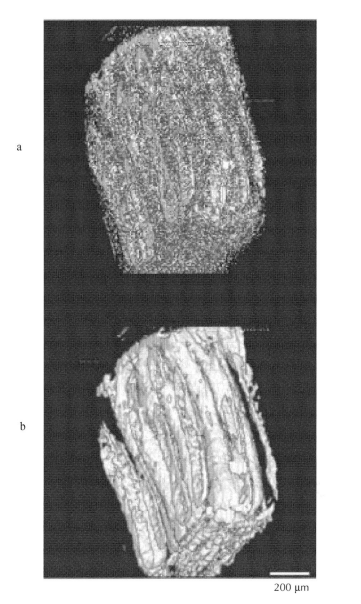

200 μm

FIGURE 17.9 Noise elimination by voxel processing. (a) A simple stacking of the serial section images; (b) the result of eliminating noise due to the removal of small-scale objects, and of subsequent smoothing with a Gaussian filter. These are represented with the solid modeling and surface shading methods. (From Baba, N., Satoh, H., and Nakamura, S., *Bioimages,* 1, 105, 1993. With permission.)

17.6.2 Rendering of 3-D Images

Once the voxel array has been obtained with the preprocessing described, several presentation modes can be used.[66,69,70] The solid body representation[71] shown in Figure 17.9 is one of them. Cutting and resectioning of the data can be performed along the principal x-, y-, and z-axes, or along arbitrary planes (Figure 17.10d). Depending on the reconstructed data, volumetric presentations can be advantageous in showing the internal structures in 3-D. This is a ray-traced method in which light is passed through the voxel array and absorbed according to the voxel values.[68] It is possible to vary the relationship with the calculated transparency, to vary the colors of the voxels, or to add light emission. In this software, object elements are brightened by the light emission

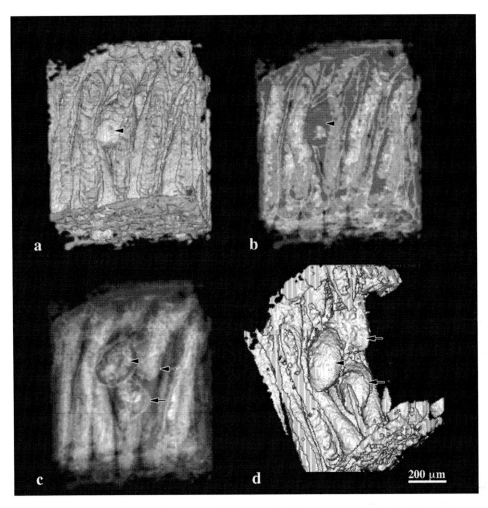

FIGURE 17.10 Volumetric presentations of a 3-D reconstruction of a single-gland adenoma (arrow) surrounded by crypts. (a, b, and c). Results increasing the transparency in order; (d), solid body representation of the cutting of the reconstructed object to compare with volumetric representations. (From Baba, N., Satoh, H., and Nakamura, S., *Bioimages*, 1, 105, 1993. With permission.)

according to a scalar product of l.grad.$g(r)$, where $g(r)$ is a 3-D density distribution (r: vector indicating a 3-D position) and l the direction of the illumination. Further, the absorption effect is simply analogized by setting a penetration range R for the light, depending on the voxel values. The light is passed through the voxel array until $\Sigma g(r) = R$, and the passing of the light is stopped at this position. Parameters for the rendering can be interactively set. This type of software is similar to those used in 3-D CT and confocal microscope imaging.

This rendering method was applied to show internal hidden parts without cutting the reconstructed object. Figures 17.10a, b, and c show the volumetric presentations of the reconstructed result, the single-gland adenoma (arrows) surrounded by crypts, in which the transparency was increased in the order A, B, C. In this representation, stained biological material is white, and unstained embedding material is black. As shown in Figure 17.10a, at low transparency, a difference in the density values was recognized between the muscularis mucosae and the other parts. As the transparency increased, the internal hidden parts were seen through outer parts. In Figure 17.10c, some hidden parts of the adenoma (arrows) were seen through crypts in the upper side, together with the surrounding crypts. However, the volumetric representation is frequently difficult to interpret because many internal details are intermingled. A combination of the volume rendering and the surface rendering of a cut result may be needed to reveal a 3-D structure. For comparison with the cutting of the object, a solid-body representation of the cut result is shown in Figure 17.10d.

As shown in this application, for the case where there is a sharp density boundary between stained and unstained materials, the serial section image reconstruction creates a satisfactory result that is simple to interpret. However, for the case where the density boundary is not sharp or obscure, although it can be observed and recognized in the section images, it may be difficult to produce a 3-D image that is easy to understand. Furthermore, in general, the stacking of serial section images produces very large amounts of data that are difficult to understand or display. To move forward with the voxel processing method in serial section reconstruction, various processing techniques must be devised to select information from a larger amount of data and to manipulate 3-D reconstructed data more easily.

17.7 CONCLUSION

Computer processing methods for serial section reconstruction are described. The entire process — from the input of section images and contour lines acquisition to the output of 3-D images — is described, in which important methods for image alignment and image deformation correction are included. With recent advances in computer technology, a personal computer with some peripherals has become capable to conveniently do the reconstruction computer work, although the preparation of serial sections and their images remains cumbersome. This convenient environment is also supported by abundant software. It is advisable for researchers to utilize the application software for appropriate processing steps.

Throughout this chapter, two basic reconstruction types are described: one is the polygon-based surface reconstruction and the other the voxel reconstruction. The former reconstructs a set of serial contour lines. In this case, when the contour lines are extracted from section images, useful image analysis is performed. Morphological information is arranged in the form of contour lines. Therefore, a reconstructed 3-D image is easy to interpret. However, because this reconstruction is based on contour lines, there is a limitation: that is, 3-D density distribution itself and an object consisting of very fine complicated parts that cannot be represented by contour lines are not reconstructed. For this case, voxel reconstruction must be adopted. Because the voxel representation is based on live image data, a realistic 3-D image can be generated. In addition, this frees researchers from the cumbersome task of tracing contour lines. However, a careless reconstruction by the voxel method using live images frequently generates a confused model due to excess image data. It is necessary to involve some processes to remove meaningless image data. It is stressed that it is important to choose the more suitable method — voxel representation or polygon-based — depending on the research purpose.

ACKNOWLEDGMENTS

The author thanks Dr. M. Baba of Japan Women's University, Dr. T. Misawa of The Jikei University School of Medicine, and Professor S. Nakamura of Iwate Medical University for much helpful advice and preparing the materials and serial sections, and also Dr. T. Dallman for preparing the 3-D figure.

REFERENCES

1. Capowski, J. J., Computer-aided reconstruction of neuron trees from several sections, *Comput. Biomed. Res.*, 10, 617, 1977.

2. Groves, P. M. and Linder, J. C., Dendro-dendritic synapses in substantia nigra: descriptions based on analysis of serial sections, *Exp. Brain Res.*, 49, 209, 1983.

3. Huijsmans, D. P., Lamers, W. H., Los, J. A., and Strackee, J., Toward computerized morphometric facilities: a review of 58 software packages for computer-aided three-dimensional reconstruction, quantification, and picture generation from parallel serial sections, *Anat. Rec.*, 216, 449, 1986.

4. Johnson, E. M. and Capowski, J. J., A system for the three-dimensional reconstruction of biological structures, *Comput. Biomed. Res.*, 16, 79, 1983.

5. Johnson, E. M. and Capowski, J. J., Principles of reconstruction and three-dimensional display of serial sections using a computer, in *The Microcomputer in Cell and Neurobiology Research*, Mize, R. R., Ed., Elsevier, New York, 1985, 249.

6. Kropf, M., Sobel, L., and Levinthai, C., Serial section reconstruction using CARTOS, in *The Microcomputer in Cell and Neurobiology Research*, Mize, R. R., Ed., Elsevier, New York, 1985, 265.

7. Levinthal, C. and Ware, R., Three-dimensional reconstruction from serial sections, *Nature*, 236, 207, 1972.

8. Macagno, E. R., Levinthal, C., and Sobel, I., Three-dimensional computer reconstruction of neurons and neuronal assemblies, *Annu. Rev. Biophys. Bioeng.*, 8, 323, 1979.

9. Moens, P. B. and Moens, T., Computer measurements and graphics of three-dimensional cellular ultrastructure, *J. Ultrastruct. Res.*, 75, 131, 1981.

10. Peachey, L. D., Damsky, C. H., and Veen, A., Computer assisted three-dimensional reconstructions from high-voltage electron micrographs of serial slices of biological material, in *Proc. 8th Int. Congr. Electron Microscopy,* Vol. 1, Australian Academy of Science, Canberra, Australia, 1974, 330.

11. Perkins, W. J. and Green, R. J., Three-dimensional reconstruction of biological sections, *J. Biomed. Eng.*, 4, 37, 1982.

12. Rakic, P., Stensas, L. J., Sayre, E. P., and Sidman, R. L., Computer-aided three-dimensional reconstructions and quantitative analysis of cells from serial electron microscopic montages of foetal monkey brain, *Nature*, 250, 31, 1974.

13. Sobel, I., Levinthal, C., and Macagno, E. R., Special techniques for the automatic computer reconstruction of neuronal structures, *Annu. Rev. Biophys. Bioeng.*, 9, 347, 1980.

14. Stevens, B. J., Variation in number and volume of the mitochondria in yeast according to growth conditions. A study based on serial sectioning and computer graphics representation, *Biol. Cell*, 28, 37, 1977.

15. Stevens, J. K., Davis, T. L., Friedman, M., and Sterling, P., A systematic approach to reconstructing microcircuitry by electron microscopy of serial sections, *Brain Res. Rev.*, 2, 265, 1980.

16. Veen, A. and Peachy, L. D., Trots: a computer graphics system for three-dimensional reconstructions from serial sections, *Comput. Graphics*, 2, 135, 1977.

17. Willey, T. J., Schultz, R. L., and Gott, A. H., Computer graphics in three dimensions for perspective reconstruction of brain ultrastructure, *IEEE Trans. Biomed. Eng.*, 20, 288, 1973.

18. Wong, Y.-M. M., Thompson, R. P., Cobb, L., and Fitzharris, T. P., Computer reconstruction of serial sections, *Comput. Biomed. Res.*, 16, 580, 1983.

19. Young, S. J., Fram, E. K., and Craig, B. A., Three-dimensional reconstruction and quantitative analysis of rat lung type II cells: a computer-based study, *Am. J. Anat.*, 174, 1, 1985.

20. Coombs, G. H., Tetley, L., Moss, V. A., and Vickerman, K., Three-dimensional structure of the Leishmania amastigote as revealed by computer-aided reconstruction from serial sections, *Parasitology,* 92, 13, 1986.

21. Braverman, M. S. and Braverman, I. M., Three-dimensional reconstruction of objects from serial sections using a microcomputer graphics system, *J. Invest. Dermatol.*, 86, 290, 1986.

22. Briarty, L. G. and Jenkins, P. H., GRIDSS: an integrated suite of microcomputer programs for three-dimensional graphical reconstruction from serial sections, *J. Microscopy*, 134, 121, 1984.

23. Baba, N., Naka, M., Muranaka, Y., Nakamura, S., Kino, L, and Kanaya, K., Computer-aided stereographic representation of an object reconstructed from micrographs of serial thin sections, *Micron Microsc. Acta*, 15, 221, 1984.

24. Gras, H. and Killman, F., NEUREC — A program package for 3-D reconstruction from serial sections using a microcomputer, *Comput. Programs Biomed.*, 17, 145, 1983.

25. Kinnamon, J. C., Three-dimensional reconstructions from serial sections using the IBM PC, *Eur. J. Cell Biol.*, 48 (Suppl. 25), 65, 1989.

26. Moss, V. A., The computation of 3-dimensional morphology from serial sections, *Eur. J. Cell Biol.*, 48 (Suppl. 25), 57, 1989.

27. Prothero, J. and Prothero, J., Three-dimensional reconstruction from serial sections. I. A portable microcomputer-based software package in Fortran, *Comput. Biomed. Res.*, 15, 598, 1982.

28. Street, C. H. and Mize, R. R., A simple microcomputer-based three-dimensional serial section reconstruction system (MICROS), *J. Neurosci. Methods*, 7, 359, 1983.

29. Sundsten, J. W. and Prothero, J. W., Three-dimensional reconstruction from serial sections II. A microcomputer-based facility for rapid data collection, *Anat. Rec.*, 207, 665, 1983.

30. Young, S. J., Royer, S. M., Groves, P. M., and Kinnamon, J. C., Three-dimensional reconstructions from serial micrographs using the IBM PC, *J. Electron Microsc. Tech.*, 6, 207, 1987.

31. Bron, C., Greminet, P., Launay, D., Jourlin, M., Gautschi, H. P., Biichi, T., and Schilpbach, J., Three-dimensional electron microscopy of entire cells, *J. Microscopy*, 157, 115, 1990.

32. Cookson, M. J., Dykes, E., Holman, J. G., and Gray, A., A microcomputer based system for generating realistic 3D shaded images reconstructed from serial sections, *Eur. J. Cell Biol.*, 48 (Suppl. 25), 69, 1989.

33. Lockhausen, J., Kristen, U., Menhardt, W., and Dallas, W. J., Three-dimensional reconstruction of a plant dictyosome from a series of ultrathin sections using computer image processing, *J. Microscopy*, 158, 197, 1989.

34. Marko, M. and Leith, A., Sterecon — Three-dimensional reconstructions from stereoscopic contouring, *J. Structural Biol.*, 116, 93, 1996.

35. Menhardt, W., Lockhausen, J., Dallas, W. J., and Kristen, U., An environment for three-dimensional shaded perspective display of cell components, *Micron Microsc. Acta*, 17, 349, 1986.

36. Nakamae, E., Harada, K., Kaneda, K., Yasuda, M., and Sato, A., Reconstruction and semi-transparent stereographic display of an object consisting of multi-surfaces, *Trans. Inf. Process. Soc. Jpn.*, 26, 181, 1985.

37. Yaegashi, H., Takahashi, T., and Kawasaki, M., Microcomputer-aided reconstruction: a system designed for the study of 3-D microstructure in histology and histopathology, *J. Microscopy*, 146, 55, 1987.

38. Baba, N., Nakamura, S., Kino, I., and Kanaya, K., Three-dimensional reconstruction from serial section images by computer graphics, *J. Electron Microsc. Tech.*, 3, 401, 1986.

39. Baba, N., Baba, M., Osumi, M., Nakamura, S., Imamura, M., Koga, M., and Kanaya, K., Computer graphics representation for a three-dimensional reconstruction from serial sections, *Sci. Form*, 2, 49, 1986.

40. Baba, N. and Kanaya, K., Three-dimensional reconstruction from serial section images by computer graphics, *Scanning Microsc.*, 2 (Suppl.), 303, 1988.

41. Baba, N., Baba, M., Imamura, M., Koga, M., Ohsumi, Y., Osumi, M., and Kanaya, K., Serial section reconstruction using a computer graphics system: applications to intracellular structures in yeast cells and the periodontal structure of dog's teeth, *J. Electron Microsc. Tech.*, 11, 16, 1989.

42. Domon, T. and Wakita, M., The three-dimensional structure of the clear zone of a cultured osteoclast, *J. Electron Microsc.*, 40, 34, 1991.

43. Sachs, L., *Angewandte Statistik,* 5th ed. Springer Verlag, Berlin, 1978.

44. Koenig, G., Mickel, W., Storl, J., Meyer, D., and Stangle, J., Digital stereophotogrammetry for processing SEM data, *Scanning*, 9, 185. 1987.

45. Durr, R., Peterhans, E., and von der Heydt, R., Correction of distorted image pairs with elastic models, *Eur. J. Cell Biol.*, 48 (Suppl. 25), 85, 1989.

46. Higdon, A., Ohlsen, E. H., Stiles, W. B., Weese, J. A., and Riley, W. F., *Mechanics of Materials*, 4th ed., John Wiley & Sons, New York, 1985.

47. Saxton, W. O., Accurate atom positions from focal and tilted beam series of high resolution electron micrographs, *Scanning Microscopy*, 2 (Suppl.), 213, 1988.

48. Baba, N., Ichise, N., and Tanaka, T., Image area extraction of biological objects from a thin section image by statistical texture analysis, *J. Electron Microsc.*, 45, 298, 1996.

49. Rosenfeld, A. and Kak, A. C., *Digital Picture Processing*, Academic Press, New York, 1982, 130, 231, 295.

50. Gonzalez, R. C. and Woods, R. E., *Digital Image Processing*, Addison-Wesley, New York, 1992, 506.

51. Weszka, J. S., Dyer, C. R., and Rosenfeld, A., A comparative study of texture measures for terrain classification, *IEEE Trans. Syst., Man., Cybern.*, SMC-6, 269, 1976.

52. Haralick, R. M., Shanmugan, K., and Dinstein, I. H., Texture features for image classification, *IEEE Trans. Syst., Man., Cybern.*, SMC-3, 610, 1973.

53. Baba, M. and Osumi, M., Transmission and scanning electron microscopic examination of intracellular organelles in freeze-substituted *Kloeckera* and *Saccharomyces cerevisiae* yeast cells, *J. Electron Microsc. Tech.*, 5, 249, 1984.

54. Takeshige, K., Baba, M., Tsuboi, S., Noda, T., and Ohsumi, Y., Autophagy in yeast demonstrated with proteinase-deficient mutants and conditions for its introduction, *J. Cell Biol.*, 119, 301, 1992.

55. Baba, M., Takeshige, K., Baba, N., and Ohsumi, Y., Ultrastructural analysis of the autophagic process in yeast: detection of autophagosomes and their characterization, *J. Cell Biol.*, 124, 903, 1994.

56. Baba, N., Baba, M., Osumi, M., Nakamura, S., and Kanaya, K., Improvements of a serial section reconstruction system to be applicable to more complicated structures, in *Proc. 9th Eur. Congr. Electron Microscopy*, Vol. 3, Dickinson, H. G. and Goodhew, P. J., Eds., IOP Publishing, York, England, 1988, 373.

57. Newman, W. M. and Sproull, R. F., *Principles of Interactive Computer Graphics*, McGraw-Hill, New York, 1975.

58. Rogers, D. F., *Procedural Elements for Computer Graphics*, McGraw-Hill, New York, 1985.

59. Ohsumi, Y. and Anraku, Y., Specific induction of Ca^{2+} transport activity in MATa cells of *Saccharomyces cerevisiae* by a mating pheromone, α factor, *J. Biol. Chem.*, 169, 10482, 1985.

60. Baba, M., Baba, N., Ohsumi, Y., Kanaya, K., and Osumi, M., Three-dimensional analysis of morphogenesis induced by mating pheromone α factor in *Saccharomyces cerevisiae*, *J. Cell Sci.*, 94, 207, 1989.

61. Chawla, S. D., Glass, L., Friewald, S., and Proctor, J. W., An interactive computer graphics system for 3-D stereoscopic reconstruction from serial sections: analysis of metastatic growth, *Comput. Biol. Med.*, 7, 223, 1982.

62. Dallman, T., Private communication.

63. Misawa, T., Hataba, Y., Baba, N., Kanaya, K., Sakurai, K., and Suzuki, T., Ultrastructural characteristics of experimental splenomegaly in portal hypertensive rats. A scanning electron microscopic study of the red pulp, in *Proc. 12th Int. Congr. Electron Microscopy*, San Francisco Press, San Francisco, 1990, 270.

64. Otten, E. and van Leeuwen, J. L., MacReco, a 3D reconstruction package from serial sections and images, *Eur. J. Cell. Biol.*, 48 (Suppl. 25), 73, 1989.

65. Zsuppan, F. and Rethelyi, M., Computer reconstruction of serial EM pictures: approximation of missing sections, in *Proc. 8th Eur. Congr. Electron Microscopy*, Vol. 2, Budapest, 1984, 1361.

66. Russ, J. C., Automatic vs. computer-assisted 3D reconstruction, *Proc. 50th E.M.S.A.*, San Francisco Press, San Francisco, 1992, 1050.

67. Nakamura, S. and Kino, I., Morphogenesis of minute adenomas in familial polyposis coli, *J. Natl. Cancer Inst.*, 7, 41, 1984.

68. Baba, N., Satoh, H., and Nakamura, S., Serial section image reconstruction by voxel processing, *Bioimages*, 1, 105, 1993.

69. Goldwasser, S. M. and Reynolds, R. A., Real-time display and manipulation of 3-D medical objects: the voxel processor architecture, *Comput. Vision Graphics Image Process.*, 39, 1, 1987.

70. Forsgren, P.-O., Visualization and coding in three-dimensional image processing, *J. Microscopy*, 159, 195, 1990.

71. Frank, J., McEwen, B. F., Radermacher, M., Turner, J. M., and Rieder, C. L., Three-dimensional tomographic reconstruction in high voltage electron microscopy, *J. Electron Microsc. Tech.*, 6, 193, 1987.

18 Automatic Counting of Round Particles in Microscopic Images by Pattern Recognition

Nico A.M. Schellart

CONTENTS

18.1 INTRODUCTION

18.1.1 SCOPE OF THE CHAPTER

Until the 1970s, counting of objects in microscopic images was performed by the microscopist, often with the aid of an ocular grid. In the early 1970s computer-assisted counting was developed.

With the availability of faster computers every year with more working and hard-disk memory, applications of automated counting are clearly increasing. Nowadays, counting software modules of computer image processing software packages are available at the market of microscopic systems and biomedical software houses. Despite this, in the last decade, counting by hand of autoradiographic grains is practiced about as much as (semi-)-automatic counting.

This chapter deals with the counting in microscopic digitized images of (basically) specimens of a single type of object

1. That is rather similar of size
2. That is roughly circular in shape
3. Light transmission or reflection (dark field) of which is rather homogeneous within its border (Thus lack a clear texture)
4. That (partly) overlaps infrequently or not

Examples of such types of objects are autoradiographic grains, cell nuclei, cell bodies, microorganisms, and colonies of microorganisms. There are also nonbiological applications, for example, in astronomy (counting of stars and galaxies) and geology. In the life sciences, the applications are most frequently found in the field of the neuroscience, where grains are the most often counted type of object. Originally, counting strategies, especially the algorithms, have mostly been developed for autoradiographic grains,[1-6] but minor modifications often make them applicable for other object types. In addition, counting procedures for other round particles have been published.[7-12] Due to the historical development, this chapter focuses especially on the methods of computerized autoradiographic grain counting.

18.1.2 HARDWARE AND SOFTWARE FOR COUNTING

In experimental biology and medical research, autoradiography[13] is a well-known technique to study (e.g., DNA replication, RNA and protein synthesis and glucose metabolism). Autoradiography is the visualization, usually on a photographic film, of the patterns of distribution of radiation emitted by the studied specimen itself. In biology and biomedicine, the specimen is usually a light-microscopic (LM) or electron-microscopic (microautoradiography) section with incorporated radioactive material. The image on the film reflects in varying degrees, depending on auto-absorption and scatter, the distribution of the radioactivity in the section. All aspects of the autoradiographic method, concerning technique as well as interpretation, are discussed in the monograph by Rogers,[13] to which the reader is referred for further information.

In the early days of this technique, autoradiographic data was at best quantified by visually (manually) counting the autoradiographic grains or by manually controlled densitometry. With present powerful PC computers, counting appears to have been developed to routine image processing procedures.[6-8,11,12,14-17] Nevertheless, counting (as well as densitometry) is afflicted with many pitfalls, as described below.

Densitometry is suitable in those cases where the particle density is sufficiently high that a considerable part of the particles, say more than 10%, are touching or overlapping. By focusing the (monochromatic) measuring beam, nowadays often a laser beam, within the plane of the particles and by automatic scanning (e.g., by computer controlled driving of the microscope stage), the intensity value of the transmitted beam can be stored and processed by the computer, together with the coordinates of the stage. In this way, areas of several millimeters (mm^2) can be analyzed with high resolution. For smaller areas ($<<1$ mm^2), the digitized image can be analyzed directly. This chapter deals minimally with these densitometric techniques; rather, it focuses on automated counting of individual particles by image analysis, mostly established with the LM. Counting by computer, nowadays hundreds of particles, has in addition to its speed the advantage that higher reliability and reproducibility are attained, since subjective (criteria) and unstable (fatigue, tedium) human factors are banished or at least strongly diminished, as discussed by Rogers (Chapter 10 of Ref. 13) and others.[1,18,19]

In short, the setup for particle counting consists of a research microscope (LM or EM), a high-quality CCD video camera mounted on the phototube of the microscope, a video interface, a PC computer, and a (laser) printer with medium resolution. Various complete setups or assemblies of parts are commercially available, often with image handling software packages included. Software with a counting facility is sometimes available in the public domain, for example, the Macintosh-based Image (NIH, Bethesda, MD)[7,12,20] which processes TIFF files made with Apple Scan[7] or with Imager™ (Appligene, Pleasanton, CA).[12] Examples of commercial image software with one or another counting module used during the last decade include Lusex 450 Particle Analyzer (Nihon Regulator Co, Ltd.),[21] ImagePlus+ (Dapple),[14] Image 1.37 (PB-500687, National Technical Information Service, Bethesda, MD),[10] Topcon IMAGEnet (Bausch and Lomb, Rochester, NY),[22] Image-Measure 5100 (Imaging Technology, Inc., Seattle, WA),[8] Discovery (Becton Dickinson, Leiden, NL),[19] and the various versions of Quantimet, the oldest system (Leica, Wetzlar, Germany)[23] with the present Windows 95-based Qwin software. Slightly younger is IBAS[6,17] with its successor VIDAS (Kontron Elektronik, Eching/Munich, Germany). In some cases, the customer must write the specific application software himself, in a standard computer language or sometimes in a dedicated language and with other systems as macros (IBAS and VIDAS, Quantimet). Also, other manufacturers of microscopes (e.g., Olympus) provide image analysis software with a particle counting module. Most commercial systems are only suitable for counting at high magnification. In recent years it has become easier for the investigator himself to assemble the hardware of the setup in order to explore a setup optimized for a specific application.[1,2,5,15,24]

The literature about computer-aided grain counting prior to 1979 has been reviewed by Rogers (Chapter 10 of Ref. 13), and the literature until 1991 by Schellart.[25] This chapter discusses counting methods developed in recent years, especially those based on particle recognizing strategies. Another main theme of this chapter is the probability theory of particle counting, a subject that has minimally been discussed in literature.

18.2 TYPES OF APPLICATIONS AND COUNTING STRATEGIES

In medical biology, zoology, and botany, there are two major fields where computerized counting of particles in microscopic images is practiced. By far, the most common is counting of autoradiographic grains in, for example, chromosomes, cell nuclei, cell bodies, or cell layers. The other one is counting of biological round objects, like bacterial colonies, small organisms, and subcellular structures, mostly made visible by immunotechniques. The reliability of automatic counting has also found its way to the clinic. Counting of corneal cells[22,26] and sperm (human and cattle)[27,28] are two well-known applications. However, their demands are too specific to discuss them here in detail.

Applications for automatic counting in autoradiographs can be divided in two types. The first one analyzes specific structures in the image, which should be recognized and defined in the image. These structures include, for example, chromosomes, cell nuclei, synapses, or whole cells in which the grains are counted. In these applications, often DNA and RNA structures, which can, for example be obtained by *in situ* hybridization,[6,15,29] are labeled by a radioisotope. The second application is grain counting in tissue studied at low magnification.[1,2]

These two types of applications have specific requirements and constraints. The first type demands high magnification, achieved by a 100x objective, often with oil immersion and bright-field illumination. Then, the image frame of interest is usually selected manually under the microscope, such that a complete set of chromosomes (or the cell of interest) is within the frame. For other applications, a special procedure is needed to define the contour of the cells or other structures to be investigated. This can be done by drawing the contour with a light-pen on a writing tablet or with a mouse on the screen. Also, a software procedure can be used to determine the contour automatically with the possibility of manual correction, after which the number of grains or the area occupied by the grains can be determined. For the first type of applications, the area of interest is on the order of 100 to 1000 μm^2, but for the second one it is 10^4 to 10^5 μm^2. Consequently, the latter applications (also non-grain particles) need low magnification, usually with a 20, 25, or most commonly 40x objective.

In those cases where a substantial number of particles (grains) are (nearly) touching or are more or less overlapping, scoring of individual particles is impossible. Now, the number of particles can be calculated when the mean particle size is estimated from a part of the image with a low particle density and the area-fraction of a defined structure (e.g., chromosome or nucleus) occupied by particles is determined.[6,15,17] This method will be referred to as the indirect counting strategy. The actual density is progressively larger, the larger the density found due to overlapping particles. A method for correction is discussed in Section 18.3. A special version of the indirect counting strategy is outlining by image segmentation the area where grains occur in an (automatic) outlined structure, for example, a cell nucleus. In a specific application of this method, the surface of the outlined grain area appeared to be proportional to the number of grains (as established by visual counting).[19]

Often, the algorithms for determination of the total area of particle occupation are ultimately based on exceeding a fixed or variable (local background correction) threshold criterion of pixel luminance. This is a completely different strategy[6,15,17] than software to count individual, recognized particles.[1,2,7,8] Recognition can be performed by the strategy of 8-connected pixels (see Section 18.5.1),[1,2] by outlining the closed border of (large) pixel clusters in which individual pixels exceed the theshold,[7] or by a self-learning algorithm (neural network computing) that recognizes particles whose parameters (such as area, major axis, shape, density) vary within certain limits.[24] A low-level semi-automatized method is visual inspection of the image and marking objects with the computer mouse.[20] Different types of objects can be marked and then counted by the computer. Obviously, this strategy of marked-object-counting can only be practiced when the number of objects is small. Unfortunately, various papers do not give the necessary details about the counting method to decide whether a real recognizing algorithm is applied or simply one or another basic type of thresholding method.

A parameter that is inherent to the biological preparation and crucial for the choice of the counting strategy is the density of the particles. Very high densities (i.e., when substantial and multiple overlap between the particles occurs, such that more than 20% of the particles are fused) make microdensitometry most suitable. The variable measured is basically the fraction of coverage. This is the normal practice in ^{14}C and ^{35}S autoradiography of, for example, brain tissue. For moderate densities (between 10 and 20% covering), the indirect counting strategy is useful. For lower densities, particle recognition is undoubtedly the most reliable strategy. Two types of radiographs can be distinguished. One is an autoradiograph with local moderate densities, so that fusion occurs fairly frequently. This may lead to areas with single grains and with clustered grains. The other one is an autoradiograph with spatially, gradually changing densities of the grains with interspersed fused grains. In these cases grains should be recognized and counted individually with a correction for coinciding grains, in the same way as done for low densities (<5 grains/100 μm^2). In addition, a correction for grain clusters, based on cluster size, will improve the accuracy of the grain count (see Sections 18.5.1.2 and 18.5.1.3).

18.3   PROBABILITY THEORY OF COUNTING

18.3.1   FACTORS INFLUENCING COUNTING STATISTICS

The interpretation of the results of particle counting faces several problems of a probabilistic nature. Obvious and important pitfalls biasing the results are:

- Two or more particles can occlude each other.
- A particle covers partly two or more pixels such that the pixel intensities become subthreshold.

Methods are described to correct for these complications. However, the impact of these procedures may be influenced by the variability in particle size and shape.

18.3.2 Coinciding Particles

To obtain statistical insight into the problem of coincidence, some restrictions have to be made before mathematically formulating this problem. It is supposed that all particles are identical and ideal, and that each particle occupies only 1 pixel. However, a pixel can contain several particles. The problem is to calculate the probability that a pixel contains 0, 1, 2, ..., m particles when n particles are randomly distributed over N pixels. For the probability (i.e., the fraction of pixels containing m particles), p(n,m) holds that:

$$p(n, m) = \frac{n!}{((n-m)!\,m!} \left(\frac{1}{N}\right)^m \left(1 - \frac{1}{N}\right)^{n-m}$$

$(n - m)!\, m!$

with $n = 0, 1, 2, \ldots, \infty$;   (18.1)

$m = 0, 1, 2, \ldots, n$;

$N = 2, 3, \ldots, \infty$.

The number of pixels with m particles are found by multiplying p by N. To calculate the values of the probability p(n,m), the simple recursion:

$$p(n, m) = p(n, m-1) \cdot (n - [m-1]) / (m[N-1])$$ (18.2)

is useful.

Figure 18.1 gives p(n, m) for N = 1024 (= 32 × 32 pixels). Along the horizontal axis, the particle density (number of occupied pixels/N) increases from low and moderate (the domain of particle counting) to intermediate (the domain of measuring the area covered by particles and particle clusters) to high (the domain of densitometry). The left parts of the curves have slopes approximately equal to m. The — for particle counting — most important part of the function space p(n,m) can be approximated by the Poisson distribution. The parts of the curves above the dashed, nearly horizontal line deviate less than 1% from the Poisson distribution. (This also holds for $n/N \leq 0.2$ as long as N is large.)

For n << N, the number of empty pixels is slightly larger than N – n. When n = 0.281N (arrow in Figure 18.1), the curve (N – n)/N (at most 1 particle/pixel) deviates 5% from p(n,0) due to multiple occupations.

Particle counts with high particle densities (>0.1) must be corrected for multiple occupations. The actual particle density D_c (multiple occupation counts multiple), for grains the density of radioactive disintegrations, is directly related to the measured density D_m (multiple occupations count as a single occupation), according to $D_c = C \cdot D_m$, where C is a correction factor (>1.0) that depends on D_m. With N and n given, the number of empty pixels is $N \cdot p(n, 0)$; so, $D_m = 1 - p(n, 0)$ and $C = n/(N \cdot D_m)$. For the theoretical model, C-1 has been calculated for $N = (2^k)^2$ pixels with k = 3, 4, ..., 10 and is plotted versus D_m in Figure 18.2 with both axes logarithmical. Figure 18.2 shows that for $D_m > 0.01$, the curves with values of $k \geq 6$ coincide. Also, for $D_m < 0.1$, correction plays only a minor role and for $D_m < 0.01$, correction is irrelevant. In contrast, the densities generally practiced in autoradiographic densitometry unequivocally need correction, the more since C increases more than linearly with D_m. However, for very high densities, the theoretical model does

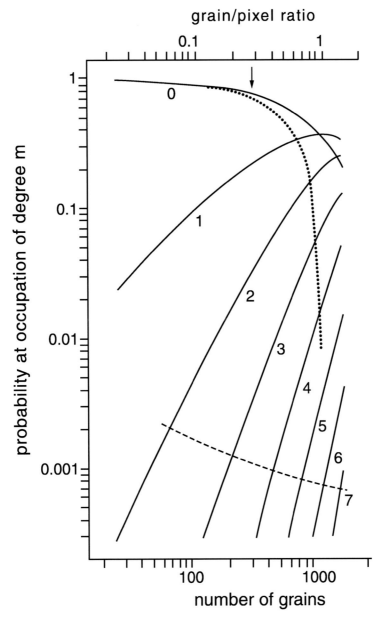

FIGURE 18.1   Probability of multiple occupations as a function of the number of particles in a 32×32 pixel matrix. The drawn curves give the probability at an m-fold occupation, indicated by the number at the curve. The part of the graph above the dashed line gives the probabilities deviating less than 1% from the Poisson distribution. The dotted curve represents $(N - n)/N$. The arrow indicates the point where this curve deviates 5% from $p(n, 0)$ (Copyright CRC Press LLC, Boca Raton, FL.)

not hold any longer due to the effect of self-screening, and since the radioactivity is not distributed completely randomly, but with some ordered pattern due to the cellular ultrastructure. This reduces the correction factor C, for example, as indicated by the heavy dotted line (intuitive approximation) in Figure 18.2. In many applications, the density is highly variable. Therefore, equidensity lines have to be determined and correction should be performed for each density class.

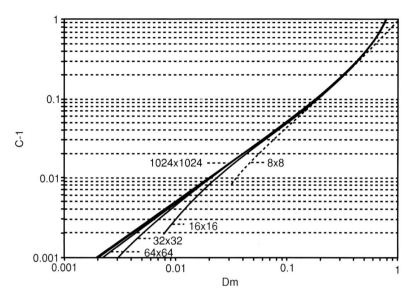

FIGURE 18.2   Correction of particle density for multiple occupations. The calculations have been made for a frame of 8×8, 16×16, …, 1024×1024 pixels. For $D_m > 0.01$, the curves practically coincide ($K > 4$). Along the ordinate the correction factor C minus one ($C - 1$) is plotted as a function of the measured particle density D_m (= occupied pixels/N). The product $(C - 1) \cdot 100\%$ gives the correction in percentage. The lower end of the curves has a D_m value corresponding to two particles within the frame.

In addition to the problem of multiple occupations, it is also possible that two particles cover adjacent pixels. It is possible to theoretically correct the score for the case that two particles cover adjacent 8-connected pixels. Suppose that one particle occupies exactly 1 pixel. Then, the change at such a twinned particle is roughly 8 times larger than at two coinciding particles (a double, m = 2). This means that for a particle density of 10% for $N \geq 1024$, 91% of the scores are single particles, 1% are doubles, and 8% are twinned. So, the score should not be corrected with C = 1.055, but with a much larger factor C' (an approximation yields $C' = 1.16$). For densities smaller than 0.2, C' changes nearly proportional). However, this correction only holds for the ideal condition that 1 pixel occupies only 1 pixel. In practice, this is often not the case. This problem of paired particles is discussed in Section 18.5.1.2.

As noticed, in practice, particles do not always just occupy only 1 pixel. Due to variability in size (related to pixel size) and position related to the pixel matrix, a particle can occupy 2 or more pixels. This problem will be treated in Section 18.3.3.

18.3.3   Influence of Particle Size, with Grain Size in Particular

The silver bromide crystals of the photographic emulsions used in autoradiography range from 0.06 to 0.34 μm (Ref. 13, appendix). After development of the crystals, the particles measure on the average from 0.2 to 0.7 μm, as found by various authors (see Reep and Greegan[2] for references). Size as well as shape of the grains are strongly dependent on the developer and developing conditions (time, temperature). The grain size is below or at the border of LM resolution with bright-field (transmission) illumination; but in dark-field illumination, the grains look 2 to 4 times larger.

An intriguing problem in the theory of particle counting is to optimize the size of pixel with respect to that of the particle. Obviously, the ratio between particle size and pixel size strongly determines the counting strategy and also influences the reliability of the results. For particles considerably smaller than the pixel size, many particles will be indistinguishable from random local intensity fluctuations. For the reverse situation, (i.e., particles much larger than the pixel size), the particle will be detected easily and reliably, but at the cost of processing speed and computer memory. In some, especially older applications, high particle:pixel area ratios have been used, mostly to quantify the labeled area of nuclei or chromosomes. However, in more recent applications concerning tissue examination, the particle:pixel area ratio is generally smaller and ranges from 1.5 to 9.[2,5,15]

Intuitively, the particle size seems to be optimal when only 1 pixel is (nearly) completely covered. The problem can be solved when the distribution of the area of the pixel covered by a particle is known. When the particle covers more than 1 pixel, only the pixel with the largest covered area (at most 100%) is relevant, since this pixel has the highest chance to be above threshold. As far as is known in geometrical probability theory, this problem cannot be solved analytically. However, the distribution can easily be approximated by calculating the area of overlap for a finite number of pixel-to-particle positions. Here, the actual calculations were made for perfect circular particles with unit radius and a two-dimensional matrix with perfect square pixels with a width w of 1.0, 1.2, 1.4, or 1.6 units. To approximate the distribution, a step size of 0.1 or 0.05 (pixel width 1.2) was used in two orthogonal directions. Figure 18.3a shows the distribution of the area of overlap of the 576 possible particle-to-pixel positions with a pixel width of 1.2. The large peak to the right is due to particles that completely or nearly completely cover a pixel. Decreasing pixel width results in a further increase of this peak and an attenuation of the peak at 0.8 (ordinate). For a pixel width $\leq \frac{1}{2}\sqrt{2}$, the distribution degenerates to a single narrow peak (delta function) at w^2, since then always at least 1 pixel is completely covered. Therefore, the particle:pixel area ratio of 2π represents the transition between meaningful ratios yielding distributions of final width and distributions degenerated to a delta-function (ratio > 2π). For increasing pixel widths, the slender peak to the right diminishes and the broad peak to the left becomes larger and dominant. For w $\geq$ 2, the maximally covered pixel area amounts to π. From here, increasing w again yields a slender peak to the right, but now at π, with a fast decreasing area of the part of the distribution to the left of this peak. Since for these values of w the ratio of overlap is always smaller than $\frac{1}{4}\pi$, w should be chosen between $\frac{1}{2}\sqrt{2}$ and 2. Figure 18.3b is helpful in estimating an optimal value. Along the ordinate, the cumulative chance is plotted that at least 50% or 75% of the area of a pixel is covered by a particle, with the particle:pixel area ratio along the abscissa. When one demands that at least 50% of the pixel be covered by the particle, and that only 5% of the particle-to-pixel positions will not fulfill this condition, then the particle:pixel area ratio should be 1.4. For the 75% and 90% criteria, this ratio is 2.5 and 3.6, respectively. When a ratio of 2 is chosen and a criterion of 50% (equivalent brightness half the value with maximal covering), then only 2% of the pixels are missed (false negatives). This appears a good compromise between accuracy and counting speed (which reduces about linear with the increase of the particle:pixel area ratio).

Up to now, variability in particle size has not been taken into account. As long as this variability is small, the results described above are not affected. However, for a large variability, a higher area ratio should be chosen such that the smallest particles are also scored. In the literature, substantial data about the size variability of grains (and other particles) is virtually absent. Small (coefficient of variation of 0.19)[1] as well as large (range of a factor of 2.4)[2] variabilities have been reported. Data about the shape of grains, besides some qualitative data (Chapter 2, Ref. 13), is also lacking. Studied at high magnification (electron microscope), their shape often appears to be very irregular and variable. However, under the light microscope (LM), they look fairly circular. To meet the demand for some quantitative data about size and shape, grains of ^{3}H-methylacrylate test plates made with Kodak NTB-2 emulsion and developed with Kodak D-19 were measured. It appears that the shape can be approximated by an ellipse. In bright-field, the ratio of the short vs. the long axis is 0.69 ± 0.11 (N = 168), whereas in dark-field, the ratio is 0.74 ± 0.085 (N = 165). Thus, dark-field grains look more circular and vary less in size than bright-field grains. The equivalent grain size (defined as the diameter of the circle with the same area) is 0.46 ± 0.076 µm and 1.95 ± 0.31 µm, respectively.

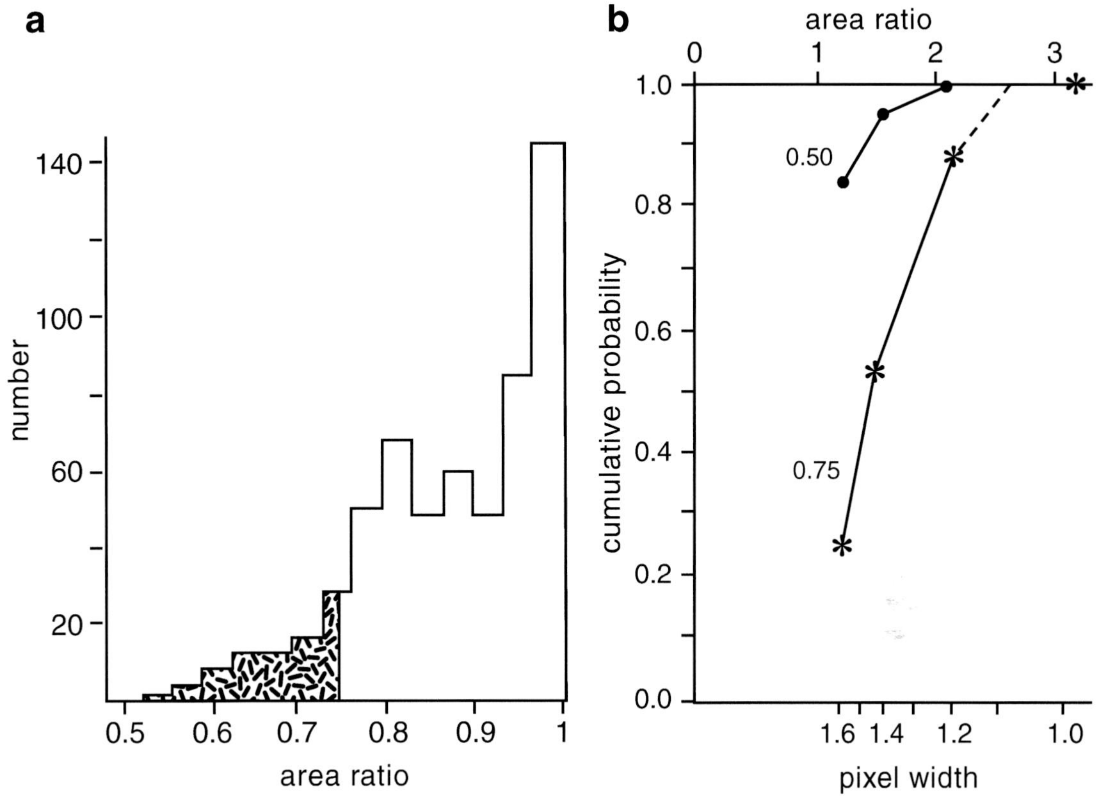

FIGURE 18.3   The influence of pixel size. (a) Distribution of the area of the pixel (1.2 × 1.2) covered by the particle (radius 1) calculated for 576 particle-to-pixel positions with a step size of 0.05 in both directions. The abscissa (area ratio) gives the area fraction of that pixel which is most covered by the grain. The non-stippled area, expressed as fraction of the total distribution (i.e., the cumulative probability), is depicted as one of the data points in b (abscissa 1.2, ordinate 0.87). (b) Cumulative probability of a coverage of at least 50% and 75%, calculated from the approximated distributions for pixel widths of 1.0, 1.2, 1.4, and 1.6. Along the abscissa, the area ratio (particle:pixel) as well as the pixel width (nonlinear scale) are indicated. (Copyright CRC Press LLC, Boca Raton, FL.)

In conclusion, taking into account the moderate variability in size and shape of the roughly circular particle, the mean particle area should be twice that of a pixel.

18.4   PRECOUNTING PROCEDURES AND BASIC APPROACH

In autoradiographic grain counting, the use of dark-field has several advantages over bright-field illumination:

- The contrast between the grains, which show up as bright specks, and the background is considerably larger.
- The small size and precise shape of the grain are less critical since they look larger and more circular.
- The magnification of the LM can be smaller than 20 to 40x.

The precounting procedures comprise the following steps:

1a. Optimize the digitized bright field-image of the biological material by adjusting the microscope (focus and alignment), and then freeze the bright-field image.

1b. In the case of grain counting, continuous acquisition of the digitized dark-field image of the frame of 1a, while displaying contrast values to optimize the grain-background contrast. This can be done by adjusting the dark-field focusing. Next, freezing of the dark-field image and reducing the influence of the background.

2. Set the parameters of counting algorithm: size of the subdivision of the field, threshold range, contrast difference.

3. Identify the particles in the image of 1b (grains), otherwise that of 1a, by pattern recognition. Store all true pixel values and set all false pixel values to zero to enable statistical analyses of the particle configuration (see Sections 18.5.2.2 and 18.5.2.3).

4. Count the particles (in each subdivision).

To get rid of variations of background intensity in order to use a constant threshold, gray values of each pixel and its neighborhood can be filtered or smoothed in some way. A simple way is, after subdividing the field, to subtract the local mean of each subfield and set negative values to zero. There exists a large variety of smoothing filters (dilatation) with which the image can be convoluted. To remove very coarse inhomogeneities, application of one or another way of low-pass filtering (2-D Fourier transformation) of the whole image is more effective. Another way is low-pass filtering or smoothing of the image and subtracting the result of the original image,[5] which results in a kind of spatial high-pass filtering. This filter method to remove the background is promising for large field analysis of low-density autoradiographs with few artifacts, digitized so that the pixel size is slightly less than the grain size. The method has also been applied to high magnification grain counting (many pixels per grain) in tissue with single grains as well as clusters of grains.[5] Since this filtering method ignores grain clusters, the image with removed background was added to a modified original image. The latter was obtained after subtracting a constant intensity, so that finally solitary particles and clusters remained. With this method, with respect to the particle clusters, there is a risk of false positives (large artifacts) and false negatives. An even more refined method to improve the contrast of the particles with respect to the background has been described by Kerrigan et al.[32] and comprises the following steps:

1. Dilatation of the image by heavily smoothing (maximum filter); this removes the particles (here grains).

2. A minimum filter (image erosion) is used to restore the contours of the original particles.

3. The original image is subtracted from that of step 2, resulting in sharp, bright contours of the particles. Background and particle-fills have moderate gray values. Since grains occupy only some pixels, they lack a fill and therefore pop out as bright specks on a nearly homogeneous background.

After the counting procedure, the particle densities can be superimposed numerically or indicated with graphic symbols in a matrix arrangement upon the bright-field section image.[1] In tissue with inevitable differences in particle densities, the identified particles can be superimposed on the section image as bright pixels. The particle density can also be visualized by calculating and displaying iso-particle density contours. A flowchart of the principal precount and count procedures can be found in Ref. 1.

18.5 STRATEGY OF PARTICLE COUNTING

18.5.1 THE CENTRAL PIXEL TRUE APPROACH

18.5.1.1 Basic Procedure

In autoradiography, especially with ^{3}H, the grain density is often low to moderate, so that in general, only 1% to 10% of the pixels are covered by the silver grains, caused by radioactive decay. When a distribution of gray values of the pixels in the digitized image is determined, the distribution does not show a peak or shoulder that can be attributed to the particles, due to the large local variability of the background intensity. This background is caused by the underlying section (see Section 18.7.1), imperfect optics (see Section 18.7.2.2), and artifacts such as fine dust particles. The lack of a distinct and predominant contribution of the particles in the pixel intensity histogram rules out the use of straightforward or more sophisticated counting (or densitometry) for low (or moderate) particle densities. Owing to the often high variability of background intensity, particle discrimination by exceeding a static criterion generally fails. However, with a dynamic criterion, being a preset difference in gray value with respect to the immediate surrounding of the pixel to be tested (the central pixel), the gray value of pixels with a definite higher intensity can be scored as particles. Although this approach can be used under bright-field as well as dark-field conditions, reliable results can be obtained more easily with dark-field.[1,2,13]

The central pixel approach has been described by Schellart et al.[1] and Reep and Greegan.[2] The procedure of Reep and Greegan[2] can be seen basically as an improvement and refinement of that of Schellart et al.[1] The method of Schellart et al.[1] is discussed first.

The digitized image is scanned from left to right and from top to bottom until a pixel is found that has an intensity value between a preset, fixed lower level and upper level (to ignore noise and bright artifacts, respectively). Such a pixel, called true (T pixel), forms the central pixel of a 3×3 matrix. The 8 pixels surrounding the central pixel are all connected by one side or by one corner with the central pixel (8-connected). The gray value of the central pixel is compared with the gray values of the 8-connected pixels. An 8-connected pixel is T when its intensity is larger than that of the central pixel minus a preset intensity difference. In this way, single particles of size 1 to 9 can be detected. However, the particle can only be scored when a 1-pixel-wide margin around the 3×3 matrix, this is a 5×5 margin, contains only false (F) pixels. If the 5×5 margin is not empty (=F), the particle candidate is considered to be an artifact (too large or touching another particle). However, by this (too limited) approach, particles consisting of a single (central) pixel with an empty 3×3 margin and a not empty 5×5 margin will be rejected. Therefore, after finding a T pixel, the first test the particle candidate has to pass is to examine whether the central pixel is a 1-pixel particle. Thus, the steps in the particle recognition procedure are:

1. Jump to the next pixel and examine whether the pixel is T. If the pixel is false (F), re-enter step 1; if not, proceed to step 2.
2. Collect the 3×3 margin of the central T pixel and convert the intensity values to T and F values. Count the number of T pixels (n_p) in the 3×3 margin. If this number is zero, then increment the number of 1-pixel particles (size 1), set the 3×3 margin to zero intensity, jump to the next pixel, and go to step 1. If not zero, proceed to step 3.
3. Read the 5×5 margin and scan this margin for a T pixel. If a T pixel is found, return to step 1; otherwise, go to step 4.
4. Increment the number of particles (size n_p). Set the 5×5 margin to zero intensity, jump two pixels to the right, and proceed with step 1.

Visual control of the counting process can be improved by adjusting the intensity of the T pixels above the upper threshold.

18.5.1.2   Evaluation of the Basic Procedure

It appears that in dark-field images, the ten-most frequently occurring configurations represent the vast majority (about 95%),[1] as illustrated in Figure 18.4. Seven of these ten configurations could have been detected with a 2×2 inner matrix and all with a 3×2 matrix. Self-evidently, as a result of scanning from left to right and from top to bottom, a T pixel in the upper row and left column of the 3×3 matrix is seldom found.

	configuration			**%**		**configuration**			**%**
(1)	F	F	F	44	(6)	F	F	F	3
	F	**T**	F			F	**T**	F	
	F	F	F			**T**	**T**	F	
(2)	F	F	F	25	(7)	F	F	F	2
	F	**T**	**T**			**T**	**T**	**T**	
	F	F	F			F	F	F	
(3)	F	F	F	6	(8)	F	F	F	2
	F	**T**	F			F	**T**	F	
	F	**T**	F			F	**T**	**T**	
(4)	F	F	F	6	(9)	F	F	F	1.5
	F	**T**	**T**			F	**T**	**T**	
	F	**T**	**T**			F	F	**T**	
(5)	F	F	F	3	(10)	F	F	F	1
	F	**T**	**T**			**T**	**T**	F	
	F	**T**	F			**T**	**T**	**T**	

FIGURE 18.4 The ten most frequently occurring configurations of particles in a 3×3 margin with the central pixel true approach. F indicates a false pixel and, T a true pixel. Due to the strategy, the pixels in the upper row are always F. (Copyright CRC Press LLC, Boca Raton, FL.)

When a particle density smaller than 5% (5% of the pixels covered) occurs, it can safely be concluded that only a minority of the 2-pixel particles are due to pairing of two particles. Most of them will be caused by (slightly larger) particles sufficiently covering 2 pixels. In the example of Figure 18.4, statistically half of the 3-pixel particles and most larger particles may be considered as artifacts.

18.5.1.3   Refinements

In a particle field, it is difficult to say whether clusters of 3 to 9 T pixels within the 3×3 matrix are clustered particles or artifacts. The procedure of Reep and Greegan[2] was designed for low particle densities. For low densities, clustered particles should be very rare. Therefore, two tests to exclude such fake clusters were developed. A contrast test checks whether the mean intensity value of a 3×3 margin is a preset intensity difference lower than the intensity of the central pixel. When this test is passed, it is determined how many T pixels in the 3×3 margin have the same intensity as the central pixel. (The very first test examines whether the central pixel has a higher intensity than any of the 3×3 margin pixels.) The chance for T pixels of equal intensity is substantial since here the particles are twice as wide as the pixels and the actual number of T intensity levels is only ca. 35.[2] When more than a preset number of pixels attains this intensity, then the central pixel is considered to be part of an artifact and is rejected as particle center. The particle candidate in the example depicted in Figure 18.5a may be rejected by one of the tests, depending on the setting of the parameters.

To examine the 5×5 margin for rejecting large artifacts, Reep and Greegan[2] introduced (nearly) the same two tests as mentioned above for the 3×3 margin. These two tests are performed before the two small artifact tests described above are carried out. The two tests for the large artifacts are a less rigorous approach compared to the 5×5 empty margin criterion of Schellart et al.[1] The tests overcome the problem that some combinations of particles with equal and/or different sizes are not scored (generally 1- and 2-pixel particles). With the plain 5×5 empty margin criterion of Schellart et al.,[1] multi-pixel particles separated by 1 F pixel will be rejected, and multi-pixel particle candidates separated by 1 F pixel from an isolated T pixel will also be rejected by the empty 5×5 margin test (Figure 18.5b, c). Since with the procedure of Schellart et al.,[1] 2-pixel particles may occur frequently (Figure 18.4), 2-pixel particle candidates were tested for an empty 4×3 or 3×4 margin,[1] a procedure severely reducing the number of false negatives.

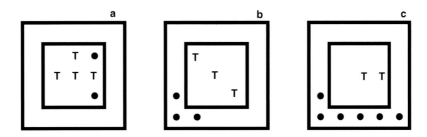

FIGURE 18.5 Examples of missed particles in complex configurations. The three configurations show a particle in the center which may be missed with the procedure of Reep and Greegan,[2] depending on the values of the two tests screening small (a) or large artifacts (b and c). In b and c, the particle candidate may be missed when in the 5×5 margin too many pixels with a dot are T. The middle and right configurations show a particle in the center which will be missed by the procedure of Schellart et al.[1] The dots in the pixels indicate pixels of which one or more are true. (Copyright CRC Press LLC, Boca Raton, FL.)

One of the basic differences between the methods of Schellart et al.[1] and Reep and Greegan[2] is that the latter allows one or some (depending on the parameter settings) T pixels in the 5×5 margin. This means that a T pixel in the 5×5 margin may be adjacent to a T pixel in the 7×7 margin. However, this has not been tested.[2] So, there is a chance, depending on the setting of the parameters, that thread-like long structures are false positive. In contrast, a 3-pixel long (double) particle with its central pixel at the end may be scored as a false negative. Figures 18.5b, c give some examples of configurations that yield false negatives in the empty 5×5 margin[4] and the nearly empty 5×5 margin[2] method.

The small inaccuracies of both procedures can simply be overcome by testing for each T pixel in the 3×3 margin, whether or not it is connected to a T pixel in the 5×5 margin. Self-evidently, this test increases analysis time substantially, although it is quite feasible with present speeds of computation.

A final refinement involves a test on the size of the particles and particle clusters, and is illustrated by the following example. For densities smaller than 100 particles/64^2 pixels and a mean particle size of 2, the maximal particle size can safely be limited to 4, irrespective of the counting strategy, since statistically it is unlikely ($p = 0.008$) that an occupied 3×3 matrix contains more than two particles. Particles of size 3 and 4 should obtain a weight of 2 (counted twice). In addition to the dependency on particle density and particle size, the size correction can also be influenced by characteristics of the biological material (e.g., the occurrence of local high particle densities). This may hamper the implementation of a simple test on particle size.

18.5.2   STATIC THRESHOLDING AND GRADIENT THRESHOLDING

The versions of the central pixel true (T) approach, based on particle size and shape, constitute a true particle recognition procedure. A clearly distinct approach of particle counting is to decide exclusively on pixel intensity or intensity gradient, whether or not it is covered by a particle. Thus, the decision whether a pixel is T or F is solely based on thresholding. This approach is most frequently chosen when the particles are heavily clustered. It is often used in labeling studies of chromosomes and cell nuclei. In these applications, the grains cover many pixels and the actual goal is to determine the area of the soma, nucleus, or chromosome covered by grains. This can be done directly with the raw digitized image, although sometimes a preprocessed (e.g., filtered) image is used.

Older procedures apply thresholding straightforwardly to estimate the field area (often a nucleus or chromosome) occupied by particles. Here, the field area is often determined by the experimenter himself (e.g., with a light-pen). Another strategy is the indirect counting strategy, as described in Section 18.2. A refinement of this method has been proposed by Masseroli et al.[6] (but his formula 2 lacks a clear derivation and, moreover, appears to be incorrect).

18.5.3   RELIABILITY AND SENSITIVITY

The reliability of counting algorithms can be tested by:

1. Comparing the particle counts obtained by visual counting and automatized counting.[1,2,7,8,10,24]
2. Comparing particle counts of the same frame, obtained by computer with different parameter settings.[1,2]

The high correlation (often >0.95) found by various authors[1,2,7,8,10,24] when comparing both ways of counting clearly shows that computerized counting is a very good and fast substitute for human counting. Remarkably, for moderate particle densities obtained at high LM magnification, the computer yields lower counts than those of visual counting,[27,29] whereas for low densities, they are the same. At low magnification, this difference was not found.[2] This effect is obviously due to the higher occurrence of adjacent (or overlapping) particles, which are scored as one particle by the computer, but double in visual counting. Judging overlapping particles by eye appears to be more accurate than by computer.[30] Reproducibility of computer countings, which is better than for visual countings, shows high correlations, even when countings with optimal and suboptimal parameter settings are compared.[1,2]

Reep and Greegan[2] did the most extensive study to test the sensitivity of a particle recognition program. Their algorithm has five parameters. The first three are the minimal intensity value of the potential particle center (Imin), the 5×5 margin mean contrast parameter (Pcon), and the number of pixels in the 5×5 margin not exceeding a small contrast difference (P5 × 5). The fourth and fifth parameters concern the 3×3 margin. The sensitivity of the parameters was tested by varying the setting of the first one with relaxed (not critical) values of the other. After optimizing the first one, the second parameter was optimized, and so on. It appeared that changing Imin by two points on a 64 intensity scale changes the particle counts at most by 5%. The same change of the setting of Pcon results in a slightly higher error, whereas a change of P5 × 5 of only 1 pixel nearly doubles the error. The fourth and fifth parameters are less critical and contribute only marginally to increasing the accuracy.

18.5.4   COMPARISON OF COUNTING SPEED

A fair comparison of the counting speeds of the various dedicated systems and the commercial image processing systems is difficult to make. This is due to large differences in computer systems, including image handling hardware and differences in software (e.g., the choice of the computer language). Detailed data about hardware and software is generally not available in the current literature and, also, precise information about counting speed is often lacking. Nevertheless, generally speaking, multipurpose commercial systems appear to have lower speeds than dedicated systems.

18.6 BACKGROUND NOISE OF GRAINS AND OTHER PARTICLES

In autoradiographs, grains show up that are not due to radiation of the incorporated radioactive material. The causes of the occurrence of these grains have been discussed extensively by Rogers (Chapter 6 in Ref. 13). Two types of background noise can be distinguished. The first one regards spontaneously developed grains, induced thermally or photically. An (approximate) estimate of the amount of this Poisson distributed background noise is given by the grain density well separated from the labeled biological material. It is generally less than 0.5 grains/100 μm,[2] so that correction is necessary only for low grain densities. A mathematical procedure to correct the mean grain count per area of variable size (e.g., nuclei) for this background has been described by Korr and Schmidt.[3,4] For low grain numbers per nucleus (<5), the result of this correction deviates substantially from the difference between the mean grain count with and without labeling.

The second type of background consists of grains that are not homogeneously distributed over the section. These grains are caused by mechanical stress, chemography (due to biological or other agents), environmental radiation, etc. Standard methods to correct for these artifacts do not exist.

18.7 DISTURBANCES AND LIMITATIONS

18.7.1 TISSUE EFFECTS

Large differences in optical density of the biological material at a large spatial scale can give rise to false positives and negatives, depending on the algorithm. A remedy is to divide the frame into a number of subfields and calculate the mean pixel intensity of each subfield. This value can be used to eliminate most of the background by subtracting it from the gray value of the individual pixels in the field (negative values are set to zero). A more sophisticated (no border effects between fields) but mathematically more complex method is to filter out the low spatial frequencies, which can be done rapidly with present computer speed. Background effects can often be reduced further by applying monochromatic light.

18.7.2 OTHER DISTURBANCES

18.7.2.1 Focusing

Focusing is generally done manually, but can also be computer-assisted[1] or automated. The effect of focusing of autoradiographic grains was thoroughly examined by Reep and Creegan.[2] By determining the total grain count versus the focusation, they found an optimal range of 6 μm (error 1.5%). By calculating the spatial correlation of the grain count in an image frame, Schellart et al.[1] obtained a range of 2 μm, and for or high magnification images, the range goes down to 1.6 μm. Thus, optimal focusing is a prerequisite for reliable counting, irrespective of the type of object.

18.7.2.2 Optical Limitations

A serious limitation of many microscope/video systems is the difficulty in obtaining a homogeneous illumination across the microscope field (object field).[1,31] In large field images, especially obtained with 18x, 25x, or 40x objectives, illumination is, at best, circular symmetric with a slight decrease of intensity toward the edge of the field. This effect is generally strengthened by the optics of the camera. Homogeneity of the illumination can be improved optically by adjusting the illumination during display of the digitized image, without a biological object and with reduced intensity. Again, the performance of the optics is improved by the use of monochromatic light, resulting in sharper particle images. The remaining inhomogeneity is eliminated by low-frequency filtering.

Another optical flaw is cushion-shaped distortion, especially at the edges, which even occurs with plan-apo objectives. The influence of the optical imperfections at the edges can be quantified by determining the particle density in the fields at the edge compared to fields in the center when using an autoradiographic test plate. By repeating this procedure at various times, the averaged found differences, being some percent, are used to make a correction table.

18.7.2.3 Imperfections of Autoradiographic Preparation

Counting inaccuracies can also be caused by fine dust particles and inhomogeneities in the layer thickness of the photographic emulsion. These and other imperfections of autoradiographic material are discussed by Rogers (Chapter 7 of Ref. 13).

18.8 CONCLUSIONS

The chosen strategy of particle counting depends on the kind of biological material (chromosomes, nuclei, tissue, microorganisms) and the density of the particles, the number of objects (e.g., nuclei, wells, frames) to be counted for the occurrence of particles, and the chosen degree of reliability. For limited material, the marked-object-count strategy is adequate. For routine purposes with a demand for reasonable accuracy, the indirect counting strategy is a good choice. However, for basic research the particle recognition procedure is recommended.

For chromosome and nuclear studies, mostly performed with high LM magnification in semi-thin sections, the static threshold approach generally suffices, as concluded from the results of reliability tests. When high particle densities cannot be avoided, for example, due to heavy autoradiographic labeling, the algorithm has to detect isolated particles as well as clustered particles. This means that false positives (small artifacts) cannot be avoided, either with the indirect, nor with the recognition strategy. Corrections for overlapping and fused particles occurring with moderate densities are a necessity.

In general, typical tissue studies demand a particle-recognizing algorithm. In order to have a sufficient resolution and depth of focus with as large as possible tissue area, the LM magnification should be between 20x and 40x. Counting grains in dark-field is more efficient than in bright-field. The particle:pixel area ratio should be about 2.0, as a compromise between reliability (<2.5% error) and recognition speed. Due to intensity fluctuations of the background (caused by the section) occurring over large areas, some type of spatial filtering is necessary. Strictly local photic background fluctuations demand dynamic thresholding (contrast criterion). Artifact rejection tests are a necessity. The central pixel true approach appears to be a successful and reliable strategy. The performance of this approach for moderate particle densities is expected to improve when the regular 5×5 margin is replaced by an irregular margin dependent on the pattern of the directly adjacent true pixels, and forming within the 5×5 matrix a 1-pixel-wide contour around the particle.

For all applications, but especially those based on static thresholding, a rigorous standardization of the various steps of the procedure is helpful to improve the performance. This involves the settings of the (stabilized) light source, the microscope, and the video camera. If the procedures of preparing and developing the histological material are also standardized, it is also profitable to standardize the values of the parameters of the algorithm.

REFERENCES

1. Schellart, N. A. M., Zweijpfenning, R. C. J. V., van Marle, J., and Huysmans, D. P., Computerized pattern recognition used for grain counting in high resolution autoradiographs with low grain densities, *Comput. Methods Programs Biomed.*, 23, 103, 1986.
2. Reep, R. L. and Greegan, W. L., An accurate method for automated counting of silver grains in autoradiographs, *Comp. Biomed. Res.*, 21, 244, 1988.
3. Korr, H. and Schmidt, H., An improved procedure for background correction in autoradiography, *Histochemistry*, 88, 407, 1988.
4. Korr, H. and Schmidt, H., A new procedure for correcting background in quantitative autoradiographic studies, *Acta Histochem.*, SXXXVII, 149, 1989.
5. Parkkinen, J. J., Paukkonen, K., Pesonen, E., Lammi, M. J., and Markkanen, S., Quantitation of autoradiographic grains in different zones of articular cartilage with image analyzer, *Histochemistry*, 93, 241, 1990.
6. Masseroli, M., Bollea, A., Bendotti, C., and Forloni, G., *In situ* hybridization histochemistry quantification: automatic count on single cell in digital image, *J. Neurosci. Meth.*, 47, 93, 1993.
7. Parry, R. L., Chin, T. W., and Donahoe, P. K., Computer-aided cell colony counting, *Biotechniques*, 10, 772, 1991.
8. Wright, M., Bakus, G. J., Ortiz, A., and Ormsby, B., Computer image processing and automatic counting and measuring of fouling organisms, *Comput. Biol. Med.*, 21, 173, 1991.
9. Wachsmuth, E. D., Germer, M., Paulus, K., and Persohn, E., Automatic counting of nuclei in series of conventional liver sections to differentiate hypertrophy and hyperplasia, *Toxicol. Appl. Pharmacol.*, 121, 264, 1993.
10. Perricone, M. A., Saldate, V., and Hyde, D. M., Quantitation of fibroblast population growth rate *in situ* using computerized image analysis, *Microsc. Res. Tech.,* 31, 257, 1995.
11. Zalewski, K. and Buchholz, R., Morphological analysis of yeast cells using an automatic image processing system, *J. Biotechnol.*, 48, 43, 1996.
12. Cui, Y. and Chang, L.-J., Computer-assisted, quantitative cytokine enzyme-linked immunospot analysis of human immune effector cell function, *Biotechniques*, 22, 1146, 1997.
13. Rogers, A. W., *Technique of Autoradiography*, 3rd ed., Elsevier/North Holland Biomedical Press, Amsterdam, 1979.
14. Doucet, G., Descarrier, L., Audet, M. A., Garcia, S., and Berger, B., Radioautographic method for quantifying regional monoamine innervations in the rat brain. Application to the cerebral cortex, *Brain Res.*, 441, 233, 1988.
15. Escot, C., Le Roy, X., Chalbos, D., Joyeyx, C., Simonsen, E., Daures, J. P., Soussaline, F., and Rochefort, H., Computer-aided quantification of RNA levels detected by *in situ* hybridization of tissue sections, *Analytic Cellular Pathol.*, 3, 215, 1991.
16. Retaux, S., Trovero, F., and Besson, M. J., Role of dopamine in the plasticity of glutamic acid decarboxylase messenger RNA in the rat frontal cortex and the nucleus accumbens, *Eur. J. Neurosci.*, 6, 1782, 1994.
17. Giorgi, S., Forloni, G., Baldi, G., and Consolo, S., Gene expression and *in vitro* release of galanin in rat hypothalamus during development, *Eur. J. Neurosci.*, 7, 944, 1995.
18. Fry, D. L., Tousimis, A. J., Talbot, T. L., and Lewis, S. J., Methods to quantify silver in autoradiographs, *Am. J. Physiol.*, 238, H414, 1980.
19. Chadwick, A. D., Diggle, S. P., Young, A. R., and Potten, C. S., The quantitation and kinetics of unscheduled (repair) DNA synthesis in ultraviolet-irradiated human skin by automated image analysis, *Br. J. Dermatol.*, 135, 516, 1996.
20. Ide, M., Jimbo, M., Yamamoto, M., and Kubo, O., Tumor cell counting using an image analysis program for MIB-1 immonohistochemistry, *Neurol. Medico. Chir.* (Tokyo), 37, 158, 1997.
21. Kekki, M., Santamäki, T., Talanti, S., and Vesikari, E., Accuracy of an automatic counting method by television-computer equipment in the quantitative autoradiography, *Acta Histochem.*, 81, 171, 1987.
22. Erickson, P., Doughty, M. J., Comstock, T. L., and Cullen, A. P., Endothelial cell density and contact lens-induced corneal swelling, *Cornea*, 17, 152, 1998.

23. Barbareschi, M., Girlando, S., Mauri, F. M., Forti, S., Eccher, C., Mauri, F. A., Togni, R., Dalla Palma, P., and Doglioni, C., Quantitative growth fraction evaluation with MIBI and Ki67 antibodies in breast carcinomas, *Am. J. Clin. Pathol.*, 102, 171, 1994.

24. Konstantinidou, A., Patsouris, E., Kavantzas, N., Pavlopoulos, P. M., Bouropoulou, V., and Davaris, P., Computerized determination of proliferating cell nuclear antigen expression in meningiomas, *Gen. Diagn. Pathol.*, 142, 311, 1996.

25. Schellart, N. A. M., Automatic grain counting in autoradiographs by computerized pattern recognition, in *Image Analysis in Biology*, Häder, D.-P., Ed., CRC Press, Boca Raton, FL, 1991, chap. 14.

26. Vecchi, M., Braccio, L., and Orsoni, J. G., The Topcon SP 1000 and Image-Net systems. A comparison of four methods for evaluating corneal endothelial cell density, *Cornea*, 15, 271, 1996.

27. Shiran, E., Stoller, J., Blumenfeld, Z., Feigin, P. D., and Makler, A., Evaluating the accuracy of different sperm counting chambers by performing strict counts of photographed beads, *J. Assist. Reprod. Genet.*, 12, 434, 1995.

28. Centola, G. M., Comparison of manual microscopic and computer-assisted methods for analysis of sperm count and motility, *Arch. Androl.*, 36, 1, 1996.

29. Mize, R. R., Thouron, C., Lucas, L., and Harlan, R., Semiautomatic image analysis for grain counting in *in situ* hybridization experiments, *Neuroimage*, 1, 163, 1994.

30. Thielmann, H. W., Popanda, O., Edler, L., Boing, A., and Jung, E. G., DNA repair synthesis following irradiation with 254-nm and 312-nm ultraviolet light is not diminished in fibroblasts from patients with dysplastic nevus syndrome, *J. Cancer Res. Clin. Oncol.*, 121, 327, 1995.

31. McCasland, J. S. and Woolsey, T. A., New high-resolution 2-deoxyglucose method featuring double labeling and automatic data collection, *J. Comp. Neurol.*, 278, 543, 1988.

32. Kerrigan, J. R., Martha, P. M., Jr., Krieg, R. J., Jr., Queen, T. A., Monahan, P. E., and Rogol, A.D., Augmented hypothalamic proopiomelanocortin gene expression with pubertal development in the male rat: evidence for an androgen receptor-independent action, *Endocrinology*, 128, 1029, 1991.

19 Calcium Imaging in Living Cells

Peter Richter and Donat-P. Häder

CONTENTS

19.1 INTRODUCTION

19.1.1 THE VITAL ROLE OF CALCIUM IN SIGNAL TRANSDUCTION

Calcium ions play an eminent role in the regulation of many cellular processes in eukaryotic as well as in prokaryotic cells. Various mechanisms have evolved to keep the cytosolic calcium concentration at a very low level of about $10^{-7} M$, which is about 4 orders of magnitude lower than the calcium concentration in the surrounding medium. This tight control of the calcium concentration is achieved by means of calcium pumps (Ca^{2+}-ATPases and Ca^{2+}/H^+ antiporters) located in the plasma membrane and in the membrane of organelles.[1-4]

Changes of the cytosolic calcium concentration mediated by activated calcium channels in the plasma membrane or in the membrane of internal calcium stores lead to a defined response of the cell.[5] Various mechanisms trigger the opening or closing of the different calcium channels. Voltage-gated channels are triggered by changes in the membrane potential and can be located in the plasma membrane, the T-tubule membrane of muscle cells, or the cilia membrane. Activated voltage-sensitive calcium channels in the T-tubules trigger the opening of RyR/Ca^{2+} release channels (ryanodine receptors), probably via mechanical coupling. Many calcium channels are activated due to binding of specific ligands such as hormones, ions, or cyclic nucleotides at binding sites at or near the channels. In these cases, calcium acts as an intracellular second messenger. Some types of calcium channels are triggered by mechanical stimulation;[6,7] for example, stretch-activated channels in membranes of various ciliates and flagellates[8-10] and transduction channels in the hair bundle of vertebrates.[11] Stretch-activated channels are also found in plant cells[12] and retinal pigment epithelial cells.[13] The pharmacology of mechanogated channels is reviewed in Refs. 14 and 15. A general overview of membrane channels can be found in Ref. 16.

The regulation of cell functions by changes in the calcium concentration is achieved in many different ways. In some cases, changes of the membrane potential due to calcium influx lead to a cellular response. A change in the membrane potential in *Paramecium*, which can be elicited by a mechanical stimulus, influences the frequency and direction of the cilia beating upon triggering voltage-sensitive calcium channels located in the cilia.[9]

In many cases, calcium binds to regulatory proteins such as calmodulin or protein kinase C, forming an active protein-calcium complex which in turn activates other proteins.[17-19] In this way, calcium is involved in many important physiological processes, such as neuronal transmission,[20,21] visual processes,[22] fertilization,[23,24] proliferation,[25] chemotaxis[26,27] and gravitaxis,[9] tip growth in higher plants[28,29] and algae,[30] chromosome segregation, and various hormonal[31] and immune-system[32] responses. In addition, the cytosolic calcium concentration is affected by exogenous factors such as toxic chemicals, temperature, or irradiation.[33-35]

Investigation of cellular calcium has become very significant in biological and medical research, as reflected by the numbers of publications increasing from year to year. This chapter describes some techniques of calcium imaging in living cells.

19.1.2 FLUORESCENCE PROBES AND OTHER NON-INVASIVE TECHNIQUES

Many biomolecules undergo short-lifetime (about 10^{-9} s) excited states after absorption of a photon in their absorption wavelength range. Returning to the ground state, the greater part of the received energy is emitted as a fluorescence photon. Due to energy dissipation in the excited state, the emitted light has a lower energy, which corresponds to a longer wavelength than the excitation light. The principle of all fluorescence techniques is the excitation of desired molecules with suitable light qualities and the detection of the resulting considerable weaker emission.[36,37]

Fluorescence probes are specially designed molecules with a specific binding site (e.g., ion-specific chelators or antibodies) coupled to a fluorescent dye (mostly a polyaromatic hydrocarbon or heterocyclic molecule[36]) that can be detected with fluorescence techniques. Binding the target substance leads to conformational changes of the dye molecule, resulting in changes in its absorption or fluorescence properties.

The combination of fluorescence techniques with other noninvasive methods developed and modified over the last decade enables cell function studies which were not previously possible. Micrometer-sized particles like cells or organelles can be micromanipulated with the help of focused light.[38] They can be grasped and moved with the help of focused infrared or near-infrared laser light, a technique better known as "optical tweezers." Microbeams of pulsed UV radiation focused into the object plane can be used for very precise microsurgery purposes.[38-41]

A very powerful tool for cell research is the development of photoactivatable (caged) probes. Reagents such as ions, chelators, hormones, second messengers, and nucleotides are bound to special caging groups (mostly derivates of *o*-nitrobenzylic compounds). They are biologically inactive as long as they are bound to this "cage."[36] The photolysis of the caged substances can be controlled spatially as well as temporally. Low-energy UV flashes can be used for controlled release of the caged substance in the test solution.[42] The internal calcium concentration, for example, can be manipulated with photolabile calcium chelators such as nitrophenyl-EGTA (NP-EGTA) or DM-nitrophen, which show a dramatic (about 12,500-fold in the case of NP-EGTA) decrease in their affinity to calcium after UV irradiation. In contrast, there are primarily inactive calcium chelators that show an increase in their affinity upon illumination.[36,43-46] Some researchers combine fluorescence measurements with electrophysiological techniques.[45,47] This allows simultaneous measurements of the membrane potential and the calcium concentration, which is very advantageous, for example, in the investigation of voltage-dependent calcium kinetics.

Confocal microscopy allows three-dimensional reconstructions of fluorescence-labeled cell structures[48] or the detection of local changes in ion activity.[23,49,50] A brief overview of the principles of confocal microscopy can be found in Section 19.2.3.

19.2 MEASUREMENT OF CYTOSOLIC CALCIUM

19.2.1 CHEMICAL AND PHYSICAL PROPERTIES OF FLUORESCENT CALCIUM INDICATORS

Different fluorescent calcium indicators with different chemical or physical properties have been developed and are commercially available (Molecular Probes, Eugene, Oregon). The calcium binding indicator is normally derived from the calcium chelators 1,2-bis(*o*-aminophenoxy)ethane-N,N,N′,N′-tetraacetic acid (BAPTA) or ethylenglycol bis(β-aminoethylether)-N,N,N′,N′-tetraacetic acid (EGTA), forming a dynamic complex with calcium ions.[50] The conformational change of the binding site upon calcium binding also affects the fluorophore. Wavelength-stable indicators such as Calcium Crimson (Figure 19.1) show an increase in the amplitude of the fluorescence. Upon calcium binding, dual-wavelength or wavelength-shifting indicators such as Fura-2 show a shift of the fluorescence excitation (Figure 19.2). In wavelength-stable dyes, the fluorescence intensity is related to the calcium concentration. In contrast, the response of wavelength-shifting dyes is determined by the ratio of the emission signal at two different excitation wavelengths. One excitation wavelength corresponds to the absorption maximum of the calcium-bound form of the dye and the second to the calcium-free form. In Fura-2 measurements, the emission at 510 nm is monitored after excitation at 340 nm (calcium-bound) and 380 nm (calcium-free). The ratiometric indicator Indo-1 is a dual-emission dye excited at 338 nm. It changes the emission maximum from 480 nm (calcium free) to 405 nm upon calcium binding.

An important parameter for the experimental design is the value of the dissociation constant (K_d), which reflects the concentration at which 50% of the calcium indicator molecules are complexed with calcium ions. Small K_d values indicate a high affinity to calcium. A K_d value of the indicator close to the physiological calcium concentration of the sample warrants a dynamic range large enough for response of the indicator due to calcium changes of the sample. The optimal range in calcium response of Calcium Crimson, for example, is indicated by the steep part of the sigmoidal curve in Figure 19.3, which is calculated from the mass action equation:

$$\left[Ca^{2+}\right] = K_d \frac{[\text{Crimson-Ca}]}{[\text{Crimson-Free}]} \tag{19.1}$$

where [Ca^{2+}] is the calcium concentration, K_d is the dissociation constant (185 n*M*), [Crimson-Ca] is the concentration of the Crimson-calcium complex, and [Crimson-free] is the concentration of the unbound Crimson. It should be noted that the K_d value can be influenced by properties of the medium (e.g., pH, ionic strength, Mg^{2+} concentration). As described below, the mass action equation allows a recalculation of the fluorescence signal to the corresponding calcium concentrations.

A number of dyes with different modifications are available. A list of common fluorescence indicators is given in Tables 19.1 and 19.2. Due to the charge of the carboxylic acid groups at its binding site, the dye molecule is lipophobic and can hardly pass the cell membrane. In contrast, the acetomethoxy (AM) ester form of the dye, in which the carboxylic groups are uncharged by coupling the acetoxymethyl, easily permeates the membrane. To avoid accumulation of the small

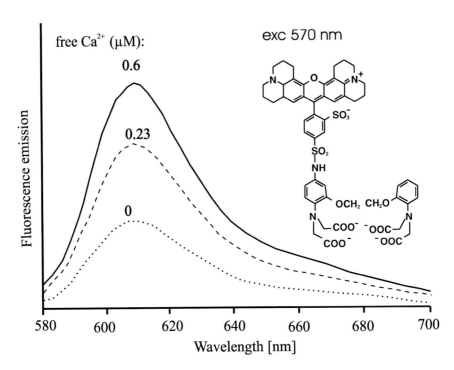

FIGURE 19.1 Fluorescence emission spectra and chemical structure of the wavelength-stable fluorescent calcium indicator Calcium Crimson. Optimal excitation at 590 nm; maximal emission at 615 nm. The amplitude of its fluorescence intensity increases upon calcium binding. (Modified from Haugland.[36] With permission.)

dye molecules inside internal compartments, some dyes are available as large dextran-bound molecules. Loading techniques are described in detail in the next section.

To measure the near-membrane Ca^{2+} concentration, which can be very different from the cytosolic Ca^{2+} concentration,[52] some calcium indicators are conjugated with a lipophilic alkyl chain anchoring the dye to the membrane.[49,53,54] Membrane-bound [18]C-Fura-2 responds several times faster to transient changes in the calcium concentration upon depolarization than cytosolic Fura-2.[53]

19.2.2 CA²⁺-FLUORESCENCE INDICATORS COMPARED WITH OTHER CALCIUM MEASUREMENT TECHNIQUES

The recently developed fluorescence lifetime measurement (FLM) technique and time-resolved lifetime microscopy (TRFLM) are designed to detect the difference in the fluorescence lifetime of the Ca^{2+}-bound and Ca^{2+}-free form of the calcium indicator.[88-90] The Ca^{2+}-free form has a prolonged fluorescence lifetime. Unlike the fluorescence signal, the lifetimes of the Ca^{2+}-bound and Ca^{2+}-free forms of the indicator are not affected by chemical properties of the environment.[88] With regard to single-wavelength dyes, the use of FLM could be advantageous compared to conventional fluorescence microscopy. Local accumulations of the dye within the sample and/or changes in the chemical properties of the solution, which can lead to artifacts in the use of fluorescence technique, have only small effects on FLM.

Another elegant method of calcium measurement is the use of bioluminescence based on Ca^{2+} indicators like aequorin.[33,36] The aequorin complex consists of an apoprotein (22,000 MW) and the luminophore coelenterazine or coelenterazine derivatives. The presence of oxygen and Ca^{2+} result in an irreversible oxidation of the luminophore and a release of blue light. The gene for the

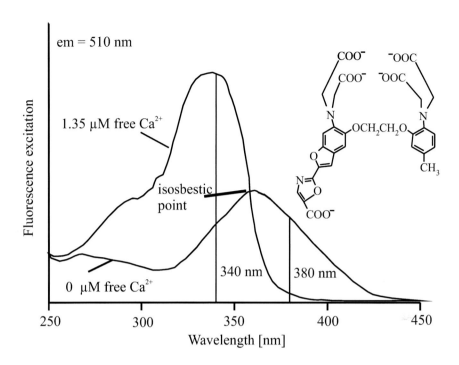

FIGURE 19.2 Fluorescence excitation spectra and chemical structure of the dual-wavelength calcium indicator Fura-2. Its excitation peak at an emission of 510 nm is shifted from 340 nm (calcium-free dye) to about 365 nm (calcium-bound dye). The isosbestic point at about 360 nm is not affected by changes in the calcium concentration. Calcium measurements are performed by ratioing the emission intensity (510 nm) at excitation wavelengths of 340 nm and 380 nm, respectively. The difference between both intensities is higher at 380 nm as for 365 nm (overlap of the excitation spectra). For details, see text. (Modified from Haugland.[36] With permission.)

apoprotein could be transformed into different cell types with the help of various vectors.[26,33,36,91-95] In this case, the membrane permeant coelenterazine must be applied to the sample separately. Since aequorins need no excitation light, they are very suitable to indicate calcium dynamics in cells with pronounced autofluorescence such as green plant cells.[94,95] Recent experimental protocols for the work with aequorins are described in Refs. 96 and 97.

Another sensitive method for calcium measurement is the work with calcium-sensitive micro-electrodes. Felle and Hepler found good coincidence of the signal obtained with an electrode compared to simultaneous Fura-2 measurements.[98]

19.2.3 HARDWARE REQUIREMENTS

The excitation and detection of the fluorescence light can be performed with an epifluorescence microscope equipped with a suitable filter set. The excitation light is preferably produced with a xenon lamp combined with an excitation filter with required transmission properties. Other possibilities are the use of suitable lasers, diodes, monochromators, or acousto-optical-tunable filters (AOTF).[49,99]

A dichroic mirror (reflection in the excitation wavelength range, transmission in the emission wavelength range) deflects the excitation light through the objective onto the sample, causing fluorescence. The fluorescence light can pass through the dichroic mirror. The emission light is selected with an emission filter transmitting in the fluorescence wavelength range (Figure 19.4). An alternating excitation with two wavelengths for wavelength-shifting dyes can be performed with

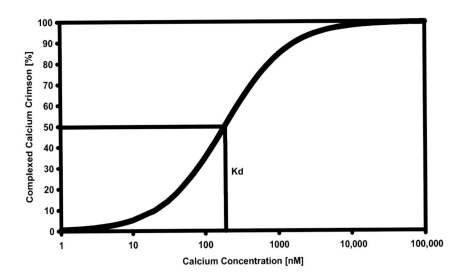

FIGURE 19.3 Binding curve of Calcium Crimson with calcium according to the calcium concentration. The curve was calculated from Eq. 19.1 with K_d 185 nM. The x-axis is the calcium concentration and the y-axis is the percentage of complexed Calcium Crimson. The dynamic range of the indicator is optimal at concentrations around the K_d value (the steep, linear part of the sigmoid curve).

TABLE 19.1
Single-Wavelength Dyes

Indicator	Exc/Em (nm)	F_{Ca}/F_{free}[a]	K_d	Recent Publications
Fluo-3	490/530	~100-200	390 nM	55–57
Rhod-2	550/580	>100	570 nM	58–61
Calcium Green-1	506/531	~14	190 nM	62–65
Calcium Green-2	503/536	~100	550 nM	66–68
Calcium Green-5N	506/532	~38	14 µM	66–69
Calcium Orange	549/576	~3	185 nM	70, 71
Calcium Orange-5N	549/582	~5	20 µM	46
Calcium Crimson	590/615	~2.5	185 NM	72–74

[a] F_{Ca}/F_{free} = Fluorescence increase upon calcium binding.

filter-wheels, monochromators, or AOTF. Excitation in the UV range requires UV-transmitting optics in the pathway of the microscope. Complete filter sets adapted to various fluorescence indicators are commercially available. Since the intensity of the fluorescence signal is normally very low (10^{-5} to 10^{-6} lx),[100] the signal needs to be recorded with a sensitive camera (intensified CCD camera, integrating camera) mounted on top of the microscope. The transfer of the camera signal to a computer equipped with a frame grabber card and suitable software allows evaluation of the fluorescence signal.

Data acquisition in confocal microscopy is different. Laser light is sent through a small aperture (pinhole) and focused as a small point into the sample. Emitted fluorescent light is focused through a confocal pinhole into a photomultiplier tube (PMT) that records the intensity of fluorescence in

TABLE 19.2
Wavelength-Shifting Dyes

Indicator	Excitation [nm]	Emission [nm]	K_d	Recent Publications
Quin-2	330 + 350	490	60 nM	75–78
Fura-2	330 + 350	510	145 nM	49, 98, 110
Mag-Fura-2	340 + 380	510	25 µM	79–81
Fura Red	420 + 480	660	140 nM	82–84
Indo-1	360	405 + 480	230 nM	85–87

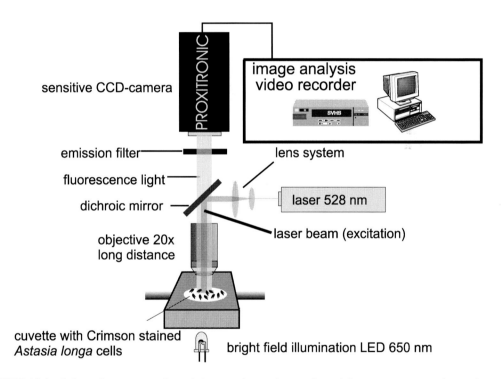

FIGURE 19.4   Schematic representation of the experimental setup for calcium measurements in moving flagellates.

this focal point. Emitted light that does not arise from the sharp focus is not focused in the PMT and does not contribute significantly to the sensor signal.

The specimen can be scanned by moving the focal point in defined steps over the sample. The calculation of the fluorescence intensity of each focal position allows a computer based reconstruction of the sample in spatial resolution. Moving the focus also in the z-direction allows three-dimensional image reconstruction. For detailed information, see literature cited in Section 19.1.2.

19.2.4   LOADING OF FLUORESCENCE DYES INTO CELLS

The chemical and structural properties of the dye determine the loading procedure. The easiest method is loading the AM ester form of the dye. This uncharged modification of the dye can permeate through the cell membrane. Internal, cell-specific esterases cleave the acetoxymethyl group from the dye, which produces the anionic and impermeant form leading to an accumulation of the dye inside the cell. The intracellular concentration of the trapped dye is several-fold higher than the concentration applied to the medium.

Non-acetoxymethylized dyes can be loaded noninvasively into pH-tolerant cells when the medium is adjusted to low pH (acid loading). In this case, the carboxylic acids are protonated, resulting in an uncharged modification of the dye. After membrane permeation, the dye encounters neutral pH values, shifting the chemical equilibrium of the dye to the anionic and cell-impermeant base form.[101]

Some dyes, especially the AM forms, often compartmentalize inside organelles where they may encounter high calcium concentrations leading to a strong unwanted fluorescence response. High-molecular-weight dextran conjugates of calcium indicators were developed to overcome this problem. The dyes are conjugated to dextran molecules (10,000 to 2,000,000 MW). The dextran conjugates are generally loaded with the help of invasive techniques such as electroporation, scrape loading, microinjection,[23,102] or membrane perforation with a UV laser microbeam.[38,103] Electroporation and scrape loading induce transient pores in the membrane of the cell, which allows the introduction of molecules from the surrounding medium. In the case of electroporation, this short time permeabilization of the membrane is achieved with the application of short electric pulses[104] or, in case of scrape loading, via application of mechanical forces.[33,105-108]

19.2.5 CALIBRATION OF THE CALCIUM SIGNAL

Although the experimental effort of calcium measurements working with single-wavelength dyes is smaller than the effort necessary for the use of wavelength-shifting dyes, some possible errors sources have to be taken into account. Wavelength-shifting dyes are relatively independent of the dye concentration because of the ratiometric evaluation. The signal obtained from single-wavelength dyes is the absolute fluorescence intensity, which depends on the calcium concentration and, in addition, on the dye concentration. An unequal distribution of the dye concentration inside a cell mimics a higher calcium concentration at the locations with higher dye concentrations. Bleaching or leakage of the dye during measurement would erroneously indicate a decrease in the calcium concentration. Thick regions of the cell indicate a higher calcium concentration than thin cell regions. Because of these problems, some scientists exclude single-wavelength dyes for calcium quantification.[109]

However, single-wavelength dyes have some advantages over dual-wavelength dyes. Besides the lower experimental effort, single-wavelength dyes with a long excitation wavelength and emission maximum, like Calcium Orange or Calcium Crimson, reduce interference of the excitation light with the sample. Autofluorescence, photodamage, as well as light scattering are reduced. Single-wavelength dyes were up until recently, more common for laser scanning microscopy[110,111] because of technical and optical limitations of UV light sources necessary for the excitation of dual-wavelength dyes. These problems have been overcome with the modification of a conventional laser scanning microscope equipped with acousto-optic tunable filters (AOTF), allowing rapid switching between different excitation wavelengths.[49] Due to the UV excitation range dual-wavelength dyes are not suitable for parallel work with caged compounds (Section 19.1.2) that are released upon UV irradiation.

19.2.5.1 Calibration of Single-Wavelength Dyes

According to Eq. 19.1, the ratio between the Ca^{2+}-bound and the Ca^{2+}-free dye (and not the absolute dye concentration) is necessary to calculate the calcium concentration. The fluorescence signal is maximal when all dye molecules are complexed with calcium at saturating concentrations, and minimal when all dye molecules are unbound in calcium-free conditions. The signal proportion of the complexed dye can be expressed with the actual fluorescence of the dye minus its minimal (zero calcium) fluorescence. The signal proportion of the unbound dye is the fluorescence of maximal calcium saturation minus the present fluorescence signal.[112] A modification of Eq. 19.1 leads to the following expression:

$$[Ca^{2+}] = K_d \frac{(F - F_{min})}{(F_{max} - F)} \qquad (19.2)$$

where F is actual fluorescence, F_{min} is the fluorescence under calcium-free conditions, F_{max} is the fluorescence at saturating calcium concentrations, and K_d is the dissociation constant.

To determine the absolute calcium concentration of the sample, it is necessary to measure F_{max} and F_{min}. The calibration can be performed with calcium calibration kits available from Molecular Probes.[36] Since the dye concentration inside the cell is generally not known and the chemical conditions of the cytoplasm may influence the K_d value or the fluorescence properties of the dye, it is advantageous to perform the calibration process directly inside the cells.

This determination method normally leads to severe damage of the cells and should therefore be performed after the actual measurement or with control cells. The cell membrane is permeabilized with detergents such as digitonin or saponin[112] or ionophores such as ionomycin, nigericin, or calcimycin. It has to be determined whether the added reagents have an autofluorescence in the excitation range. Calcimycin (A23187), for example, shows interference with the longer wavelength (380 nm) of Fura-2 excitation.[45] This problem can be eliminated using its autofluorescence-free analogue 4-bromo-A23187. The permeabilization of the cell membrane allows externally added calcium or EGTA to enter the cell. After addition of high calcium concentrations, the F_{max} value can be determined. Excessive concentrations of EGTA or BAPTA, which complex calcium[51] (K_d about 10^{-8} M), allow determination of the calcium-free fluorescence F_{min}. From the binding curve (Figure 19.3), it can be seen that the application of 10^{-3} M Ca^{2+} and downregulation to 10^{-8} M Ca^{2+} is sufficient to obtain F_{max} and F_{min}, respectively, of Calcium Crimson.

Many cells show a more or less pronounced autofluorescence upon excitation. To obtain the amount of cellular autofluorescence, it is necessary to eliminate the fluorescence of the dye, which is in most cases not zero even for calcium-free conditions (see Section 19.2.1). This can be achieved by measuring and recording the cellular autofluorescence before staining the cells. Another possibility is dye quenching with $MnCl_2$[113,114] after membrane permeabilization, which can be reversed with DTPA (diethylenetriamine pentaacetic acid); the method is described in detail by Thomas.[112]

19.2.5.2 Calibration of Wavelength-Shifting Dyes

A variety of commercial ratioing imaging systems are available, as well, many descriptions and protocols for the calibration of wavelength-shifting dyes have been published. Methods for 3-D imaging of cells by confocal microscopy are described, for example, by Rizzuto and Fasolato,[115] Nitschke et al.,[49] and others. Methods for ratioing by means of conventional fluorescence microscopy were published, among others, by Williams and Fay,[116] Thomas,[112] Mason et al.[45] (in combination with electrophysiological methods), and Webb et al.[1]

The calibration of wavelength-shifting dyes is performed with the Grynkiewitz equation,[117] which is similar to Eq. 19.1:

$$[Ca^{2+}] = K_d \frac{(R - R_{min})}{(R_{max} - R)} \times \frac{S_{f_2}}{S_{b_2}} \qquad (19.3)$$

where R = present ratio, R_{max} = ratio at saturating calcium concentration, R_{min} = ratio at calcium-free conditions, and S_{f2}/S_{b2} = ratio of the calcium-free corresponding wavelength Ca^{2+}-bound/Ca^{2+}-free conditions. The ratios were calculated from the fluorescence of the wavelength according to calcium-bound conditions divided by the fluorescence of the wavelength indicating calcium-free conditions.

In the case of Fura-2, the ratios were obtained by ratioing 340 nm to 380 nm. S_{f2}/S_{b2} is the ratio of the calcium saturated to calcium-free signal at 380 nm (the calcium-free corresponding wavelength). A decrease of the signal at the isosbestic point at 360 nm indicates loss of dye due to leakage or bleaching. The calibration procedure corresponds to the methods described in Section 19.2.5.1.

19.3 CALCIUM IMAGING IN MOVING FLAGELLATES

Euglena gracilis is a unicellular freshwater flagellate that orients itself driven by a pronounced negative gravitaxis toward the water surface. Previous experiments revealed that the orientation is based on an active, calcium-dependent process, most likely triggered by membranal stretch-sensitive calcium channels elicited by forces to the membrane applied by sedimentation of cell body.[118,119] To investigate changes of the cytosolic calcium concentration due to gravity, the calcium signal of moving cells was monitored during a sounding rocket flight.

19.3.1 EXPERIMENTAL PROCEDURE

19.3.1.1 Cells and Loading Procedure

The experiments were performed with *Astasia longa,* a close relative of *Euglena gracilis.* The chlorophyll-free *Astasia* cells show only a negligible autofluorescence in contrast to the green *Euglena* cells and are not influenced by the excitation light like colorless *Euglena* strains (e.g., 1F or 9B), which showed a pronounced photophobic response due to the laser illumination.

The cells were concentrated to 10^7 cells per milliliter. Calcium Crimson dextran[36] (Molecular Probes; 10,000 MW) was added to an end concentration of 2 μM and incubated for several hours. The loading was performed by electroporation (Peqlab; single pulse 4 ms, 500 V, two repetitions). The cells were washed several times in fresh medium immediately after electroporation. After confirming their sufficient fluorescence level, the cells were filled in custom-made stainless steel cuvettes and mounted into an experimental module of a sounding rocket (see next section).

19.3.1.2 Space Experiment

The experiment consisted of two identical setups (Figure 19.5). Two epifluorescence microscopes equipped with 20X long-distance objectives were mounted on a centrifuge. The excitation was performed with 20-mW lasers (LCM-LL-11qp; Laser Compact Co. Ltd., Russia) at 538 nm. The intensity of the lasers was reduced with 90 % neutral glass filters. The red part of the laser radiation resulting from the pump diode was removed with filters. The laser beam was expanded by f10 and f40 lenses (Spindler & Hoyer, Germany). A suitable beam splitter deflected the laser beam onto the sample. The fluorescence of the samples was additionally filtered with a suitable emission filter (interference filter, 610 nm) and a 550-nm cutoff filter (the beam splitter and the emission filter were from AMP, Thanesch, Germany; and the cutoff and bandpass filters from Zeiss, Oberkochen, Germany). The fluorescence signal of each of the two microscopes was recorded with sensitive CCD cameras (Proxitronic HL4 S NIR 2+; Proxitronic, Germany), transmitted to the ground control center, and recorded with SVHS-video recorders (Philips VR948/02M, Germany). Weak bright-field illumination was performed with 620-nm LEDs. Telemetry was available to control the experiment from the ground. Focusing and lateral movement of the cuvette was possible by joystick

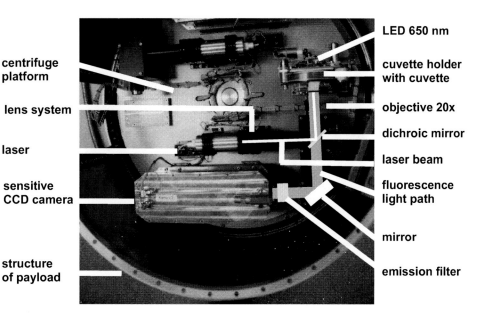

FIGURE 19.5 Experimental setup of the MAXUS 3 space experiment. Crimson-loaded *Astasia longa* cells are excited with 538-nm laser light. The fluorescence signal from the cells is detected with an image intensifier CCD camera. The cuvette with the cells (top right) can be revolved.

control in order to find the optimal monitoring position. The centrifuge could be adjusted to defined accelerations. During the start, the cuvette was rotated (about 1 Hz) to avoid cell sedimentation or swimming out of the field of vision.

The total microgravity time ($<10^{-4}$ g) was about 840 s. During this time, different acceleration forces were applied to the cells by centrifugation in various combinations corresponding to 0 g, 0.1 g, 0.2 g, and 0.3 g.

The analysis of the calcium signals was performed later from the video recordings with a fast Pentium computer equipped with a frame grabber card (Matrox, Dorral, Quebec, Canada) and a modified version of the image analysis software WinTrack 2000 (developed by D.-P. Häder and M. Lebert; see Chapter 20). This modification was necessary because of the small signal-to-noise ratio of the cells and the unequal illumination in the focal plane. Statistical interpretations whether the calcium signal differs in dependence of the acceleration needs the calculation of a huge number of single cell tracks because of the free movement of the cells into and out of the focus. The signal of cells in the focal plane (about 6 μm, cuvette thickness = 100 μm) is much brighter and sharper than the image of cells outside the focus. In addition, the calcium signal can differ from cell to cell due to different dye loadings or to physiological parameters concerning the cytosolic calcium concentration.

In ground experiments performed with the experimental setup described above, the cuvette with oriented, upward swimming *Astasia* cells was rotated around the short horizontal axis of the vertically oriented cuvette by an angle of 90°. Upon reorientation, the cells always showed the same kinetics of the calcium fluorescence intensity. After a lag phase of about 10 s, the portion of cells showing a higher calcium signal increased. About 30 s after the turn, the average calcium signal was maximal, followed by a fast (about 10 s) decrease back to the intensity level before the perturbation. The kinetics of the calcium fluorescence corresponds to the reorientation of the cells (Figure 19.6).

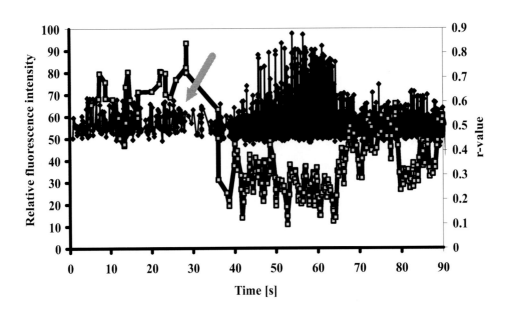

FIGURE 19.6 Kinetics of the calcium fluorescence and reorientation of *Astasia longa*. The cuvette with oriented, upward swimming cells was turned by an angle of 90°. The lower line (squares) shows the kinetics of the reorientation, represented by the r-value (increasing r-values indicate better gravitactic orientation of the cells; see Chapter 20), and the upper line indicates the fluorescence kinetic of the cells. Each data point represents a single cell track.

ACKNOWLEDGMENTS

The authors gratefully acknowledge financial support from DLR (No. JW-50 WB 9827) and thank M. Lebert and H. Tahedl for fruitful discussions and N. Steffens for excellent technical assistance.

REFERENCES

1. Webb, A. A. R., McAinsh, M. R., Taylor, J. E., and Hetherington, A. M., Calcium ions as intracellular second messengers in higher plants, *Adv. Bot. Res.*, 22, 45, 1996.
2. Gangola, P. and Rosen, B. P., Maintenance of intracellular calcium in *Escherichia coli*, *J. Biol. Chem.*, 262, 12570, 1987.
3. Mac Lennan, D. H., Rice, W. J., and Green, N. M., The mechanism of Ca^{2+} transport by sarcoplasmic reticulum Ca^{2+}-ATPases, *J. Biol. Chem.*, 272, 28815, 1997.
4. Norris, V., Grant, S., Freestone, P., Canvin, J., Sheikh, F. N., Toth, I., Trinel, M., Modha, K., and Norman, R. I., Calcium signalling in bacteria, *J. Bact.*, 178, 3677, 1996.
5. Golovian, V. A. and Blaustein, M. P., Spatially and functionally distinct Ca^{2+} stores in sarcoplasmic and endoplasmic reticulum, *Science*, 275, 1643, 1997.
6. Petrov, G. and Usherwood, N. R., Mechanosensitivity of cell membranes, *Eur. Biophys. J.*, 23, 1, 1994.
7. Sachs, F. and Morris, C. E., Mechanosensitive ion channels in nonspecialized cells, in *Reviews of Physiology and Biochemistry and Pharmacology*, Springer, Berlin, 1998, 1.
8. Machemer, H. and Bräucker, R., Gravireception and graviresponse in ciliates, *Acta Protozool.*, 31, 185, 1992.
9. Machemer, H. and Teunis, P. F. M., *Sensory Coupling and Motor Responses*, Gustav Fischer Verlag, Stuttgart, Jena, New York, 1996.

10. Yoshimura, K., A novel type of mechanoreception by the flagella of *Chlamydomonas*, *J. Exp. Biol.*, 199, 295, 1996.

11. Howard, J., Roberts, W. M., and Hudspeth, A. J., Mechanoelectrical transduction by hair cells, *Annu. Rev. Biophys. Biophys. Chem.*, 17, 99, 1988.

12. Shimmen, F., Studies of mechano-perception in characeae: effects of external Ca^{2+} and Cl^-, *Plant Cell Physiol.*, 38, 691, 1997.

13. Stalmans, P. and Himpens, B., Confocal imaging of Ca^{2+} signaling in cultured rat retinal pigment epithelial cells during mechanical and pharmacologic stimulation, *Invest. Ophthalmol. Vis. Sci.*, 38, 176, 1997.

14. Hamill, O. P. and Mc Bride, D. W., The pharmacology of mechanogated membrane ion channels, *Pharmacol. Rev.*, 48, 231, 1996.

15. Banes, A. J., Tsuzaki, M., Yamamoto, J., Fischer, T., Brigman, B., Brown, T., and Miller, L., Mechanoreception at the cellular level: the detection, interpretation, and diversity of responses to mechanical signals, *Biochem. Cell Biol.*, 73, 349, 1995.

16. Peracchia, C., *Handbook of Membrane Channels Molecular and Cellular Physiology*, Academic Press, San Diego, 1994.

17. Zielinski, R. E., Calmodulin and calmodulin-binding proteins in plants, *Annu. Rev. Plant Physiol. Plant Mol. Biol.*, 49, 697, 1998.

18. Rampersaud, A., Utsumit, R., Delgado, J., Forst, S. A., and Inouye, M., Ca^{2+}-enhanced phosporylation of a chimeric protein kinase involved with bacterial signal transduction, *J. Biol. Chem.*, 266, 7633, 1991.

19. Nishizuka, Y., The role of protein kinase C in cell surface signal transduction and tumor promotion, *Nature*, 308, 693, 1984.

20. Ghosh, A. and Greenberg, M. E., Calcium signaling in neurons: molecular mechanisms and cellular consequences, *Science*, 268, 239, 1995.

21. Berridge, M. J., Neuronal calcium signaling, *Neuron*, 21, 13, 1998.

22. Müller, K. and Kaupp, U. B., Signaltransduktion in Sehzellen, *Naturwissenschaften*, 85, 49, 1998.

23. Stricker, S. A., Intracellular injections of a soluble sperm factor trigger calcium oscillations and meiotic maturation in unfertilized oocytes of a marine worm, *Developm. Biol.*, 186, 185, 1997.

24. Wiesener, B., Weiner, J., Middendorff, R., Hagen, V., Kaupp, U. B., and Weyland, I., Cyclic nucleotide-gated channels on the flagellum control Ca^{2+} entry into sperm, *J. Cell Biol.*, 142, 474, 1998.

25. Hoffman, S., Gopalakrishna, R., Gundimeda, U., Murata, T., Spee, C., Ryan, S. J., and Hinton, D. R., Verapamil inhibits proliferation, migration and proteinase C activity in human retinal pigment epithelial cells, *Exp. Eye Res.*, 67, 45, 1998.

26. Nebl, T. and Fischer, P. R., Intracellular Ca^{2+} signals in *Dictyostelium* chemotaxis are mediated exclusively by Ca^{2+} influx, *J. Cell Sci.*, 110, 2845, 1997.

27. Tisa, F. S., Baldomero, M. O., and Adler, J., Inhibition of *Escherichia coli* chemotaxis by ω-conotoxin, a calcium ion channel blocker, *J. Bacteriology*, 175, 1235, 1994.

28. Messerli, M. and Robinson, K. R., Tip localized Ca^{2+} pulses are coincident with peak pulsatile growth rates in pollen tubes of *Lilium longiflorum*, *J. Cell Sci.*, 110, 1269, 1997.

29. Zhenbiao, Y., Signaling tip growth in plants, *Curr. Opinion Plant Biol.*, 1, 525, 1998.

30. Shaw, S. L., Quatrano, R. S., Polar localization of a dihydropyridine receptor on living *Fucus* zygotes, *J. Cell. Sci.*, 109, 335, 1996.

31. Woods, N. M., Cuthbertson, K. S. R., and Cobbold, P. H., Repetitive transient rises in cytoplasmic free calcium in hormone-stimulated hepatocytes, *Nature*, 319, 600, 1986.

32. Trapani, J. A., Dual mechanisms of apoptosis induction by cytotoxic lymphocytes, *Int. Rev. Cytol.*, 182, 111, 1998.

33. Borle, A. B., Ca^{2+}-bioluminescent indicators, *Methods Toxicol.*, 19, 315, 1994.

34. Gong, M., van der Luit, A. H., Knight, M. R., and Trewavas, A. J., Heat-shock-induced changes in intracellular Ca^{2+} level in tobacco seedling in relation to thermotolerance, *Plant Physiol.*, 116, 429, 1998.

35. Jones, D. L., Gilroy, S., Larsen, P. B., Howell, S. H., and Kochian, L. V., Effect of aluminium on cytoplasmic Ca^{2+} homeostasis in root hairs of *Arabidopsis thaliana* (L.), *Planta*, 206, 378, 1998.

36. Haugland, R. P., *Handbook of Fluorescent Probes and Research Chemicals*, Molecular Probes, Eugene, Oregon, 1997.

37. Tevini, M. and Häder, D.-P., *Allgemeine Photobiologie*, Georg Thieme Verlag, Stuttgart, 1985.
38. Greulich, K. O. and Pilzarczyk, G., Laser tweezers and optical microsurgery in cellular and molecular biology. Working principles and selected applications, *Cell. Mol. Biol.*, 44, 701, 1998.
39. Pool, R., Making light work of cell surgery, *Science*, 248, 29, 1990.
40. Schütze, K. and Clement-Sengewald, A., Catch and move-cut or fuse, *Nature*, 368, 667, 1994.
41. Block, S. M., Making light work with optical tweezers, *Nature*, 360, 493.
42. Gilroy, S., Read, N. D., and Trewavas, A. J., Elevation of cytoplasmatic calcium by caged calcium or caged inositol triphosphate initiates stomatal closure, *Nature*, 346, 769, 1990.
43. Kaplan, J. H., Graham, C. R., and Ellis-Davies, C. R., Photolabile chelators for the rapid photorelease of divalent cations, *Proc. Natl. Acad. Sci.*, *U.S.A.*, 85, 6571, 1988.
44. Graham, C. R. and Kaplan, J. H., Nitrophenyl-EGTA, a photolabile chelator that selectively binds Ca^{2+} with high affinity and releases it rapidly upon photolysis, *Proc. Natl. Acad. Sci.*, *U.S.A.*, 91, 187, 1993.
45. Mason, W. T., Dempster, J., Zorec, R., Hoyland, J., and Lledo, P. M., Fluorescence measurements of cytosolic calcium: combined photometry with electrophysiology, *Methods Neurosci.*, 27, 81, 1995.
46. Ellis-Davies, C. R., Kaplan, J. H., and Barsotti, R. J., Laser photolysis of caged calcium: rates of calcium release by nitrophenyl-EGTA and DM-nitrophen, *Biophys. J.*, 70, 1006, 1996.
47. Helmchen, F., Gerard, J., Borst, G., and Sakmann B., Calcium dynamics associated with a single action potential in a CNS presynaptic terminal, *Biophys. J.*, 72, 1458, 1997.
48. Braun, M. and Wasteneys, G. O., Distribution and dynamics of the cytoskeleton in graviresponding protonemata and rhizoid of characean algae: exclusion of microtubles and a convergence of actin filaments in the apex suggest an actin mediated gravitropism, *Planta*, 205, 39, 1998.
49. Nitschke, R., Wilhelm, S., Borlinghaus, R., Leipziger, J., Bindels, R., and Greger, R., A modified confocal laser scanning microscope allows fast ultraviolet ratio imaging of intracellular Ca^{2+} activity using Fura-2, *Pflügers Arch. – Eur. J. Physiol.*, 433, 653, 1997.
50. Lipp, P., Lüscher, C., and Niggli, E., Photolysis of caged compounds characterized by ratiometric confocal microscopy: a new approach to homogeneously control and measure the calcium concentration in cardiac myocytes, *Cell Calcium*, 19, 255, 1996.
51. Gurney, A. M. and Bates S. E., Use of chelators and photoactivatable caged compounds to manipulate cytosolic calcium, *Methods Neurosci.*, 27, 123, 1995.
52. Augustine, G. J. and Neher, E., Calcium requirements for secretion in bovine chromaffin cells, *J. Physiol.*, 450, 247, 1992.
53. Etter, E. F., Kuhn, M. A., and Fay, F. S., Detection of changes in near-membrane Ca^{2+}-concentration using a novel membrane-associated Ca^{2+}-indicator, *J. Biol. Chem.*, 269, 10141, 1994.
54. Etter, E. F., Minta, A., Poenie, M., and Fay, F. S., Near-membrane $[Ca^{2+}]$ transients resolved using the Ca^{2+} indicator FFP18, *Proc. Natl. Acad. Sci.*, *U.S.A.*, 93, 5368, 1996.
55. Fleet, A., Ellis-Davies, G., and Bolsover, S., Calcium buffering capacity of neuronal cell cytosol by flash photolysis of calcium buffer NP-EGTA, *Biochem. Biophys. Res. Commun.*, 250, 786, 1998.
56. Venance, L., Premont, J., Glowinski, J., and Giaume, C., Gap junctional communication and pharmacological heterogeneity in astrocytes cultured from the rat striatum, *J. Physiol. (London)*, 510, 429, 1998.
57. Grondahl, T. O., Hablitz, J. J., and Langmoen, I. A., Depletion of intracellular Ca^{2+} stores or lowering extracellular calcium alters intracellular Ca^{2+} changes during cerebral energy deprivation, *Brain Res.*, 796, 125, 1998.
58. Bowser, D. N., Minamikawa, T., Nagley, P., and Williams, D. A., Role of mitochondria in calcium regulatory of spontaneously contracting cardiac muscle cells, *Biophys. J.*, 75, 2004, 1998.
59. Kamiya, H. and Ozawa, S., Kainate receptor-mediated inhibition of presynaptic Ca^{2+} influx and EPSP in area CA1 of the rat hippocampus, *J. Physiol. (London)*, 509, 833, 1998.
60. Del Nido, P. J., Glynn, P., Buenaventura, P., Salama, G., and Koretsky, A. P., Fluorescence measurement of calcium transients in perfused rabbit heart using Rhod-2, *Am J. Physiol.*, 274, 728, 1998.
61. Babock, D. F., Herrington, J., Goodwin, P. C., Park, Y. B., and Hille, B., Mitochondrial participation in the intracellular Ca^{2+} network, *J. Cell Biol.*, 136, 833, 1997.
62. Dearnaley, J. D., Levina, N. N., Lew, R. R., Heath, I. B., and Goring, D. R., Interrelations between cytoplasmatic Ca^{2+} peaks, pollen hydration and plasma membrane conductance during compatible and incompatible pollinations of *Brassica napus* papillae, *Plant Cell Physiol.*, 38, 985, 1997.

63. Ricci, A. J. and Fettiplace, R., Calcium permeation of the turtle hair cell mechanotransducer channel and its relation to the composition of endolymph, *J. Phys. (London)*, 506, 159, 1998.

64. Prajer, M., Fleury, A., and Laurent, M., Dynamics of calcium regulation in *Paramecium* and possible morphogenetic implications, *J. Cell Sci.*, 110, 529, 1997.

65. Young, R. C. and Hession, R. O., Paracrine and intracellular signaling mechanisms of calcium waves in cultured human uterin myocytes, *Obstet. Gynecol.*, 90, 928, 1997.

66. David, G., Barret, J. N., and Barret, E. F., Stimulation-induced changes in [Ca^{2+}] in lizard motor nerve terminals, *J. Phys. (London)*, 504, 83, 1997.

67. Srivastava, S. K., Wang, L. F., Ansari, N. H., and Bhatnagar, A., Calcium homeostasis of isolated single cortical fibers of rat lens, *Invest. Ophthalmol. Vis. Sci.*, 38, 2300, 1997.

68. Gilchrist, J. S., Palahniuk, C., and Bose, R., Spectroscopic determination of sarcoplasmic reticulum Ca^{2+} uptake and Ca^{2+} release, *Mol. Cell Biochem.*, 172, 159, 1997.

69. Ukhanov, K. and Payne, R., Rapid coupling of calcium release to depolarization in *Limulus polyphemus* ventral photoreceptor as revealed by microphotolysis and confocal microscopy, *J. Neurosci.*, 17, 1701, 1997.

70. Basarsky, T. A., Duffy, S. N., Andrew, R. D., and Mac Vicar, B. A., Image spreading depression and associated intracellular calcium waves in brain slices, *J. Neurosci.*, 18, 7189, 1998.

71. Geibel, J., Abraham, R., Modlin, I., and Sachs, G., Gastrin-stimulated changes in Ca^{2+} concentration in parietal cells depends on adenosine 3',5'-cyclic monophosphate levels, *Gastroenterology*, 109, 1060, 1995.

72. Pu, R. and Robinson, K. R., Cytoplasmic calcium gradients and calmodulin in the early development of the fucoid alga *Pelvetia compressa*, *J. Cell. Sci.*, 111, 3197, 1998.

73. Umbach, J. A., Grasso, A., Zurcher, S. D., Kornblum, H. I., Mastragiacomo, A., and Gundersen, C. B., Electrical and optical monitoring of alpha-latroxin action at *Drosophila* neuromuscular junctions, *J. Neurosci.*, 87, 913, 1998.

74. Richter, P., Krywult, M., Sinha, R. P., and Häder, D.-P., Calcium signals from heterocysts of *Anabaena* sp. after UV radiation, *J. Plant Physiol.*, 154, 137, 1999.

75. Hawrth, R. A., Redon, D., Biggs, A. V., and Potter, K. T., Ca uptake by heart cells. II. Most entering Ca appears to leave without mixing with the sarcoplasmic reticulum Ca pool, *Cell Calcium*, 23, 199, 1998.

76. Dalledonne, I., Milzani, A., and Colombo, R., Effect of replacement of tightly bound Ca^{2+} by Ba^{2+} on actin polymerization, *Arch. Biochem. Biophys.*, 351, 141, 1998.

77. Rowin, M. E., Whatley, R. E., Yednock, T., and Bohnsack, J. F., Intracellular calcium requirements for betal integrin activation, *J. Cell. Physiol.*, 175, 193, 1998.

78. Lijen, P., Echevaria-Vaquez, D., Fagard, R., and Petrov, V., Protein kinase C induced changes in erythrocyte Na^+/H^+ exchange and cytosolic free calcium in humans, *Am. J. Hypertens.*, 11, 81, 1998.

79. Martinez-Zaguilan, R., Paranami, J., and Martinez, G. M., Mag-fura-2 (furaptra) exhibits both low (μM) and high (nM) affinity for Ca^{2+}, *Cell. Physiol. Biochem.*, 8, 158, 1998.

80. Speake, T. and Elliot, A. C., Modulation of calcium signals by intracellular pH in isolated rat pancreatic cells, *J. Physiol. (London)*, 506, 415, 1998.

81. Tojyo, Y., Tanimura, A., and Matsumoto, Y., Monitoring of Ca^{2+} release from intracellular stores in permeabilized rat parotid acinar cells using the fluorescent indicators Mag-fura-2 and calcium green C18, *Biochem. Biophys. Res. Commun.*, 240, 189, 1997.

82. Wu, Y. and Clusin, W. T., Calcium transients alternans in blood-perfused ischemic hearts: observations with fluorescent indicator Fura Red, *Am. J. Physiol.*, 273, 2161, 1997.

83. Sipido, K. R., Stankovicova, T., Flameng, W., Vanhaecke, J., and Verdonck, F., Frequency dependence of Ca^{2+} release from the sarcoplasmic reticulum in human ventricular myocytes from end-stage heart failure, *Cardiovasc. Res.*, 37, 478, 1998.

84. Mayer, C., Quasthoff, S., and Grafe, P., Differences in the sensitivity to purinergic stimulation of myelinating and non-myelinating Schwann cells in peripheral human and rat nerve, *Glia*, 23, 374, 1998.

85. Brust-Mascher, I. and Webb, W. W., Calcium waves induced by large voltage pulses in fish keratocytes, *Biophys J.*, 75, 1669, 1998.

86. Binah, O., Liu, C. C., Young, J. D., and Berke, G., Channel formation and $[Ca^{2+}]_i$ accumulation induced by perforin N-terminus peptides: comparison with purified perforin and whole lytic granules, *Biochem. Biophys. Res. Commun.*, 240, 647, 1997.

87. Ju, Y. K. and Allen, D. G., Intracellular calcium and Na^+-Ca^{2+} exchange current in isolated toad pacemaker cells, *J. Physiol. (London)*, 508, 153, 1998.

88. Herman, B., Wodnicki, P., Kwon, S., Periasamy, A., Gordon, G. W., Mahajan, N., and Wang, X. F., Recent developments in monitoring calcium and protein interactions in cells using fluorescence lifetime microscopy, *J. Fluorescence*, 7, 1, 1997.

89. Lakowicz, J. R., Szmacinski, H., and Johnson, M. L., Calcium imaging using fluorescence lifetimes and long-wavelength probes, *J. Fluorescence*, 2, 47, 1992.

90. Lakowicz, J. R., Emerging applications of fluorescence spectroscopy to cellular imaging: lifetime imaging, metal-ligand probes, multi-photon excitation and light quenching, *Scanning Microsc. Suppl.*, 10, 213, 1996.

91. Tsien, R. Y., The green fluorescent protein, *Annu. Rev. Biochem.*, 67, 509, 1998.

92. Cubitt, A. B., Firtel, R. A., Fischer, G., Jaffe, L. F., and Miller, A. L., Patterns of free calcium in multicellular stages of *Dictyostelium* expressing jellyfish apoaequorin, *Development*, 121, 2291, 1995.

93. Schaap, P., Nebl, T., and Fisher, P. R., A slow sustained increase in cytosolic Ca^{2+} levels mediates stalk gene induction by differentiation inducing factor in *Dictyostelium*, *EMBO J.*, 15, 5177, 1996.

94. Johnson, C. H., Knight, M. R., Kondo, T., Masson, P., Sedbrook, J., Haley, A., and Trewavas, A., Circadian oscillations of cytosolic and chloroplastic free calcium in plants, *Science*, 269, 1863, 1995.

95. Knight, M. R., Campell, A. K., Smith, S. M., and Trewavas, A. J., Transgenic plant aequorin reports the effects of touch and cold-shock and elictors on cytoplasmatic calcium, *Nature*, 352, 524, 1991.

96. Brini, M., Pinton, P., and Bastinianutto, C., Targeting, expressing and calibrating recombinant aequorin, in *Imaging Living Cells*, Rizzuto, R. and Fasolato, C., Eds. Springer, Berlin, 263, 1999.

97. Rutter, G. A., Imaging Ca^{2+} in small mammalian cells with recombinant targeted aequorin, *Imaging Living Cells*, Rizzuto, R. and Fasolato, C., Eds., Springer, Berlin, 284, 1999.

98. Felle, H. H. and Hepler, P. K., The cytosolic Ca^{2+} concentration gradient of *Sinapis alba* root hairs as revealed by Ca^{2+}-selective microelectrode tests and fura-dextran ratio imaging, *Plant Physiol.*, 114, 39, 1997.

99. Xiaolu, W. and Lewis, E. N., Acousto-optic tunable filters and their application in spectroscopic imaging and microscopy, in *Fluorescence Imaging Spectroscopy and Microscopy*, Xue, F. W. and Herman, B., Eds., John Wiley & Sons, New York, 125, 1999.

100. Mason, W. T., Hoyland, J., McCann, T. J., Somasundaram, B., and O'Brien, W., Strategies for quantitative digital imaging of biological activity in living cells with ion-sensitive fluorescence probes, *Imaging Living Cells*, Rizzuto, R. and Fasolato, C., Eds., Springer, Berlin, 3, 1999.

101. Legué, V., Blancaflor, E., Wymer, C., Perbal, G., Fantin, D., and Gilroy, S., Cytoplasmic free Ca^{2+} in *Arabidopsis* roots changes in response to touch but not gravity, *Plant Physiol.*, 114, 789, 1997.

102. Stricker, S. A., Repetitive calcium waves induced by fertilization in the nemertean worm *Cerebratus lacteus*, *Dev. Biol.*, 176, 496, 1996.

103. Weber, G., Microperforation of plant tissue with a UV laser microbeam and injection of DNA into cells, *Naturwissenschaften*, 75, 35, 1988.

104. Weaver, J. C., Electroporation: a general phenomenon for manipulating cells and tissues, *J. Cell. Biochem.*, 51, 426, 1993.

105. Lee, G., Delohery, T. M., Ronai, Z., Brandt-Rauf, P. W., Pincus, M. R., Murphy, R. B., and Weinstein, I. B., A comparison of techniques for introducing macromolecules into living cells, *Cytometry*, 14, 265, 1993.

106. El Fouly, M. H., Trosko, J. E., and Chang, C. C., Scrape-loading and dye transfer. A rapid and simple technique to study gap junctional intercellular communication, *Exp. Cell. Res.*, 168, 422, 1987.

107. Mc Neil, P. L., Murphy, R. F., and Taylor, D. L., A method for incorporating macromolecules into adhaerens cells, *J. Cell Biol.*, 98, 1556, 1984.

108. Schlatterer, C., Knoll, G., and Malchow, D., Intracellular calcium during chemotaxis of *Dictyostelium discoideum*: a new Fura-2 derivate avoids sequestration of the indicator and allows long-term calcium measurements, *Eur. J. Cell. Biol.*, 58, 172, 1992.

109. Trewavas, A. J. and Malho, R., Ca^{2+} signalling in plant cells: the big network, *Curr. Opinion Plant Biol.*, 1, 428, 1998.

110. Klauke, N. and Plattner, H., Imaging of Ca^{2+} transients induced in *Paramecium* cells by apolyamine secretagogue, *J. Cell Sci.*, 110, 975, 1997.
111. Zimprich, F., Ashworth, R., and Bolsover, S., Real-time measurements of calcium dynamics in neurons developing *in situ* within zebrafish embryos, *Pflügers Arch.*, 436, 489, 1998.
112. Thomas, A. T., Cell function studies using fluorescent Ca^{2+} indicators, *Methods in Toxicology*, Academic Press, New York, 1B, 287, 1994.
113. Plieth, C., Sattelmacher, B., Hansen, U.-P., and Thiel, G., The action potential in *Chara*: Ca^{2+} release from internal stores visualized by Mn^{2+}-induced quenching of Fura dextran, *Plant J.*, 13, 167, 1998.
114. Merrit, J. E., Jacob, R., and Hallam, T. J., Use of manganese to discriminate between calcium influx and mobilization from internal stores in stimulated human neutrophils, *J. Biol. Chem.*, 264, 1522, 1989.
115. Rizzuto, R. and Fasolato, C., Imaging living cells, *Springer Lab Manual*, 1999.
116. Williams, D. A. and Fay, F. S., Intracellular calibration of the fluorescent calcium indicator Fura-2, *Cell Calcium*, 11, 75, 1990.
117. Grynkiewicz, G., Poenie, M., and Tsien, R. Y., A new generation of fluorescent calcium indicators with greatly improved fluorescence properties, *J. Biol. Chem.*, 260, 3440, 1985.
118. Lebert, M. and Häder, D.-P., How *Euglena* tells up from down, *Nature*, 379, 590, 1996.
119. Lebert, M., Richter, P., Porst, M., and Häder, D.-P., Mechanism of gravitaxis in the flagellate *Euglena gracilis*, *Proc. C.E.B.A.S. Workshop. Annual Issue 1996*, 225, 1996.

Movement Analysis

CONTENTS

20.1 INTRODUCTION

Quantitative analysis by real-time computer image analysis has gained wide acceptance in many fields, ranging from cytology,[1-3] microbiology,[4,5] plant biology,[6-8] zoology,[9] microscopy,[10-12] and medicine[13-15] to meteorology,[16] oceanography,[17-20] ecology,[21,22] and earth sciences.[23] The tasks include the tracking of organisms,[24-26] the analysis of cloud movements, and the determination of animal migrations. Computer-aided analysis of motility and orientation has become an important tool in space research.[27-30] In medical research, cell tracking is used to follow the movements of lymphocytes[31] and sperm cells.[32] One ambitious goal is to recognize cancer cells and determine their aberrant behavior from their altered motility patterns.[33,34] The tracking algorithm is widely independent of the size of the objects, which can range from microscopic to celestial objects.

The rapid development of video technology and the even more dramatic improvements in computer hardware are the basis for real-time analysis of large numbers of tracks, which warrants high statistical significance and detailed data analysis.[35-37] In parallel, the development of mathematical approaches and computer algorithms allow researchers a more and more detailed extraction of relevant movement and orientation parameters.[38,39] Recently, a robot was constructed, which is capable of recognizing and touching moving objects; this task requires real-time movement analysis. The biologically inspired control systems produce strangely life-like movements.[40]

Although the human visual apparatus and brain are far superior to even the fastest computers, the latter have a number of advantages that make machine vision an attractive tool. In contrast to human observers, the hardware and software dedicated to cell tracking does not tire. In addition, machine vision is characterized by high objectivity. In a class experiment, students were asked to analyze the directionality of flagellate movements. For this purpose, one group of students was handed a videotape with cell tracks and asked to draw short movement vectors on an acetate sheet fixed on the video monitor. They were told that the cells show phototaxis and the light had impinged from the left side of the screen. At the end of the day, the students had produced a large amount of data clearly proving phototactic orientation of the cell population with respect to the light direction. This experiment was repeated with another group of students, who were asked to evaluate the same videotape and given the same explanation — with the only difference that the light source was said to have been on the right. Again, the results showed a high statistical significance for a positive phototaxis — only to the other side. Even in double-blind experiment, it is difficult for an experimenter to arrive at an objective analysis.

Computer-aided cell tracking, in contrast, is free of subjective bias. In addition, modern hardware and software are capable of following hundreds of objects in parallel and evaluating many cell form and movement parameters.[41] A new tracking software (WinTrack 2000, Real Time Computer, Möhrendorf, Germany) has been shown to track more than 12,000 cells per minute. This was of importance in the evaluation of as many as possible cell tracks from recent space experiments on parabolic rocket flights with limited duration on the order of a few minutes.[42-44] The parameters extracted from the analyzed objects and their paths include the size, shape, and position of the objects,[45,46] their linear and angular velocity, and their deviation from a predefined direction.

20.2 HARDWARE

The movement of a motile object can be recorded by a number of devices. The traditional tool is a video camera with a suitable objective. For light-sensitive objects, infrared cameras can be used.[47] Today, charge-coupled device (CCD) cameras are usually used to record the video image, the spectral sensitivity of which extends far into the infrared.[48] For unicellular organisms or microorganisms, the recording device is mounted on a light microscope. Most research microscopes can be equipped with a so-called C mount that facilitates the connection of standard CCD cameras (Figure 20.1). In order to increase the optical contrast of the organisms against the background, dark-field microscopy can be used, in which the organisms appear bright on a dark background.[49-51] In the case of larger microorganisms (such as ciliates) or multicellular organisms, the movement tracks can be followed with a macro lens. In most cases, cellular details are not of interest and only the position of the cells needs to be determined.

The spatial resolution should be at least as high as that of the subsequent digitizer (see below). Another limitation is the video frequency: full frames are recorded at a video frequency of 25 Hz in most parts of the world (30 Hz is the American standard). Each full frame consists of two half frames with interlaced video lines that are transmitted at double the frequency. During online analysis, each video frame must be digitized and analyzed in no more than 40 ms (33 ms, U.S. standard). Thus, the analysis is limited to events that are slower than this video frequency. For example, it is not possible to follow the beating pattern of flagella or cilia that beat at a frequency of about 50 Hz. One way around this limitation is to use faster cameras and to store the images either electronically or on tape. The analysis is then performed offline from the recorded images.

In very fast movements, analysis can be difficult because the translocation of the objects can be so far that the identification of the object fails, especially at high densities of moving objects. A correspondence can only be determined if the movement vectors are smaller than the mean distance between the objects.[52] If the object distances are statistically distributed, misinterpretations cannot be excluded.

FIGURE 20.1 Schematic setup for tracking of microorganisms. The image of the moving cells is recorded by a CCD camera on top of a microscope and digitized in a frame grabber plugged into a slot of a PC that has access to the digitized image in memory.

Another limitation is the dynamic range of the recording device, that is, the brightness difference between the brightest and darkest parts in the image. In the time domain, this problem is partially offset by the automatic gain control (AGC), which electronically changes the amplification when the image is too bright or too dark. This automatic adjustment can also be a problem, for example, when a few dark objects are recorded on a bright background. Then, the AGC circuit lowers the amplification so that some structure is seen in the bright background, which may not be desired, while details in the objects of interest are lost.

Whatever the source of the video image may be, in the next step the sequence of images needs to be digitized. For this purpose, the image is divided into an array of equally sized elements called pixels. Typical spatial resolutions range from 512×512 to 2048×2048 pixels. The spatial resolution is defined by the digitization frequency. In order to produce a 512×512 pixel resolution, a frequency of 10 MHz is necessary, for 768×512 pixel resolution, the frequency needs to be 15 MHz. The image information must be stored in electronic form at least during analysis by the software. An image of 1024×1024 pixels needs a memory of 1 Megabyte (MB). Therefore, it is easy to see that a substantial amount of memory is necessary to record longer video sequences. The brightness in each pixel is assigned a digital value. This is done in a video analog/digital (A/D) flash converted (digitizer). The digitizer has also a limited dynamic range of, for example, 64 or 256 gray values.[53] The darkest black is arbitrarily set to a gray value of zero, and the brightest to the maximal value (e.g., 255). Likewise, color can be digitized using a combination of three colors, which results in three times larger image files.[54]

In the beginning of video digitization, these devices were large stand-alone electronic instruments, and they took up to several minutes to digitize an individual image, which is of course prohibitive in real-time image analysis.[55] Today, digitizers have shrunk to the size of a card that fits into the slot of a standard PC (Figure 20.2) These frame grabber cards are commercially available, for example, from Matrox (Dorval, Quebec, Canada), Imaging Technology, Inc. (Woburn, MA), Delta-T Devices (Cambridge, England), Matrix (Oppenweiler, Germany), Synopsis (Cambridge, England), Tecnom (Medon-la-Fôret, France), Digihurst Ltd. (Newark Close, Royston, Britain), and Bio-Rad (Düsseldorf, Germany). Complete image analysis systems are available from,

FIGURE 20.2   The PCimage-SG/SGVS frame grabber card. (Matrix Vixion GmbH, Oppenweiler, Germany. With permission.)

for example, Real Time Computer (Möhrendorf, Germany, Universal Imaging Corporation (West Chester, PA), Clemex (Quebec, Canada), AI Tektron (Düsseldorf, Germany), and AVS/UNIRAS (Birkenrod, Denmark). While in most cases, cell tracking can be performed in black-and-white, most modern frame grabbers also perform video digitization of color video.

20.3   IMAGE MANIPULATION

Before analysis, the image can be enhanced, which is facilitated by the numerical representation of the image in memory. Table 20.1A shows part of a digitized image. The content of each individual memory cell and thus the image can easily be manipulated. The brightness can be increased or decreased by simply adding or subtracting a constant number to or from each gray value, respectively. Similarly, a horizontal and/or vertical ramp can be defined to compensate for a linear spatial offset in the image. Multiplication or division of each pixel with a constant enhances or reduces the contrast. Thresholding is another useful technique to suppress the background below or above a predefined value, which removes noise.

A multitude of mathematical filters have been developed to improve image quality.[56,57] Smoothing can be achieved by replacing each pixel P by the mean value P' calculated from the original value and that of its eight neighbors, which smoothes sharp boundaries and reduces noise.[58] This calculation is repeated for each pixel in the image. The outer pixels, which do not have eight neighbors, are copied into the new image.

$$
\begin{array}{ccc}
A & B & C \\
D & P & E \\
F & G & H
\end{array}
\qquad (20.1)
$$

$$
P' = (P + A + B + C + D + E + F + G + H)/9 \qquad (20.2)
$$

TABLE 20.1A
Part of a Digitized Image Represented by Gray Values, Which Corresponds to the Indicated Part in Figure 20.3a

73	60	54	51	67	62	50	67	75	79	72	57	78	63	54	60	52	58	76	
59	66	60	68	78	69	75	70	59	59	75	69	74	69	57	70	77	77	77	
49	47	46	63	67	53	65	69	64	75	61	69	81	71	64	66	78	67	72	
74	88	85	255	73	37	26	42	59	55	52	53	69	74	63	81	84	86	96	
80	70	62	60	57	57	56	68	76	59	35	59	73	71	30	50	78	83	89	
75	91	82	55	61	57	63	82	88	74	60	67	74	73	51	71	89	77	75	
88	74	61	51	56	75	75	92	75	69	60	82	78	57	67	90	97	91	79	
66	80	94	83	58	50	56	75	67	66	57	68	75	83	81	90	82	72	69	
71	63	54	41	48	71	56	66	67	64	59	53	61	57	72	79	72	67	75	
81	71	52	43	50	44	52	70	69	74	48	59	59	59	55	61	86	87	85	
70	72	61	52	52	50	43	65	71	57	48	53	70	75	64	48	59	58	42	
59	56	50	51	72	64	49	36	61	92	82	68	56	47	48	51	66	60	34	
73	64	79	60	68	71	58	51	50	79	72	51	54	70	76	77	56	24	0	
82	91	67	45	81	89	71	69	69	69	63	54	68	47	25	20	35	8	0	
93	57	49	52	73	64	53	69	85	69	67	60	39	0	0	0	0	0	0	
91	71	71	73	81	65	59	56	63	57	63	50	11	0	0	0	0	0	0	
87	84	72	67	76	59	45	47	47	60	84	54	20	0	0	0	0	0	0	
89	64	47	30	52	60	74	79	85	76	61	38	0	0	0	0	0	0	0	
90	76	65	56	73	97	103	84	93	81	48	11	0	0	0	0	0	0	0	
94	77	54	67	88	106	122	98	59	32	20	17	0	0	0	0	0	0	0	
58	53	60	44	72	86	67	51	24	0	9	3	0	0	0	0	0	0	0	
22	30	52	50	61	75	57	24	0	0	0	0	0	0	0	0	0	0	0	
0	0	30	39	62	67	33	20	0	0	0	0	0	0	0	0	0	0	0	
0	0	0	17	62	48	25	0	0	0	5	0	0	0	0	0	0	0	0	
0	0	0	0	0	0	0	0	0	0	0	0	0	0	0	0	0	0	0	
17	12	0	0	0	3	3	0	1	0	0	0	0	0	0	0	0	0	0	
7	0	0	0	11	24	32	25	10	2	0	0	0	0	0	0	0	0	0	
45	10	0	0	0	31	32	43	37	18	7	19	8	0	0	0	0	0	0	
44	27	0	0	0	0	23	33	37	24	14	16	0	0	0	0	0	0	0	
57	43	44	14	0	0	4	41	51	31	7	0	0	0	0	0	0	0	0	

Note: Note the error in Row 4, Column 4.

Each number represents the gray value at this image position. Figure 20.3a shows the corresponding image represented by the pixel matrix in Table 20.1A. There is an error in the image due to pixel noise in row 4, column 4 which shows up as a bright pixel in the image.

Table 20.1B shows the altered gray-value matrix after smoothing, and Figure 20.3b shows the result of the operation. The pixel error has been smoothed but the sharp borderline between dark and bright areas is also smoothed. A variant of this technique is a hat filter, which weights the pixels before smoothing according to the following scheme:

$$\begin{matrix} 1 & 2 & 1 \\ 2 & 4 & 2 \\ 1 & 2 & 1 \end{matrix} \qquad (20.3)$$

Laplace or Sobel algorithms are more advanced mathematical techniques based on calculations also involving 3×3 (or higher) matrices of pixels gliding over the image.[59] Each pixel in the derived image is calculated from its original value and that of its neighbors. During Laplace filtering,

TABLE 20.1B

Same Part of a Digitized Image After a Smoothing Operation, Which Corresponds to the Indicated Part in Figure 20.3

61	58	59	64	68	69	67	67	69	73	75	76	69	64	62	65	68	71	73
59	57	57	61	64	65	64	66	68	68	68	70	70	67	63	64	67	70	69
62	63	86	88	84	60	56	58	61	62	63	67	69	69	68	71	76	79	75
69	66	86	85	80	54	52	58	63	59	57	61	68	66	63	66	74	81	79
79	78	94	87	79	54	54	62	67	62	57	60	68	64	62	66	77	84	83
80	75	67	60	58	61	69	75	75	66	62	65	70	63	62	69	80	84	79
77	79	74	66	60	61	69	74	76	68	67	69	73	71	73	79	84	81	72
75	72	66	60	59	60	68	69	71	64	64	65	68	70	75	81	82	78	69
73	70	64	58	54	53	60	64	68	63	60	59	63	66	70	75	77	77	68
73	66	56	50	50	51	57	62	67	61	57	56	60	63	63	66	68	70	57
70	63	56	53	53	52	52	57	66	66	64	60	60	59	56	59	64	64	46
68	64	60	60	60	58	54	53	62	68	66	61	60	62	61	60	55	44	24
71	69	62	63	66	69	62	57	64	70	70	63	57	54	51	50	44	31	14
75	72	62	63	67	69	66	63	67	69	64	58	49	42	35	32	24	13	3
78	74	64	65	69	70	66	66	67	67	61	52	36	21	10	8	7	4	0
80	75	66	68	67	63	57	58	61	66	62	49	26	7	0	0	0	0	0
83	75	64	63	62	63	60	61	63	66	60	42	19	3	0	0	0	0	0
82	74	62	59	63	71	72	73	72	70	57	35	13	2	0	0	0	0	0
77	72	59	59	69	86	91	88	76	61	42	21	7	0	0	0	0	0	0
66	69	61	64	76	90	90	77	58	40	24	12	3	0	0	0	0	0	0
47	55	54	60	72	81	76	55	32	16	9	5	2	0	0	0	0	0	0
24	33	39	52	61	64	53	30	13	3	1	1	0	0	0	0	0	0	0
8	14	24	41	53	54	38	17	4	0	0	0	0	0	0	0	0	0	0
0	3	9	23	32	33	21	8	2	0	0	0	0	0	0	0	0	0	0
4	3	3	8	14	15	8	3	0	0	0	0	0	0	0	0	0	0	0
11	4	1	1	4	8	9	7	4	1	0	0	0	0	0	0	0	0	0
26	10	2	1	7	15	21	20	15	8	5	3	3	0	0	0	0	0	0
36	14	4	1	7	17	27	30	25	16	11	7	4	0	0	0	0	0	0
48	30	15	6	5	10	23	33	35	25	15	7	4	0	0	0	0	0	0
61	50	36	22	9	6	13	26	33	29	20	10	4	0	0	0	0	0	0

Note: The error in Row 4, Column 4 has been smoothed out.

each pixel is multiplied by a specific number (kernel), which is 8 for the most common algorithm, and the values of all its surrounding eight neighbors are subtracted from this value (Table 20.1C).

$$
\begin{matrix}
-1 & -1 & -1 \\
-1 & 8 & -1 \\
-1 & -1 & -1
\end{matrix}
\tag{20.4}
$$

Laplace filtering results in a reduction in brightness of all areas with similar gray levels, while sharp boundaries show up as bright lines. Other kernel numbers result in completely different manipulations. The Sobel algorithm calculates the difference between the gray values of the upper and lower neighbors of a pixel, that of the left and right neighbors as well as for the diagonals, and replaces each pixel by the highest difference in the gray values of opposite pixels. By this method, the Sobel algorithm highlights not only horizontal and vertical but also diagonal edges (Figure 20.3c).

TABLE 20.1C
Same Part of a Digitized Image After Applying the Sobel Algorithm, Which Corresponds to the Indicated Part in Figure 20.3c

50	42	26	42	52	62	46	42	46	48	86	78	58	60	30	40	52	32	30
58	58	34	34	44	46	28	46	52	52	48	34	44	62	34	28	28	26	62
42	42	156	148	150	82	72	58	46	46	58	38	44	64	32	36	40	42	60
54	46	255	42	194	74	30	34	40	74	52	50	38	84	52	58	38	30	50
50	64	132	190	182	38	64	70	58	64	28	46	26	90	58	70	36	56	72
54	56	78	58	26	24	40	30	56	68	38	48	46	66	42	78	44	64	78
54	66	76	56	28	30	46	52	84	80	34	32	56	52	38	46	58	76	82
62	54	60	66	46	52	42	60	74	64	50	44	60	38	42	56	92	84	90
58	74	102	90	44	32	30	34	46	62	50	42	54	62	54	58	64	44	72
64	68	62	30	26	50	38	38	48	68	56	24	26	40	52	48	44	66	118
66	74	62	22	34	36	30	56	38	56	48	26	34	50	60	38	52	104	152
62	48	52	28	30	46	46	30	56	24	52	50	48	52	40	26	48	100	106
40	50	42	22	44	50	64	42	56	44	98	68	38	38	46	58	108	120	78
48	70	92	34	44	64	52	32	28	46	70	54	76	112	148	138	104	76	20
64	96	60	32	32	88	66	36	54	62	62	90	120	102	60	52	42	28	4
64	66	44	26	50	76	52	48	62	46	60	106	108	46	0	0	0	0	0
74	88	86	64	68	70	38	40	44	26	60	118	94	20	0	0	0	0	0
86	98	90	42	42	62	88	94	70	56	114	132	74	12	0	0	0	0	0
68	84	50	52	80	86	74	70	90	124	126	98	44	0	0	0	0	0	0
72	96	70	40	82	50	106	152	162	136	94	52	20	0	0	0	0	0	0
102	82	60	42	84	94	158	174	138	74	46	30	12	0	0	0	0	0	0
92	96	62	38	64	86	124	116	66	22	8	8	2	0	0	0	0	0	0
44	74	94	84	56	94	120	76	28	2	2	2	0	0	0	0	0	0	0
4	20	56	108	112	110	98	52	12	2	0	2	0	0	0	0	0	0	0
26	18	2	52	82	86	48	18	0	2	2	2	0	0	0	0	0	0	0
44	16	8	6	22	44	54	44	24	8	0	0	0	0	0	0	0	0	0
102	46	14	6	38	38	66	72	68	42	28	22	18	4	0	0	0	0	0
108	76	24	6	44	58	30	18	54	58	36	20	22	4	0	0	0	0	0
90	90	76	38	12	56	70	44	38	64	26	26	28	4	0	0	0	0	0
84	112	120	108	58	4	64	66	32	48	44	36	26	0	0	0	0	0	0

Note: All uniform areas are converted to dark and all edges are outlined in white.

20.3.1 LOOKUP TABLES

Figure 20.4a shows part of a black-and-white image. The pixel matrix can be manipulated by modifying the numeric values in memory. For this purpose, lookup tables (LUT) can be defined[60] according to which the individual gray values are substituted (Table 20.2). Row A shows one application to convert a positive image into its negative: each 0 is substituted by a 255, each 1 by 254, etc. The result of this procedure — applied to the image in Figure 20.4a — is shown in Figure 20.4b. In most cases, the LUT is defined in software but the manipulation of the gray values is done by hardware to allow real-time manipulation.

Another example is binarization (Table 20.2B): each gray value below a certain threshold is set to zero (black) and the rest is set to 255 (white). This procedure increases the contrast and can be used to segment the image, that is, extract the objects (bright) from the background (dark). Figure 20.4c shows the result of this procedure. The human visual apparatus cannot easily distinguish similar gray values. Therefore, a certain range of gray values of interest can be spread out, while background gray values are set to 0 or 255 (Table 20.3C, Figure 20.4d).

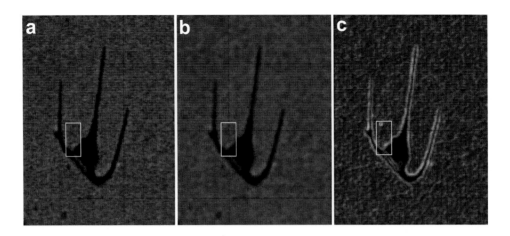

FIGURE 20.3   (a) Black-and-white image of the dinoflagellate *Ceratium*. The indicated part of the image is represented in Table 20.1A in numerical form. (b) Result of a smoothing operation using a 3 × 3 matrix. (c) Result of a Sobel operation; subsequently, the figure was smoothed and the contrast enhanced.

The human eye can distinguish colors more easily. Therefore, a pseudo-color technique is often used (Table 20.3D). In this approach, certain gray values are mapped to the color channels of the monitor. For example, the values 0 to 85 are sent to the blue channel with increasing color intensity. The values from 86 to 170 are sent to the green channel, and the remainder (bright pixels) to the red channel. This is obviously only one possible color combination. The color ranges can overlap to produce, e.g., yellow or orange. Using an 8-bit definition for each color channel allows a combination of over 16 million colors. True color digitization requires separate treatment of the three colors. Of course, these can also be manipulated by LUTs, which are used to modify the color presentation of photographic images using specific image software.

20.4   SOFTWARE

Figure 20.5 shows the initial screen of a modern tracking software (WinTrack 2000, Real Time Computer, Möhrendorf, Germany). This software was developed in Visual Basic and takes advantage of all the Windows features, including 3-D appearance, communication boxes, buttons, etc.

Most of the screen is devoted to the display of the digitized image, which can show the stream of images or a snapshot that can be frozen at any time by clicking on a toggle in the Video Settings menu. In the snapshot mode, the user can binarize the image to test which areas of the image are currently recognized as objects and which as background. Another toggle allows selection of either bright-field mode (the software detects dark objects on bright background) or dark-field mode (bright objects on dark background).

Initially, the image is adjusted in focus and brightness using the optical controls on the microscope. In addition, the video settings can be adjusted by sliding the brightness control and the contrast control or by entering numerical values. The result of this procedure can be followed visually on screen or more precisely by calculating a gray-value histogram (see below). This histogram is updated online so that one can see the change in the histogram as one adjusts the brightness and control sliders.

A zoom function is implemented to visualize details of the objects. The image can be magnified by integer factors of up to 16 times or made smaller by the same factor. This procedure does not change the pixel resolution (Figure 20.6), nor does it affect the evaluation of movement parameters which is performed on the full image regardless of whether it is displayed fully or only partially.

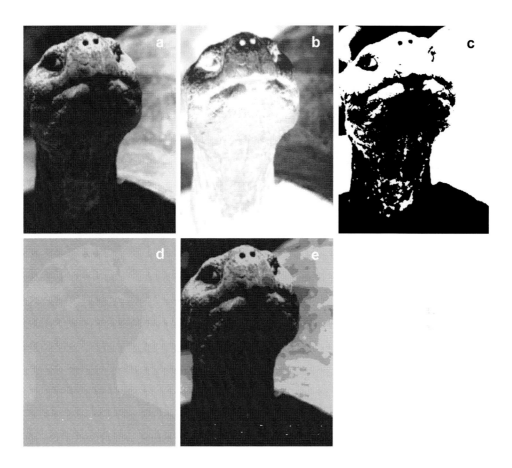

FIGURE 20.4   (a) Black-and-white image of a turtle head. (b) The same image after inversion by application of the LUT defined in Table 20.2A. (c) Result of a binarization using the LUT defined in Table 20.2B. (d) The same image taken with very poor contrast; and (e) after stretching the central range of gray values using the LUT defined in Table 20.2C.

Another feature of the program is to save images on the hard disk of the computer under a user-selected name or retrieve images for subsequent analysis.

20.4.1 CALIBRATION

Most tracking systems operate on the basis of pixels.[61] To measure sizes and velocities in physical units, the system must be calibrated. For this purpose, a calibration tool such as an object micrometer or yardstick is inserted into the field of view. Two positions of the calibration device are marked by a mouse click and the computer calculates the distance in pixels. This is compared with the corresponding physical distance, from which a calibration factor is derived. This calibration factor is stored internally and used in the next calibration. In addition, the depth of the measurement cuvette can be indicated, so that the program can calculate, for example, true cell densities per milliliter. A scale bar is automatically inserted in the digitized video image and the description of the hardware configuration (e.g., microscope and objective magnification) is listed in the status bar

TABLE 20.2
Lookup Tables Used to Substitute the Gray Values in an Image
by a Different One

	A	B	C	D	
0	255	0	0	0	
1	254	0	0	3	
2	253	0	0	6	Blue
3	251	0	0	9	
4	250	0	0	12	
⋮	⋮	⋮	⋮	⋮	
124	131	0	31	118	
125	130	0	63	121	
126	129	0	95	124	
127	128	255	137	127	Green
128	127	255	169	130	
129	126	255	201	133	
130	125	255	233	136	
131	124	255	255	139	
⋮	⋮	⋮	⋮	⋮	
251	4	255	255	243	
252	3	255	255	246	
253	2	255	255	249	Red
254	1	255	255	252	
255	0	255	255	255	

Note: A: The values of the digitized image (Figure 20.4A) are inverted (result in Figure 20.4B. B: Binarization of the image by setting all gray values below 127 to 0 and all others to 255 (result in Figure 20.4C). C: Stretching the central gray values (124–130) over a larger range results in an increase in image contrast; all other gray values are set to 0 or 255, respectively. D: Sending low, intermediate, and high gray values to different color channels results in a pseudo-color representation of the b/w image.

at the bottom of the screen. The status bar also indicates the current data and time and the path for the storage.

The video image is not square, but has a ratio of 4:3 for width and height. If the full screen is mapped on a square pixel matrix, the individual pixels are not square — which can present a problem in the analysis of the form, the velocity, and the angular deviation of moving objects. The digitizer card (Meteor PPB, Matrox, Dorval, Quebec, Canada) used for the WinTrack program operates with square pixels; thus, the measured distances in x- and y-directions should be equal. However, there may be distortions by the video camera used. For example, a circle may be distorted to an ellipse. To test for video distortions, a square or circular object is recorded and the diameters in horizontal and vertical directions are measured. If they are not identical, the software calculates a correction factor that is used to correct the X/Y ratio.

20.4.2 IDENTIFICATION OF OBJECTS

A simple, fast, and therefore — especially in real-time image analysis — successful method is the segmentation of an image using a pixel-oriented threshold technique. This is based on the simple assumption that there are two regions in the image that can be attributed to the objects and the

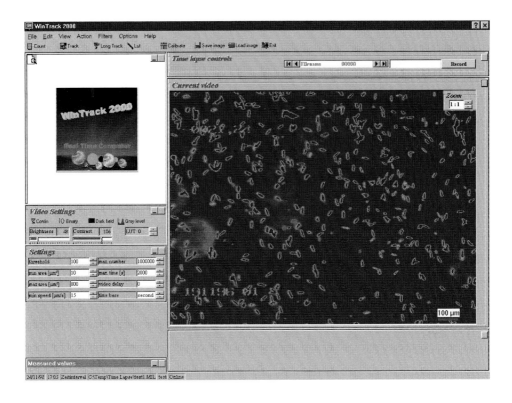

FIGURE 20.5 Initial screen of WinTrack 2000 (Real Time Computer, Möhrendorf, Germany).

background, respectively. Ideally, the gray values of the two regions follow a normal distribution and show two local maxima in the gray-value histogram (Figure 20.7).

The minimum between the two maxima represents the gray value that separates those of the objects (low gray values in bright-field mode) and the background (high gray values). This local minimum between these two maxima can be used as a threshold; all pixels with gray values below the threshold are regarded as belonging to objects, and all pixels with gray values above the threshold as belonging to the background (in bright-field mode). This distinction results in a binary image in which all object pixels can be assigned the value 1, and all background pixels the value 0. This technique can be applied to all images with an even background and clear object boundaries.

If the background is not homogeneous, a technique using a dynamically adjusted background is employed. A human observer recognizes an object by its contrast to the background in its vicinity. To mimic this approach, the algorithm searches for an abrupt change in the gray value as it scans the image, indicating the presence of an object. Then it determines the gray value in the vicinity of the object and calculates a threshold to identify an object in this area of the image. This threshold is adaptively changed as the software identifies objects in other areas. In addition, the evenness of a region can be improved by smoothing operations. Another approach is to subtract an initial image from the subsequent images to remove the uneven background. By this technique in particular, moving objects are recognized.

There are several methods to identify the area and contour of an object in the image.[62-64] Once a gray-value threshold is defined on the basis of a gray-value histogram, the image is scanned line by line from the top left corner until a first pixel is hit, the gray value of which is above the predefined threshold.[65,66] Assuming that this is an edge pixel of an organism, the 8 neighbor pixels are tested for being object pixels. The x- and y-coordinates of those pixels identified as object pixels

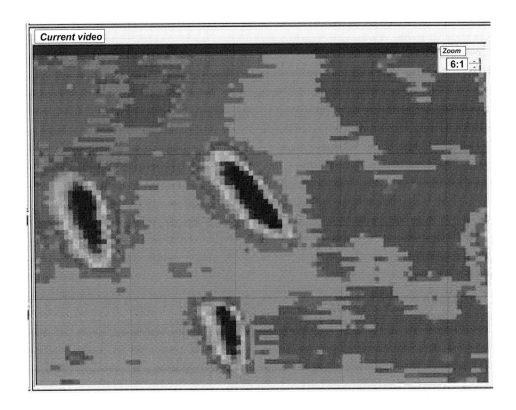

FIGURE 20.6 Zoomed image of flagellates showing the pixel structure.

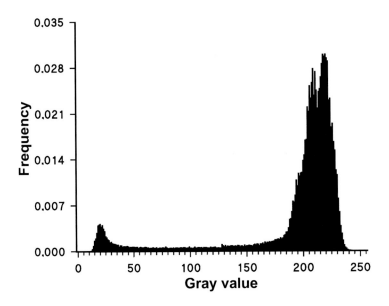

FIGURE 20.7 Histogram of the gray-value distribution in an image of dark organisms on a bright background.

are recorded and, for each new object pixel, this operation is repeated (Figure 20.8). This procedure warrants that only those pixels are identified that are connected, forming one unique object. Identified object pixels are set to a reserved gray value or color so that they are not queried again. The process is continued until the growing area of identified pixels has completely filled the area of the object. Especially with larger objects, this method is time-consuming.

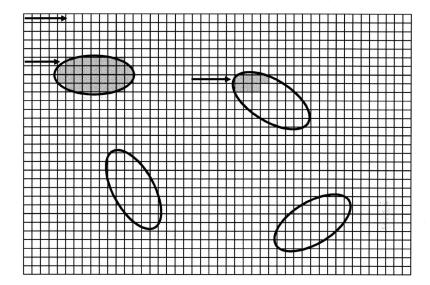

FIGURE 20.8   Filling algorithm to determine the area of objects.

Alternative methods utilize edge detection algorithms.[67,68] This simple technique also starts by scanning the image line by line until an organism pixel has been found. The x- and y-coodinates of this edge pixel are stored and the neighboring edge pixel in clockwise direction is identified by analyzing the eight neighbors of this first one, also in clockwise direction. An edge pixel is defined by the fact that in a number of connected pixels, there is at least one neighbor that does not belong to the object. A transition from a background pixel to an object pixel signals that the next edge pixel has been found. This procedure is repeated until the first edge pixel is hit again and thus the outline has been determined. During calculation, the extreme x- and y-coordinates in the outline are established, from which the geometric center of gravity can be calculated.[69]

A faster variant of this approach is the chain code algorithm.[70] A chain code is a sequence of short contour elements coded in numerical form that uniquely describes the outline of an object in a digital image. The directions from the central pixel P to is eight neighbors are defined by the numbers 0 through 7 in a counterclockwise fashion, starting to the right of the pixel.

The technique also starts with a line-by-line search until an edge pixel P has been detected (Figure 20.9). Starting from the current scan direction rotated two positions counterclockwise, the neighboring pixel is tested for being an edge pixel in a clockwise direction. This is performed by a modulo 8 addition of 2. Assuming that the first chain code element was defined as 1, the next tested (possible) neighbor is searched in direction 3. If the tested neighbor is not an organism pixel, the shift is decremented (modulo 8) and the search is continued in a clockwise fashion until an organism pixel is found. Now the new chain code element is calculated by adding this shift (modulo 8) to the previous value. In a clockwise direction, the maximal direction change is a shift of 4 (180°). If the shift value falls below 4, there are no neighbors. The whole outline is scanned until

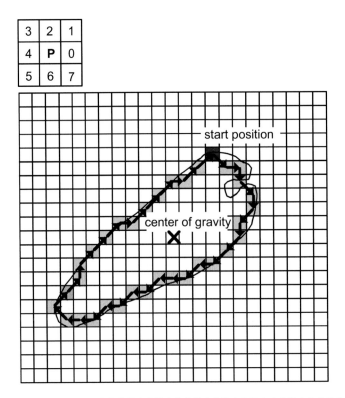

Chain: 7,0,6,7,6,5,6,5,5,4,5,4,4,5,4,5,4,5,4,4,3,1,1,2,1,1,1,0,1,1,1,0,1,1

FIGURE 20.9 Schematic diagram exemplifying the chain code algorithm used to analyze the outline of an object.

the first edge point is found again, with this routine performed iteratively. The determined directions to the next edge points are stored in the growing chain of elements.[71] The advantage of this robust technique is that it is very fast and allows for a drastic reduction in the amount of information. One disadvantage is that, in contrast to the object fill method, "holes" in objects are not detected.

20.4.3 AREA AND POSITION DETERMINATION

From the chain code, the length of the outline, the area of the enclosed object, and the coordinates of the gravicenter can be calculated using equations developed for this purpose. The length of the outline is given by:

$$L = T\left(n_e + n_o\sqrt{2}\right) \tag{20.5}$$

where n_e is the number of even links, n_o is the number of odd links, and T is a scaling factor that depends on the optical magnification. The area A of the two-dimensional projection of the organism can be determined from the integration of the chain elements with respect to the x-axis:

$$A = \sum_{i=1}^{n} a_{ix}\left(y_{i-1} + 0.5a_{iy}\right) \tag{20.6}$$

In order to calculate the gravicenter, the first moments about the *x*- and *y*-axes are needed; about the *x*-axis,

$$M_1^x = \sum_{i=1}^{n} 0.5 a_{ix} \left[\left(y_{i-1} \right)^2 + a_{iy} \left(y_{i-1} + 1/3\, a_{ix} \right) \right] \tag{20.7}$$

The first moment about the *y*-axis is determined by rotating the chain by $\pi/2$ about the origin and then employing the same equation. The centroid with the coordinates x_c and y_c is calculated using the first moments and the area:

$$x_c = -M_1^y / A \tag{20.8}$$

$$y_c = -M_1^x / A \tag{20.9}$$

All calculations can be performed as integer operations, which facilitates real-time analysis. All edge pixels are arbitrarily set to 255 (bright white) after they have been identified. Thus, they serve as markers for already-found organisms. Starting from the first found edge pixel of the previously calculated organism, the search is continued until the whole frame has been analyzed and all organisms have been found.

20.4.4 CELL COUNTING

Cell counting is routinely performed in many medical and biological laboratories to count colonies, microorganisms, or macroscopic objects such as leaves.[72] The WinTrack 2000 software contains a module to count objects (Figure 20.10). The cell suspension is transferred into a cuvette of known depth. Motile cells do not need to be fixed or immobilized for this purpose because digitization is done in real-time. Small holes in the objects can be filled before counting. The calibration procedure described above warrants that areas and lengths be calculated in the appropriate physical units, and even cell densities are calculated. The software provides a Video Settings window in which the user can define a lower and an upper limit for the area of objects. Objects that do not fit in this size range are ignored. All others are outlined in a different color in the overlay over the image to allow the user to control if all desired objects are found. The software then determines a size distribution of the identified objects in ten equal size classes. Multiple counting is possible with user-defined number of counts. Each count is acknowledged by the user, so that he/she has time to, for example, change the position of the sample in the cuvette. The measured parameters are number of objects, mean area, and mean form factor, and are shown in the Measured Values window. The form factor is a parameter to distinguish, for example, round from long objects. It is calculated from the ratio of the squared perimeter L and the area A adjusted to the area of a circle:

$$\left(2\pi r \right)^2 / \pi r^2 = 4\pi \tag{20.10}$$

$$F = \frac{L^2}{A4\pi} \tag{20.11}$$

In addition, the length and width parameters can be selected optionally. The data can be stored in an ASCII file for further analysis using spreadsheet programs (e.g., Excel). The user can also enter a headline into the data file to describe the individual experiment. Here, the individual data

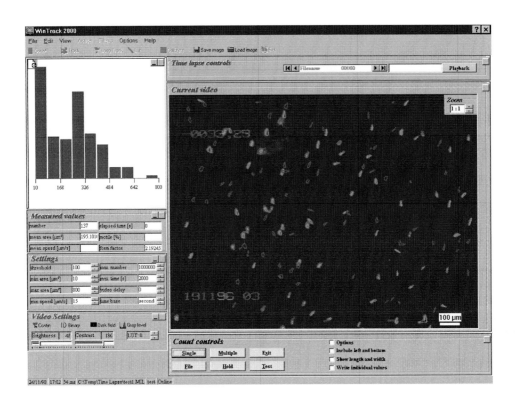

FIGURE 20.10   Count module in the WinTrack 2000 software. A size distribution of the selected objects is shown in a dedicated window.

for each object is listed. In addition, the data and time, the selected video settings (brightness, contrast, bright/dark field selection, LUT, threshold, selected minimal and maximal size), as well as the cell density per milliliter (ml) are recorded. During manual counting using, for example, a Thoma chamber, it is conventional to ignore those objects that touch the left side or the bottom of a cell.

When objects are in direct contact with others, errors in counting occur.[73] When the average size of the objects is known, bigger ones can be rejected or the area can be divided when it is outside predefined boundaries, which may still be inaccurate — especially in dense suspensions. To solve this problem, an erosion technique is applied: shells 1 pixel wide are removed and set to a specific gray value around the circumference of each object. This process is repeated until connecting areas are separated at the isthmus. Subsequently, the shells are blown up (dilated) again to their original value. However, a gap at least 1 pixel wide is left between adjacent objects. Objects can also be separated manually before analysis using the cursor controls, a mouse, a digitizing tablet or a touch-sensitive screen.

Cell counting was recently applied in an ecological approach to determine the densities in populations of marine phytoplankton. For this purpose, the cell suspensions were filled in a 3-m Plexiglas column and samples were taken from 18 evenly spaced lateral outlets using a peristaltic pump capable of handling 18 samples in parallel. Samples were taken at regular time intervals to determine the distribution of the population over the day (Figure 20.11). The position of the cells was controlled by light and gravity and followed an endogenous rhythm.

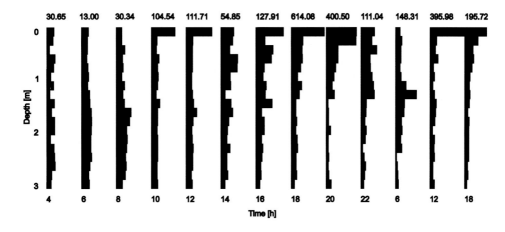

FIGURE 20.11 Linear histograms of the cell density distribution in the water column during 2 days. The numbers above the histograms indicate the Chi-square values. (After Häder and Eggersdorfer, 1991.[18])

20.4.5 ORGANISM TRACKING

Movement of objects relative to the recording device can be recorded as a change in the gray value or color at a given position with time. The following discussion concentrates on gray images; however, the same algorithms can be used for color images. The gray value is thus a function of time and space.

$$f(x, y, t) = f(x + dx, y + dy, t + dt) \tag{20.12}$$

Movement can also be regarded as a stream of gray-scale patterns through the image, which has been termed "optical flow."[74] Keep in mind that the image is only a two-dimensional representation of three-dimensional objects. If one disregards rotational and deformational changes, one is left with translational movements, which can be detected in the two-dimensional projection by the change in the position (x, y-coordinates) of the centers of gravity (centroids).[75]

Another problem occurs when objects meet or partially or totally occlude each other. A related problem occurs when touching objects move apart; in this case, no correlation to previous frames can be found. These difficulties can be aggravated by shadow images, nonuniform irradiation in the image, and/or movement perpendicular to the plane of projection.

Two different methods can be distinguished for following the movement of objects in a sequence of images:[76]

- Determination of the optical flux by analysis of the difference between subsequent images
- Calculation of corresponding image segments

In the first approach, which has been used to study street scenes,[77] an image is digitized and stored in the video memory. Then the next image is digitized and the first is subtracted from the second one on a pixel-by-pixel basis.[78] Provided no changes have occurred between the two frames, the difference image contains only zero gray values (i.e., it is black).[79] In reality, there are always small differences due to thermal noise and digitization errors, but basically the image information is lost. If a change in image information has occurred between the two frames, the objects that changed position show up as bright areas. In practice, this procedure is modified by initializing the whole image matrix with zero values. An element in the matrix is incremented when the gray values of corresponding pixels differ between subsequent frames. The calculated first-order difference picture (FODP) shows the traces of the moving objects in the full width of the object. The tracks can be extracted by the technique of skeletoning, during which shells of 1-pixel width are removed iteratively until only a line 1 pixel wide remains. From the FODP, a second-order difference picture (SODP) can be calculated by defining those elements in the pixel array as 1 where elements of different values are found in the FODP. The SODP can therefore be used to determine the start and end positions of tracks. Due to the large number of pixels in the array, these calculations that involve all picture elements in the image are rather time-consuming and can be performed in real-time only with dedicated hardware.

The second method is based on the determination of corresponding image segments in subsequent frames. The position and form of each object (or a limited number of objects for faster calculation) are determined in each frame of the sequence and stored for the following analysis.[80,81] Using complex search algorithms allows one to identify each object in subsequent frames so that the movement vectors can be determined.[82] One method for identifying the individual objects is the cross-correlation method by Smith and Phillips,[83] which, however, has the disadvantage that it is rather time-consuming. Takahashi and Kiobatake[84] have increased the computational speed by reducing the correlation matrix to 32×32 pixels between subsequent frames, centered around the object found in the first image. The maximum of the correlation matrix is equal to the dislocation of the object between two images. In a method developed by Nobel and Levine,[85] the positions of objects in subsequent frames are stored in an array. Then the movement vectors are calculated by the nearest neighbor algorithm. Aggrawal and Duda[86] describe a method for the movement analysis of polygons that can even overlap partially, based on the detection of corresponding vortices. An algorithm to determine the translation, rotation, and dilation of moving objects was developed by Schalkoff and McVey.[87] Only in complicated cases with, for example, overlapping filamentous organisms is interactive image analysis necessary.[88]

Using subsequent frames recorded at frequent and regular time intervals allows to follow movement vectors of organisms in the time domain.[89-94] Likewise, the growth of an organism or organelle can be followed.[95-99] The objects can range from individual cells such as flagellates or ciliates[100-104] to multicellular organisms.[105]

20.4.6   QUANTIFICATION OF ORIENTATION AND MOTILITY

This section concentrates on the movements of whole organisms and disregards the analysis of ciliary and flagellar beat patterns,[106-109] which are covered in other chapters in this volume.

The track module in the WinTrack 2000 software determines the movement parameters of motile objects on the basis of short track segments (minimally 160 ms). For this purpose, five subsequent frames from the incoming video sequence are digitized and stored in memory. Next, all objects in the first image are identified using the algorithms described above for object detection. Then the positions of the cells are determined in the second image. However, in this case, the image is not scanned line by line, but the organisms are searched starting at the gravicenters found in the first image. If an object has been identified at or near the position of the corresponding object in the first image, it is accepted provided there are no large discrepancies in the areas. A considerable increase in the area is a sign that two organisms meet during the process of tracking and a sudden decrease indicates that the object has moved out of the focus. If there is no or more than one object in the search area, the organism is regarded as lost and not considered for the rest of the calculation. The procedure is repeated for the third, fourth, and fifth frames. For each object, vectors are calculated from the gravicenters in the first and fifth frames (Figure 20.12). These vectors are used to calculate the distance s the objects have moved:

$$s = \sqrt{\left(y_2 - y_1\right)^2 + \left(x_2 - x_1\right)^2}$$

(20.13)

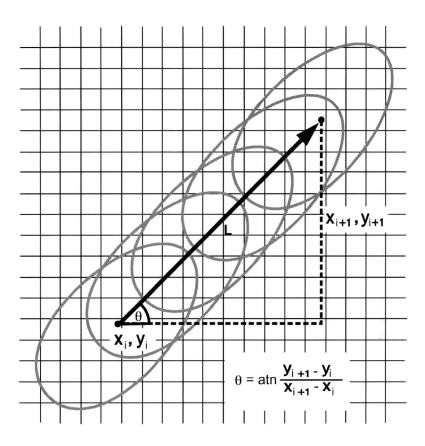

FIGURE 20.12 Track analysis from a short sequence of video frames. The length of the vector connecting the centroid in the first and final image can be used to determine the linear velocity of the object. The angular deviation from a predefined direction can be used to determine the precision of orientation of a population.

where x and y are the x- and y-coordinates of an object in the first and final frame, respectively. The velocity of the objects is determined by dividing the distance that the object moved during the time interval $t_2 - t_1$ needed for this displacement (read from the hardware clock of the computer):

$$v = k \frac{\sqrt{(y_2 - y_1)^2 + (x_2 - x_1)^2}}{t_1 - t_2} \tag{20.14}$$

where k is the calibration factor that allows the velocity to be automatically shown in the correct dimension.

If slower objects need to be tracked, a video delay can be selected. Each delay unit corresponds to 40 ms (European norm; 33 ms is the American norm).

The angular deviation α from the zero direction (top of the screen) is also determined from the movement vectors:

$$\alpha = \arctan \left| \frac{y_1 - y_2}{x_1 - x_2} \right| \tag{20.15}$$

Furthermore, the area (also with the correct dimension [e.g., μm^2]) and form factor of each object are calculated. When all objects in the field of view are evaluated, the entire sequence is repeated, starting with recording the next five frames from the incoming stream of video frames. This procedure is terminated when either the maximal time or the maximal number of evaluated tracks determined in the settings window is exceeded or the user terminates it. The positions of the organisms can be indicated by an overlay during the tracking sequence. However, this option considerably slows the process.

After each cycle, the results are updated in the Measured Values window, utilizing all tracks evaluated up to this point of time; the number of evaluated tracks, the mean area and mean form factor of the found objects, the mean speed of movement, the elapsed time, and the percentage of motile cells are displayed. With the exception of the percentage of motile cells, all values are based only on the objects that meet the selection criteria (minimal and maximal area and speed).

A histogram is plotted in a separate window and updated each cycle; it shows the direction of movement of the population binned in 72 sectors. The upward direction in the histogram ($0°$) corresponds to the upward direction in the video window. A statistical value for the precision of orientation, the r-value, is calculated according to the following equation[110,111]

$$r = \frac{\sqrt{\left(\sum \sin \alpha\right)^2 + \left(\sum \cos \alpha\right)^2}}{n} \tag{20.16}$$

where n is the number of tracks and α is the angular deviation of each track segment defined above. The r-value runs from 0 for a completely random orientation to 1 for an ideal orientation of all objects in the same direction. The Rayleigh test is applied to determine whether or not the distribution is random. The critical value for the error probability P is read from a table. For measurements with more than 100 tracks, the value $2nr^2$ is calculated from the r-value and the number of tracks n and compared with the table entry.

In addition, the alignment is calculated. The alignment is a measure of how well the tracks are aligned with the x- or y-axis and calculated from $(\sum |\sin \alpha| - \sum |\cos \alpha|)/n$, where α is the angular deviation of each track and n is the total number of tracks. Thus, the value ranges between +1 (all tracks parallel to the y-axis) and –1 (all tracks parallel to the x-axis); a movement on the 45° line results in a value of 0. Another orientation parameter is the k value, which is calculated by ($\sum$ Upward swimming cells $-\sum$ Downward swimming cells $\times$ 100/n). This value varies between 100 (all cells swim in the upward semicircle) and –100 (all cells swim downward); 0 indicates random orientation.

The mean direction of movement (θ) is calculated, printed numerically, and indicated by a green bar in the histogram.

$$\theta = \arctan \frac{\left(\sum \sin \alpha\right)^2}{\left(\sum \cos \alpha\right)^2} \tag{20.17}$$

The orientation histogram can be copied to the clipboard and inserted into other application programs (Figure 20.13). The direction histogram can be replaced by a velocity histogram that shows the velocity distribution of the motile objects in dependence of their direction (with the correct units, according to the calibration procedure). Also, the mean velocities are calculated for the upward, downward, and sideward moving objects (moving in 120° sectors each). This is of interest when objects move with different velocities in dependence of their direction (gravitaxis,

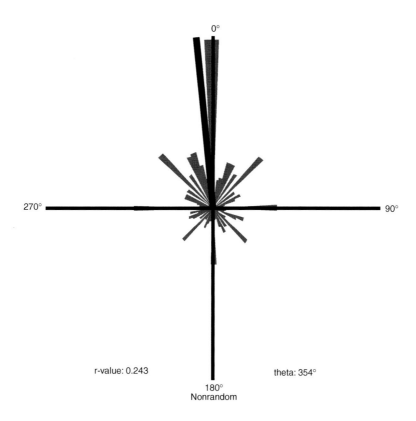

FIGURE 20.13 Orientation histogram showing the direction of movement of a population of microorganisms. The inserted bar to the left of 0° indicates the mean direction of the population ($\theta = 354°$). The r-value is a measure of the precision of orientation. The Rayleigh test is used to determine whether or not the distribution is random.

chemotaxis).[112] The Windows technology allows one to perform the calculations in parallel with the video frame acquisition (multithreading).

Another option of the tracking module is to display the kinetics of the orientation parameters over the current measurement period. In this mode, the kinetics of the precision of orientation, mean direction of movement, mean speed in the upward, downward, and sideward directions, overall mean speed, as well as the mean form factor and the mean area are shown in Figure 20.14. The minimal time can be defined when a new value is introduced in the kinetics panels, and the minimal number of organisms defines the minimal number of tracks to be analyzed before the kinetic values are updated. The data can be stored in ASCII files, including a headline with date and time as well as a description of the experiment. All relevant information concerning the settings are also recorded, as for the Count module. The software can be programmed by the user to create sequences of files at predefined time intervals or containing a predefined number of tracks.

20.4.7 Long Tracking

Often, it is of interest to follow organisms over considerable time spans rather than only short time segments. This type of analysis is required, for example, for the determination of the free path

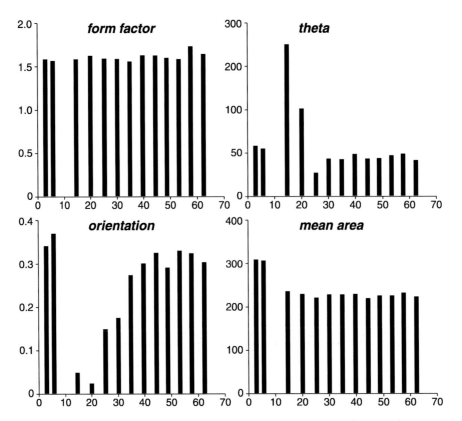

FIGURE 20.14   Diagrams showing the kinetics of the mean form factor, mean direction of movement (theta), precision of orientation, and mean area in a population of flagellates during gravitactic orientation.

length of the organisms.[113] For this purpose, a true real-time analysis is inevitable because only a limited number of frames can be stored in memory. Long tracking differs from the tracking software described in the previous section in that the algorithm tries to follow the objects as long as they are visible on screen. During long tracking, chances are high that an object will be lost. If no organism is found at the previous position, the neighborhood is searched using a dynamically adjusted area, the size of which can be adjusted in dependence of the mean path length of the organism within the time interval and their density. When the long tracking procedure starts, all objects are selected which meet the selection criteria (minimal and maximal area and velocity) and are followed until they are lost. This occurs as a result of one of the following conditions:

- The object leaves the area of view.
- The object hits another object. When this happens, another algorithm takes over. This assumes that the crossing particles will basically follow their initial path. Therefore, the software waits a short time, defined by size and velocity of the object, and then tries again to find the objects in a "look ahead sector" in continuation of the original path.
- The area changes by more than 50% because it leaves the focus.

An object is regarded as not motile when a maximal predefined number of frames is reached during which this specific object has not moved. The number of currently tracked objects is indicated in the measured values window in addition to the other measured parameters. Another calculated parameter is the linearity, which indicates the ratio of the linear distance between the beginning and end point of a track, divided by the actual path length which follows all curves and bends.

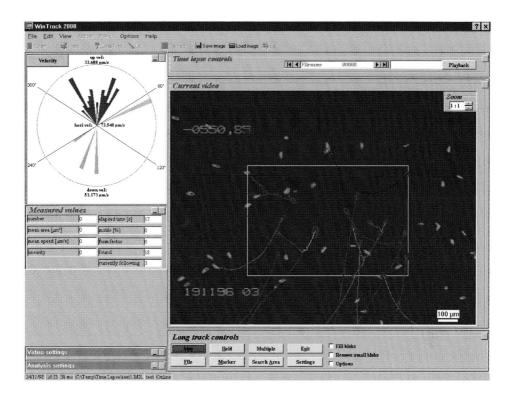

FIGURE 20.15 Long track module in the WinTrack 2000 software. The rectangle on the digitized image selects the organisms to be tracked.

When an object moves on a straight path, the linearity is 1; a small value indicates a high degree of meandering.

The long track algorithm continues until all tracked objects are lost and then starts again. The number of objects can be restricted by selecting a search area that allows the user to define a rectangular area on screen with the mouse (Figure 20.15). The search area can be defined small enough to follow a single object.

The tracks are drawn as an overlay on screen in different colors in order to reconstruct the track paths. The histogram area shows the same histograms as the track module (orientation, velocity distribution). In addition, a panel can be selected to show the tracks recorded up to that time (Figure 20.16). All beginning points are mathematically moved to the center of the diagram.

The maximal number of organisms to be tracked simultaneously can be defined, as well as the minimal time to follow an object before a track is recorded as successful. Also, the time interval (in seconds) after which an object is regarded as immotile can be defined when it has not changed position during that time. The maximal number of frames to follow an object can be selected.

20.4.8 Time-Lapse Recorder

The WinTrack 2000 software has a built-in electronic time-lapse recorder. This software device stores full video frames on the hard disk. The speed of modern hard disks allows the system to store files at a frequency of faster than six frames per second. The capacity of today's hard disks allows for the storage of long video sequences. A 4-GB disk has a capacity for over 9000 frames, which translates to a video sequence of more than 2.5 h duration at one frame per second.

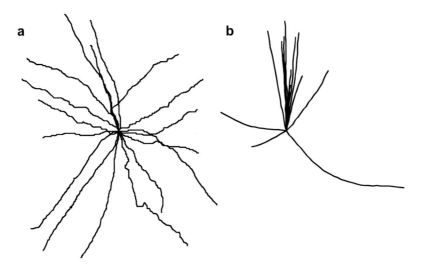

FIGURE 20.16 Projected long tracks of moving objects. The origins have been mathematically moved to the center of the diagram.

The user selects a file name to which a running number for each frame is appended. The user can select the time interval at which frames are recorded. The delay time can be up to 100 h. The frame sequence can be played back in any analysis mode. A slide control allows one to restore a selected frame. Another field indicates the time-stamp of this frame, which is the time elapsed after the start of the series recording. Recorded series can be evaluated by the Count, Track, and Long track modules in a similar manner as online video sequences.

20.4.9   THREE-DIMENSIONAL TRACKING

In most cases, an experimental setup can be developed in which the two-dimensional tracking suffices. In other applications, three-dimensional (3-D) tracking is indispensable. This increases the computational requirements considerably. Other methods to follow objects in three dimensions include defocusing.[114] Figure 20.17 shows the setup to track motile objects in 3-D using two cameras oriented perpendicular to each other. The cameras aim at the same volume in a swimming aquarium.[115,116] A laser beam can be used to align the cameras as well as the irradiation beams. The images of the two cameras are combined on a split screen using a video wiper and external sync generator for the two cameras.

The software is designed to first identify an object in one camera view and determine its x- and y-coordinates. Then it tries to identify a corresponding object in the other half image at the same y-coordinate and determine the z-coordinate. The subsequent calculation of the direction of movement is done in a van Mises space using similar mathematics as for two-dimensional analysis expanded for three dimensions.

20.5   APPLICATIONS

Microorganisms use a number of external factors to orient in their environment.[117,118] While some motile microorganisms have been found to orient in the water column with the aid of thermal[119,120] and chemical[121,122] gradients, the magnetic field of the earth,[123,124] and even electrical currents.[125]

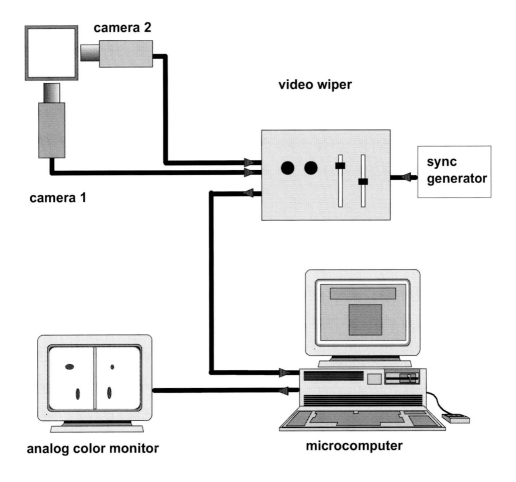

FIGURE 20.17   Schematic diagram for a 3-D tracking device. Two cameras positioned perpendicularly to each other focus on the same spot in a swimming cuvette. The two video images are combined in one by a video viper and the program evaluates both half images.

The flagellate *Euglena gracilis* and other photosynthetic and nonphotosynthetic organisms mainly orient with respect to light[126-128] and gravity.[129,130]

In addition to phototaxis, the cells orient themselves precisely with respect to the gravity vector. Under microgravity conditions during a parabolic flight of a rocket, the cells show random orientation, which proves the assumption that gravity is the signal for orientation under terrestrial conditions. On a recent space flight on the American Shuttle Columbia, a series of experiments was conducted designed to determine the threshold for gravity perception. For this purpose, the cells were kept in cuvettes that were either stored under microgravity or on a 1-g reference centrifuge. For the experiment, the astronauts transferred the cuvettes into a slowly rotating centrifuge microscope, which allowed viewing of the swimming cells under increasing centrifugal acceleration.[131] Accelerations up to 0.08g, no orientation could be seen. But at 0.12g, the cells started to show negative gravitaxis. The number of applications is virtually unlimited. The computer does not care if it follows moving microorganisms, people, automobiles, or celestial objects. Many other examples are described in other chapters of this volume.

REFERENCES

1. van Opstal, W.Y.X., Ranger, C., Lejeune, O., Forgez, P., Boudin, H., Bisconte, J.C., and Rostene, W., Automated image analyzing system for the quantitative study of living cells in culture, *Microsc. Res. Techn.*, 28, 440, 1994.

2. Belyaev, N., Paavilainen, S., and Korpela, T., Characterization of bacterial growth on solid medium with image analysis, *J. Biochem. Biophys. Meth.*, 25, 125, 1992.

3. Zicha, D. and Dunn, G.A., An image processing system for cell behavior studies in subconfluent cultures, *J. Microscopy*, 179, 11, 1995.

4. Gualtieri, P., Colombetti, G., and Lenci, F., Automatic analysis of the motion of microorganisms, *J. Microscopy*, 139, 57, 1985.

5. Lopez-de-Victoria, G., Zimmer-Faust, R.K., and Lovell, C.R., Computer-assisted video motion analysis: a powerful technique for investigating motility and chemotaxis, *J. Microbiol. Meth.*, 23, 329, 1995.

6. Rodrigo, J., Rivas, E., and Herrero, M., Starch determination in plant tissues using a computerized image analysis system, *Physiol. Plant*, 99, 105, 1997.

7. Baker, B., Olszyk, D.M., and Tingey, D., Digital image analysis to estimate leaf area, *J. Plant Physiol.*, 148, 530, 1996.

8. Tendel, J. and Häder, D.-P., Effects of ultraviolet radiation on orientation movements of higher plants, *J. Photochem. Photobiol. B.: Biology*, 27, 67, 1995.

9. Colmorgen, M. and Paul, R.J., Imaging of physiological functions in transparent animals (*Agonus cataphractus, Daphnia magna, Pholcus phalangioides*) by video microscopy and digital image processing, *Comp. Biochem. Physiol.*, 111A, 583, 1995.

10. Walter, R.J. and Berns, M.W., Computer-enhanced video microscopy: digitally processed microscope images can be produced in real-time, *Proc. Natl. Acad. Sci, U.S.A.*, 78, 6927, 1981.

11. Mineyuki, Y., Yamada, M., Takagi, M., Wada, M., and Furuya, M., A digital image processing technique for the analysis of particle movements: its application to organelle movements during mitosis in *Adiantum* protonemata, *Plant Cell Physiol.*, 24, 225, 1983.

12. Carragher, B. and Smith, P.R., Advances in computational image processing for microscopy, *J. Struct. Biol.*, 116, 2, 1996.

13. Firestone, L.M., Preston, K., Jr., and Nathwani, B.N., Continuous class pattern recognition for pathology, with applications to non-Hodgkin's follicular lymphomas, *Pattern Recogn.*, 29, 2061, 1996.

14. Preston, K., Jr., Joe, B., Siderits, R., and Welling, J., Three-dimensional reconstruction of the human renal glomerulus, *J. Microscopy*, 177, 7, 1995.

15. Hall, R.W., Image processing algorithms for eye movement monitoring, *Computer, Biomed. Res.*, 16, 563, 1983.

16. Endlich, R.M., Wolf, D.E., Hall, D.J., and Brain, A.E., Use of pattern recognition technique for determining cloud motions from sequences of satellite photographs, *J. Appl. Meteorol.*, 10, 105, 1071.

17. Eggersdorfer, B. and Häder, D.-P., Phototaxis, gravitaxis and vertical migrations in the marine dinoflagellate, *Prorocentrum micans, FEMS Microbiol. Ecol.*, 85, 319, 1991.

18. Eggersdorfer, B. and Häder, D.-P., Phototaxis, gravitaxis and vertical migrations in the marine dinoflagellate, *Peridinium faeroense* and *Amphidinium caterii, Acta Protozool.*, 30, 63, 1991.

19. Häder, D.-P., Novel method to determine vertical distributions of phytoplankton in marine water columns, *Env. Exp. Bot.*, 35, 547, 1995.

20. Hüller, R., Glossner, E., Schaub, S., Weingärtner, J., and Kachel, V., The macro flow planktometer: a new device for volume and fluorescence analysis of macro plankton including triggered video imaging in flow, *Cytometry*, 17, 109, 1994.

21. Stallwitz, E. and Häder, D.-P., Effects of heavy metals on motility and gravitactic orientation of the flagellate, *Euglena gracilis, Eur. J. Protistol.*, 30, 18, 1994.

22. Tahedl, H. and Häder, D.-P., Fast examination of water quality using the automatic biotest ecotox based on the movement behavior of a freshwater flagellate, *Water Res.*, 33, 426, 1999.

23. D'Errico, F., Image analysis and 3D optical surface profiling of upper palaeolithic mobiliary art, *Eur. Microsc. Anlys.*, 39, 29, 1996.

24. Hemmersbach-Krause, R., Briegleb, W., Vogel, K., and Häder, D.-P., Swimming velocity of *Paramecium* under the conditions of weightlessness, *Acta Protozool.*, 32, 229, 1993.

25. Hemmersbach-Krause, R., Briegleb, W., Häder, D.-P., Vogel, K., Klein, S., and Mulisch, M., Protozoa as model systems for the study of cellular responses to altered gravity conditions, *Adv. Space Res.*, 14, 49, 1994.

26. Nilsson, G.E., Rosen, P., and Johansson, D., Anoxic depression of spontaneous locomotor activity in crucian carp quantified by a computerized imaging technique, *J. Exp. Biol.*, 180, 153, 1993.

27. Häder, D.-P., Lebert, M., and Richter, P., Gravitaxis and graviperception in *Euglena gracilis*, *Adv. Space Res.*, 21, 1277, 1998.

28. Häder, D.-P., Rosum, A., Schäfer, J., and Hemmersbach, R., Gravitaxis in the flagellate *Euglena gracilis* is controlled by an active gravireceptor, *J. Plant Physiol.*, 146, 474, 1995.

29. Hemmersbach, R., Voormanns, R., and Häder, D.-P., Graviresponses in *Paramecium biaurelia* under different accelerations: studies on the ground and in space, *J. Exp. Biol.*, 190, 2199, 1996.

30. Kühnel-Kratz, C., Schäfer, J., and Häder, D.-P., Phototaxis in the flagellate, *Euglena gracilis*, under the effect of microgravity, *Microgr. Sci. Technol.*, 4, 188, 1993.

31. Levine, M.D., Youssef, Y.M., Noble, P.B., and Boyarsky, A., The quantification of blood cell motion by a method of automatic digital picture processing, *IEEE Trans. Pattern Anal. Machine Intell.*, 2, 444, 1980.

32. Leidl, W., Riemke, P., and Schröppel, I., Computergesteuerte Videomikrographie-Auswertung zur Bestimmung der Spermienmotilität am Modell des Bullen, *Deutsche Tierärztl. Wochenschr.*, 8, 461, 1987.

33. Schor, S.L., Schor, A.M., and Brazill, G.W., The effects of fibronectin on the migration of human foreskin fibroblasts and Syrian hamster melanoma cells into three-dimensional collagen fibres, *J. Cell Sci.*, 48, 301, 1981.

34. Link, N.G., Nathwani, B.N., and Preston, K., Jr., Subclassification of follicular lymphomas by computerized image processing, *Analyt. Quent. Cytol. Histol.*, 11, 119, 1989.

35. Clark, K.D. and Nelson, D.L., An automated assay for quantifying the swimming behavior of *Paramecium* and its use to study cation responses, *Cell Motil. Cytoskel.*, 19, 91, 1991.

36. Häder, D.-P. and Vogel, K., Real time tracking of microorganisms, in *Image Analysis in Biology*, Häder, D.-P., Ed., CRC Press, Boca Raton, FL, 1991, 289.

37. Häder, D.-P. and Vogel, K., Simultaneous tracking of flagellates in real time by image analysis, *J. Math. Biol.*, 30, 63, 1991.

38. Taylor, C.J., Cootes, T.F., Lanitis, A., Edwards, G., Smyth, P., and Kotcheff, A.C.W., Model-based interpretation of complex and variable images, *Phil. Trans. Roy. Soc. Lond.*, B, 352, 1267, 1997.

39. Bengtsson, E., Nordin, B., and Pedersen, F., MUSE — A new tool for interactive image analysis and segmentation based on multivariate statistics, *Comput. Meth. Prog. Biomed.*, 42, 181, 1994.

40. Beardsley, T., Here's looking at you, *Sci. Am.*, 280, 22, Jan. 1999.

41. Lebert, M. and Häder, D.-P., How *Euglena* tells up from down, *Nature*, 379, 590, 1996.

42. Häder, D.-P., Gravitaxis in the flagellate *Euglena gracilis* — Results from NIZEMI, clinostat and sounding rocket flights, *J. Gravit. Physiol.*, 1, P-82, 1994.

43. Häder, D.-P., Rosum, A., Schäfer, J., and Hemmersbach, R., Graviperception in the flagellate *Euglena gracilis* during a shuttle space flight, *J. Biotechnol.*, 47, 261, 1996.

44. Häder, D.-P., Vogel, K., and Schäfer, J., Responses of the photosynthetic flagellate, *Euglena gracilis*, to microgravity, *Microgravity Sci. Technol.*, III, 110, 1990.

45. Serra, J., Digitalization, *Mikroskopie*, 37(Suppl.), 109, 1980.

46. Kokubo, Y. and Hardy, W.H., Digital image processing, a path to better pictures, *Ultramicroscopy*, 8, 277, 1982.

47. Häder, D.-P., Effects of solar and artificial UV irradiation on motility and phototaxis in the flagellate, *Euglena gracilis, Photochem. Photobiol.*, 44, 651, 1986.

48. Vogel, K., Hemmersbach-Krause, R., Kühnel, C., and Häder, D.-P., Swimming behavior of the unicellular flagellate, *Euglena gracilis*, in simulated and real microgravity, *Micrograv. Sci. Technol.*, 5, 232, 1993.

49. Häder, D.-P., Polarotaxis, gravitaxis and vertical phototaxis in the green flagellate, *Euglena gracilis,* *Arch. Microbiol.*, 147, 179, 1987.

50. Häder, D.-P. and Häder, M., Ultraviolet-B inhibition of motility in green and dark bleached *Euglena gracilis* Curr. Microbiol., 17, 215, 1988.

51. Häder, D.-P. and Häder, M., Effects of solar radiation on photoorientation, motility and pigmentation in a freshwater *Cryptomonas, Botanica Acta*, 102, 236, 1989.

52. Jähne, B., *Digitale Bildverarbeitung*, Springer-Verlag, Berlin, 1989.

53. Castillo, X., Yorkgitis, D., and Preston, K., Jr., A study of multidimensional multicolor images, *IEEE Trans. Biomed. Eng.*, 29, 111, 1982.

54. Knecht, D.A., Preparation of figures for publication using a digital color printer, *Biotechnology*, 14, 1006, 1993.

55. Alleman, A., Kim, Y., Milton, S., and Bush, J., Microcomputer-based system for real-time optical imaging, *Med. Biol. Eng. Comput.*, 33, 728, 1995.

56. Malik, N.R. and Huang, G., Integer filters for imaging processing, *Med. Biol. Eng. Comput.*, 26, 62, 1988.

57. Malik, N.R., Microcomputer realisations of Lynn's fast digital-filtering designs, *Med. Biol. Comput.*, 18, 638, 1980.

58. Julez, B. and Harmon, L.D., Noise and recognizability of coarse quantized images, *Nature*, 308, 211, 1984.

59. Voss, K., Differentialgeometrie und digitale Bildverarbeitung, *Bild und Ton*, 43, 165, 1990.

60. Adler, J., The use and abuse of look up tables, *Eur. Microsc. Analys.*, 39, 7, 1996.

61. Saxton, M.J., Single-particle tracking: models of directed transport, *Biophys. J.*, 67, 2110, 1994.

62. Vaija, J., Lagaude, A., and Ghommidh, C., Evaluation of image analysis and laser granulometry for microbial cell sizing, *Antonie van Leeuwenhoek*, 67, 139, 1995.

63. Whalen, T.A., Volume and surface area estimation from microscopic images, *J. Microscopy*, 188, 93, 1997.

64. Wu, H.-S. and Barba, J., An efficient semi-automatic algorithm for cell contour extraction, *J. Microscopy*, 179, 270, 1995.

65. Häder, D.-P., Automatic area calculation by microcomputer-controlled video analysis, *EDV Med. Biol.*, 18, 33, 1987.

66. Häder, D.-P., Advanced techniques in photobehavioral studies: computer-aided studies, in *Light in Biology and Medicine*, Douglas, R.H., Moan, J., and Dall'Acqua, F., Eds., Plenum Press, New York, 1988, 385.

67. Grant, G. and Reid, A.F., An efficient algorithm for boundary tracing and feature extraction, *Computer Graph. Imag. Process.*, 17, 225, 1981.

68. Pavlidis, T., *Graphics and Image Processing*, Springer-Verlag, Berlin, 1982.

69. Jobbagy, A. and Furnee, E.H., Marker centre estimation algorithms in CCD camera-based motion analysis, *Med. Biol. Eng. Comput.*, 32, 85, 1994.

70. Freeman, H., Analysis and manipulation of lineal map data, in *Map Data Processing*, Academic Press, 1980, 151.

71. Kincaid, D.T. and Schneider, R.B., Quantification of leaf shape with a microcomputer and Fourier transform, *Can. J. Bot.*, 61, 2333, 1983.

72. Häder, D.-P. and Griebenow, K., Versatile digital image analysis by microcomputer to count microorganisms, *EDV Med. Biol.*, 18, 37, 1987.

73. Häder, D.-P. and Griebenow, K., Orientation of the green flagellate, *Euglena gracilis*, in a vertical column of water, *FEMS Microbiol. Ecol.*, 53, 159, 1988.

74. Gibson, J.J., *The Perception of the Visual World*, Houghton and Mifflin, Boston, MA, 1950.

75. Roach, J.W. and Aggarwal, J.K., Determining the movement of objects from a sequence of images, *IEEE Trans. Pattern Anal. Machine Intell.*, PAMI-2, 554, 1980.

76. Jain, R., Extraction of motion information from peripheral processes, *IEEE Trans. Pattern Anal. Machine Intell.*, PAMI-3, 489, 1981.

77. Jain, R. and Nagel, H.H., On the analysis of accumulative difference pictures from image sequences or real world scenes, *IEEE Trans. Pattern Anal. Machine Intell.*, PAMI-1, 206, 1979.

78. Kenny, P.A., Dowsett, D.J., Vernon, D., and Ennis, J.T., A technique for digital image registration used prior to subtraction of lung images in nuclear medicine, *Phys. Med. Biol.*, 35, 679, 1990.

79. Gualtieri, P. and Coltelli, P., A digital microscope for real time detection of moving microorganisms, *Micron Microsc. Acta*, 20, 99, 1989.

80. Häder, D.-P., Real-time tracking of microorganisms, *Binary*, 6, 81, 1994.

81. Häder, D.-P., Tracking of flagellates by image analysis, in *Biological Motion Proceedings Königswinter 1989*, Alt, W. and Hoffmann, G., Eds., Springer-Verlag, Berlin, 1990, 343.

82. Häder, D.-P., Use of image analysis in photobiology, in *Photobiology: The Science and Its Applications*, Riklis, E., Ed., Plenum Press, New York, 1991, 329.

83. Smith, E.A. and Phillips, D.R., Automated cloud tracking using precisely aligned digital ATS pictures, *IEEE Trans. Comput.*, C-21, 715, 1972.

84. Takahashi, T. and Kobatake, Y., Computer-linked automated method for measurement of the reversal frequency in phototaxis of *Halobacterium halobium*, *Cell Struct. Funct.*, 7, 183, 1982.

85. Noble, P.B. and Levine, M.D., *Computer Assisted Analysis of Cell Locomotion and Chemotaxis*, CRC Press, Boca Raton, FL, 1986.

86. Aggarwal, J.K. and Duda, R.O., Computer analysis of moving polygonal images, *IEEE Trans. Comput.*, C-24, 966, 1975.

87. Schalkoff, R.J. and McVey, E.S., A model and tracking algorithm for a class of video targets, *IEEE Trans. Pattern Anal. Machine Intell.*, PAMI-4, 2, 1982.

88. Häder, D.-P. and Vogel, K., Interactive image analysis system to determine the motility and velocity of cyanobacterial filaments, *J. Biochem. Biophys. Meth.*, 22, 289, 1991.

89. Allen, R.D., New directions and refinements in video-enhanced microscopy applied to problems in cell motility, in *Advances in Microscopy*, Alan R. Liss, New York, 1985, 3.

90. Amos, L., Movements made visible by microchip technology, *Nature*, 330, 211, 1987.

91. Coates, T.D., Harman, J.T., and McGuire, W.A., A microcomputer-based program for video analysis of chemotaxis under agarose, *Computer Meth. Programs Biomed.*, 21, 195, 1985.

92. Rikmenspoel, R. and Isles, C.A., Digitized precision measurements of the movement of sea urchin sperm flagella, *Biophys. J.*, 47, 395, 1985.

93. Burton, J.L., Law, P., and Bank, H.L., Video analysis of chemotactic locomotion of stored human polymorphonuclear leukocytes, *Cell Motility Cytoskel.*, 6, 485, 1986.

94. Gordon, D.C., MacDonald, I.R., Hart, J.W., and Berg, A., Image analysis of geo-induced inhibition, compression, and promotion of growth in an inverted *Helianthus annuus* L. seedling, *Plant Physiol.*, 76, 589, 1984.

95. Omasa, K. and Onoe, M., Measurement of stomatal aperture by digital image processing, *Plant Cell Physiol.*, 25, 1379, 1984.

96. Jaffe, M.J., Wakefield, A.H., Telewski, F., Gulley, E., and Biro, R., Computer-assisted image analysis of plant growth, thigmomorphogenesis, and gravitropism, *Plant Physiol.*, 77, 722, 1985.

97. Omassa, K. and Aiga, I., Environmental measurement, image instrumentation for evaluating pollution effects on plants, in *System & Control Encyclopedia*, Vol. 2, Singh, M.G., Ed., Pergamon Press, Oxford, 1987, 1516.

98. Popescu, T., Zängler, F., Sturm, B., and Fukshansky, L., Image analyzer used for data acquisition in phototropism studies, *Photochem. Photobiol.*, 50, 701, 1989.

99. Häder, D.-P. and Lebert, M., The photoreceptor for phototaxis in the photosynthetic flagellate *Euglena gracilis*, *Photochem. Photobiol.*, 68, 260, 1998.

100. Häder, D.-P., Gravitaxis and phototaxis in the flagellate *Euglena* studied on TEXUS missions, in *Life Science Experiments Performed on Sounding Rockets (1985–1994)*, Cogoli, A., Friedrich, U., Mesland, D., and Demets, R., Eds., ESA Publications Division, ESTEC, Noordwijk, The Netherlands, 1997, 77.

101. Häder, D.-P. and Hemmersbach, R., Graviperception and graviorientation in flagellates, *Planta*, 203, S7, 1997.

102. Hemmersbach, R., Voormanns, R., Briegleb, W., Rieder, N., and Häder, D.-P., Influence of accelerations on the spatial orientation of *Loxodes* and *Paramecium*, *J. Biotechnol.*, 47, 271, 1996.

103. Häder, D.-P., Rhiel, E., and Wehrmeyer, W., Ecological consequences of photomovement and photobleaching in the marine flagellate *Cryptomonas maculata*, *FEMS Microbiol. Ecol.*, 53, 9, 1988.

104. Dusenbery, D.B., Using a microcomputer and videocamera to simultaneously track 25 animals, *Comput. Biol. Med.*, 15, 169, 1985b.

105. Sanderson, M.J. and Dirksen, E.R., A versatile and quantitative computer assisted photoelectronic technique used for the analysis of ciliary beat cycles, *Cell Motil.*, 5, 267, 1985.

106. Baba, S.A. and Mogami, Y., An approach to digital image analysis of bending shapes of eukaryotic flagella and cilia, *Cell Motil.*, 5, 475, 1985.

107. Omoto, C.K. and Brokaw, C.J., Bending patterns of *Chlamydomonas* flagella. II. Calcium effects on reactivated *Chlamydomonas* flagella, *Cell Motil.*, 5, 53, 1985.

108. Cantatore, G., Ascoli, C., Colombetti, G., and Frediani, C., Doppler velocimetry measurements of phototactic response in flagellated algae, *Biosci. Rep.*, 9, 475, 1989.

109. Häder, D.-P. and Lebert, M., Real time computer controlled tracking of motile microorganisms, *Photochem. Photobiol.*, 42, 509, 1985.

110. Batschelet, E., *Circular Statistics in Biology*, Academic Press, London, 1981.

111. Mardia, K.V., *Statistics of Directional Data*, Academic Press, London, 1972.

112. Machemer, H. and Bräucker, R., Gravireception and graviresponses in ciliates, *Acta Protozool.*, 31, 185, 1992.

113. Glazzard, A.N., Hirons, M.R., Mellor, J.S., and Holwill, M.E.J., The computer assisted analysis of television images as applied to the study of cell motility, *J. Submicrosc. Cytol.*, 15, 305, 1983.

114. Bianco, B. and Diaspro, A., Analysis of three-dimensional cell imaging obtained with optical microscopy techniques based on defocusing, *Cell Biophysics*, 15, 189, 1989.

115. Kühnel-Kratz, C. and Häder, D.-P., Real time three-dimensional tracking of ciliates, *J. Photochem. Photobiol.*, 19, 193, 1993.

116. Kühnel-Kratz, C. and Häder, D.-P., Light reactions of the ciliate *Stentor coeruleus* — A three-dimensional analysis, *Photochem. Photobiol.*, 59, 257, 1994.

117. Häder, D.-P., Ecological consequences of photomovement in microorganisms, *Photochem. Photobiol., B: Biol.*, 1, 385, 1988.

118. Mizuno, T., Maeda, K., and Imae, Y., Thermosensory transduction in *Escherichia coli*, in *Transmembrane Signaling and Sensation*, Oosawa, F., Yoshioka, T., and Hayashi, H., Eds., Japan Sci. Soc. Press, Tokyo and VNU Sci. Press BV, Netherlands, 1984, 147.

119. Poff, K.L., Temperature sensing in microorganisms, in *Sensory Perception and Transduction in Aneural Organisms*, Colombetti, G., Lenci, F., and Song, P.-S., Eds., Plenum Press, New York, London, 1985, 299.

120. Berg, H.C., Physics of bacterial chemotaxis, in *Sensory Perception and Transduction in Aneural Organisms*, Colombetti, G., Lenci, F., and Song, P.-S., Eds., Plenum Press, New York, 1985, 19.

121. MacNab, R.M., Biochemistry of sensory transduction in bacteria, in *Sensory Perception and Transduction in Aneural Organisms*, Colombetti, G., Lenci, F., and Song, P.-S., Eds., Plenum Press, New York, 1985, 31.

122. Frankel, R.B., Magnetic guidance of organisms, *Annu. Rev. Biophys. Bioeng.*, 13, 85, 1984.

123. Esquivel, D.M.S. and de Barros, H.G.P.L., Motion of magnetotactic microorganisms, *J. Exp. Biol.*, 121, 153, 1986.

124. Mast, S.O., *Light and Behavior of Organisms*, John Wiley & Sons, New York; Chapman & Hall, London, 1911.

125. Diehn, B., Feinleib, M., Haupt, W., Hildebrand, E., Lenci, F., and Nultsch, W., Terminology of behavioral responses of motile microorganisms, *Photochem. Photobiol.*, 26, 559, 1977.

126. Häder, D.-P., Colombetti, G., Lenci, F., and Quaglia, M., Phototaxis in the flagellates, *Euglena gracilis* and *Ochromonas danica, Arch. Microbiol.*, 130, 78, 1981.

127. Colombetti, G., Häder, D.-P., Lenci, F., and Quaglia, M., Phototaxis in *Euglena gracilis*: effect of sodium azide and triphenylmethyl phosphonium ion on the photosensory transduction chain, *Curr. Microbiol.*, 7, 281, 1982.

128. Lenci, F., Colombetti, G., and Häder, D.-P., Role of flavin quenchers and inhibitors in the sensory transduction of the negative phototaxis in the flagellate, *Euglena gracilis, Curr. Microbiol.*, 9, 285, 1983.

129. Batschelet, E., Statistical methods for the analysis of problems in animal orientation and certain biological rhythms, in *Animal Orientation and Navigation*, Galles, S.R., Schmidt-Koenig, K., Jacobs, G.J., and Belleville, R.F., Eds., NASA, Washington, 1965, 61.

130. Hemmersbach-Krause, R., Briegleb, W., Häder, D.-P., Vogel, K., Klein, S., and Mulisch, M., Protozoa as model systems for the study of cellular responses to altered gravity conditions, *Adv. Space Res.*, 14, 49, 1994.

131. Häder, D.-P., NIZEMI — Experiments on the slow rotating centrifuge microscope during the IML-2 mission, *J. Biotechnol.*, 47, 223, 1996.

21 Computer-Aided Analysis of Movement Responses of Microorganisms

Tetsuo Takahashi

CONTENTS

21.1 INTRODUCTION

Motile bacteria migrate upward the concentration gradient of a chemoattractant and swim away from a repellent, but they detect temporal gradient stimuli rather than a spatial gradient.[1-3] This behavior is called bacterial chemotaxis and has been studied extensively since the late l960s.[4-6] In earlier experiments, chemotaxis was quantitatively described simply by counting the number of bacterial cells entered into a capillary tube that has been filled with attractant or repellent chemicals.[4] Three-dimensional, single-cell tracking, developed by Berg and Brown,[7] revealed that the following basic nature in bacterial behavior leads to the net migration in a spatial gradient:

1. The swimming patterns of enteric bacteria are determined by periods of smooth runs interrupted by a brief period of erratic motion (called tumbling) during which the cell changes its directional axis.
2. The length of smooth run is determined stochastically (i.e., tumbling occurs spontaneously at random).
3. Bacteria swimming upward and downward the concentration gradient of a chemoattractant show longer and shorter periods of smooth runs, respectively, than those under

normal unstimulated conditions. In other words, the frequency of directional changes is modulated by the time derivative of the local concentration of the chemoeffectors surrounding the cell.

A more frequently used technique for quantitating chemotaxis of enteric bacteria is to "tether" cells by their flagellum to a microscope slide to observe the rotation of the cell body, reflecting tumbling (clockwise rotation of flagellum) or smooth-swimming (counterclockwise) behavior.[8] Also used is long-exposure photography of trajectories of swimming cells in a buffer solution after rapid mixing with chemoeffectors.[9,10] These time-resolved methods established that two fundamental processes are involved in chemotaxis: the excitation process governing the increase or the decrease in tumbling frequency, and the adaptation process through which the frequency gradually returns to its prestimulus level.[11]

Currently, both excitation and adaptation processes can be measured at high time resolution by computer-assisted image analysis.[12] Over the past decade, computerized methods have been widely applied to this field and in many laboratories image analysis was substituted for time-consuming manual assay by eye on photographic or videotape recording of the output behavior of bacterial chemotaxis or photobehavioral responses. This chapter focuses on the software techniques and criteria for detecting behavioral responses of bacteria and also of flagellated algae.

21.2   BASIC STRATEGIES

Two different approaches are briefly described in this section. One is to develop a compact software program for the limited memory space of a host computer and therefore suitable for relatively small systems having home made or custom-built hardware configurations. The other is to use a general-purpose motion analyzer system that recently became commercially available. In the latter case, the major concerns should involve application procedures and efficacy in detecting responses when adopting software packages provided by manufacturers or other commercial resources.

21.2.1   CORRELATION MATRIX AND DISPLACEMENT VECTORS

Takahashi and Kobatake[13] first reported an automated system for determining frequency of reversals of the swimming direction of halobacterial cells, using a small computer system combined with a commercially available frame grabber. In this system, a video field was digitized to two gray-scale (black-and-white) 256×256 pixel data fields. Data was captured every 200 to 700 ms and then transferred sequentially to the main memory of the host computer for further analysis. Due to limitation in memory space, only 6 (in later version up to 16) frames were stored and analyzed. Moving objects were searched for in the starting frame using the chain code algorithm.[14-16] The X and Y coordinates of all the dots representing a single moving object were stored for subsequent calculation of the displacement from the current frame to the subsequent frame using the correlation matrix method.[17]

Use of a correlation matrix is advantageous if each cell has its own characteristic shape and when the orientation of the cell axis is preserved during swimming. This is because the correlation matrix shows a sharp maximum with its highest value when the shape and size of the object are identical between frames.[13,17] If one assumes an upper limit of the velocity of the objects, the size of the correlation matrix should be reduced for the sake of fast processing. When the matrix has two obvious maxima, it may reflect the fact that two objects come close together and thus the object being tracked can be abandoned. The displacement of each object between two consecutive frames is actually a two-dimensional vector, herein referred to as a displacement vector. Further analyses, such as detecting swimming directions or reversal responses, are all based on these displacement vectors and are discussed in the first two topics in Section 21.3.

21.2.2 "PATH" DATA AND OPERATION THROUGH FILES

Another approach to detect behavioral responses is to use relatively long sequences of data in the time domain, and is useful for the acquisition of time-resolved information. The typical behavioral response of bacteria to a temporal stimulus persists for several seconds to minutes.[18] To record the whole sequence of a response with ca. 70-ms time resolution requires more than 200 digitized images, which is too large to be stored in the main memory of standard laboratory computers. In such cases, it is beneficial to start with a file containing time series of X and Y coordinates of the centroid(s) of single or multiple objects. This type of file is commonly available from many commercial motion analyzer systems (e.g., Image Tracker PTV, InterQuest, Osaka, Japan; ExpertVision, MotionAnalysis Corp., Santa Rosa, CA). Detection of behavioral responses from these files is discussed as the last topic of the next section.

21.3 MOVEMENT ANALYSIS TECHNIQUE FOR THE DETECTION OF REVERSAL, STOP, AND/OR TUMBLING RESPONSES

21.3.1 CASE 1: REVERSAL RESPONSES OF *HALOBACTERIUM SALINARUM*

Halobacterium salinarum (formerly referred to as *H. halobium* and as *H. salinarum*) is a rod-shaped, monopolarly or bipolarly flagellated archaebacterium whose size ranges from 1.5 to 4 μm in length. Bipolarly flagellated cells appear predominantly in the stationary growth phase. If two flagella bundles at both poles rotate in different directions, they push and pull the cell body to produce a smooth run.[19] In this case, reversal of rotation on one flagella bundle causes the cell to stop migration. This switching event occurs independently and stochastically at each pole, but a short inactive period before the subsequent reversal of the same flagellar bundle[20,21] produces zig-zag trajectories of cells under normal unstimulated conditions (Figure 21.1).

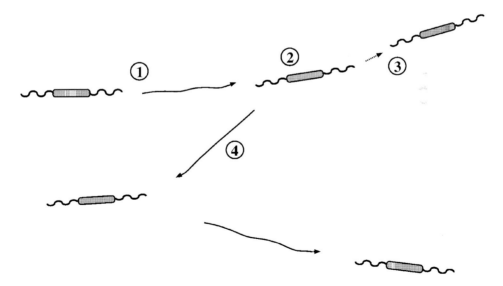

FIGURE 21.1 Schematic representation of motility of *Halobacterium salinarum*. The cell swims smoothly when the rotational senses of two flagellar bundles are opposite to each other (1). The cell stops swimming when either one of the flagellar bundles changes its rotational sense and fluctuates its cell axis (2). The cell restarts swimming after the change in rotational sense of the same flagellar bundle (3) or reverse the swimming direction after that of the other (4).

Delivery of repellent stimuli induces reversal of flagellar rotation, causing the cells to reverse the swimming direction to stop swimming (Figure 21.2). This has been regarded as a good measure of halobacterial sensitivity to visible light.[22] Conversely, attractant stimuli, such as a sudden turning off of blue-UV illumination or a step-like increase in the intensity of orange light, or a transfer from anaerobic to aerobic conditions, inhibits switching of flagellar rotation for several seconds to more than a minute.[23,24]

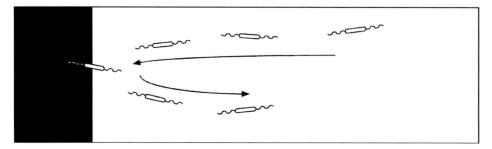

FIGURE 21.2 Schematic representation of the step-down photophobic response of *Halobacterium salinarum*.[19] On entering into the dark area, the halobacterial cell shows stop or reversal response to be trapped in the region where attractant light is illuminated. The responses are also elicited by a step-like decrease in intensity of attractant light.

Figure 21.3 illustrates the earlier criterion for the detection of stop and reversal responses from two subsequent displacement vectors.[13] A smooth run was identified when the displacement of the object from the current to the subsequent frame is close to that from the previous frame to the current frame. Reversals and stop responses were identified when the displacement vector pointed to areas away from the previous vector or the displacement was small. Owing to the feature of the movement responses of *Halobacterium salinarum* as described, this criterion was satisfactorily sensitive. At the saturated level of photorepellent stimuli, the fraction of cells showing reversal or stop response during a 3-s period was shown to be as much as 80% of the cell population (Figure 21.4). When preferentially smooth swimming mutant cells were tracked without photostimuli, the automated method recorded a value less than 10%.[25] This could be regarded as the lowest limit of the detection method.

21.3.2 CASE 2: PHOTOPHOBIC RESPONSE OF *EUGLENA* CELLS

Photophobic responses of *Euglena* were regarded as elementary reactions mediating phototaxis.[26] Also quantitated were the responses using the basically identical hardware setup, but in this case more precise criteria were necessary.[27] The response of *Euglena* cells to a sudden increase or decrease in light intensity includes gyration of the cell axis, and the direction of the cell axis is no longer preserved after the response. Trajectories of cells showing these responses (Figure 21.5) are comparable to those of bacteria exhibiting tumbling (cf. Figure 21.7) but in *Euglena*, the swimming direction fluctuates only slightly, even during a smooth run.

An adjustable parameter was introduced to evaluate the relation between two consecutive displacement vectors of *Euglena* cells. This parameter determines the shape of a hyperbola, whose axis was chosen to match the direction of the displacement vector of a cell from one frame to the subsequent frame, and its focus coincided with the end point of the vector (Figure 21.6). The vector representing further displacement from this frame to the next frame was compared with the "old" vector and examined with this hyperbolic curve: if the next displacement vector points to the region outside the hyperbola, the cell is assumed to exhibit the response. The following relations were actually used in the software program:

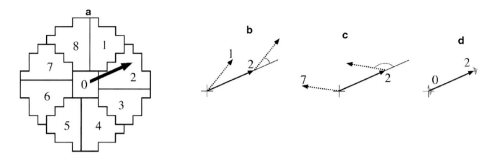

FIGURE 21.3 Principle for detection of reversal and stop responses of *Halobacterium salinarum*.[13] (a) The thick arrow represents the displacement vector of an object from one frame to the subsequent frame. In this example, the vector pointing to the region 2 is shown. When the next vector (i.e., representing displacement from the second frame to the third frame) points to the same or adjacent regions except the region 0, the swimming trajectory is regarded as a smooth path. (b to d) Relation between the consecutive displacement vectors. The broken arrows and gray arrowhead indicate the second vectors. The number at the end of each vector indicates corresponding regions shown in (a). Because the first vector is pointing to region 2, second vectors pointing to the regions 1 (b), 7 (c), and 0 (d) indicate smooth run, reversal, and stop response, respectively.

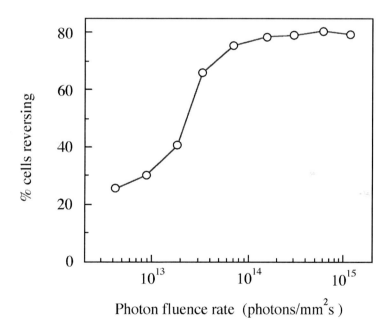

Photon fluence rate (photons/mm^{2}s)

FIGURE 21.4 Fluence rate-response curve for step-down photophobic responses of *Halobacterium salinarum* measured by automated tracking. Fractions of cell tracks that were recognized by the software program as to be showing reversal or stop response upon a sudden turning off of the actinic light were plotted against the initial actinic intensity of the 580-nm attractant light. Each point represents data from a population consisting of 100 to 200 cell tracks.

$$|\mathbf{a} + \mathbf{b}| - |\mathbf{a} - \mathbf{b}| > k \text{ (smooth run)}$$

$$|\mathbf{a} + \mathbf{b}| - |\mathbf{a} - \mathbf{b}| \leq k \text{ (tumbling)}$$

where **a** and **b** represent the displacement from the current frame to the next and the one from the previous frame to the current frame, respectively.

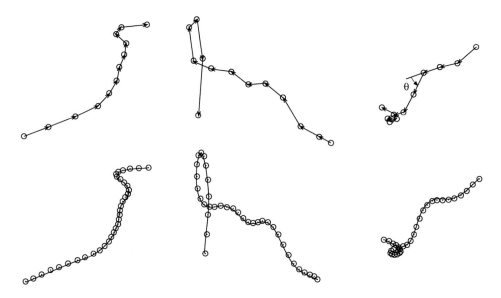

FIGURE 21.5 Trajectories of *Euglena* cells showing photophobic responses. (Upper) Three trajectories of the centroids captured every 67 ms. (Lower) To visualize the relation between displacement vectors that can be detected with the hardware setup used, the positions of the centroids were replotted at 400-ms intervals. θ defines the angular velocity along a path.

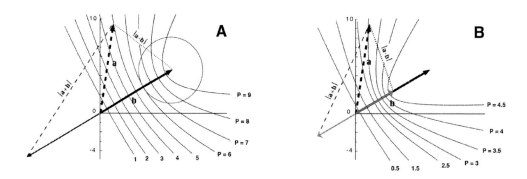

FIGURE 21.6 Principle for the detection of phobic responses of *Euglena* cells. (A) The newly found displacement vector, **a** (broken arrow), was evaluated using the displacement from the previous frame to the current frame as the reference vector, **b** (thick arrow). When the length of the broken line, $|\mathbf{a+b}|$, was larger by a predetermined value P multiplied by one tenth of $|\mathbf{b}|$ than that of the dotted line, $|\mathbf{a-b}|$, **a** was regarded as indicating a smooth run. The parameter P characterizes the shape of the hyperbolic curve that defines the border between the smooth run and the response. To ensure that the vector whose direction was almost identical but its absolute value was smaller than that of the previous vector, runs were regarded as smooth. The circle around the previous vector was also used as another criterion, namely, the condition for smooth run was: $|\mathbf{a-b}|$ < r, and $|\mathbf{a+b}|-|\mathbf{a-b}|$ > k, where r was the radius of the circle. (B) Alternatively, it is possible to use **b**/2 as the reference vector.

Since the elements of all the vectors are integers and the program was written in assembly language, a triangular translation table was used for the calculation of the absolute values of vectors instead of performing floating-point calculations. This reduced the total size of the software as well as processing time.

When the parameter k was large, combination with another criterion was also useful to rescue the smooth paths whose current vector had smaller absolute values. This was carried out by introducing a circle around the end point of the previous vector **b** (Figure 21.6). The program recognizes a smooth run when the new vector **a** satisfies the relation:

$$|\mathbf{a} - \mathbf{b}| < r$$

where r is the second adjustable parameter.[27]

21.3.3 Case 3: Detection of the Tumbling Response from "Path" Data

The time series of X-Y coordinates representing the trajectory of an object or of its centroid is called a *path* (ExpertVision System, MotionAnalysis Corp., Santa Rosa, CA), or a "trace" (Image Tracker PTV, InterQuest, Osaka, Japan). Tracking for a long period of time produces path data too large to be stored in the memory space, and it becomes inevitable to save single or multiple *paths* in a disk file for further analysis. The file can be processed using a set of operational commands provided by the manufacturers or handled through general-purpose spreadsheet software. Sometimes it is necessary, before extracting features of each trajectory, to smooth the path by setting a window in the time domain (usually 3 to 7 consecutive data points were used). Figure 21.7 shows the X-Y projection of a typical path of a bacterium (*Ectothiorhodospira halophila*) captured every 67 ms and smoothed three times using three successive data points as window.

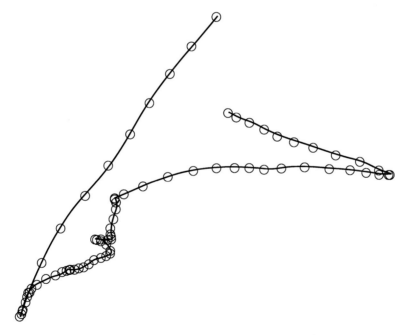

FIGURE 21.7 Typical path of a bacterium showing smooth run and tumbling.

From the *path* file, various parameters can be calculated also as time series, which may reflect a feature of the swimming behavior. These include distance between consecutive frames, linear velocity, acceleration (first derivative of linear velocity), angle of the movement vector with respect to the X-axis, and angular velocity (see Figure 21.5) and its derivative. Among them, linear velocity and the absolute value of angular velocity are first choices to distinguish the behavioral responses from a smooth run.[28-30] The latter is also referred to as "rate of change of direction" (RCDI).[29]

Sager et al.[31] reported that the ensemble average of angular speed recorded at 15 frames per second for a population of *Escherichia coli* (*E. coli*) increased from ca. 45 to 70 degrees per frame and decreased to ca. 20 degrees per frame after mixing with repellent and attractant chemicals, respectively. Time-resolved experiments recently made by Khan et al.[12,32] also showed that RCDI values are suitable for monitoring excitation kinetics of chemotactic responses in *E. coli*. They combined motion analysis techniques with flash photolysis of caged chemoeffectors to attain high time resolution and indicated that kinetic constants up to 16 s^{-1} were measurable for freely swimming bacterial suspensions.[12] As a consequence of the higher frame rate (30 s^{-1}), their RCDI values were almost one half of those reported by Sager et al.,[31] that is, ca. 20 degrees per frame under nonstimulating condition, ca. 40 degrees per frame after repellent stimulation, and ca. 11 degrees per frame after attractant stimulation, suggesting sufficient reproducibility for the use of ensemble averages of the RCDI value.[30,32]

In *Halobacterium salinarum*, it is possible to derive the time-dependent reversal status of a single individual cell from a record of RCDI, or RCDI divided by its linear velocity, if one assumes a certain threshold value that discriminates the reversal and stop responses from a smooth run.[17,29,30]

21.4 OTHER APPLICATIONS

21.4.1 CASE 4: PHOTOTACTIC ORIENTATION OF *CHLAMYDOMONAS*: SIMULTANEOUS MEASUREMENTS OF TAXIS AND PHOBIC RESPONSES

For a cell population, displacement vectors often show a biased distribution regarding their direction. When it occurs in a sufficiently large population, it is regarded as tactic response with respect to the direction of a stimulus. Figure 21.8 shows a custom-made chamber (designed by Dr. Watanabe of National Institute of Basic Biology, Okazaki, Japan)[33] for the measurement of phototaxis in flagellated algae to the light beam incident in the horizontal direction. For recording tactic orientation in Takahashi's system, each displacement vector was accumulated in a counter whose dimension was the same as that of the correlation matrix.[34-36] An example of the output of the counter is shown in Figure 21.9 as a three-dimensional histogram. When it is necessary to quantitate the extent of orientation, the following phototactic indices are usually employed:

$$\text{Positive phototaxis index} = \{|X^+| - 1/4\} \times (4/3),$$

$$\text{Negative phototaxis index} = \{|X^-| - 1/4\} \times (4/3),$$

where X^+ and X^- represent the fraction of cells swimming toward or away from the stimulus light source, respectively, as illustrated in Figure 21.9.[34]

FIGURE 21.8 A microscope chamber for the measurement of phototaxis.[44]

Simultaneous measurements of tactic orientation and photophobic (stop) responses in a retinal-deficient strain of the unicellular green alga *Chlamydomonas* indicated that these two responses appeared to be strongly correlated in fluence rate dependence and also in dependence of the concentration of the chromophore retinal. It was inferred from this, the currently accepted view, that these two responses are mediated by a single photoreceptor species.[34]

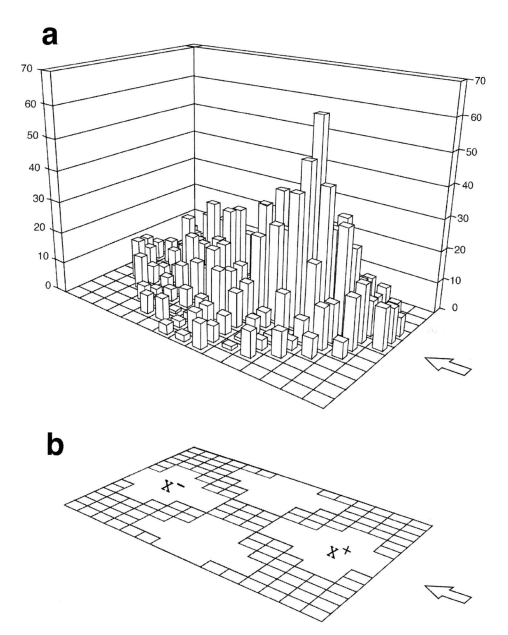

FIGURE 21.9 (a) 3-D histogram of displacement vectors showing a biased orientation due to phototaxis. The arrow indicates the direction of the actinic light. (b) Selected areas for calculating phototactic indices with respect to the direction of the light (white arrow) are indicated as white regions on the correlation matrix (see text).

Tactic response can also be measured by calculating ensemble averages of the angle of the movement axis with respect to x-axis from commercially available motion analyzer systems[37-38] as well as phobic responses in *Chlamydomonas*.[39,40]

21.4.2 CASE 5: ACTION SPECTROSCOPY

Computer-aided automated analysis is especially useful in experiments that need accumulation of large data, such as action spectroscopy.[41,42] Takahashi, in collaboration with Dr. Watanabe of National Institute of Basic Biology, have developed five identical sets of motion analyzer systems for general use at the Okazaki Large Spectrograph (Figure 21.10). Applications of these systems to a variety of microalgal species are described elsewhere.[27,43,44]

FIGURE 21.10   Five identical sets of motion analyzer systems being operated at the Okazaki Large Spectrograph.

21.5   CONCLUSIONS

Video image analysis techniques for the detection of behavioral responses of bacteria and microalgae are described. Real-time analysis using a relatively small custom-made system has allowed fast and sufficiently accurate quantitation of behavioral responses, while application of commercially available systems to this research area enabled time-resolved analysis of behavioral data. In *Halobacterium salinarum* and *Euglena*, phobic responses to step-like photostimuli are detected as deviations in displacement vectors of individual cells. Tumbling/smooth run behavioral switch in enteric bacteria can be detected as changes in RCDI or RCDI divided by the linear velocity at the population level. Parallel execution with several movement analyzer systems with a spectrograph provides a great advantage for action spectroscopic study.

ACKNOWLEDGMENTS

The author wishes to express thanks to Professor Masakatsu Watanabe, National Institute for Basic Biology, Okazaki, Japan, for continuous collaboration and also for providing a nice photograph for this book, Professor John Spudich for arranging a first opportunity to use the MotionAnalysis system. Mamoru Kubota, Professors Koji Nakanishi, Akimori Wada, and Masayoshi Ito are also acknowledged for their kind collaboration and Professor Donat-P. Häder for improvement of the manuscript.

REFERENCES

1. Stryer, L., *Biochemistry*, 4th ed., W. H. Freeman, New York, 1995.
2. Berg, H.C., A physicist looks at bacterial chemotaxis, *Cold Spring Harbor Symp. Quant. Biol.*, 53, 1, 1988.
3. Koshland, D.E., Jr., *Bacterial Chemotaxis as a Model Behavioral System*, Raven Press, New York, 1980.
4. Pfeffer, W., Über chemotaktische Bewegungen von Bacterien, Flagellaten und Volvocineen, *Untersuch. Bot. Inst. Tübingen*, 2, 582, 1888.
5. Adler, J., Chemotaxis in *Escherichia coli*, *Cold Spring Harbor Symp. Quant. Biol.*, 30, 289, 1966.
6. Adler, J., Chemoreceptors in bacteria, *Science*, 166, 1588, 1969.
7. Berg, H.C. and Brown, D.A., Chemotaxis in *Escherichia coli* analyzed by three dimensional tracking, *Nature*, 239, 500, 1972.
8. Silverman, M. and Simon, M., Flagellar rotation and the mechanism of bacterial motility, *Nature*, 249, 73, 1974.
9. Macnab, R.M. and Koshland, D.E., Jr., The gradient sensing mechanism in bacterial chemotaxis, *Proc. Natl. Acad. Sci., U.S.A.*, 69, 2509, 1972.
10. Spudich, J.L. and Koshland, D.E., Jr., Quantitation of the sensory response in bacterial chemotaxis, *Proc. Natl. Acad. Sci., U.S.A.*, 72, 710, 1975.
11. Springer, M.S., Goy, M.F., and Adler, J., Protein methylation in behavioral control mechanisms and in signal transduction, *Nature*, 280, 279, 1979.
12. Jasuja, R., Keyoung, J., Reid, G.P., Trentham, D.R., and Khan, S., Chemotactic responses of *Eschericha coli* to small jumps of photoreleased *L*-asparate, *Biophys. J.*, 76, 1706, 1999.
13. Takahashi, T. and Kobatake, Y., Computer-linked automated method for measurement of the reversal frequency in phototaxis of *Halobacterium halobium*, *Cell Struct. Funct.*, 7, 183, 1982.
14. Häder, D.-P., Real-time tracking of microorganisms, in *Image Analysis in Biology*, CRC Press, Boca Raton, FL, 1991, 289.
15. Freeman, H., Computer processing of line-drawing images, *Comput. Surv.*, 6, 57, 1974.
16. Freeman, H., Analysis and manipulation of lineal map data, in *Map Data Processing*, Academic Press, New York, 1980, 151.
17. Takahashi, T., Automated measurement of movement responses in halobacteria, in *Image Analysis in Biology*, CRC Press, Boca Raton, FL, 1991, 315.
18. Hazelbauer, G.L., The bacterial chemosensory system, *Can. J. Microbiol.*, 34, 466, 1988.
19. Alam, M. and Oesterhelt, D., Morphology, function and isolation of halobacterial flagella, *J. Mol. Biol.*, 176, 459, 1984.
20. McCain, D.A., Amici, L.A., and Spudich, J.L., Kinetically resolved state of the *Halobacterium halobium* flagellar motor switch and modulation of the switch by sensory rhodopsin I, *J. Bacteriol.*, 169, 4750, 1987.
21. Schimz, A. and Hildebrand, E., Response regulation and sensory control in *Halobacterium halobium* based on an oscillator, *Nature*, 317, 641, 1985.
22. Dencher, N.A. and Hildebrand, E., Two photosystems controlling behavioral responses of *Halobacterium halobium*, *Nature*, 257, 46, 1975.
23. Spudich, J.L. and Stoeckenius, W., Photosensory and chemosensory behavior of *Halobacterium halobium*, *Photobiochem. Photobiophys.*, 1, 43, 1979.
24. Lindbeck, J.C., Goulbourne, E.A., Johnson, M.S., and Taylor, B.L., Aerotaxis in *Halobacterium salinarium* is methylation-dependent, *Microbiology*, 141, 2945, 1995.
25. Takahashi, T., Tomioka, T., Kamo, N., and Kobatake, Y., A photosystem other than PS370 also mediates the negative phototaxis of *Halobacterium halobium*, *FEMS Microbiol. Lett.*, 28, 161, 1985.
26. Diehn, B., Phototaxis and sensory transduction in *Euglena*, *Science*, 181, 1009, 1973.

27. Matsunaga, S., Hori, T., Takahashi, T., Kubota, M., Watanabe, M., Okamoto, K., Masuda, K., and Sugai, M., Discovery of signaling effect of UV-B/C light in the extended UVA/blue-type action spectra for step-down and step-up photophobic responses in the unicellular flagellate alga, *Euglena gracilis, Protoplasma*, 201, 45, 1998.

28. Sundberg, S.A., Bogomolni, R., and Spudich, J.L., Selection and properties of phototaxis-deficient mutants of *Halobacterium halobium, J. Bacteriol.*, 164, 282, 1985.

29. Sundberg, S.A., Alam, M., and Spudich, J.L., Excitation signal processing times in *Halobacterium halobium* phototaxis, *Biophys. J.*, 50, 895, 1986.

30. Marwan, W., Schäfer, W., and Oesterhelt, D., Signal transduction in *Halobacterium halobium* depends on fumarate, *EMBO, J.*, 9, 355, 1990.

31. Sager, B.M., Sekelsky, J.J., Matsumura, P., and Adler, J., Use of a computer to assay motility in bacteria, *Anal. Biochem.*, 173, 271, 1988.

32. Khan, S., Spudich, J.L., McCray, J.A., and Trentham, D.R., Chemotactic signal integration in bacteria, *Proc. Natl. Acad. Sci., U.S.A.*, 92, 9757, 1995.

33. Watanabe, M. and Furuya, M., Phototactic behavior of individual cells of *Cryptomonas* sp. in response to continuous and intermittent light stimuli, *Photochem. Photobiol.*, 35, 559, 1982.

34. Takahashi, T., Yoshihara, K., Watanabe, M., Kubota, M., Johnson, R., Derguini, F., and Nakanishi, K., Photoisomerization of retinal at 13-ene is important for phototaxis of *Chlamydomonas reinhardtii*: simultaneous measurements of phototactic and photophobic responses, *Biochem. Biophys. Res. Commun.*, 178, 1273, 1991.

35. Takahashi, T., Kubota, M., Watanabe, M., Yoshihara, K., Derguini, F., and Nakanishi, K., Diversion of the sign of phototaxis in a *Chlamydomonas reinhardtii* mutant incorporated with retinal and its analog, *FEBS Lett.*, 314, 275, 1992.

36. Takahashi, T. and Watanabe, M., Photosynthesis modulates the sign of phototaxis of wild-type *Chlamydomonas reinhardtii*. Effects of background illumination and 3-(3′,4′-dichlorophenyl)-1,1-dimethylurea, *FEBS Lett.*, 33, 516, 1993.

37. Zacks, D.N., Derguini, F., Nakanishi, K., and Spudich, J.L., Comparative study of phototactic and photophobic receptor chromophore properties in *Chlamydomonas reinhardtii*, *Biophys. J.*, 65, 508, 1993.

38. Sakamoto, M., Wada, A., Akai, A., Ito, M., Goshima, T., and Takahashi, T., Evidence for the archae-bacterial-type conformation about the bond between the β-ionone ring and the polyene chain of the chromophore retinal in chlamyrhodopsin, *FEBS Lett.*, 434, 335, 1998.

39. Hegemann, P. and Bruck, B., Light-induced stop response in *Chlamydomonas reinhardtii*: occurrence and adaptation phenomena, *Cell Motil. Cytoskel.*, 14, 501, 1989.

40. Lawson, M.A., Zacks, D.N., Derguini, F., Nakanishi, K., and Spudich, J.L., Retinal analog reconstitution of photophobic responses in a blind *Chlamydomonas reinhardtii* mutant. Evidence for an archaebacterial like chromophore in a eukaryotic rhodopsin, *Biophys. J.*, 60, 1490, 1991.

41. Tomioka, H., Takahashi, T., Kamo, N., and Kobatake, Y., Action spectrum of the photoattractant response of *Halobacterium halobium* in early logarithmic growth phase and the role of sensory rhodopsin, *Biochim. Biophys. Acta*, 884, 578, 1986.

42. Takahashi, T., Tomioka, H., Nakamori, Y., Kamo, N., and Kobatake, Y., Phototaxis and the second sensory pigment in *Halobacterium halobium*, in *Primary Processes in Photobiology*, Kobayashi, T., ed., Springer-Verlag, Berlin, 1987, 101.

43. Erata, M., Kubota, M., Takahashi, T., Inoue, I., and Watanabe, M., Ultrastructure and phototactic action spectra of two genera of cryptophyte flagellate algae, *Cryptomonas* and *Chroomonas, Protoplasma*, 188, 258, 1995.

44. Horiguchi, T., Kawai, H., Kubota, M., Takahashi, T., and Watanabe, M., Phototactic responses of four marine dinoflagellates with different types of eyespot and chloroplast, *Phycol. Res.*, 47, 101, 1999.

22 Simultaneous Tracking of Multiple Subjects

David B. Dusenbery

CONTENTS

22.1 INTRODUCTION

Studies of the behavior of all types of organisms are plagued by large variations in individual behavior. Usually, one is most interested in the average behavior of a population and much less in individual variation. In many instances, the information sought concerns relatively simple questions about the movements of the subjects. When this is the case and a suitable group of subjects can be viewed simultaneously by a camera, it is possible to have a computer automatically track the movements of the individuals and average together appropriate data about their behavior. The summary data can then be reported continuously within seconds of its occurrence. This data could be used directly or used to control some event in real-time.

This chapter focuses on experience in the author's laboratory in developing and using systems for simultaneous tracking of the locomotion of many individual subjects and real-time reporting of statistical data on their behavior. Although experience with this technique is mostly limited to use with nematodes (round worms), it should be applicable to a much broader range of organisms.

22.2 METHODS

22.2.1 STRATEGIES

22.2.1.1 Simple Recognition

To simultaneously track multiple subjects, it is important to identify their positions rapidly. Consequently, a simple strategy has been used. In the original system, the location of each subject is determined by the location of the first pixel that is found to be above threshold. During each round of tracking, the video frame is examined and if the pixel at the previous position of a subject is still above threshold, no movement is recorded. If the pixel is now below threshold, a movement is recorded for the subject and a search is initiated.

During the search, adjacent pixels in an ever-widening circle are examined out to a limit of a radius of 9 pixels. The first pixel found to be above threshold is recorded as the new location of the subject. If no pixels above threshold are found within the radius of 9, the search is abandoned. This search area, the magnification, and the time between rounds of tracking are adjusted so that the fastest moving subjects are not likely to get out of the search area before a new position is defined. With this tracking strategy, it is actually the trailing edge of the subject that is tracked.

This crude method does not detect all movements. If a subject reverses direction of locomotion, it will appear to be stationary until its head passes the point where the tail was located when the change of direction occurred. A more precise detection of movement can be obtained by tracking fewer subjects and determining their centers; but for many purposes, it is advantageous to track more subjects.

To provide more flexibility, a recent version of the software provides optional search strategies that may find a position closer to the center of the subject. One strategy simply looks for the pixel within the search area that most closely matches the ideal intensity of the subject. This slows down the search because all pixels within the search area must be examined every time. Another search strategy follows intensity gradients to a maximum or minimum. This strategy can potentially find a target further away and faster because fewer pixels must be examined, but it requires a subject image with smoothly varying intensities.

22.2.1.2 Visual Feedback

Early experience in developing this technique demonstrated that it is very useful to have a video screen that overlays the image from the camera with indications from the computer of where the system has recorded the position of each subject. This allows the user to determine if the system is seeing dirt or noise rather than the appropriate subjects and if it is tracking the subjects properly. To indicate the positions, markers were originally used that were single pixels that differed greatly in intensity or color from the camera image. The most recent version of the software allows other kinds of markers, including labels that allow subjects on the screen to be identified with recorded data.

22.2.2 HARDWARE

22.2.2.1 Computer

The personal computers that are now so widely available have more than sufficient capacity to perform the basic functions necessary for simultaneous tracking. The first system[1] used a computer based on the Motorola 6809 microprocessor with 72 KB of memory and the OS-9 operating system. The second-generation system[2] used an IBM XT computer. And the third-generation system[3] used an IBM AT computer. The most recent version employs Apple Power Macintosh computers.

If a particular application requires that subjects be viewed in an environment that makes them difficult to distinguish from the background or other objects, it may be helpful and important to use a real-time image processor to enhance contrast before the tracking software searches for the subjects. Such processors are generally quite expensive but are becoming increasingly available. If contrast is a potential problem, it would be wise to select a computer that can accommodate such a processor.

22.2.2.2 Camera Interface

The electronic equipment to interface the camera to the computer is the most specialized and possibly the most expensive hardware requirement. In assembling a system from scratch, the availability of an appropriate camera interface should be the first consideration. Although hardware to interface video cameras to personal computers is now widely available, when the author's investigations started in the early 1980s, the interface was built in the lab (by Bill Goolsby). This interface digitized the video in a single scan in 1/60 second into an array of 240×256 pixels with 1 bit of intensity. A second slower commercial interface was used to generate markers on a video screen to indicate where the computer was tracking the subjects.

The interface used with the IBM computers was the PCVISION frame grabber from Imaging Technology, Inc. (600 West Commings Park, Woburn, MA 01801). It digitizes a standard RS-170 video signal into 6 bits of intensity and stores a video frame of 512×480 pixels in its own memory in 1/30 second. In addition to its greater resolution, it contains a color video output with a programmable lookup table for mapping bytes of video memory to any of 16 million colors. This output is connected to an RGB color monitor. This system allows markers to be displayed in colors different from the camera input and also allows the use of color to distinguish between input intensities that are above and below threshold.

A Power Macintosh computer with a Scion LG-3 video digitizer is currently being employed. The Scion board provides a great deal of flexibility in adjusting gain and contrast. This makes it much easier to obtain appropriate lighting conditions.

22.2.2.3 Camera

The requirements of the camera are relatively simple. Although color cameras are widely available, color is rarely necessary and the inexpensive closed-circuit television (CCTV) cameras used in surveillance are quite satisfactory. Even if color were useful in distinguishing the subjects from background, a color filter would probably be a better solution than a color camera. If high sensitivity is required in order to work in low light, very sensitive surveillance cameras are available, although they cost thousands rather than hundreds of dollars.

If the subjects are small, the camera may need to be fitted with a macro lens or a microscope. Maximum flexibility is provided by a camera that accepts C-mount lenses, which most surveillance cameras do — in contrast to home video cameras. For subjects on the order of a millimeter in size, an enlarger lens may be good because it is highly corrected for a flat field at appropriate working distances. The author uses a Computar f 1.9, 55 mm lens for small nematodes.[2]

22.2.2.4 Illumination

A key element in making this technique fast and easy is to have the subjects stand out in high contrast from the background so that a simple threshold intensity criterion reliably distinguishes them. The appropriate strategy depends on the subjects and their environment. People viewed from above against a concrete courtyard might have sufficient contrast, but it could probably be improved by the use of color filters to select the color of light that gives the best contrast. Black mice should be viewed against a white background and vice versa. In some circumstances, it might be appropriate to attach retro-reflectors (e.g., glass beads as used on highway signs) to the subjects and illuminate them with a beam of light from the direction of the camera. For small organisms that scatter light, dark-field illumination is ideal. The author has assembled such a system for use with nematodes (1 mm long and translucent).

22.2.3 SOFTWARE

22.2.3.1 Language

In principle, any computer language can be used as long as it can address the video memory. However, most applications will probably require a compiled language (unlike regular Basic or Java) in order to have sufficient speed to track multiple subjects. The original system used BASIC09, a partially compiled and partially interpreted language, and some time-critical parts had to be written in assembly language. The later systems used Microsoft Pascal, and assembly-language modules were used only to provide access to I/O ports not accessible from the higher level language.

The most recent version of the software is built on the public-domain program NIH Image written by Wayne Rasband (http://rsb.info.nih.gov/nih-image/about.html) in Pascal for the Code Warrior compiler. This software base provides many more options for image manipulation, such as subject markers with labels (Figure 22.1), and 35 parameters can be specified by the user. Data on individual subjects can also be recorded for later analysis. However, it is slower than previous versions because the whole image is copied to the computer memory. Nonetheless, nearly 100 subjects have been tracked on a Mac 7500 with a 150-MHz 604e processor upgrade, and much faster computers are now available.

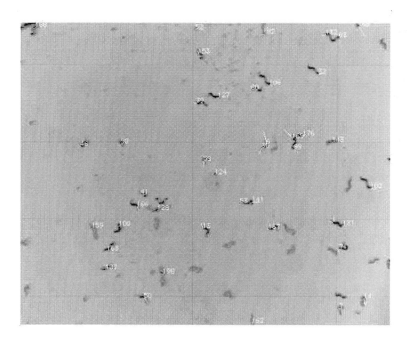

FIGURE 22.1 Image from the screen of the latest version of the software. Intensities above threshold are dark. The white numbers are labels that associate specific subjects in the image with specific subjects in the recorded data.

Another advantage of this version is that it has the capability of using simple programs written in a macro language to control tracking. The macro capability has been used to combine tracking with stimulus control.

22.2.3.2 Search Pattern

A key element in obtaining sufficient speed to make this system work is to avoid calculating the next nearest pixel to check during a search. Instead, a lookup table is generated that provides the relative positions of the pixels to be searched in order of increasing distance from the starting pixel. Thus, the search program merely has to work down the table until an appropriate pixel is located.

The generation of the search pattern starts with a table that contains pairs of positive integers (i,j) in order of increasing value of $i^2 + j^2$ up to but not including the pair 0,9. Each pair of numbers is then used to generate 4 or 8 sets of x and y offsets by assigning $x = \pm i$ and $y = \pm j$ in all combinations of sign and repeating with x and y reversed when $x \neq y$. The result is a table of x and y offsets that is in order of increasing distance from the origin. It includes every point inside a circle of radius 9.

Of course, different size areas of search patterns could be used, but 9 was considered to be a good compromise. In considering searching to larger distances, one should keep in mind that the time to search the whole area will increase with the square of the radius.

When the newest version of the program was written, optional search strategies were included that may be useful in some circumstances. With a nearest-best-intensity search, the program looks for the most extreme intensity (light or dark, as specified) within the search area. Pixels may also be weighted according to their distance from the focus of the search. With a gradient-following search, the program follows the intensity gradient from the old position as far as it goes. These strategies can also be combined to do gradient following after a threshold or best-pixel search.

22.2.3.3 Measures of Behavior

In each tracking cycle, the search procedure provides the X,Y coordinates of each subject. These are compared against the previous values to determine the changed coordinates ΔX and ΔY, which are the raw data used to describe the behavior.

The most basic measure of behavior is the rate of locomotion, which is the distance moved divided by the time between tracking cycles. The distance would normally be calculated from $(\Delta X^2 + \Delta Y^2)^{1/2}$. However, calculating the squares and particularly the square root takes a lot of micro-processor time. Originally, this delay was avoided by simply using the sum $(\Delta X + \Delta Y)$ for an approximation of the distance moved. (Assuming a large number of subjects are moving in random directions, this simplification gives a constant overestimate of 27%, which is no problem for many applications.) In more recent computers with built-in numeric processors, the calculation of $(\Delta X^2 + \Delta Y^2)^{1/2}$ is not a problem.

In using this procedure to study chemotaxis, the author wanted to measure the frequency with which the subjects changed direction of locomotion. However, a straightforward geometrical mea-sure would have required far too much computer time. Consequently, a simple logical test was employed, based on the current and previous signs of ΔX and ΔY and whether either of the current displacements was zero.

A change in the direction of locomotion was identified when there was a change in direction along both axes, or when there was a change in direction along one axis and no change in position along the other. This strategy avoided signaling a change in direction from a gradually curving path.

22.2.3.4 Finding Subjects

Some thought needs to be given to the procedure for identifying new subjects to track. Do you want the probability of a subject being tracked to be evenly distributed over the area viewed? Do you want to favor moving subjects or subjects moving toward the center? For this author's purposes, an even distribution was the main consideration. To obtain it, a position in the field of view is chosen at random, and if that pixel is not above threshold, adjacent pixels are checked until a pixel above threshold is found or the time allotted for looking for new subjects is elapsed. If the image of the subjects includes several pixels, it is expedient not to check each adjacent pixel but rather to skip over a few.

When a pixel with the appropriate intensity for a subject is found, the next step is to determine if it belongs to the image of a subject that is already being tracked. This is done by checking its position against the current positions of all the subjects being tracked. If it is within a certain distance of any of them, it is assumed to belong to a subject already being tracked and is not used. The distance used to make this decision is based on the size of the images of the subjects.

22.2.3.5   Collisions

When two subjects come together, there is no reliable way of determining which is which when they separate. In fact, with the procedures described thus far, if both subjects are being tracked, when they separate, the computer may follow only one of the two images as two subjects. To avoid this problem, in each tracking cycle that determines new positions for each of the subjects being tracked, the position of each is compared against all the others; and if any are identical, one of the two is abandoned. The choice as to which of the two to drop is arbitrary unless one has just registered a change of direction, in which case it is the one abandoned.

In most applications, it will be important to keep the density of the subjects sufficiently low that collisions occur infrequently. If there are so many subjects that collisions are frequent and many of the subjects are not tracked, the measures of the frequency of change of direction will be exaggerated by the system switching from one subject to another when they come together on paths of opposite direction and recording this as a change of direction, when in reality neither subject changed direction.

22.3   APPLICATIONS

22.3.1   Possible Range of Applications

The technique of simultaneous tracking has a very wide range of application. It should be applicable to bacteria that can be viewed by a video camera mounted on a compound microscope. When studies of locomotion in two dimensions is sufficient, the approach would be straight-forward. At the other extreme, simultaneous tracking of large animals should be possible if a camera can be positioned at an appropriate vantage point. For example, it should be possible to study the movement of people walking across a courtyard or automobiles with a camera looking down from a building.

In many cases, it would be desirable to track subjects in three dimensions. This is much more difficult and few people have done simultaneous tracking of biological subjects in three dimensions. The simplest approach would be to view the volume with two cameras. The principal difficulty is to identify corresponding images in the two cameras.

Simultaneous tracking of microscopic organisms in three dimensions faces additional problems because high resolution requires objective lenses of high numerical aperture and there will be limits on how large the numerical aperture can be and still get the lenses sufficiently close to the subjects and at an angle to one another.

The possible range of applications is illustrated by presenting several examples of applications that have employed nematodes as subjects.

22.3.2 Stimuli

22.3.2.1 Chemical Stimuli

The technique of simultaneous tracking was originally developed to study the response of nematodes to controlled changes in the concentrations of chemical stimuli. The previous technique, which revealed the need for simultaneous tracking, involved holding a single nematode by the tail with a suction pipette while various solutions were pumped past it and its shadow was projected on an array of light detectors wired to a multichannel recorder.[4] The chart was then examined manually to identify behavior that normally leads to a change in the direction of locomotion. Testing for response to a chemical involved switching back and forth between two different concentrations of the chemical in the flowing solution and determining if behavior differed between the different exposures. This technique required many hours of recording to obtain statistically significant data for weak stimuli such as oxygen.[5]

This experience led to thinking about ways of obtaining data from many subjects simultaneously and ultimately to the development of the simultaneous tracking system. The nematodes are allowed to move around on the surface of a thin layer of agar across which a stream of hydrated air blows to carry volatile stimuli. Dark-field illumination provides a high-contrast image of the translucent nematodes. This technique provided the first method by which data on the behavior of a group of nematodes exposed to a controlled chemical stimulus could be obtained.[1] The original system could simultaneously track 25 animals with the positions established once a second. Within a few minutes, statistically significant data could be acquired that had previously required many hours to obtain with one nematode at a time.[6] The results (Figure 22.2) confirmed responses to oxygen and carbon dioxide and revealed very different patterns of adaptation, which is consistent with the suggestion that oxygen is sensed internally.[5] Responses to very low concentrations have also been documented.[7]

Techniques were also developed to measure responses to spatial gradients. A parallel flow from two sources with different concentrations of the test chemical was established (Fig. 22.3). Diffusion across the flow was sufficient to establish a nearly linear gradient in the vicinity of the nematodes. The software was modified to separately measure movements across and parallel to the flow. The results clearly demonstrated that the nematodes moved toward the side with higher concentrations of carbon dioxide.[2]

22.3.2.2 Thermal Stimuli

Plant-parasitic nematodes migrate in very shallow temperature gradients[8] toward a preferred temperature.[9] They can respond to temperature gradients as low as $0.001°C/cm$.[10] Questions about the mechanism of the response led to the use of simultaneous tracking experiments in which the temperature was altered by shining an incandescent lamp on the agar containing the nematodes.[11] Turning the lamp on and off generated temperature changes that were estimated to be on the order of $10^{-4}°C/s$. These experiments employed the second-generation system and it could simultaneously track 200 nematodes with positions established every 3 s.

As seen in Figure 22.4, behavior changed within about 10 s of the lamp being turned on or off. Consequently, the nematodes responded to a temperature change on the order of $0.001°C$. Temperature changes away from the preferred temperature resulted in decreases in the rate of movement and increases in the rate of change of direction, whether the changes were toward warmer or cooler temperatures. These behavioral changes lasted about 30 s. Temperature changes toward the preferred temperature caused the response rates to change in the opposite direction. In this case, the behavioral changes persisted for several minutes. These results demonstrate that nematodes can respond to a purely temporal thermal stimulus in a manner consistent with efficient indirect orientation or klinokinesis. They further demonstrated, very clearly, different rates of adaptation

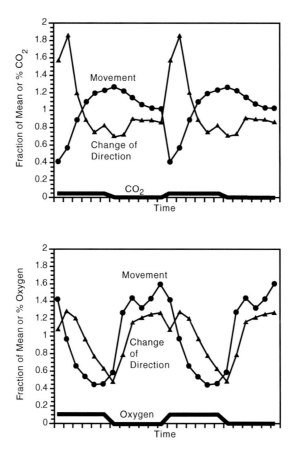

FIGURE 22.2 Behavioral responses to chemical stimulation. Average rates of locomotion and frequency of change of direction in response to cyclic changes in the concentrations of carbon dioxide or oxygen. In the upper graph, the concentration was changed back and forth between 0 and 4% carbon dioxide at 1-min intervals. In the lower graph, the oxygen concentration alternated between 0 and 10% on the same schedule. Shown are values averaged over 50 such stimulus cycles.[6]

for increases and decreases in stimulation, which has been suggested to lead to more efficient responses.[12]

22.3.3 BEHAVIORAL TOXICITY

Toxicity testing of environmental samples and synthetic chemicals is a problem of growing concern. Most such tests use survival after a period of exposure as a measure of toxicity. However, it is important to determine whether significant non-lethal effects occur at lower levels than those that are lethal. One of the most important areas of sub-lethal effects involves effects on the nervous system and behavior. Behavioral endpoints are desirable but because of the large variability in behavior, they are difficult to carry out. Nematodes have been found useful in toxicity testing.[13] They are as sensitive as other organisms used in testing water toxicity.[14] It seemed that simultaneous tracking offered an opportunity to obtain large amounts of behavioral information in a short time and relatively inexpensively.

The nematodes were exposed to the test chemicals for 24 h and then transferred to a thin pool of agar under the video camera. The third-generation system, which could simultaneously track up

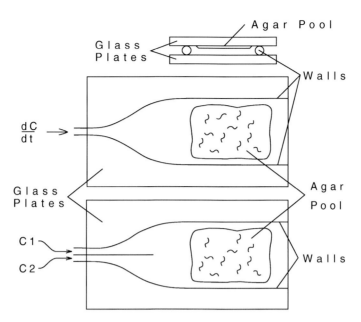

FIGURE 22.3 Diagram of the two types of flow chambers used with nematores. The top figure is a cross-section of either chamber. The middle figure is the chamber used to generate a temporal, spatially uniform stimulus. The lower figure is the chamber used to generate a steady spatial gradient. (From Pline, M. and Dusenbery, D.B., *J. Chem. Ecol.*, 13, 873, 1987. With permission.)

to 300 subjects with positions established once a second, was employed. The average rate of locomotion was determined for animals exposed to the chemical being tested and to a parallel control group that was treated similarly except for being exposed to the chemical under test. The ratio of the rate of locomotion in the exposed animals to that of the control group was used as the basic indication of neurobehavioral toxicity. This parameter was determined for a number of concentrations of the chemical to develop a dose-response curve, which was then compared to the survival curve.

Initial testing with metallic ions[3,15] demonstrated behavioral effects at concentrations well below those that caused lethality for the neurotoxins lead and mercury, but not for the non-neurotoxin copper. Surprisingly, mercury produced hyperactivity in the nematodes, suggestive of its effects on humans.

22.3.4 GAS CHROMATOGRAPHY DETECTION

One of the strengths of simultaneous tracking is that statistically significant behavioral data can be obtained in a few seconds. This makes it attractive as a detector for gas chromatography. This application has been developed in order to search for volatile stimuli that plant-parasitic nematodes might use to locate the roots of host plants (Fig. 22.5).[16]

A sample of gas was collected from around the roots and injected into the gas chromatograph. A temperature program was used to separate components of a wide range of volatility. As they eluted from the column, they passed through a thermal conductivity detector and were then carried in an air stream over nematodes on an agar slab. Data on the average distance moved, the number of changes of direction detected, and from the thermal conductivity detector were collected by the computer and plotted together. These experiments indicated that carbon dioxide was the only volatile stimulus released by the roots, although the nematodes also responded to oxygen depletion.[16]

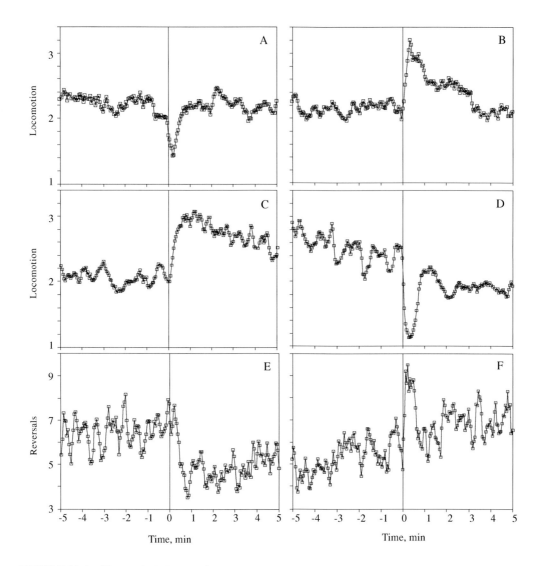

FIGURE 22.4 Changes in behavior of second-stage juveniles of *Meloidogyne incognita* caused by changes in temperature. The nematodes were grown at 23°C and consequently tended to move toward a "preferred" temperature of about 26°C.[9] Temperature changes were generated by turning a red glowing incandescent lamp on (left-side plots) or off (right-side plots) at zero time. Experiments A and B were conducted at an ambient temperature of 29°C, and the others were conducted at 23°C.[11] Notice that a change in temperature toward the preferred temperature (B, C, E) resulted in an increase in the rate of locomotion and decrease in the frequency of changes of direction and that these changes persisted for several minutes. In contrast, a change away from the preferred temperature (A, D, F) resulted in a decrease in locomotion, an increase in change of direction, and these changes in behavior lasted less than a minute. Generally, about 200 nematodes were tracked simultaneously with the position of each established every 3 s. Each plot is the average of three experiments smoothed with a running average of three. Locomotion (A-D) is the average per nematode of the sum of the number of pixels moved in the X and Y directions during the 3-s interval. At the magnifications used, pixels are separated by about 33 μm. Number of reversals, changes of direction of more than 90°, recorded in each interval per 100 nematodes is plotted (E, F).

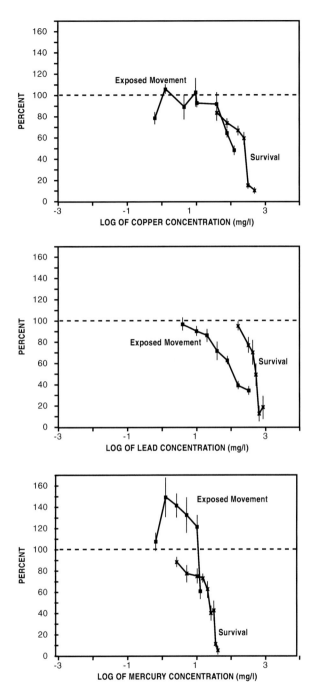

FIGURE 22.5 Rate of locomotion and survival of nematodes exposed to common salts of three different metals: copper, lead, and mercury. Values plotted are the percent survival or the average rate of locomotion of the nematodes exposed to the metal salt as a percent of the locomotion of nematodes from a parallel control that were not exposed. (From Williams, P.L., and Dusenbery, D.B., *Toxicol. Ind. Health*, 6, 425, 1990. With permission).

22.4 SUMMARY

Simultaneous tracking is a technique that can be implemented at relatively low cost on personal computers. It provides a method of obtaining large amounts of behavioral data in short time periods. This permits automatic processing by the computer and presentation of results in nearly real-time.

A wide range of applications has already been demonstrated. Future developments might include tracking in three dimensions or use of the behavioral data to control experimental parameters in real-time.

REFERENCES

1. Dusenbery, D. B., Using a microcomputer and video camera to simultaneously track 25 animals, *Comput. Biol. Med.*, 15, 169, 1985.
2. Pline, M. and Dusenbery, D. B., Responses of the plant-parasitic nematode *Meloidogyne incognita* to carbon dioxide determined by video camera-computer tracking, *J. Chem. Ecol.*, 13, 1617, 1987.
3. Williams, P. L. and Dusenbery, D. B., A promising indicator of neurobehavioral toxicity using the nematode *Caenorhabditis elegans* and computer tracking, *Toxicol. Industrial Health,* 6, 425, 1990.
4. Dusenbery, D. B., Responses of the nematode *Caenorhabditis elegans* to controlled chemical stimulation, *J. Comp. Physiol.*, 136, 327, 1980.
5. Dusenbery, D. B., Appetitive response of the nematode *Caenorhabditis elegans* to oxygen, *J. Comp. Physiol.*, 136, 333, 1980.
6. Dusenbery, D. B., Video camera-computer tracking of the nematode *Caenorhabditis elegans* to record behavioral responses, *J. Chem. Ecol.*, 11, 1239, 1985.
7. Terrill, W. F. and Dusenbery, D. B., Threshold chemosensitivity and hypothetical chemoreceptor function of the nematode *Caenorhabditis elegans*, *J. Chem. Ecol.*, 22, 1463, 1996.
8. El-Sherif, M. and Mai, W. F., Thermotactic response of some plant parasitic nematodes, *J. Nematol.*, 1, 43, 1969.
9. Diez, J. A. and Dusenbery, D. B., Preferred temperature of *Meloidogyne incognita*, *J. Nematol.*, 21, 99, 1989.
10. Pline, M., Diez, J. A., and Dusenbery, D. B., Extremely sensitive thermotaxis of the nematode *Meloidogyne incognita*, *J. Nematol.*, 20, 605, 1988.
11. Dusenbery, D. B., Behavioral responses of *Meloidogyne incognita* to small temperature changes, *J. Nematol.*, 20, 351, 1988.
12. Dusenbery, D. B., The value of asymmetric signal processing in klinokinesis, *Biol. Cybernetics*, 61, 401, 1989.
13. Samoiloff, M. R. and Bogaert, T., The use of nematodes in marine ecotoxicology, in *Ecotoxicological Testing for the Marine Environment*, Persoone, G., Jaspers, E., and Claus, C., Eds., State Univ. Ghent and Inst. Mar. Scient. Res., Bredene, Belgium, 1984, 407.
14. Williams, P. L. and Dusenbery, D. B., Aquatic toxicity testing using the nematode, *Caenorhabditis elegans*, *Envir. Toxicol. Chem.*, 9, 1285, 1990.
15. Williams, P. L. and Dusenbery, D. B., Screening test for neurotoxins using *Caenorhabditis elegans*, in *Model Systems in Neurotoxicology*, Shahar, A. and Goldberg, A. M., Eds., Alan R. Liss, New York, 1986, 163.
16. McCallum, M. E. and Dusenbery, D. B., Computer tracking as a behavioral GC detector: nematode responses to vapor of host roots, *J. Chem. Ecol.*, 18, 585, 1992.

23 The Use of Image Analysis in Ecotoxicology

Harald Tahedl and Donat-P. Häder

CONTENTS

23.1 INTRODUCTION

Ecotoxicology investigates the impact of toxic substances on the biota. This science has become more important as pollution increased in parallel to economic development. About 5 million different chemicals are known and about 80,000 of them are in use.[1] Information on the interaction of chemicals with plants, animals, and microorganisms is required to evaluate the toxic influence of these substances. The presence or absence of species in ecosystems can be used as an environmental quality indicator. In addition to the observation of populations or communities, many efforts have been made to culture and test organisms under laboratory conditions. This helps to predict possible future effects of pollutants and to evaluate observed effects in natural ecosystems. The direct effect of toxic substances on single species is used as a base for the evaluation of their ecotoxicological effects on the environment. Many toxicity tests have been developed to determine the effects of chemicals on single species under laboratory conditions. All biological toxicity tests are based on the same principle: living organisms are exposed to toxic agents and their response depending on the dose (exposure level of the toxic agent) is measured. Test organisms, realization, and conditions for different toxicity tests are defined in different guidelines such as the OECD (Organization of Economic Cooperation and Development) guidelines for testing chemicals[2] in Europe or the U.S. EPA (United States Environmental Protection Agency) guideline[3] in the United States.

Acute toxicity tests are short-term tests that most often measure survival after a 24- to 96-h period. In chronic toxicity tests, the organisms are exposed over a significant portion of their life cycle, usually between 21 and 28 days. Chronic tests measure effects on sensitive response parameters (endpoints) such as reproduction growth, or sublethal effects like behavioral, physiological and biochemical effects. Especially the measurement of sublethal effects leads to the development and implementation of new detection systems. Image analysis is a powerful tool for the analysis of motile behavior. It not only allows the evaluation of known response parameters in a standardized and automatic way, but also the inclusion of new and interesting endpoints in behavior.

23.2   REAL-TIME TRACKING

Fish have been often used as aquatic test organisms for the evaluation of ecotoxicological responses of chemicals.[4,5] Usually, the survival in acute tests or the behavior in chronic tests is determined to assess toxicity. Automated fish tests are presently used to measure acute toxicity in a flow-through system.[6] The fish swim against the stream in the flow-through chamber and the number of impulses, triggered by contact of the fish with a reaction lattice at the backside of the chamber, is counted and used as the response parameter. If toxic substances, which inhibit the fish's movements, are present, it is more difficult for the fish to swim against the stream and the number of impulses rises. The search for new and more sensitive endpoints led to the application of the image analysis system BehavioQuant (Spieser, München) in fish toxicity measurements (Figure 23.1).[7,8] This system uses a multiplexer to switch between up to 16 CCD cameras and connect them to the frame grabber card BHQ-1. The movement of usually six to ten objects in the observation chamber is recorded by video imaging, and the traces are analyzed. A background image, which is calculated as the mean of 50 to 100 digitized images before the actual measurement, is used as reference. Each pixel of the digitized video images is compared with the background image. Differences in the brightness level, exceeding a defined gray level, are used to differentiate between object pixels and background. Edge coordinates are established to calculate the mean coordinates of the found objects and are saved on hard disk. The time interval between two consecutive images is 40 ms, and the nearest neighbor algorithm is used to define the movement vector. The mean coordinates and vectors are used to calculate all behavior parameters.[9] In addition to the velocity, the relative depth from the surface and the number of turns are determined, as well as two more complex endpoints:[10-12] the preference of light and dark habitats and the mean distance of the organisms to each other. The latter describes the shoal behavior of the fish.

Chronic toxicity tests with the herbicide atrazine showed that the frequency distribution of the zebrafish *Brachydanio rerio* is a sensitive endpoint for this toxin. While standard toxicity tests with various fish species revealed lethal concentrations for 50% of the tested organisms (LC_{50}) between 8.8 and 76 mg/l, the frequency distribution of *Brachydanio rerio* changed during the first week of atrazine exposure at a concentration of 5 μg/l. The fish showed a more pronounced preference for dark habitats in the presence of atrazine.[10]

In addition to fish, the crustacean *Daphnia magna* (with a size of 5 to 6 mm) is a commonly used biologic test organism to monitor water quality by measuring acute (mortality) and chronic (reproduction and mortality) toxicity.[13-16] In 1978, Knie developed a dynamic *Daphnia* test,[17] where a defined number of organisms are observed in a flow-through system. Their swimming behavior is analyzed by light barriers, which measure the response parameter "pulse per minute." The use of image analysis instead of light barriers enables the evaluation of more detailed behavior parameters. First, a dynamic *Daphnia* test was equipped with the BehavioQuant system.[18] Two chambers, illuminated from one side by a electroluminescence foil and containing ten *Daphnia* each, were monitored by one CCD camera each. The behavior of the organisms was recorded for 2 minutes by video image and analyzed afterwards. The velocity, relative path depth from the surface, and the number of turns were calculated and used as biological endpoints. Adding the chemical Lindan at a final concentration of 1 μg/1 to the described flow-through system led to a change of all measured parameters compared to a 2-h control measurement indicating the sensitivity of the new endpoints.

Baillieul et al.[19] developed another image analysis system based on the frame grabber card DT-2862 (Data Translation 1080E, Marlboro, MA) and a PC-486 computer to detect and analyze the movement behavior of up to 30 *Daphnia magna* in a flow-through system in real-time (Figure 23.2). The authors created binary images (each 40 ms) by setting a gray threshold, to differentiate between a white background (binary 0) and black objects (binary 1). Two consecutive binary images are compared by an OR operation: if a pixel is an object pixel (binary 1) in one or both images, the result is also an object pixel. The resulting image is stored in a frame buffer and compared in

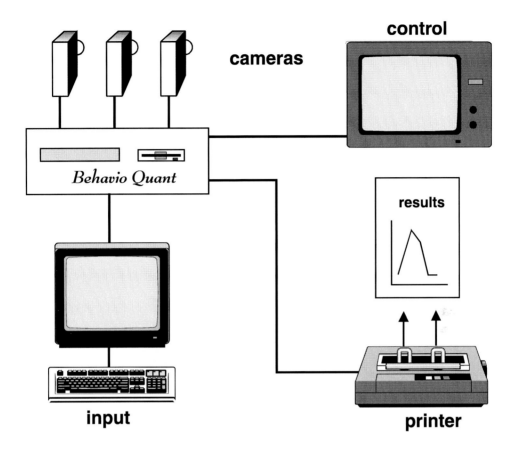

FIGURE 23.1 Schematic view of the image analysis system BehavioQuant. (Reprinted from Steinberg, C.E.W. et al., *Brachydanio rerio, Water Research*, 29, 981, 1995, with permission from Elsevier Science.)

the same way with the next binary image in the sequence. The result of this algorithm is a binary reconstruction of the trajectories of all moving objects (Figure 23.3). The average velocity of the objects is then calculated using statistical analysis of the trajectory images of each sequence, which takes about half a second per analyzed sequence. In addition, the program calculates the total number of motile objects in the active window and the number of objects in two subwindows set as the upper and lower half of the active window to obtain information on the spatial distribution of the daphnids.

Toxicity measurements were carried out using an observation chamber, which is illuminated from the top and one side by different light sources. Because of the phototactic orientation of *Daphnia*,[20] the distribution of the organisms in the chamber is not random and can be used as a response parameter. When adapting the organisms for 2 h in the chamber using a flow-through system, they showed a constant swimming behavior. Experiments with cadmium showed that all calculated parameters were influenced during a 20-h measurement procedure. The velocity and the swimming activity increased after adding cadmium at a final concentration of 3 μM to the flow-through medium, and the distribution of the organisms in the chamber changed: there were less animals in the upper part.[21] A disadvantage of the algorithm employed is the limitation to slowly moving organisms. If an organism moves faster than its own diameter between two consecutive images, then there are no more overlapping object pixels. In this case, no trajectories can be found and movement parameters cannot be calculated.[19]

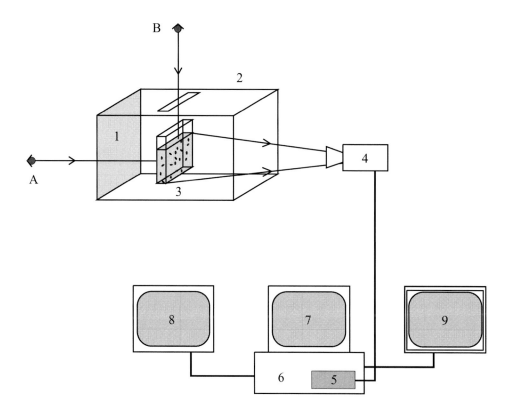

FIGURE 23.2 Biomonitoring system measuring the swimming activity of 30 simultaneously swimming daphnias. A and B light sources, 1 opal Plexiglas screen, 2 black tunnel, 3 flow-through measuring cell with daphnias (flow-through circulation and thermostatic regulation is not shown), 4 video camera, 5 frame grabber, 6 PC, 7 monitor showing swimming activity, 8 monitor showing calculated data, 9 monitor showing curves. (Reprinted from Wolf, G. et al., *Comp. Biochem. Physiol. A*, 120, 99, 1998. With permission.)

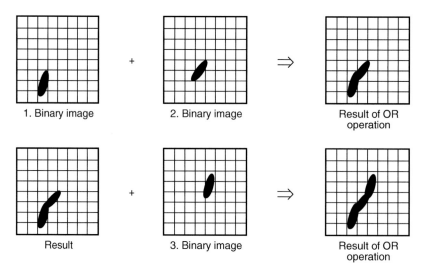

FIGURE 23.3 Strategy for the binary reconstruction of trajectories of moving objects used by Baillieul and Scheunders.

A more complicated tracking strategy is used when calculating the movement vectors by the object displacement during a longer time period than that in between two images (40 ms).[22] The centroids (centers of gravity) of all objects in a first image are calculated and compared with those of the consecutive frame. Only those objects are accepted as organisms, that fulfill a number of conditions such as limits for the area. If one centroid in the second image was found in a defined search area around a centroid in the first image, both centroids should belong to the same organism. The procedure is repeated with the following frames and can be stopped, for example, at the fifth frame (after a total time of 160 ms). The movement vector of each individual organism is determined from the displacement of the centroids between the first and the last frame (Figure 23.4). This leads to vectors longer than calculated by comparison of two consecutive frames (time interval 40 ms) and therefore enhances the calculation precision of the vector angles and lengths, which is determined by the limited pixel resolution of the digitized images. Vector analysis can be done in real-time in parallel, calculating the orientation behavior of the tracked organisms as response parameters. This has been done with the motile unicellular flagellate *Euglena gracilis* (length ≈ 80 μm). The cells orient themselves using a number of external factors, including light and gravity.[23-25] In darkness, the organisms swim upward (negative gravitaxis) toward the water surface. At low irradiances, the cells move toward the light source (positive phototaxis), while at high irradiances they show negative phototaxis.[26,27]

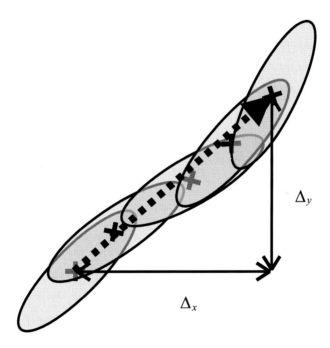

FIGURE 23.4 Calculation of the movement vector (dotted arrow) by the displacement of the centroids (crosses) of the moving object. The vector is determined from the first and final centroids.

Chronic toxicity tests with different heavy metals showed that not only growth but also gravitactic and phototactic orientation behavior of the cells measured by real-time image analysis are affected at low concentrations.[28,29] A fully automatic early warning system called ECOTOX, based on the described real-time image analysis system (Figure 23.5a), uses up to nine different movement parameters of *Euglena gracilis* as endpoints in parallel.[30] The system includes a miniaturized microscope with an observation cuvette, valves, and pumps, and has a size of 29 × 20 × 11 cm. In addition to the number of motile cells and the mean speed, some new response parameters like the precision of gravitactic orientation and the cells form are calculated and described by

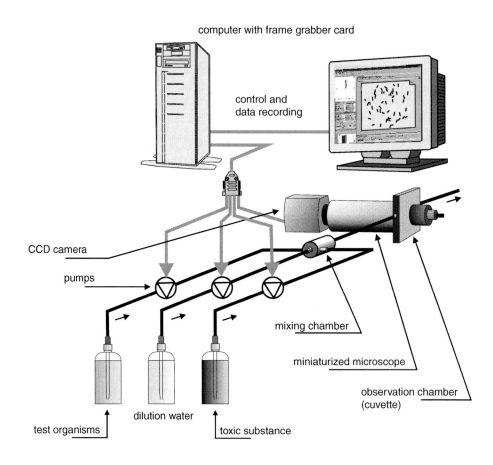

(a)

FIGURE 23.5 (a) Schematic view of the automatic early warning system ECOTOX. The software measures different movement parameters by real-time tracking of the organisms and calculates the decrease of these parameters due to the toxic concentration. (b) The user interface shows the tracked objects and plots all calculated parameters in real-time. (c) The results can be used to determine dose-effect relationships and EC_{50} values (effect concentration at which 50% of the inhibitory effect occurs.)

different statistic parameters. A measurement with a known or unknown toxic sample is compared with a control measurement, and the inhibition of the response parameters is calculated and shown in a result window in real-time. In addition, object movement is displayed on the monitor during the tracking (Figure 23.5b). In contrast to the tests with fish or *Daphnia*, the system is not built as a dynamic test in a flow-through chamber, but as a short-term static test, where test organisms and the water sample are pumped once into the observation cuvette, and measurement starts immediately for a period of, for example, 5 min. There are about 6×10^4 cells in the observation cuvette and about 50 to 100 objects in the measurement window. However, up to 1500 objects can be tracked in parallel. Due to automatic dilution of the toxic sample, dose-effect relationships can easily be calculated (Figure 23.5c). A disadvantage of this type of system can be the lower sensitivity, due to the short exposure time.[30] The big advantages are the compactness, the short response time, the calculation of many endpoints in parallel, and the high number of observed objects.

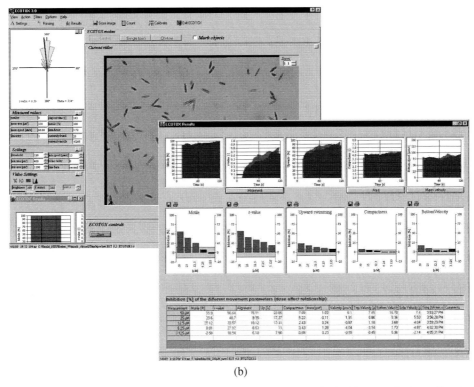

(b)

FIGURE 23.5 (Continued) (b) The user interface shows the tracked objects and plots all calculated parameters in real-time.

Real-time tracking is not only utilized in ecotoxicology but also in the closely related toxicology, the science where the direct effects of chemicals on man are the focus of attention. For measuring the toxicity of opioids and chloramphenicol, Wu et al.[31,32] used the mean speed of the ciliated protozoa *Tetrahymena pyriformis* as endpoint, measured by a computerized image analysis system. *Tetrahymena pyriformis* is an organism widely used as an alternative to animal models for the investigation of narcotic toxicity in environmental toxicology.[33] IC_{50} values (inhibition concentration at which 50% of the inhibitory effect occurs) for the tested pharmaceuticals were in between 0.27 and 59.8 mM, and the inhibitory effect was attributed to a hydrophobic interaction between drugs and membrane components.

A different model was developed to measure the toxic effects of penetration enhancers administered internally. As a biological endpoint, the mucociliary transport velocity in the nasal delivery system or in a comparable system can be used.[34,35] The frog palate epithelium is ciliated with numerous mucus-producing and -secreting glands, similar to mammalian nasal and tracheo-bronchial mucosa, and the transport rates and mechanisms of the two systems are similar.[36] To determine the ciliary transport velocity in such a frog palate system, Aspden et al.[37] measured the transport velocity of graphite particles (diameter: 50 to 100 µm) placed on the palate after the application of the pharmaceuticals. An automatic image analysis system enabled the tracking of the particles by analyzing a sequence of images taken at 1-s intervals. Because particles tend to move in a straight line with a relatively stable velocity, the new location of each particle in each new image can be estimated by extrapolating the new location from the known location in the two preceding images. The actual new location is then obtained by searching in the immediate area around the predicted location. However, this type of tracking strategy leads to problems when particles merge, disintegrate, or when stray particles move into the extrapolated search area. The authors tested

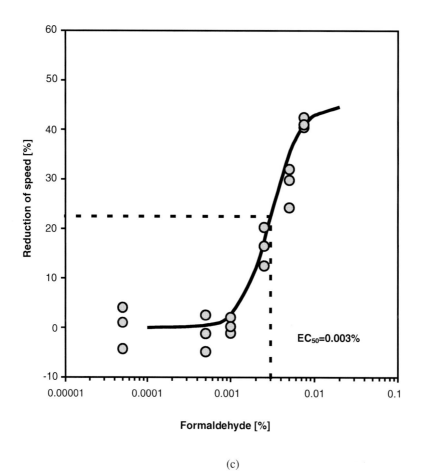

(c)

FIGURE 23.5 (Continued)   (c) The results can be used to determine dose-effect relationships and EC_{50} values (effect concentration at which 50% of the inhibitory effect occurs.)

several chitosans for their toxic potential on the frog palate model. Chitosans, cationic polysaccharides with high molecular weight, are used for a wide range of applications, for example, as a pharmaceutical excepient in oral drug formulations and have potential as a nasal delivery system. Most of the tested chitosans only briefly affected the mucociliary transport velocity and can therefore be considered to have no adverse effect on the mucociliary clearance mechanism. The last examples show that real-time tracking is not only an extension and facilitation in ecotoxicology, but also in pharmacology and toxicology.

Most of the described systems measure toxicity of chemicals using the velocity of their test organisms as the biological endpoint. Some of the systems also calculate different, more complex behavioral parameters, which can give a more detailed picture of the biological effect of the samples. Table 23.1 gives a survey of behavioral response parameters in ecotoxicological (and toxicological) risk assessment as measured by real-time tracking.

23.3   STATIC IMAGE ANALYSIS

Not only the movement behavior of test organisms can be analyzed by image analysis, but also slowly changing response parameters such as growth or reproduction. In this case, no real-time tracking is necessary. Images can be recorded and the detected objects can be analyzed with respect to their number, length, area, or form. Image analysis has been used to assess the effects of chemicals

TABLE 23.1
Behavioral Response Parameters in Toxicology Testing Methods:
Measurement by Implementation of Real-Time Image Analysis System

Measured Parameter	Ref.
Velocity	7, 8, 18, 19, 21, 28, 29, 30, 31, 32, 37
Relative depth from surface	7, 8, 18
Turning	7, 8, 18
Distance of organisms to each other (shoal behavior)	7, 8
Preference of light and dark habitats	10, 11, 12
Distribution in a light gradient	19, 21
Number of motile objects	19, 21, 28, 29, 30
Precision of orientation (phototaxis, gravitaxis)	28, 29, 30
Number of upward swimming cells (gravitaxis)	29, 30
Cells form	Unpublished, this contribution

released by fish predators on the morphology of *Daphnia lumholtzi*,[38] and Johnson et al. used image analysis to determine sub-lethal effects of chemicals on the growth of *Daphnia magna*. They measured the length of the organisms using the non-invasive image analysis technique instead of determining the growth by measuring the dry weight,[39] so that repeated measurements could be done throughout the test.

The most commonly used vascular plant in toxicity tests in *Lemna*.[40-42] Usually, a test duration of 4 to 10 days is used in these assays, with biomass yield and growth rate as response parameters. Both are determined by counting the number of leaves (frond number) in a defined area.[43] Currently, the implementation of image analysis into a *Lemna* test method for OECD (Organization of Economic Cooperation and Development) is being considered. This could automate growth measurement by analyzing frond area and/or number.

All the tests described thus far measure the effects of toxic substances indirectly, by looking at changes in growth, reproduction, or behavioral parameters. Some tests have been developed to obtain information on direct effects of toxic chemicals at the molecular or cellular level, especially to measure the primary effects of genotoxic substances on the idioplasm.[44-47] A commonly used genotoxicity test is the micronucleus test, a simple and rapid measure of induced structural and numerical chromosomal aberrations.[48] The endpoint of this test is the evaluation of polychromatic erythrocytes from bone marrow for the presence of micronuclei, which are acentric chromosome fragments or whole chromosomes. In recent years, image analysis has been implemented in this micronuclei detection. Parton et al.[49,50] were the first to use a semi-automatic image analysis system that differentiates between polychromatic erythrocytes and normochromatic erythrocytes based on the color, size, and shape of the cells. The number of micronuclei in the detected polychromatic erythrocytes is counted manually because of the difficult visual criteria for the identification of micronuclei. The authors found similar micronuclei counts calculated using the automatic system and in manual analysis. The big advantage of using image analysis is the reduction in time for measurement from 40 h to approximately 5 to 8 h.

23.4   SUMMARY

This chapter describes the implementation of image analysis in biological test systems for ecotoxicological and toxicological risk assessment. Different systems with different strategies for imaging, tracking, and analysis of the behavior of test organisms are presented. The use of image analysis in biological test systems has increased in recent years due to the advantages of this technique. New biological response parameters (endpoints), such as organism speed, distribution in a light gradient, photo- and gravitaxis, or the distance of organisms to each other, can be detected and used to measure toxic effects. In addition to these parameters, previously used endpoints such as growth or reproduction are determined much faster and in a more standardized way. The savings in time and manpower increase productivity. In particular, the measurement of growth and reproduction with the image analysis technique is non-invasive, so that repetitive measurements can be done throughout the tests. Real-time tracking systems enable the tracking of many objects in parallel and the calculation of all movement parameters of interest without any delay. This arises the statistical significance of the results and reduces the total time for measurement, which is an important advantage for biological test systems used online as early warning systems. Image analysis can help to automate bioassays and reduce the necessary equipment to minimize the size of the test instruments.

Because of the rapid development of hardware and software during the last few years, real-time tracking is no longer limited with respect to organism numbers or expensive hardware. Standard image analysis software can be used but must be modified (or developed from scratch) if automation and the measurement of special response parameters are to be realized. Currently, several fully automatic online biological test systems, using image analysis, such as BehavioQuant or ECOTOX, are available.

REFERENCES

1. Schwarzenbach, R.P., Gschwend, P.M., and Imboden, D.M., *Environmental Organic Chemistry*, John Wiley & Sons, New York, 1993.
2. OECD, *Guidelines for Testing of Chemicals*, OECD, Paris, 1993.
3. U.S. EPA, *Quality Criteria for Water*, U.S. Environmental Protection Agency, Washington, D.C., 1971.
4. OECD, *Guidelines for Testing of Chemicals*, OECD, Section 203–204, Paris, 1992.
5. William, J.A., Aquatic toxicology testing methods, in *Handbook of Ecotoxicology*, Hoffman D.J., Rattner, B.A., Burton, G.A., Jr., and Cairns, J. Jr., Eds., CRC Press, Boca Raton, FL, 25, 1995.
6. Ermisch, R. and Juhnke, I., Automatische Nachweisvorrichtung für akut toxische Einwirkungen in Strömungstests, *Gewässer und Abwasser*, 52, 16, 1973.
7. Spieser, O.H. and Scholz, W., Verfahren zur quantitativen Bewegungsanalyse von mehreren Objekten im selben Medium, Deutsche Patentschrift P. 4224750.0, 1992.
8. Blübaum-Gronau, E., Spieser, O.H., and Krebs, F., Bewertungskriterien für einen Verhaltensfischtest zur kontinuierlichen Gewässerüberwachung, *Schriftenr. Ver. Wasser Boden Lufthyg.*, 89, 333, 1992.
9. Lorenz, R., Spieser, O.H., and Steinberg, C., New ways to ecotoxicology: quantitative recording of behavior of fish as toxicity endpoint, *Acta Hydrochim. Hydrobiol.*, 23, 197, 1995.
10. Steinberg, C.E.W., Lorenz, R., and Spieser, O.H., Effects of atrazine on swimming behavior of zebrafish, *Brachydanio rerio, Water Res.*, 29, 981, 1995.
11. Lorenz, R., Brüggemann, R., Steinberg, C.E.W., and Spieser, O.H., Humic material changes effects of terbutylazine on behavior of zebrafish (*Brachydanio rerio*), *Chemosphere*, 33, 2145, 1996.
12. Baganz, D., Staaks, G., and Steinberg, C., Impact of the cyanobacteria toxin, microcystin-LR on behavior of zebrafish, *Danio rerio, Water Res.*, 32, 948, 1998.
13. Adema, D.M.M., *Daphnia magna* as a test animal in acute and chronic toxicity tests, *Hydrobiologia*, 59, 125, 1978.
14. Leeuwangh, P., Toxicity tests with Dahnids: its application in the management of water quality, *Hydrobiologia*, 59, 145, 1978.
15. Prater, B.L. and Anderson, M.A., A 96 hour bioassay of Otter Creek, Ohio, USA, *J. Water Poll. Control Fed.*, 49, 2099, 1977.
16. OECD, *Guidelines for Testing of Chemicals*, OECD, Section 202, Paris, 1984.
17. Knie, J., Der dynamische Daphnientest — ein automatischer Biomonitor zur Überwachung von Gewässern und Abwässern, *Wasser, Boden*, 12, 310, 1988.
18. Blühbaum-Gronau, E. and Hoffmann, M., Enhanced sensitivity of continuous *Daphnia* test by consideration of more behavior parameters, *Vom Wasser*, 89, 163, 1997.

19. Baillieul, M. and Scheunders, P., On-line determination of the velocity of simultaneously moving organisms by image analysis for the detection of sublethal toxicity, *Water Res.*, 21, 1027, 1998.

20. Ringelberg, J., The positive phototactic reaction of *Daphnia magna* Strauss, a contribution to the understanding of diurnal vertical migration, *Neth. J. Sea Res.*, 2, 319, 1964.

21. Wolf, G., Scheunders, O., and Selens, M., Evaluation of the swimming activity of *Daphnia magna* by image analysis after administration of sublethal cadmium concentrations, *Comp. Biochem. Physiol A*, 120, 99, 1998.

22. Vogel, K. and Häder, D.-P., Simultaneous tracking of flagellates in real-time by image analysis, *Proc. Fourth Eur. Symp. Life Science Res. Space*, (ESA SP-307), 1990, 541.

23. Brinkmann, K., Keine Geotaxis bei *Euglena, Z. Pflanzenphysiol.*, 59, 12, 1968.

24. Häder, D.-P., Phototaxis and gravitaxis in *Euglena gracilis*, in *Biophysics of Photoreceptors and Photomovements in Microorganisms*, Lenci, F., Ghetti, F., Colombetti, G., Häder, D.-P. and Song, P.-S., Eds., Plenum, New York, 1991, 203.

25. Häder, D.-P., Porst, M., Tahedl, H., Richter, P., and Lebert, M., Gravitactic orientation in the flagellate *Euglena gracilis, Micrograv. Sci. Technol.*, 10, 53, 1997.

26. Häder, D.-P., Colombetti, G., Lenci, F., and Quaglia, M., Phototaxis in the flagellates, *Euglena gracilis* and *Ochromonas danica, Arch. Microbiol.*, 130, 78, 1981.

27. Lenci, F., Colombetti, G., and Häder, D.-P., Role of flavin quenchers and inhibitors in the sensory transduction of the negative phototaxis in the flagellate *Euglena gracilis, Curr. Microbiol.*, 9, 285, 1983.

28. Stallwitz, E. and Häder, D.-P., Motility and phototactic orientation of the flagellate *Euglena gracilis* impaired by heavy metal ions, *Photochem. Photobiol., B: Biol.*, 18, 67, 1993.

29. Stallwitz, E. and Häder, D.-P., Effects of heavy metals on motility and gravitactic orientation of the flagellate, *Euglena gracilis, Eur. J. Protistol.*, 30, 18, 1994.

30. Tahedl, H. and Häder, D.-P., Fast examination of water quality using the automatic biotest ECOTOX based on the movement behavior of a freshwater flagellate, *Water Res.*, 33, 426, 1999.

31. Wu, C., Fry, C.H.F., and Henry, J.A., Membrane toxicity of opioids measured by protozoan motility, *Toxicology*, 117, 35, 1995.

32. Wu, C., Clift, P., and Fry, C. H., Membrane action of chloramphenicol measured by protozoan motility inhibition, *Arch. Toxicol.*, 70, 850, 1996.

33. Schultz, T.W., Wyatt, N.L., and Lin, D.T., Structure toxicity relationship for nonpolar narcotics: a comparison of data from the *Tetrahymena*, photobacterium and pimephales systems, *Bull. Environ. Contam. Toxicol.*, 44, 67, 1990.

34. Andersen, I.B., Camner, P., Philipson, K., and Proctor, D.F., Nasal clearance in monozygotic twins, *Am. Rev. Respir. Dis.*, 110, 301, 1974.

35. Passali, D., Bellusi, M., Ciampoli, B., and De Seta, E., Experiences in the determination of nasal mucociliary transport time, *Acta Otolaryngol.*, 97, 319, 1984.

36. Puchelle, E., Zahm, M., and Sadoul, P., Mucociliary frequency of frog palate epithelium, *Am. J. Physiol.*, 242, 321, 1982.

37. Aspden, T.J., Adler, J., Davis, S.S., Skaugrud, Ø., and Illum, L., Chitosan as a nasal delivery system: evaluation of the effect of chitosan on mucociliary clearance rate in the frog palate model, *Int. J. Pharm.*, 122, 69, 1995.

38. Tollrain, R., Fish-kairomone induced morphological change in *Daphnia lumholtzi* (Sars), *Arch. Hydrobiol.*, 130, 69, 1994.

39. Johnson, I. and Delany, P., Development of a 7-day *Daphnia magna* growth test using image analysis, *Bull. Environ. Contam. Toxicol.*, 61, 355, 1998.

40. American Society for Testing and Materials, *Standard Guide for Conducting Static Toxicity Test with Lemna gibba G3*, E 1415-91, 1991.

41. U.S. EPA, OPPTS 850.4400 Aquatic Plant Toxicity Test Using *Lemna* spp., Public draft, EPA 712-C6-156, 1996.

42. Fairchild, J.F., Ruessler, D.S., Haverland, P.S., and Carlson, A.R., Comparative sensitivity of *Selenastrum capriconutum* and *Lemna minor* to sixteen herbicides, *Arch. Environ. Contam. Toxicol.*, 32, 353, 1997.

43. Huebert, D.B. and Shay, J.M., Considerations in the assessment of toxicity using duckweeds, *Environ. Toxicol. Chem.*, 12, 481, 1993.

44. Ames, B.N., Durston, W.E., Yamasaki, E., and Lee, F.D., Carcinogens are mutagens: a single test system combining liver homogenate for activation and bacteria for detection, *Proc. Natl. Acad. Sci., U.S.A.*, 70, 2281, 1973.

45. De Flora, S., Vigano, L., D'Agostini, F., Camoirano, A., et al., Multiple genotoxicity biomarkers in fish exposed *in situ* to polluted river water, *Mutat. Res.*, 319, 167, 1993.

46. Reifferscheid, G., Heil, J., and Zahn, R.K., Die Erfassung von Genotoxinen in Wasserproben mit dem *umu*-Milrotest, *Vom Wasser*, 76, 153, 1991.

47. Schmid, W., The micronucleus test, *Mutat. Res.*, 31, 9, 1975.

48. Heddle, J.A., Cimino, M.C., Hayashi, M., Romagns, F., Shelby, M.D., Tucker, J.D., Vanparys, P., and MacGregor, J.T., Micronuclei as an index of cytogenetic damage: past, present, and future, *Environ. Mol. Mutagen.*, 18, 277, 1991.

49. Parton, J.W., Hoffman, W.P., and Garriott, M.L., Validation of an automated image micronucleus scoring system, *Mutat. Res.*, 370, 65, 1996.

50. Wolf, T. and Luepke, N.-P., Formation of micronuclei in incubated hen's eggs as a measure of genotoxicity, *Mutat. Res.*, 394, 163, 1997.

Index

A

A/D conversion, 145
Acridine Orange, 78, 125, 190
actin 14f, 127, 191, 197, 210, 212, 214, 229f, 279, 457
algorithm
    compression, 119
    dithering, 119
alignment, 12, 14, 146, 279, 282, 285, 302, 313, 317, 330–334, 347f, 351, 363, 412
annealing, 168–171, 175–177
    mixed, 179
    simulated, 178f
antibodies, 145, 189, 211, 227, 320, 374
    fluorescence labeled, 218
antioxidants, 189
area, 3, 15, 23–30, 32, 42–52, 55f, 59–63, 68, 94, 101, 144, 148, 196, 211, 348, 403, 406–408, 412–414, 432, 439
    change, 410
    cotyledon, 262, 268
    filling, 147
    leaf, 260, 262, 265
    limits, 407
    mean, 412, 414
    morphometric, 54
    of view, 414
    particles, 359
    search, 415, 436
    spot, 147, 149
    tissue, 370
    variability, 362
Astasia, 143, 382–384
autofluorescence, 142, 193, 218, 377, 381f
automatic gain control, 145, 215, 395
autophagic bodies, 339–341
autoradiography, 127, 142, 144, 356, 358, 361, 364
    grains, 356f, 369
axoneme, 241–252
    mutant, 249

B

background, 23, 25–28, 30f, 36, 45f, 48, 50, 56, 69, 77, 96, 111, 114, 119, 134, 144, 147f, 213, 217, 262, 363f, 369, 394f, 399f, 403, 437f
    color, 108f
    digital, 215
    gray level, 219

image, 111, 114
local, 358

C

chlorophyll, 278, 382, 457
chondriocytes, 63
chroma, 121, 125
chromatofocusing, 141
chromosomes, 50, 197, 200, 357, 362, 368, 370
    human, 230
ciliate, 125, 374, 394, 410
cilium, 3, 241–245, 249, 252f, 318–320, 373f, 394
Clifford-Hammersley theorem, 164
CLOSE operation, 34, 36–40, 44
co-dimension, 68f, 71
collagen, 9–19
    fibers, 14, 18f
color image, 125, 218, 260, 262–264, 268, 271
color processing, 25
color space, 110, 120, 124
    CIE*Luv*, 120
    HBS, 120
    RGB, 120
confocal
    fluorescence, 14
    reflection, 10–19
        contrast, 12f, 19
        imaging, 13
        microscopy, 10f, 18f
conical tilt, 300
    method, 279
        random, 301
        series, 296, 302–304, 309f, 313–318, 322
contour, 26, 39, 258, 329, 330f, 337f, 340–347, 351, 357, 370, 403, 405
    lines, 330, 331, 341, 343, 346, 351
    polygonation, 194
contrast, 3, 10–14, 17, 19, 23, 25, 42, 53, 56f, 100–102, 192, 211, 215, 217f, 363, 366, 374, 382, 396, 402f, 408, 438, 452, 457
    adjustment, 400
    Allen video enhanced, 228
    amplitude, 277
    difference, 364
    enhancement, 213, 215f, 250, 252, 400f
    image, 193
    optical, 394
    phase, 187, 281
    replication, 356

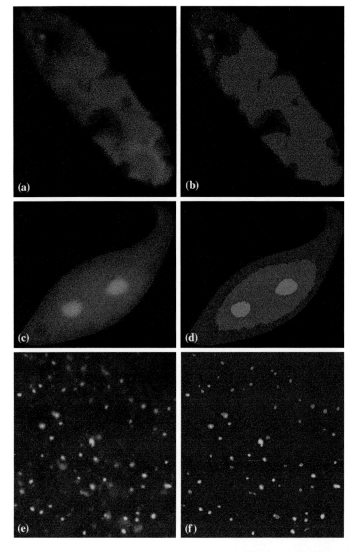

Chapter 7, Color Figure 1. (a) Fluorescence image of the flagellate *Euglena gracilis*. (c) Phase of the digestive process of the raptorial feeder ciliate *Litonotus lamella*. (e) Emission of isolated *Euglena gracilis* photoreceptors. All these images possess thousands of colors. Figures 7.1b, d, and f are the corresponding images after color reduction.

Chapter 14, Color Figure 6. Focused color image composed by repetitive interpolation (through use of Eq. 14.4) at all image points of RGB images. (From Omasa, K. and Kouda, M., *Environment Control in Biology*, 36, 4, 1998. With permission.)

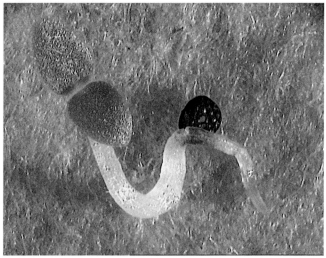

1.0 mm

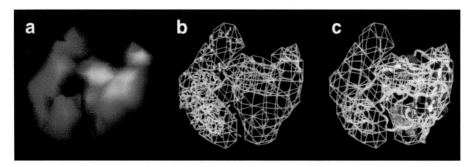

Chapter 15, Color Figure 4. Three-dimensional reconstruction of mammalian signal recognition particle, 54-kDa subunit. This unit of SRP is responsible for both signal-sequence recognition and binding, and GTP-binding and hydrolysis. High-resolution atomic structures of related biological macromolecules and their domains were computationally fitted into related domains in the SRP structure. (a) Shaded surface representation of the reconstruction. (b) Reconstruction of SRP54 in wire mesh representation has been fitted with the peptide-binding methionine-rich domain of calmodulin. This calmodulin domain is related in structure and function to the SRP54 signal-sequence binding domain. (c) Representation of p21-RAS and bound GDP, which is related in terms of amino acid sequence to the GTP-binding domain of SRP 54. The p21 structure was computationally fitted into the SRP54 electron density map as for the calmodulin domain in (b). α-Helices are colored red and β-strands and loops are given in yellow. Blue spheres demark the location of the bound GDP in p21-RAS and display atoms with radii equal to 0.5 of the van der Waals radii of the constituent atoms. (From Czarnota, G. J. et al., *J. Struct. Biol.*, 113, 35–46, 1994. With permission.)

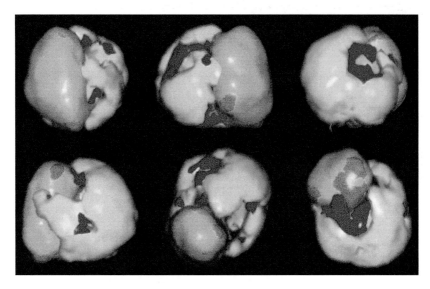

Chapter 15, Color Figure 7. Three-dimensional reconstructions of *Escherichia coli* large and small ribosomal subunits. The subunits were reconstructed using the combined mass images at 150 ± 8 eV of protein and rRNA (rendered as volumes with a blue or green solid surface for the large and small subunits, respectively). The two subunits (large and small) were reconstructed independently and merged to produce the volume shown here from various orientations. The net phosphorus distributions were reconstructed separately using the same images in conjunction with their 100-eV counterparts. These phosphorus distributions were represented as red and orange solid surfaces for the large and small subunits, respectively, and then placed back inside the total mass reconstructions at their corresponding positions. In these volumes, the rRNA appears to form a dense central core, reaching the surface at several locations (red and orange regions in these figures). The rRNA is believed to have an enzymatic role in the process of protein synthesis. Both subunits are required for peptide synthesis, and it is believed that this process occurs at the interface between the two subunits. These reconstructions provide support for the availability of rRNA at the interface, where it is believed to have an active role in protein synthesis. (From Beniac, D. R. et al., *J. Microscopy*, 88, 29, 1997. With permission.)

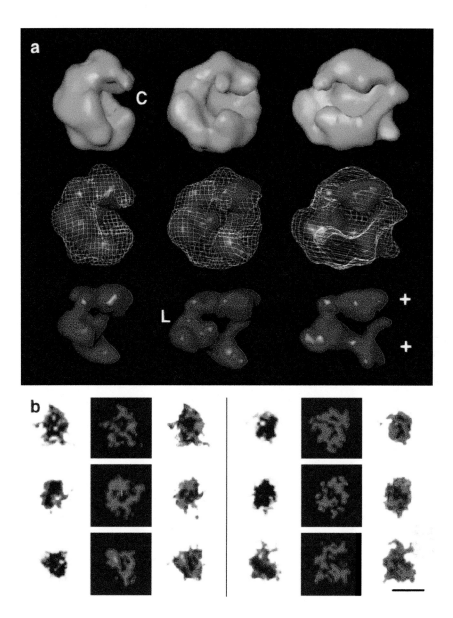

Chapter 15, Color Figure 9. Three-dimensional reconstructions of nucleosomes extracted from transcriptionally active chromatin. (a) The top row shows three views of the 155-eV energy-loss reconstruction for the nucleosome particle at a contour which corresponds to the theoretical volume of combined protein and nucleic acid components of the nucleosome. The middle row shows the 3-D difference map (phosphorus content) fitted inside a wire-mesh representation of the 155-eV energy-loss nucleosome reconstruction. Over 95% of this map lies within the 155-eV reconstruction. Each view from left to right is related to the previous by a rotation of 45° toward the left about a vertical axis in the plane of the page. L indicates a slender phosphorus signal linker that joins two lobes of phosphorus signal denoted to the right by +. C indicates a cleft between the upper and lower domains of the structure. (b) Individual electron spectroscopic images are shown here for nucleosomes associated with actively transcribing chromatin. For each nucleosome, the black-and-white image is the phosphorus contrast-enhanced nucleosome image obtained at an energy loss of 155 eV. Subtraction of a 120-eV energy-loss reference image (not shown), after a background normalization procedure, results in the difference image (red on black), interpreted as an elemental map of nucleosomal DNA phosphorus. An overlay of the difference image on top of the corresponding 155-eV energy-loss image is shown as the third image in each set (red, black, and white). Scale bar, 10 nm. (From Czarnota G. J. et al., *Micron*, 26, No. 6, 426, 1997. With permission from Elsevier Press.)

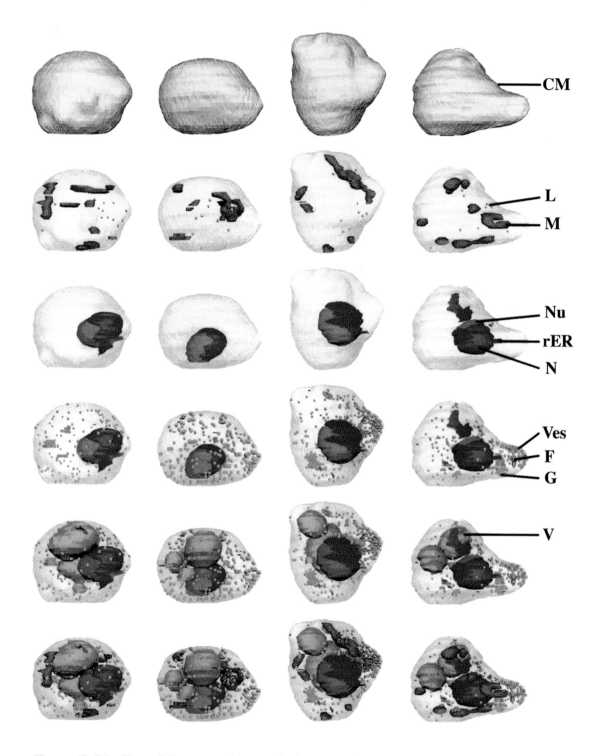

Chapter 17, Color Figure 7. Reconstructed images of α-factor-treated cells. Four typical cells corresponding to four stages in projection formation are shown from left (stage 1) to right (stage 4). The following cell organelles are seen through the cell membranes: lipid droplet (L), mitochondrion (M), nucleus (N), nucleolus (Nu), vacuole (V), vesicle (Ves), Golgi body (G), rough endoplasmic reticulum (rER), and filasome (F), in which vesicles and lipid droplets are represented by simple, small spheres in the computer graphics. (These images are rendered with SYNU applications software by Dallman. With permission. The data are taken from Baba, M. et al., *J. Cell. Sci.*, 94, 207, 1989.)